Sequential Stochastic Optimization

WILEY SERIES IN PROBABILITY AND STATISTICS

Established by WALTER A. SHEWHART and SAMUEL S. WILKS

A complete list of the titles in this series appears at the end of this volume

Sequential Stochastic Optimization

R. CAIROLI

Département de Mathématiques
Ecole Polytechnique Fédérale
Lausanne, Switzerland

ROBERT C. DALANG

Department of Mathematics
Tufts University
Medford, Massachusetts

A Wiley-Interscience Publication
JOHN WILEY & SONS, INC.
New York • Chichester • Brisbane • Toronto • Singapore

This text is printed on acid-free paper.

Library of Congress Cataloging-in-Publication Data:

Cairoli, R.
Sequential stochastic optimization / by R. Cairoli and Robert C. Dalang.
p. cm.—(Wiley series in probability and statistics. Probability and statistics)
"A Wiley-Interscience publication."
Includes bibliographical references (p.314–319) and index.
ISBN 0-471-57754-5 (cloth : acid-free paper) : $69.95
1. Optimal stopping (Mathematical statistics) 2. Dynamic programming. 3. Stochastic control theory. I. Dalang, Robert C., 1961– . II. Title. III. Series.
QA279.7.C35 1995
519.2—dc20 94-39134

Printed in the United States of America

10 9 8 7 6 5 4 3 2 1

Contents

Preface

This book presents a unified mathematical theory of optimal stopping and optimal sequential control of stochastic processes, together with several applications including sequential statistical tests involving several populations and multiarmed bandit problems. Much of the material presented here is either new or appears in a book for the first time.

The book is addressed to graduate students and mathematicians working in probability theory as well as to more applied researchers who need a well-developed framework within which their problems can be formulated and solved. Many of the results presented here will be of interest to experts in the field, as well as to researchers and students from other fields who wish to learn about this topic. The exposition is pedagogically organized so that little prior knowledge of the subject is required, and in fact, the only prerequisites are a standard graduate course in probability theory and some basic knowledge of discrete-parameter martingales. For these, the reader can consult books such as Breiman (1968), Chung (1974), and Chow and Teicher (1988).

The framework for optimal stopping and optimal sequential control problems presented here uses processes indexed by $\mathbb{N}^d$ and, more generally, by (partially) ordered sets. Within this framework, both optimal stopping and optimal sequential control can be treated using similar methods, and the theory we develop is general enough to cover a wide variety of applications. No prior knowledge of the classical theory of optimal stopping and control of discrete-time stochastic processes is assumed. In fact, the classical case expounded in books and articles such as Wald and Wolfowitz (1948, 1950), Arrow, Blackwell, and Girshick (1949), Snell (1952), Dynkin and Yushkevitch (1969), Chow, Robbins, and Siegmund (1971), Neveu (1975), and Shiryayev (1978), is treated together with the general case in a unified fashion. The vast subject of stochastic optimization in continuous time is not discussed here: Books on this subject include Krylov (1980), Kushner and Dupuis (1992), and Fleming and Soner (1993).

This book is organized as follows. Chapter 1 presents the necessary fundamentals, including the notions of essential supremum, stopping points, accessibility, martingales, and supermartingales indexed by $\mathbb{N}^d$.

Chapter 2 is devoted first of all to the study of conditions that ensure the integrability of certain suprema of partial sums of arrays of independent random variables, and second, to the strong law of large numbers for such arrays.

Chapter 3 presents the general theory of optimal stopping for processes indexed by $\mathbb{N}^d$. Consider, for example, the situation in which a gambler receives a random reward X_T if he quits the game at stage T. The problem is to determine which stages to choose and when to quit, in such a way as to maximize the expected reward. A clean mathematical formulation is provided, within which a constructive solution is given under minimal assumptions using a generalization of Snell's method.

Chapter 4 examines the possibility of reducing certain problems involving arrays of random variables to related problems for processes indexed by $\mathbb{N}$. This is shown in particular to be possible when the reward process is given by averages from an array of exchangeable random variables.

Chapter 5 presents structural properties of information flows (filtrations), highlighting the complexity of the information structure in dimensions greater than 1. Necessary and sufficient conditions for accessibility of stopping points are given. Some techniques from graph theory are recalled and used in this section. The relationship between the optimal stopping problem and classical linear programming is also explored.

Chapter 6 focuses on a sequential sampling problem involving several populations. The canonical example concerns two new medical treatments with distinct (but unknown) probabilities of curing a given illness. As each patient arrives, it is necessary to decide, based on previous observations, which treatment to administer. Weighing the total cost of treatment with a cost for making the wrong decision, the problem is to determine an optimal strategy for deciding which treatment has the highest probability of curing the patient. The mathematical model for this problem is elegantly formulated as an optimal stopping problem for a specific process indexed by $\mathbb{N}^d$. A description of the optimal strategy is given. Three concrete examples involving Bernoulli and Gaussian populations are discussed.

Chapter 7 presents a theory of optimal sequential control using ideas similar to those employed in the theory of optimal stopping. The problems differ in that optimal stopping only considers a terminal reward, whereas optimal control also allows a running reward. The optimal control problem is reduced to an optimal stopping problem involving stopping points taking values in a particular directed set. Existence of optimal strategies is proved under minimal assumptions.

As an application of the results of Chapter 7, Chapter 8 is concerned with the multiarmed bandit problem. This problem is easily formulated and the optimality of index strategies is conveniently proved in this framework. Two concrete examples are solved explicitly.

Chapter 9 specializes the results of previous chapters to Markov chains. Two problems are considered: (a) a reward that only depends on the state of the chain is obtained when the observations stops, and (b) a reward is obtained the first time the Markov chain encounters a fixed subset of the state space. The first problem falls in the setting of optimal stopping and the second in that of optimal control. The solutions to these problems involve computing the superharmonic envelope of the reward function.

In Chapter 10, this problem is specialized to the case of the optimal switching problem for two independent random walks on an interval with absorption at the endpoints. At each stage, the player can choose to let one of the random walks evolve while freezing the other. This leads to a switched process that evolves in a rectangle, and the player receives a reward when the process reaches the boundary. The structure of the solution is described explicitly via a decomposition of the state space and a geometric construction of this decomposition is given.

The material is accompanied by a variety of problems and exercises, the aim of which is to increase the reader's understanding of the subject. These problems are generally not needed for subsequent chapters.

Credit for original work along with the sources used in the preparation of this book are given in historical notes at the end of each chapter.

In some sentences, brackets [...] are used to indicate a possible alternative reading of the text.

Several chapters of this book were used by the second author in a graduate course at Tufts University.

Acknowledgments

The authors would like to thank John B. Walsh, who carefully read a preliminary version of this book and made numerous useful and detailed comments, and Xiaowen Zhou, who checked most of the exercises. They also thank S. D. Chatterji for several interesting discussions.

It is a pleasure to acknowledge the generous financial support of P.-F. Pittet, secrétaire général de l'Ecole Polytechnique Fédérale de Lausanne. The research of the second author was partially supported by grants from the National Science Foundation and the Army Research Office.

Finally, the authors thank Madame E. Gindraux for her expert typing of the manuscript and Mrs. Marie-Alix Dalang-Secrétan for her significant contributions to the preparation of the final version of this document.

R. CAIROLI
ROBERT C. DALANG

In Memoriam
Renzo Cairoli (1931–1994)

My teacher, colleague, and friend
died on July 14, 1994, as the very last lines of this book
were being written. May this book remain
as an expression of Renzo Cairoli's lifelong dedication
to the beautiful art of mathematics.

Robert C. Dalang
March 1995

Notation and Conventions

Gathered here are the notation and conventions that will be used throughout the book without comment.

Real Numbers

$\mathbb{R}$ is the real line and $\overline{\mathbb{R}}$ is the set obtained by adjoining to $\mathbb{R}$ two new elements $-\infty$ and ∞. The elements of $\overline{\mathbb{R}}$ [$\mathbb{R}$] are termed numbers [finite or real numbers]. The order structure of $\mathbb{R}$ is extended to $\overline{\mathbb{R}}$ by setting $-\infty < \infty$ and $-\infty < x < \infty$ for all $x \in \mathbb{R}$. The relation $x \leq y$ [$x < y$] reads "x less than or equal to y" or "y greater than or equal to x" ["x less than y" or "y greater than x"]. We will say that an element $x \in \overline{\mathbb{R}}$ is *nonnegative* [*positive*] if $x \geq 0$ [$x > 0$]. $\mathbb{R}_+$ [$\overline{\mathbb{R}}_+$] is the set of $x \geq 0$ in $\mathbb{R}$ [$\overline{\mathbb{R}}$] and $\mathbb{R}_+^*$ [$\overline{\mathbb{R}}_+^*$] is the set $\mathbb{R}_+ - \{0\}$ [$\overline{\mathbb{R}}_+ - \{0\}$]. The intervals of $\mathbb{R}$ or $\overline{\mathbb{R}}$ are denoted by the usual symbols $[x, y]$, $]x, y]$, ... $\mathbb{R}$ is endowed with the usual topology and $\overline{\mathbb{R}}$ with the topology generated by the open sets of $\mathbb{R}$ and the intervals $[-\infty, x[$ and $]x, \infty]$, where x belongs to $\mathbb{R}$ (with this topology, $\overline{\mathbb{R}}$ is compact). The supremum [infimum] of a subset B of $\overline{\mathbb{R}}$ will be denoted $\sup B$ [$\inf B$]. By convention, $\sup \varnothing = -\infty$ and $\inf \varnothing = \infty$, where $\varnothing$ denotes the empty set. The larger [smaller] of two elements x and y of $\overline{\mathbb{R}}$ will be denoted $\sup(x, y)$ or $x \vee y$ [$\inf(x, y)$ or $x \wedge y$]. We write x^+ (positive part of x) [x^- (negative part of x)] instead of $\sup(x, 0)$ [$-\inf(x, 0)$]; $|x| = x^+ + x^-$ is the absolute value of x.

Integers

$\mathbb{Z}$ is the set of integers, $\mathbb{N}$ the set of nonnegative integers, $\overline{\mathbb{N}}$ the set $\mathbb{N} \cup \{\infty\}$, and $\mathbb{N}^*$ [$\overline{\mathbb{N}}^*$] the set $\mathbb{N} - \{0\}$ [$\overline{\mathbb{N}} - \{0\}$]. $\overline{\mathbb{N}}$ is endowed with the order structure and with the topology induced by $\overline{\mathbb{R}}$. Elements of $\mathbb{N}$ will usually be denoted by the letters k, m, n (with precedence to n). Occasionally, these letters may also refer to the element ∞. For all $n \in \mathbb{N}$, $\mathbb{N}_n$ will denote the set $\{m \in \mathbb{N}: m \leq n\}$. The product $1 \cdots n$ is denoted $n!$.

The Set $\mathbb{I}$

Throughout this book, $\mathbb{I}$ will denote the set $\mathbb{N}^d$, where d is a given positive integer, generally > 1 but possibly equal to 1. Elements of $\mathbb{I}$ will usually be denoted by the letters s, t, u (with precedence to t); t^i is the ith coordinate of t; 0 denotes the element $(0, \ldots, 0)$, the zero element of $\mathbb{I}$, and e_i denotes the element for which the ith coordinate is 1 and all other coordinates vanish. The set $\mathbb{I}$ is equipped with the (partial) order

$$s = \left(s^1, \ldots, s^d\right) \leq t = \left(t^1, \ldots, t^d\right),$$

which reads "s less than or equal to t" or "t greater than or equal to s", and means

$$s^1 \leq t^1, \ldots, s^d \leq t^d.$$

The relation $s < t$ is equivalent to $s \leq t$ and $s \neq t$. When $d > 1$, we write $s \ll t$ to express that

$$s^1 < t^1, \ldots, s^d < t^d.$$

We conform to general usage concerning the notation that relates to order relations and we will sometimes write $t \geq s$ for $s \leq t$, $t > s$ for $s < t$, and $t \gg s$ for $s \ll t$. Two elements s and t of $\mathbb{I}$ such that $s \nleq t$ and $t \nleq s$ are said to be *incomparable*. $\mathbb{I}^*$ is the set $\mathbb{I} - \{0\}$. For all $t \in \mathbb{I}$, $\mathbb{I}_t$ will denote the set $\{s \in \mathbb{I}\colon s \leq t\}$, and we set

$$|t| = t^1 + \cdots + t^d.$$

The *direct predecessors* of $t \in \mathbb{I}^*$ are the elements s of $\mathbb{I}$ such that $s < t$ and $|t - s| = 1$. The *direct successors* of $t \in \mathbb{I}$ are the elements u of $\mathbb{I}$ such that $u > t$ and $|u - t| = 1$. The set of direct successors of t will be denoted $\mathbb{D}_t$.

The Set $\bar{\mathbb{I}}$

We use the symbol $\bar{\mathbb{I}}$ to denote the set obtained by adjoining to $\mathbb{I}$ one infinite element $\bowtie$. The symbol $\bowtie$ should not be confused with ∞. The order structure of $\mathbb{I}$ is extended to $\bar{\mathbb{I}}$ by setting $t < \bowtie$ for all $t \in \mathbb{I}$. By convention, $|\bowtie| = \infty$. The letters s, t, and u will also occasionally designate the element $\bowtie$, and in this case, in order to indicate that $t \neq \bowtie$, will sometimes say that t is finite. The supremum [infimum] of two elements s and t of $\bar{\mathbb{I}}$ will be denoted $s \vee t$ [$s \wedge t$]. Obviously, $s \vee t = (s^1 \vee t^1, \ldots, s^d \vee t^d)$ and $s \wedge t = (s^1 \wedge t^1, \ldots, s^d \wedge t^d)$. The supremum [infimum] of a family $(t_l,\ l \in \mathbb{L})$ of elements of $\bar{\mathbb{I}}$ will be denoted by $\bigvee_l t_l$ [$\bigwedge_l t_l$]. A function $t \mapsto x_t$ defined on $\bar{\mathbb{I}}$, $\overline{\mathbb{N}}$ or $\overline{\mathbb{R}}$ and with values in a (partially) ordered set (or the family (x_t) that this function determines) will be termed *nondecreasing* [*nonincreasing*] if $x_s \leq x_t$ [$x_s \geq x_t$] for all couples (s, t) such that $s \leq t$.

The Two Forms of Convergence to $\bowtie$

If t and t_n are elements of $\mathbb{I}$, we will write $t \to \bowtie$ $[t_n \to \bowtie]$ to express that $|t| \to \infty$ $[|t_n| \to \infty]$, and $t \twoheadrightarrow \bowtie$ $[t_n \twoheadrightarrow \bowtie]$ to express that $t^1 \wedge \cdots \wedge t^d \to \infty$ $[t_n^1 \wedge \cdots \wedge t_n^d \to \infty]$. If $t \mapsto x_t$ is a function defined on $\mathbb{I}$ and with values in $\overline{\mathbb{R}}$, then $\limsup_{t \to \bowtie} x_t$ and $\limsup_{t \twoheadrightarrow \bowtie} x_t$ denote, respectively, $\inf_{t \in \mathbb{I}} \sup_{u \in \mathbb{I} - \mathbb{I}_t} x_u$ and $\inf_{t \in \mathbb{I}} \sup_{u \in \mathbb{I}:\, u \geq t} x_u$. Obviously, in the case where $d = 1$, both symbols represent the same number. The symbols $\liminf_{t \to \bowtie} x_t$ and $\liminf_{t \twoheadrightarrow \bowtie} x_t$ are defined analogously, and we write $\lim_{t \to \bowtie} x_t$ and $\lim_{t \twoheadrightarrow \bowtie} x_t$ when liminf and limsup are equal.

Convergence Criterion

The reader will easily check the simple criterion that follows: $\limsup_{t \to \bowtie} x_t \leq x$ $[\limsup_{t \twoheadrightarrow \bowtie} x_t \leq x]$ if and only if $\limsup_{n \to \infty} x_{t_n} \leq x$ for all nondecreasing sequences $(t_n, n \in \mathbb{N})$ of elements of $\mathbb{I}$ such that $t_n \to \bowtie$ $[t_n \twoheadrightarrow \bowtie]$ (cf. Exercise 1 of Chapter 1).

Spatial Interpretation of $\mathbb{I}$

The set $\mathbb{I}$ $(= \mathbb{N}^d)$ will mainly be used as an index set. We will often refer to the exponent d as the dimension of $\mathbb{I}$. For example, the expression "in dimension 1" ["in dimension greater than 1"] means that $d = 1$ $[d > 1]$. The elements of $\mathbb{I}$ will thus be given a spatial rather than temporal interpretation and, in particular, we will call them "points" rather than "times." We will for instance say "at the point t" or "at stage t" rather than "at time t," "beyond t" rather than "after t."

Conventions About Functions

From now on, unless a particular range is specified, the term *function* will always refer to a function with values in $\overline{\mathbb{R}}$. We will say that a function is *finite* or *real valued* [*nonnegative*] if its range is contained in $\mathbb{R}$ $[\overline{\mathbb{R}}_+]$, and *bounded* if its range is a bounded subset of $\mathbb{R}$. We will conform to general usage concerning the notation for operations and relations concerning functions. Unless otherwise mentioned, a *Borel set* is a Borel subset of $\overline{\mathbb{R}}$ and a *Borel function* is a measurable function defined on some measurable space. Often the domain of such a function is $\overline{\mathbb{R}}$ or $\overline{\mathbb{R}}^n$ or a Borel subset of one of these two spaces. When the context clearly determines the domain, it may not be mentioned explicitly.

CHAPTER 1

Preliminaries

In this chapter, we develop the concepts and theorems that constitute the foundation for a theory centered on problems of sequential stochastic optimization. These problems typically arise in the situation where an observer (or controller) observes the evolution of a random phenomenon, or process. At each stage of the evolution, the observer can choose from a finite number of actions. Assuming that there is a cost to taking each action, or that each action brings in a reward, the controller would like to act in such a way as to minimize the cost or maximize the reward.

The fundamental assumption, common to all mathematical models that describe random phenomena, is that the behavior of the process, that is, of all the random events that characterize it, is entirely determined by a parameter ω, whose value is unknown. The controller is aware of the set Ω of all possible values of this parameter, of the probability that this parameter belongs to a particular subset of Ω, and of the set of actions that may be taken at each stage of the observation. The controller is also aware of what information concerning the value of the parameter ω will be made available to him following each action. The mathematical setup that formalizes this assumption is the starting point to the theory.

1.1 FILTERED PROBABILITY SPACES

Most of this book is built upon notions that depend on the presence of an underlying probability space $(\Omega, \mathscr{F}, \mathrm{P})$, that is, a measurable space $(\Omega, \mathscr{F})$ and a probability measure P on $\mathscr{F}$ (a nonnegative measure on $\mathscr{F}$ such that $\mathrm{P}(\Omega) = 1$). The generic element of Ω is denoted ω. The elements of the sigma field $\mathscr{F}$ represent the *observable events*. A set F in $\mathscr{F}$ is a *null set* if $\mathrm{P}(F) = 0$. We will assume that $(\Omega, \mathscr{F}, \mathrm{P})$ is *complete*, that is, that the subsets of null sets of $\mathscr{F}$ belong to $\mathscr{F}$.

Filtrations

A family $(\mathscr{F}_t,\ t \in \mathbb{I})$ of sub-sigma fields of $\mathscr{F}$ such that

a. $\mathscr{F}_0$ contains all null sets of $\mathscr{F}$, and
b. $\mathscr{F}_s \subset \mathscr{F}_t$ if $s < t$

is termed a *filtration*. A probability space is *filtered* if it is equipped with a filtration.

We will also use filtrations indexed by (partially) ordered sets other than $\mathbb{I}$. The definition remains the same: It suffices to replace $\mathbb{I}$ by the specified index set. When no confusion is possible, we will write $(\mathscr{F}_t)$ instead of $(\mathscr{F}_t,\ t \in \mathbb{I})$. Moreover, we will denote by $\mathscr{F}_{\bowtie -}$ the sigma field $\bigvee_{t \in \mathbb{I}} \mathscr{F}_t$ (that is, the smallest sub-sigma field of $\mathscr{F}$ that contains all the $\mathscr{F}_t$).

Elements of the sigma field $\mathscr{F}_t$ represent the events that are observable at stage t or, equivalently, the information available at stage t on the value of the parameter ω.

Convention

A complete probability space $(\Omega, \mathscr{F}, \mathrm{P})$ is given and fixed once and for all. Unless otherwise mentioned, we assume that this space is equipped with a filtration $(\mathscr{F}_t)$ accompanied by an additional sigma field $\mathscr{F}_{\bowtie}$ such that $\mathscr{F}_{\bowtie -} \subset \mathscr{F}_{\bowtie} \subset \mathscr{F}$. When necessary, this sigma field will be considered as incorporated into $(\mathscr{F}_t)$. If a particular filtration is used in an example or an application and the sigma field $\mathscr{F}_{\bowtie}$ is not described explicitly, then $\mathscr{F}_{\bowtie} = \mathscr{F}_{\bowtie -}$ by convention.

Modeling Stochastic Systems

In a given application, the probability space is used to describe the set of values of the parameter ω and the probability that this parameter belongs to a particular subset. The set of actions available to the observer is described by a (partially) ordered set, and the information on the value of the parameter ω after each action by the filtration. To explain why this description is appropriate, notice that the observer may initially have little or no information concerning the value of the parameter ω, but, based on the available information, will choose an action from the set of actions available at that stage. If $(a_1, \ldots, a_n)$ is the sequence of actions that the observer has taken up to time n, he will have to choose the next action from a set that may depend on $(a_1, \ldots, a_n)$, or decide to terminate the observation of the process. Of course, the information available to the observer may also depend on $(a_1, \ldots, a_n)$.

It is clearly natural to use finite and infinite sequences $(a_1, \ldots, a_n)$ and $(a_1, a_2, \ldots)$ to index the information and costs. This set is naturally endowed with a (partial) order: One sequence precedes the other if the first is obtained from the second by truncation. Abstracting from this situation leads to processes indexed by (partially) ordered sets.

A special case of particular interest is when at each stage, there is a fixed set of d actions $\{A_1, \ldots, A_d\}$ to choose from, and the information and costs do not depend explicitly on the order in which actions have been taken, but only on the number t^i of times each action A_i has been chosen. Indeed, many theoretical and practical applications arise in this form. In this case, it is natural to parametrize actions and costs by the set $\mathbb{I}$ $(= \mathbb{N}^d)$ of d-tuples $t = (t^1, \ldots, t^d)$ of integers. Since the information contained in $\mathscr{F}_t$ can only be acquired sequentially, that is, by passing from one stage to another that is a direct successor, it appears that the novelty in dimension $d > 1$ compared to dimension 1 comes from the fact that there are several possible ways to proceed, each consisting of the choice of a succession of stages, which defines a strategy or path leading to t. We shall concentrate on this situation in the first six chapters of this book.

The general case is studied in subsequent chapters. It will turn out that the general situation can be solved with only a small additional investment compared to the special case just described.

Although we will generally assume that the filtered probability space is given, we will also examine (in Chapter 6) how this probability space can be constructed from an informal description of a control problem.

An Example

Suppose that a gambler has the choice of playing one of d games at each stage. Assume that for $i = 1, \ldots, d$, the information on the parameter ω that is available after t^i plays of game i is represented by a sigma field $\mathscr{F}^i_{t^i}$, and that $\mathscr{F}^i_{t^i} \subset \mathscr{F}^i_{t^i+1}$ for all $t^i \in \mathbb{N}$. If after n games the gambler has played game i exactly t^i times $(t^1 + \cdots + t^d = n)$, then the information on the value of the parameter ω available to the gambler at the stage $t = (t^1, \ldots, t^d)$ is represented by the sigma field $\mathscr{F}_t$ equal to the smallest sigma field that contains $\mathscr{F}^1_{t^1}, \ldots, \mathscr{F}^d_{t^d}$. Notice that this information does not depend on the order in which the plays are made. Typically, the gambler will use this information to decide which game to play next. The family $(\mathscr{F}_t)$ of sigma fields thus defined is a filtration.

1.2 RANDOM VARIABLES

A *random variable* is a measurable function defined on Ω with values in a measurable space B. The sigma field of measurable sets of B is denoted $\mathscr{B}$. When B is a topological space [a countable set], then unless another choice is specified, $\mathscr{B}$ is the Borel sigma field of B, that is, the sigma field generated by the open sets of B [the sigma field $\mathscr{P}(\mathsf{B})$ of all subsets of B]. When B is not specified, it is agreed that $\mathsf{B} = \overline{\mathbb{R}}$. In agreement with the terminology we have adopted for functions, we will say that a random variable is finite or real valued [nonnegative] if its range is contained in $\mathbb{R}$ [$\overline{\mathbb{R}}_+$] and bounded if its range is a bounded subset of $\mathbb{R}$.

The set on which a random variable X satisfies some set C of conditions is frequently denoted by $\{C\}$. For example, $\{X > 0\}$ denotes the set of ω such that $X(\omega) > 0$. We will write $\mathrm{P}\{C\}$ rather than $\mathrm{P}(\{C\})$.

If F is a set in $\mathscr{F}$, then 1_F denotes the random variable that is equal to 1 on F and to 0 on $\Omega - F$. The set $\Omega - F$ is often denoted F^c.

For convenience of exposition, we will not make any distinction between a random variable X and the equivalence class of all random variables which are almost surely (a.s.) equal to X (that is, equal to X except on a null set). Accordingly, unless otherwise indicated, relations between random variables are to be interpreted as relations between their respective equivalence classes, that is, as relations holding a.s. For example, "$X \leq Y$" is equivalent to "$X \leq Y$ a.s.", "$X < Y$ on F" is equivalent to "$X < Y$ a.s. on F". In the same spirit, expressions such as "finite or nonnegative on F" have the meaning "finite or nonnegative a.s. on F." Notice that "$X < \infty$" means "$X < \infty$ a.s." Of course, when a random variable is being defined, for example, through a formula, it is often preferable to think in terms of random variables rather than equivalence classes, in other words, to consider that the formula holds everywhere on F or on Ω. Be that as it may, no confusion should arise from this slight abuse of notation.

Upwards-Directed Families

A family $(X_l, l \in \mathbb{L})$ of random variables is termed *upwards directed* if for all couples (l', l'') of elements of $\mathbb{L}$, there exists an element l of $\mathbb{L}$ such that $X_l \geq \sup(X_{l'}, X_{l''})$.

Limits

All limits (lim), upper limits (limsup), or lower limits (liminf) of a family of random variables are to be understood in the sense of a.s. convergence, unless some other mode of convergence is explicitly indicated.

Sigma Fields Generated by Random Variables, Supremum of Sigma Fields

In agreement with the convention above, the *sigma field generated by a family* $(X_l, l \in \mathbb{L})$ of *random variables* is the sigma field generated by all sets of the form $\{X_l \in B\}$, where B is measurable, and by all null sets.

The smallest sigma field $\mathscr{G}$ that contains the sigma fields $\mathscr{G}_1, \ldots, \mathscr{G}_n$ is denoted $\mathscr{G}_1 \vee \cdots \vee \mathscr{G}_n$.

Expectation, Conditional Expectation, and Conditional Probability

The *expectation* or *mean* $E(X)$ of an integrable random variable X is the integral of X. The integral of X over a set F in $\mathscr{F}$ will sometimes be denoted $E(X; F)$ ($E(X; Y \in B)$ if $F = \{Y \in B\}$). This notation will also be used for *semi-integrable* random variables, that is random variables X such that X^+ or X^- is integrable. In this case, $E(X; F)$ may equal ∞ or $-\infty$.

The space of all random variables X such that $E(|X|^p) < \infty$ $(p \geq 1)$ will be denoted by L^p.

The conditional expectation of an integrable or semi-integrable random variable X with respect to a sub-sigma field $\mathscr{G}$ of $\mathscr{F}$ will be denoted $E(X|\mathscr{G})$; if $\mathscr{G}$ is the sigma field generated by a family $(X_l,\ l \in \mathbb{L})$ of random variables, then $E(X|\mathscr{G})$ will be denoted $E(X|X_l,\ l \in \mathbb{L})$.

The *conditional probability* of a set $F \in \mathscr{F}$ given $\mathscr{G}$ (that is, the conditional expectation of 1_F given $\mathscr{G}$) will be denoted $P(F|\mathscr{G})$, or $P(X \in B|\mathscr{G})$ if $F = \{X \in B\}$.

Essential Supremum

When $\mathbb{L}$ is not countable, it is necessary to replace the notion of supremum of a family $(X_l,\ l \in \mathbb{L})$ of random variables by that of *essential supremum*. This random variable, denoted $\operatorname{esssup}_l X_l$, is the unique random variable X such that

a. $X \geq X_l$ for all $l \in \mathbb{L}$, and
b. $X \leq Y$ for all random variables Y such that $Y \geq X_l$ for all $l \in \mathbb{L}$.

It is clear that there cannot be more than one random variable X for which properties a and b are satisfied. The theorem below shows that one such variable does indeed exist. Notice that this variable is generally not equal to $\sup_l X_l$ (cf. Exercise 2).

Theorem (Existence of the essential supremum). *Let $(X_l,\ l \in \mathbb{L})$ be an arbitrary family of random variables. There exists a sequence $(l_n,\ n \in \mathbb{N})$ of elements of $\mathbb{L}$ such that the random variable* $\sup_n X_{l_n}$ *satisfies properties* a *and* b *above. If the family is upwards directed, then the sequence $(l_n,\ n \in \mathbb{N})$ can be chosen such that $X_{l_n} \leq X_{l_{n+1}}$ for all $n \in \mathbb{N}$. If in addition, $X_{l'}$ is integrable for some element l' of $\mathbb{L}$, then this same sequence can be chosen so that $l_0 = l'$; in this case,*

$$E\left(\operatorname*{esssup}_l X_l|\mathscr{G}\right) = \sup_n E\left(X_{l_n}|\mathscr{G}\right) = \operatorname*{esssup}_l E(X_l|\mathscr{G})$$

for any sub-sigma field $\mathscr{G}$ of $\mathscr{F}$; in particular,

$$E\left(\operatorname*{esssup}_l X_l\right) = \sup_n E(X_{l_n}) = \sup_l E(X_l).$$

Proof. By replacing each random variable X_l by arctan X_l, we can assume that $|X_l| \leq c$ for all $l \in \mathbb{L}$, where c is a finite constant. Let $\mathscr{D}$ be the set of all countable subsets of $\mathbb{L}$. Put $\alpha = \sup_{D \in \mathscr{D}} \mathrm{E}(\sup_{l \in D} X_l)$ and choose a sequence $(D_n, n \in \mathbb{N})$ of elements of $\mathscr{D}$ such that $\alpha = \sup_n \mathrm{E}(\sup_{l \in D_n} X_l)$. Set $D_0 = \bigcup_n D_n$ and $X = \sup_{l \in D_0} X_l$. By the definition of α, $\mathrm{E}(X) \leq \alpha$. On the other hand, $\mathrm{E}(X) \geq \alpha$, because $X \geq \sup_{l \in D_n} X_l$ for all $n \in \mathbb{N}$. It follows that $\mathrm{E}(X) = \alpha$, and therefore also that $\mathrm{E}(\sup(X, X_l)) = \alpha$ for all $l \in \mathbb{L}$. This shows that property a holds for X. Since property b also clearly holds for X, an arbitrary enumeration $(l_n, n \in \mathbb{N})$ of the elements of D_0 furnishes the desired sequence of elements of $\mathbb{L}$. In the case where the family $(X_l, l \in \mathbb{L})$ is upwards directed, set $l'_0 = l_0$ and choose by induction l'_{n+1} in $\mathbb{L}$ such that $X_{l'_{n+1}} \geq \sup(X_{l'_n}, X_{l_{n+1}})$. This determines a sequence $(l'_n, n \in \mathbb{N})$ such that $X_{l'_n} \leq X_{l'_{n+1}}$ for all $n \in \mathbb{N}$. Since we can add l' to D_0, and since the choice of l_0 is arbitrary, the last part of the proposition follows from the monotone convergence theorem for conditional expectations. Indeed, if $X_{l_n} \leq X_{l_{n+1}}$ for all $n \in \mathbb{N}$ and if X_{l_0} is integrable, then

$$\sup_n \mathrm{E}\left(X_{l_n} | \mathscr{G}\right) = \mathrm{E}\left(\sup_n X_{l_n} | \mathscr{G}\right)$$

$$= \mathrm{E}\left(\operatorname*{esssup}_l X_l | \mathscr{G}\right) \geq \operatorname*{esssup}_l \mathrm{E}\left(X_l | \mathscr{G}\right) \geq \sup_n \mathrm{E}\left(X_{l_n} | \mathscr{G}\right).$$

Remarks

1. It is easy to see that if $F \in \mathscr{F}$, then

$$\left(\operatorname*{esssup}_l X_l\right) 1_F = \operatorname*{esssup}_l \left(X_l 1_F\right)$$

(cf. Exercise 3).
2. The *essential infimum* of $(X_l, l \in \mathbb{L})$, denoted $\operatorname{essinf}_l X_l$, is the random variable $-\operatorname{esssup}_l(-X_l)$.
3. If $(F_l, l \in \mathbb{L})$ is a family of elements of $\mathscr{F}$, then the essential supremum [infimum] of this family, denoted $\operatorname{esssup}_l F_l$ [$\operatorname{essinf}_l F_l$], is the set F whose indicator function 1_F satisfies $1_F = \operatorname{esssup}_l 1_{F_l}$ [$1_F = \operatorname{essinf}_l 1_{F_l}$].
4. Any set L of random variables can be considered to be a family indexed by L itself. In this case, the essential supremum [infimum] is simply denoted esssup L [essinf L].

Uniform Integrability

We will end this section by briefly recalling some facts concerning uniform integrability. For the omitted proofs, we refer the reader to the book of Dellacherie and Meyer (1978).

A family $(X_l, l \in \mathbb{L})$ of random variables is *uniformly integrable* if

$$\lim_{k\to\infty} \sup_l \int_{\{|X_l|>k\}} |X_l| \, dP = 0.$$

A sequence of random variables converges in L^1 if and only if it is uniformly integrable and it converges in probability.

The closure in L^1 of the convex set generated by the elements of a uniformly integrable family is uniformly integrable. If $X \in L^1$ and if $(\mathscr{G}_l, l \in \mathbb{L})$ is a family of sub-sigma fields of $\mathscr{F}$, then $(E(X|\mathscr{G}_l), l \in \mathbb{L})$ is uniformly integrable.

A *test function* for uniform integrability is any function f defined on $\overline{\mathbb{R}}_+$ such that $f(\infty) = \infty$ and whose restriction to $\mathbb{R}_+$ is finite, convex, increasing and satisfies $\lim_{x\to\infty} f(x)/x = \infty$. Such a test function is necessarily continuous.

In order that a family $(X_l, l \in \mathbb{L})$ of random variables be uniformly integrable, it is necessary and sufficient that there exist a test function for uniform integrability f such that $\sup_l E(f|X_l|)) < \infty$.

In particular, if $(X_l, l \in \mathbb{L})$ is a uniformly integrable family, then $\sup_l E(|X_l|) < \infty$, and if $\sup_l E(|X_l|^p) < \infty$ for some $p > 1$, then the family $(X_l, l \in \mathbb{L})$ is uniformly integrable.

Here is an extension of Fatou's Lemma to semi-integrable random variables.

Fatou's Lemma. *If the positive part $(X_n^+, n \in \mathbb{N})$ of a sequence $(X_n, n \in \mathbb{N})$ of random variables is uniformly integrable and if* $\limsup_{n\to\infty} X_n$ *is semi-integrable, then*

$$E\left(\limsup_{n\to\infty} X_n\right) \geq \limsup_{n\to\infty} E(X_n).$$

Proof. Set $X_n^k = \inf(X_n, k)$. Clearly,

$$E\left(\limsup_{n\to\infty} X_n^k\right) = \lim_{n\to\infty} E\left(\sup_{m\geq n} X_m^k\right) \geq \limsup_{n\to\infty} E(X_n^k),$$

therefore, because $X_n \geq X_n^k$ and $\limsup_{n\to\infty} X_n$ is semi-integrable,

$$E\left(\limsup_{n\to\infty} X_n\right) \geq \limsup_{n\to\infty} E(X_n^k).$$

Take any $\varepsilon > 0$ and choose $k > 0$ such that

$$\sup_n E(X_n^+; X_n^+ > k) \leq \varepsilon.$$

Then for all n,

$$\begin{aligned} E(X_n^k) &\geq E(X_n; X_n \leq k) = E(X_n) - E(X_n; X_n > k) \\ &= E(X_n) - E(X_n^+; X_n^+ > k) \geq E(X_n) - \varepsilon. \end{aligned}$$

Consequently,

$$\mathrm{E}\Big(\limsup_{n\to\infty} X_n\Big) \geq \limsup_{n\to\infty} \mathrm{E}\big(X_n^k\big) \geq \limsup_{n\to\infty} \mathrm{E}(X_n) - \varepsilon,$$

which establishes the lemma, since ε is arbitrary.

1.3 STOPPING POINTS

In view of our spatial interpretation of the index set, the concept analogous to that of stopping time in dimension 1 will here be termed stopping point.

More precisely, we will term *stopping point* (relative to the filtration $(\mathscr{F}_t, t \in \mathbb{I})$) any function T defined on Ω with values in $\bar{\mathbb{I}}$ such that

$$\{T = t\} \in \mathscr{F}_t, \qquad \text{for all } t \in \mathbb{I}. \tag{1.1}$$

Clearly, property (1.1) can be expressed equivalently by

$$\{T \leq t\} \in \mathscr{F}_t, \qquad \text{for all } t \in \mathbb{I}. \tag{1.2}$$

Because $\{T = \bowtie\} = \{T < \bowtie\}^c = (\bigcup_{t \in \mathbb{I}} \{T = t\})^c$, if T is a stopping point, then $\{T = \bowtie\}$ belongs to $\mathscr{F}_{\bowtie -}$. However, notice that $\{T > t\}$ does not belong to $\mathscr{F}_t$ in general, since its complement is not $\{T \leq t\}$ unless the dimension d is equal to 1.

If T is a constant function, then T is a stopping point. All stopping points are random variables with values in $\bar{\mathbb{I}}$ (equipped with thc sigma field $\mathscr{P}(\bar{\mathbb{I}})$). If a function defined on Ω with values in $\bar{\mathbb{I}}$ is a.s. equal to a stopping point, then this function is also a stopping point. The definition of stopping point is therefore compatible with the convention that identifies two random variables that are equal a.s.

The set of stopping points will be considered ordered by the order relation $\leq$ induced by $\bar{\mathbb{I}}$.

In agreement with our conventions, a stopping point is finite if it is a.s. finite. However, depending on the context, some finite stopping points will in fact be finite everywhere, but no specific indication will be given, because we do not distinguish between two random variables equal a.s. A stopping point T will be termed *bounded* if there exists $t \in \mathbb{I}$ such that $T \leq t$.

In dimension 1, the definition of stopping point reduces to the classical definition of stopping time (relative to the filtration $(\mathscr{F}_n, n \in \mathbb{N})$). Stopping times will usually be denoted by the letters $\tau, \sigma, \ldots$.

Sigma Fields Associated with Stopping Points

We will term *sigma field associated* with a stopping point T, and will denote by $\mathscr{F}_T$, the sigma field of all sets F in $\mathscr{F}_{\bowtie}$ such that

$$F \cap \{T = t\} \in \mathscr{F}_t, \qquad \text{for all } t \in \mathbb{I}. \tag{1.3}$$

Clearly, property (1.3) can be expressed equivalently by

$$F \cap \{T \leq t\} \in \mathscr{F}_t, \qquad \text{for all } t \in \mathbb{I}. \tag{1.4}$$

Notice that $\mathscr{F}_T$ is the sigma field formed by the subsets F of Ω that satisfy (1.3) or (1.4) with $\bar{\mathbb{I}}$ instead of $\mathbb{I}$.

Clearly, $\mathscr{F}_T$ contains all null sets, and if two stopping points are a.s. equal, then their respective associated sigma fields are equal. The identification of two stopping points equal a.s. is therefore compatible with the notion of associated sigma field. If T is equal to the constant t, it is clear that $\mathscr{F}_T = \mathscr{F}_t$.

Properties of Stopping Points and Their Associated Sigma Fields

Here are some properties that we will use further on. The proofs which are straightforward are left to the reader (cf. Exercise 6).

1. If T is a stopping point, then T is $\mathscr{F}_T$-measurable, and if U is a random variable with values in $\bar{\mathbb{I}}$ which is $\mathscr{F}_T$-measurable and such that $U \geq T$, then U is a stopping point. Moreover, for all $t \in \bar{\mathbb{I}}$ and for all integrable random variables X, $\mathrm{E}(X|\mathscr{F}_T) = \mathrm{E}(X|\mathscr{F}_t)$ on $\{T = t\}$.
2. If S and T are two stopping points, then $S \vee T$ is a stopping point, because $\{S \vee T \leq t\} = \{S \leq t\} \cap \{T \leq t\} \in \mathscr{F}_t$. However, $S \wedge T$ is not a stopping point in general even if S or T is constant. But it is a stopping point if

$$\{S < T\} \cup \{S \geq T\} = \Omega, \tag{1.5}$$

because in this case $\{S \wedge T \leq t\} = \{S \leq t\} \cup \{T \leq t\} \in \mathscr{F}_t$. Notice that (1.5) always holds when $d = 1$.

A simple example of a stopping point T such that $s \wedge T$ is not a stopping point is the following. Assume that $d = 2$ and that F is an element of $\mathscr{F}$ such that $0 < \mathrm{P}(F) < 1$. Let

$$\mathscr{F}_t \text{ be the sigma field generated by} \begin{cases} \text{the null sets,} & \text{if } t = 0, \\ F \text{ and the null sets,} & \text{if } t \neq 0. \end{cases}$$

The family $(\mathscr{F}_t)$ thus defined is a filtration and $T = (0,1)\,1_F + (1,1)\,1_{F^c}$ is a stopping point relative to this filtration. However, $(1,0) \wedge T$ is not a stopping point, because $\{(1,0) \wedge T = 0\} = F \notin \mathscr{F}_0$.
3. If S and T are two stopping points such that $S \leq T$, then $\mathscr{F}_S \subset \mathscr{F}_T$.
4. If S and T are two stopping points, then the set $\{S = T\}$ belongs to $\mathscr{F}_S \cap \mathscr{F}_T$, but the set $\{S < T\}$ only belongs to $\mathscr{F}_T$, in general. This is because $\{S < T\}$ is not the complement of $\{S \geq T\}$ if $d > 1$. Under the hypothesis (1.5), $\{S < T\}$ also belongs to $\mathscr{F}_S$.

5. If S and T are two stopping points and if $F \in \mathscr{F}_S \cap \mathscr{F}_T$, then the function U defined by

$$U = \begin{cases} S \text{ on } F, \\ T \text{ on } F^c, \end{cases} \tag{1.6}$$

is a stopping point.

6. If $(T_n,\ n \in \mathbb{N})$ is a sequence [nonincreasing sequence] of stopping points, then $\bigvee_n T_n$ $[\bigwedge_n T_n]$ is a stopping point. Indeed,

$$\left\{ \bigvee_n T_n \leq t \right\} = \bigcap_n \{T_n \leq t\} \quad \left[\left\{ \bigwedge_n T_n \leq t \right\} = \bigcup_n \{T_n \leq t\} \right] \in \mathscr{F}_t.$$

It follows that for all sequences $(T_n,\ n \in \mathbb{N})$ of stopping points, $\limsup_{n \to \infty} T_n = \bigwedge_n \bigvee_{m>n} T_m$ is a stopping point.

7. If $(T_n,\ n \in \mathbb{N})$ is a nonincreasing sequence of stopping points and $T = \bigwedge_n T_n$, then $\mathscr{F}_T = \bigcap_n \mathscr{F}_{T_n}$. Indeed, if $F \in \mathscr{F}_{T_n}$ for all $n \in \mathbb{N}$, then

$$F \cap \{T \leq t\} = F \cap \left(\bigcup_n \{T_n \leq t\} \right) = \bigcup_n (F \cap \{T_n \leq t\}) \in \mathscr{F}_t,$$

which shows that $F \in \mathscr{F}_T$ and therefore that $\bigcap_n \mathscr{F}_{T_n} \subset \mathscr{F}_T$. Since the converse inclusion is an immediate consequence of item 3 above, equality of the two sigma fields follows.

8. If $(T_n,\ n \in \mathbb{N})$ is a nondecreasing sequence of stopping points and $T = \bigvee_n T_n$ is finite, then $\mathscr{F}_T = \bigvee_n \mathscr{F}_{T_n}$. Indeed, if $F \in \mathscr{F}_T$, then $F \cap \{T_n = T\} \in \mathscr{F}_{T_n}$, because

$$F \cap \{T_n = T\} \cap \{T_n = t\} = F \cap \{T = t\} \cap \{T_n = t\} \in \mathscr{F}_t.$$

However,

$$F = F \cap \left(\bigcup_n \{T_n = T\} \right) = \bigcup_n (F \cap \{T_n = T\});$$

therefore $F \in \bigvee_n \mathscr{F}_{T_n}$ and thus $\mathscr{F}_T \subset \bigvee_n \mathscr{F}_{T_n}$. Since the converse inclusion is a direct consequence of item 3, the two sigma fields are equal. Notice that this conclusion fails trivially when $\bigvee_n T_n$ is not finite. Indeed, when $d = 2$, this typically occurs for $T_n = (n, 0)$.

1.4 INCREASING PATHS AND ACCESSIBLE STOPPING POINTS

In dimension greater than 1 and unless the filtration satisfies certain conditions, the notion of stopping point has only limited applicability. This should not be surprising given that in dimension 1, a stopping time is typically the instant where an event occurs for the first time, and in dimension greater than 1, there is no useful interpretation of the expression "for the first time."

In many problems, the class of stopping points is advantageously replaced by the more restricted class of accessible stopping points. In essence, these are stopping points that can be reached by following certain random increasing paths. We begin by defining these paths.

Optional and Predictable Increasing Paths

We term *optional increasing path* any sequence $(\Gamma_n,\ n \in \mathbb{N})$ of finite stopping points such that

$$\Gamma_0 = 0 \quad \text{and} \quad \Gamma_{n+1} \in \mathbb{D}_{\Gamma_n}\,(\text{a.s.}), \qquad \text{for all } n \in \mathbb{N}.$$

In other words, an optional increasing path is an increasing sequence $(\Gamma_n,\ n \in \mathbb{N})$ of stopping points such that

$$|\Gamma_n| = n, \qquad \text{for all } n \in \mathbb{N}. \tag{1.7}$$

We term *predictable increasing path* any optional increasing path $(\Gamma_n,\ n \in \mathbb{N})$ such that

$$\Gamma_{n+1} \text{ is } \mathscr{F}_{\Gamma_n}\text{-measurable}, \qquad \text{for all } n \in \mathbb{N}. \tag{1.8}$$

Clearly, in order that a sequence $(\Gamma_n,\ n \in \mathbb{N})$ of functions defined on Ω with values in $\mathbb{I}$ be a predictable increasing path, it is necessary and sufficient that

$$\Gamma_n \le \Gamma_{n+1}, \quad |\Gamma_n| = n, \quad \text{and} \quad \{\Gamma_n = s\} \cap \{\Gamma_{n+1} = t\} \in \mathscr{F}_s,$$

for all $n \in \mathbb{N}$ and all couples (s, t) of elements of $\mathbb{I}$ such that $|s| = n$ and $t \in \mathbb{D}_s$.

The simplest examples of predictable increasing paths are *deterministic increasing paths*, that is, increasing sequences $(\gamma_n,\ n \in \mathbb{N})$ of elements of $\mathbb{I}$ such that $|\gamma_n| = n$ for all $n \in \mathbb{N}$.

If $(\Gamma_n,\ n \in \mathbb{N})$ is an optional or a predictable increasing path, then the *filtration associated* with this path is the sequence of sigma fields $(\mathscr{F}_{\Gamma_n},\ n \in \mathbb{N})$.

From now on, in order to avoid double lower indices, we will write $\mathscr{F}_n^\Gamma$ instead of $\mathscr{F}_{\Gamma_n}$. Moreover, in order to denote an optional or a predictable increasing path and its associated filtration, we will also use the abbreviated symbols (Γ_n) and $(\mathscr{F}_n^\Gamma)$, or even Γ and $\mathscr{F}^\Gamma$. For deterministic increasing paths, we will use the letter γ instead of Γ.

In the sequel, for all optional or predictable increasing paths Γ, we will set $\Gamma_\infty = \bowtie$ by convention. In agreement with this convention, $\mathscr{F}_\infty^\Gamma$ will be the sigma field $\mathscr{F}_\bowtie$ and will play the role of the additional sigma field that accompanies $\mathscr{F}^\Gamma$. When necessary, this sigma field will be considered incorporated into $\mathscr{F}^\Gamma$. The sigma field $\bigvee_n \mathscr{F}_n^\Gamma$ will be denoted by $\mathscr{F}_{\infty-}^\Gamma$.

A predictable increasing path can be thought of as a strategy to be used to acquire the information contained in the filtration $(\mathscr{F}_t)$, in the sense that it provides at each stage of the observation a way of deciding on the choice of the next stage. In this strategy, the index n has a temporal interpretation. We will therefore say, for instance, that Γ_n is the stage at time n, or that $\mathscr{F}_n^\Gamma$ contains the information available at time n, or that a random variable τ is a stopping time relative to the filtration $\mathscr{F}^\Gamma$.

Notice that if Γ is a predictable increasing path, or even only an optional increasing path, and if τ is a stopping time relative to the filtration $\mathscr{F}^\Gamma$, then Γ_τ is a stopping point. Indeed, for all $t \in \mathbb{I}$,

$$\{\Gamma_\tau = t\} = \{\Gamma_{|t|} = t\} \cap \{\tau = |t|\},$$

which shows that $\{\Gamma_\tau = t\} \in \mathscr{F}_t$, because $\{\tau = |t|\} \in \mathscr{F}_{|t|}^\Gamma$. It is easy to check that $\mathscr{F}_\tau^\Gamma = \mathscr{F}_{\Gamma_\tau}$.

In dimension 1, the concepts that we have just defined have little interest. In fact, in this case there is only one optional or predictable increasing path, namely, the deterministic increasing path defined by $\Gamma_n = n$ for all $n \in \mathbb{N}$. There is, however, no reason to exclude dimension 1 from our discussions, since when interpreted with the help of this path, the conclusions we will reach will give some of the results that are classical in dimension 1.

The Debut of a Set Along an Increasing Path

A subset H of $\bar{\mathbb{I}} \times \Omega$ $[\mathbb{I} \times \Omega]$ is *adapted* if its t-section $\{\omega: (t, \omega) \in H\}$ belongs to $\mathscr{F}_t$ for all $t \in \bar{\mathbb{I}}$ $[\mathbb{I}]$.

Let H be an adapted subset of $\mathbb{I} \times \Omega$ $[\mathbb{I} \times \Omega]$ and Γ an optional increasing path. For all ω, set

$$\tau(\omega) = \inf\{n \in \bar{\mathbb{N}}\ [\mathbb{N}]: (\Gamma_n(\omega), \omega) \in H\}. \tag{1.9}$$

The function τ thus defined is a stopping time relative to the filtration $\mathscr{F}^\Gamma$. Indeed, for all $n \in \mathbb{N}$ and all $t \in \mathbb{I}$, $\{\omega: (\Gamma_n(\omega), \omega) \in H\} \cap \{\Gamma_n = t\} \in \mathscr{F}_t$; therefore $\{\omega: (\Gamma_n(\omega), \omega) \in H\} \in \mathscr{F}_n^\Gamma$ and so $\{\tau \leq n\} \in \mathscr{F}_n^\Gamma$, because $\{\tau \leq n\} = \bigcup_{m \leq n} \{\omega: (\Gamma_m(\omega), \omega) \in H\}$. It follows that Γ_τ is a stopping point and we will term this stopping point the *debut of H along Γ*.

Increasing Paths that Pass Through a Stopping Point

We will say that an optional increasing path Γ *passes through* a stopping point T, or that T is *located on* Γ, if

$$\Gamma_{|T|} = T. \tag{1.10}$$

Notice that this relation holds trivially on $\{T = \bowtie\}$, because $|T| = \infty$ on this set and $\Gamma_\infty = \bowtie$. On the other hand, together with (1.7), it implies that $\{|T| = n\} = \{T = \Gamma_n\}$ for all $n \in \mathbb{N}$, and therefore that $|T|$ is a stopping time relative to the filtration $\mathscr{F}^\Gamma$, because $\{T = \Gamma_n\} \cap \{\Gamma_n = t\} = \{T = t\} \cap \{\Gamma_n = t\} \in \mathscr{F}_t$ for all $t \in \mathbb{I}$.

It thus appears that Γ passes through T if and only if there exists a stopping time τ relative to the filtration $\mathscr{F}^\Gamma$ such that $\Gamma_\tau = T$.

Accessible and Inaccessible Stopping Points

We will say that a stopping point T is *accessible* if there exists a predictable increasing path which passes through T. If a stopping point is not accessible, we will sometimes say that it is *inaccessible*.

All constant stopping points are accessible. In dimension 1, all stopping points are accessible. The debut of an adapted subset $\bar{\mathbb{I}} \times \Omega$ along a predictable increasing path is an accessible stopping point. If T is an accessible stopping point and Γ a predictable increasing path that passes through T, then $T \vee \Gamma_n$ and $T \wedge \Gamma_n$ are accessible stopping points for all $n \in \mathbb{N}$. This follows from the fact that $T \vee \Gamma_n = \Gamma_{|T| \vee n}$, $T \wedge \Gamma_n = \Gamma_{|T| \wedge n}$, and that $|T| \vee n$ and $|T| \wedge n$ are stopping times relative to the filtration $\mathscr{F}^\Gamma$. Note that $\bigvee_n (T \wedge \Gamma_n) = T$, so that every accessible stopping point is the supremum of a sequence of bounded accessible stopping points. Since the stopping point T in the example described in item 2 of Section 1.3 is accessible (along a deterministic increasing path), it appears that even in the case where T is accessible, $s \wedge T$ is not a stopping point in general.

Examples of Inaccessible Stopping Points

Consider again the filtration $(\mathscr{F}_t)$ introduced in item 2 of Section 1.3 and set $T = (1,0)\,1_F + (0,1)\,1_{F^c}$. If Γ is a predictable increasing path, then $\Gamma_{|T|} = \Gamma_1$ is $\mathscr{F}_0$-measurable, therefore constant. Relation (1.10) is therefore not satisfied, which allows us to conclude that T is not an accessible stopping point.

We will see further on (in Chapter 5) that under certain conditions relating to the structure of the filtration $(\mathscr{F}_t)$, all stopping points are accessible. This is in particular true in dimension 2 when the sigma fields $\mathscr{F}_t$ are generated by a family of independent random variables. In dimension greater than 2, however, the fact that the sigma fields $\mathscr{F}_t$ are generated by independent random variables does not necessarily imply that all stopping points are accessible. Here is an example that justifies this assertion (see also Exercise 7).

Let X_1, X_2, and X_3 be three independent random variables that take the values 0 and 1 with positive probability. For $i = 1, 2, 3$, let $\mathscr{F}_0^i$ be the sigma field generated by the null sets, and for all $t^i \in \mathbb{N}^*$, let $\mathscr{F}_{t^i}^i$ be the sigma field generated by X_i. For $t = (t^1, t^2, t^3) \in \mathbb{I}$ $(= \mathbb{N}^3)$, put $\mathscr{F}_t = \mathscr{F}_{t^1}^1 \vee \mathscr{F}_{t^2}^2 \vee \mathscr{F}_{t^3}^3$

and set

$$F_1 = \{X_2 = 1, X_3 = 1\}, \quad F_2 = \{X_1 = 0, X_3 = 0\},$$
$$F_3 = \{X_1 = 1, X_2 = 0\}.$$

Moreover, set

$$T = \begin{cases} (0,1,1) \text{ on } F_1, \\ (1,0,1) \text{ on } F_2, \\ (1,1,0) \text{ on } F_3, \\ (1,1,1) \text{ on } (F_1 \cup F_2 \cup F_3)^c. \end{cases}$$

Clearly, T is a stopping point. Suppose that there exists a predictable increasing path Γ that passes through T. Fix i such that $\mathrm{P}\{\Gamma_1 = e_i\} > 0$. From the definition of T, $\{T \not\geq e_i\} = F_i$. Because $\{\Gamma_1 = e_i\} \in \mathscr{F}_0$, $\mathrm{P}\{\Gamma_1 = e_i\}$ is equal to 1. Therefore

$$\mathrm{P}\{\Gamma_1 = e_i, T \not\geq e_i\} = \mathrm{P}(F_i) > 0.$$

Consequently, $\mathrm{P}\{T \not\geq \Gamma_1\} > 0$, which is impossible because $\mathrm{P}\{T \geq \Gamma_1\} = 1$, since $T = \Gamma_{|T|}$ and $|T| \geq 2$.

Unicity of the Path that Leads to an Accessible Stopping Point

In general, there are several predictable increasing paths Γ that pass through a given accessible stopping point T. It is interesting to know whether there exist accessible stopping points $T > 0$ for which there is only one path from 0 to T, in other words, stopping points T that determine $(\Gamma_{n \wedge |T|})$. The following example shows that such points do indeed exist, even with $|T|$ arbitrarily large and with arbitrarily small probability that one of the coordinates of T equals zero. Let $d = 2$ and let $(X_t, t \in \mathbb{I})$ be a family of independent random variables that take only the values 0 and 1, both with positive probability. For all $t \in \mathbb{I}$, let $\mathscr{F}_t$ denote the sigma field generated by the family $(X_s, s \in \mathbb{I}_t)$. Define Γ_n by setting $\Gamma_0 = 0$ and, by induction on n,

$$\Gamma_{n+1} = \begin{cases} \Gamma_n + (1,0), & \text{if } X_{\Gamma_n} = 1, \\ \Gamma_n + (0,1), & \text{if } X_{\Gamma_n} = 0. \end{cases}$$

Clearly, $\Gamma = (\Gamma_n)$ is a predictable increasing path (relative to the filtration $(\mathscr{F}_t)$). Set $T = \Gamma_m$, where m is an arbitrary positive integer. We are going to prove that if S is a stopping point such that $S \leq T$, then $\Gamma_{|S|} = S$, which implies the unicity of $(\Gamma_{n \wedge |T|})$. To prove this, assume that $\mathrm{P}\{\Gamma_{|S|} \neq S\} > 0$ and show that this leads to a contradiction. Because $\{|S| = m\} \subset \{S = T\} \subset \{\Gamma_{|S|} = S\}$, we deduce that $\{\Gamma_{|S|} \neq S\} \subset \{|S| < m\}$; therefore there exists

$s \in \mathbb{I}$ such that $|s| < m$ and $P\{S = s, \Gamma_{|s|} \neq s\} > 0$. But $\{\Gamma_{|s|} \neq s\}$ is the union of the two disjoint sets $\{\Gamma^1_{|s|} < s^1\}$ and $\{\Gamma^2_{|s|} < s^2\}$. Consequently, $P\{S = s, \Gamma^1_{|s|} < s^1\} > 0$ or $P\{S = s, \Gamma^2_{|s|} < s^2\} > 0$. Assume that $P\{S = s, \Gamma^1_{|s|} < s^1\} > 0$. By the definition of Γ,

$$\left\{S = s, \Gamma^1_{|s|} < s^1\right\} = \bigcup_{r^1 \in \mathbb{N},\, r^1 < s^1} \left\{S = s, \Gamma_{r^1 + s^2} = (r^1, s^2), X_{r^1, s^2} = 0\right\};$$

therefore there exists $r^1 \in \mathbb{N}$ such that $r^1 < s^1$ and

$$P(F \cap \{X_{r^1, s^2} = 0\}) > 0, \qquad \text{where } F = \left\{S = s, \Gamma_{r^1 + s^2} = (r^1, s^2)\right\}.$$

On the other hand, by the independence of the family $(X_t,\ t \in \mathbb{I})$ and because F belongs to $\mathcal{F}_s$,

$$\begin{aligned}
&P\{S = s, T = (r^1, m - r^1)\} \\
&\quad \geq P(F \cap \{T = (r^1, m - r^1)\}) \\
&\quad = P(F \cap \{X_{r^1, s^2} = 0, X_{r^1, s^2+1} = 0, \ldots, X_{r^1, m - r^1 - 1} = 0\}) \\
&\quad = P(F \cap \{X_{r^1, s^2} = 0\}) P\{X_{r^1, s^2+1} = 0\} \cdots P\{X_{r^1, m - r^1 - 1} = 0\} \\
&\quad > 0,
\end{aligned}$$

which contradicts the relation $S \leq T$.

In this example, the range of the stopping point T is the set $\{t \in \mathbb{I}: |t| = m\}$ and is therefore transversally totally ordered, so to speak. In the same vein, it is possible to provide examples showing that the unicity of $(\Gamma_{n \wedge |T|})$ can also occur in the case where the range of T is totally ordered by $\leq$ (cf. Exercise 8).

Absence of a Section Theorem

When H is an adapted subset of $\mathbb{I} \times \Omega$, one can ask whether there exists a section of H by a stopping point, in other words a stopping point T such that $\{\omega: (T(\omega), \omega) \in H\} = \{\omega: H_\omega \neq \varnothing\}$, where $H_\omega = \{t \in \mathbb{I}: (t, \omega) \in H\}$ denotes the section of H along ω. We will see in Exercises 9 and 10 that the answer to this question is negative (except in dimension 1), even in the case where all stopping points are accessible.

1.5 SOME OPERATIONS ON ACCESSIBLE STOPPING POINTS

The results of this section are not surprising, but their proof requires more than a little effort.

Proposition (Accessibility of the maximum). *If T_1 and T_2 are two accessible stopping points, then $T_1 \vee T_2$ is also an accessible stopping point. More* *precisely, $T_1 \vee T_2$ is accessible through T_1; in other words, there exists a predictable increasing path Γ that passes through T_1 and $T_1 \vee T_2$. Moreover, if ${}^1\Gamma$ is a predictable increasing path that passes through T_1, then Γ can be chosen in such a way that $(\Gamma_{n \wedge |T_1|}) = ({}^1\Gamma_{n \wedge |T_1|})$.*

Proof. Let ${}^1\Gamma$ and ${}^2\Gamma$ be two predictable increasing paths that pass, respectively, through T_1 and T_2. Clearly, we can assume that for all ω, $({}^1\Gamma_n(\omega))$ and $({}^2\Gamma_n(\omega))$ are deterministic increasing paths such that ${}^1\Gamma_{|T_1(\omega)|} = T_1(\omega)$ and ${}^2\Gamma_{|T_2(\omega)|} = T_2(\omega)$. For all ω, set $\tau(\omega) = |T_1(\omega)|$ and let $(\Gamma_n(\omega))$ denote the sequence defined by induction as follows:

$$\Gamma_n(\omega) = \begin{cases} {}^1\Gamma_n(\omega), & \text{if } n \le \tau(\omega) < \infty \\ & \text{or } \quad \tau(\omega) = \infty, \\ \Gamma_{n-1}(\omega) + {}^2\Gamma_{\tau_n(\omega)}(\omega) - {}^2\Gamma_{\tau_n(\omega)-1}(\omega), & \text{if } n > \tau(\omega), \end{cases}$$

where $\tau_n(\omega) = \inf\{m \in \mathbb{N}: {}^2\Gamma_m(\omega) \not\leq \Gamma_{n-1}(\omega)\}$. This sequence is a deterministic increasing path that passes through $T_1(\omega)$ and $T_1(\omega) \vee T_2(\omega)$, because $\Gamma_n(\omega) = {}^1\Gamma_n(\omega)$ if $n \le \tau(\omega) < \infty$ or $\tau(\omega) = \infty$, and as soon as $n = \tau(\omega)$, the successive increments of $\Gamma_n(\omega)$ are equal to those of ${}^2\Gamma_n(\omega)$ that correspond to increments of coordinates that are at least equal to the corresponding coordinates of $T_1(\omega)$. Let us prove that Γ_n is a stopping point. Since $\Gamma_0 = 0$, it suffices to establish this assertion in the case where $n > 0$ and Γ_{n-1} is assumed to be a stopping point. In fact, we are going to show that $\{\Gamma_n = t\}$ belongs to $\mathscr{F}_t \cap \mathscr{F}^{\Gamma}_{n-1}$ for all $t \in \mathbb{I}$, which will show at the same time that (Γ_n) is a predictable increasing path. Observe to begin with that

$$\begin{aligned} \{\Gamma_n = t\} &= (\{\tau > n-1\} \cap \{\Gamma_n = t\}) \cup (\{\tau \le n-1\} \cap \{\Gamma_n = t\}) \\ &= \left(\{\tau > n-1\} \cap \{{}^1\Gamma_n = t\}\right) \\ &\quad \cup \left(\{\tau \le n-1\} \cap \left\{\Gamma_{n-1} + {}^2\Gamma_{\tau_n} - {}^2\Gamma_{\tau_n - 1} = t\right\}\right). \end{aligned}$$

However, $\{\tau > n-1\}$ belongs to $\mathscr{F}^{{}^1\Gamma}_{n-1}$ and ${}^1\Gamma_n$ is $\mathscr{F}^{{}^1\Gamma}_{n-1}$-measurable, which implies that $\{\tau > n-1\} \cap \{{}^1\Gamma_n = t\}$ belongs to $\mathscr{F}_t \cap \mathscr{F}^{{}^1\Gamma}_{n-1}$, therefore to $\mathscr{F}_t \cap \mathscr{F}^{\Gamma}_{n-1}$, since ${}^1\Gamma_{n-1} = \Gamma_{n-1}$ on $\{\tau > n-1\}$. On the other hand, because

$$\begin{aligned} &\{\tau \le n-1\} \cap \left\{\Gamma_{n-1} + {}^2\Gamma_{\tau_n} - {}^2\Gamma_{\tau_n - 1} = t\right\} \\ &\quad = \bigcup_{i=1}^{d} \left(\{\tau \le n-1\} \cap \left\{{}^2\Gamma_{\tau_n} - {}^2\Gamma_{\tau_n-1} = e_i\right\} \cap \{\Gamma_{n-1} = t - e_i\}\right), \end{aligned}$$

and because $\{\tau \le n-1\}$ belongs to $\mathscr{F}^{\Gamma}_{n-1}$, we will be able to conclude that the right-hand side belongs to $\mathscr{F}_t \cap \mathscr{F}^{\Gamma}_{n-1}$ once we have proved that ${}^2\Gamma_{\tau_n} - {}^2\Gamma_{\tau_n-1}$ is $\mathscr{F}^{\Gamma}_{n-1}$-measurable. Let e denote one of the elements $e_1, \ldots, e_d$ and write

$$\left\{{}^2\Gamma_{\tau_n} - {}^2\Gamma_{\tau_n-1} = e\right\} = \bigcup_{m\geq 1}\left(\{\tau_n = m\} \cap \left\{{}^2\Gamma_m - {}^2\Gamma_{m-1} = e\right\}\right).$$

By the definition of τ_n,

$$\{\tau_n = m\} \cap \{\Gamma_{n-1} = t\} = \left\{{}^2\Gamma_0 \leq t, \ldots, {}^2\Gamma_{m-1} \leq t, {}^2\Gamma_m \nleq t\right\} \cap \{\Gamma_{n-1} = t\};$$

therefore $\{\tau_n = m\}$ belongs to $\mathscr{F}^{\Gamma}_{n-1}$, because the right-hand side belongs to $\mathscr{F}_t$ for all $t \in \mathbb{I}$. On the other hand, ${}^2\Gamma_m - {}^2\Gamma_{m-1}$ is $\mathscr{F}^{2\Gamma}_{m-1}$-measurable and ${}^2\Gamma_{m-1} \leq \Gamma_{n-1}$ on $\{\tau_n = m\}$, therefore

$$\begin{aligned}\{\tau_n = m\} &\cap \left\{{}^2\Gamma_m - {}^2\Gamma_{m-1} = e\right\} \cap \{\Gamma_{n-1} = t\} \\ &= \{\tau_n = m\} \cap \{\Gamma_{n-1} = t\} \cap \left\{{}^2\Gamma_m - {}^2\Gamma_{m-1} = e\right\} \cap \left\{{}^2\Gamma_{m-1} \leq t\right\}.\end{aligned}$$

Since the right-hand side belongs to $\mathscr{F}_t$ for all $t \in \mathbb{I}$, we conclude that $\{\tau_n = m\} \cap \{{}^2\Gamma_m - {}^2\Gamma_{m-1} = e\}$ belongs to $\mathscr{F}^{\Gamma}_{n-1}$, and consequently the same is true of $\{{}^2\Gamma_{\tau_n} - {}^2\Gamma_{\tau_n-1} = e\}$, which completes the proof.

Corollary 1 (Accessibility of a split point). *If T_1, T_2, and S are accessible stopping points such that $T_1 \wedge T_2 \geq S$ and if $F \in \mathscr{F}_S$, then the stopping point defined by*

$$T = \begin{cases} T_1 & \text{on } F, \\ T_2 & \text{on } F^c, \end{cases}$$

is accessible through S.

Proof. The fact that T is a stopping point follows from the property in item 5 of Section 1.3. By the proposition, there exists a predictable increasing path ${}^k\Gamma$ that passes through S and T_k $(k = 1, 2)$. Set

$$\Gamma_n = {}^1\Gamma_n \text{ on } \{n < |S|\} \quad \text{and} \quad \Gamma_n = \begin{cases} {}^1\Gamma_n & \text{on } F \cap \{n \geq |S|\}, \\ {}^2\Gamma_n & \text{on } F^c \cap \{n \geq |S|\}. \end{cases}$$

Clearly, $\Gamma_n \leq \Gamma_{n+1}$ and $|\Gamma_n| = n$. Moreover, for all couples (s, t) of elements of $\mathbb{I}$ such that $|s| = n$ and $t \in \mathbb{D}_s$,

$$\begin{aligned}\{\Gamma_n = s\} \cap \{\Gamma_{n+1} = t\} = &\left(\left\{{}^1\Gamma_n = s\right\} \cap \left\{{}^1\Gamma_{n+1} = t\right\} \cap \{n < |S|\}\right) \\ &\cup \bigcup_{u \in \mathbb{I}_s}\left(\left\{{}^1\Gamma_n = s\right\} \cap \left\{{}^1\Gamma_{n+1} = t\right\} \cap F \cap \{S = u\}\right) \\ &\cup \bigcup_{u \in \mathbb{I}_s}\left(\left\{{}^2\Gamma_n = s\right\} \cap \left\{{}^2\Gamma_{n+1} = t\right\} \cap F^c \cap \{S = u\}\right).\end{aligned}$$

Since $\{n < |S|\}$ belongs to $\mathscr{F}_{n-1}^{1\Gamma}$ and ${}^1\Gamma_{n+1}$ is $\mathscr{F}_{n-1}^{1\Gamma}$-measurable, the first set in parentheses on the right-hand side belongs to $\mathscr{F}_s$. The same is true of the two unions over u, because ${}^k\Gamma_{n+1}$ is $\mathscr{F}_{n-1}^{k\Gamma}$-measurable ($k = 1, 2$) and $F \cap \{S = u\}$ and $F^c \cap \{S = u\}$ belong to $\mathscr{F}_u \subset \mathscr{F}_s$. Therefore $\{\Gamma_n = s\} \cap \{\Gamma_{n+1} = t\}$ belongs to $\mathscr{F}_s$. It follows that (Γ_n) is a predictable increasing path, which establishes the corollary, since this path clearly passes through S and T.

Corollary 2 (Accessibility of the supremum). *Let $(T_k,\ k \in \mathbb{N})$ be a nondecreasing sequence of accessible stopping points. There exists a predictable increasing path that passes through all the T_k, therefore also through $T = \bigvee_k T_k$.*

Proof. By the proposition, we can define inductively a sequence $({}^k\Gamma,\ k \in \mathbb{N})$ of predictable increasing paths such that ${}^k\Gamma$ passes through T_k and ${}^{k+1}\Gamma_n = {}^k\Gamma_n$ on $\{n \leq |T_k|\}$. In order to obtain the desired path (Γ_n), it suffices to set

$$\Gamma_n = {}^k\Gamma_n \quad \text{on } \{n \leq |T_k|\} \cup \big(\{n > |T_k|\} \cap \{T_{k-1} < T = T_k\}\big).$$

Indeed, on each of these sets, $\Gamma_n = {}^k\Gamma_n$ and $\Gamma_{n+1} = {}^k\Gamma_{n+1}$ for an appropriate k; therefore $\Gamma_n \leq \Gamma_{n+1}$ and $|\Gamma_n| = n$. Moreover, $\Gamma_{|T_k|} = {}^k\Gamma_{|T_k|} = T_k$, which shows that (Γ_n) passes through all the T_k. Finally, in order to check that $\{\Gamma_n = s\} \cap \{\Gamma_{n+1} = t\} \in \mathscr{F}_s$ for all couples (s, t) of elements of $\mathbb{I}$ such that $t \in \mathbb{D}_s$, it suffices to observe that this set is equal to

$$\bigcup_k \big(\{{}^k\Gamma_n = s\} \cap \{{}^k\Gamma_{n+1} = t\} \cap \{n \leq |T_k|\}\big)$$
$$\cup \bigcup_k \big(\{{}^k\Gamma_n = s\} \cap \{{}^k\Gamma_{n+1} = t\} \cap \{n > |T_k|\} \cap \{T_{k-1} < T = T_k\}\big)$$

and to observe that $\{{}^k\Gamma_{n+1} = t\}$, $\{n \leq |T_k|\}$, and $\{n > |T_k|\} \cap \{T_{k-1} < T = T_k\}$ belong to $\mathscr{F}_{n-1}^{k\Gamma}$.

1.6 STOCHASTIC PROCESSES AND MARTINGALES

A *stochastic process*, or simply a *process*, is a family $(X_t,\ t \in \mathbb{J})$ of random variables (defined on Ω) indexed by a subset $\mathbb{J}$ of $\bar{\mathbb{I}}$ (most often by $\bar{\mathbb{I}}$ itself, or by $\mathbb{I}$). Occasionally, other index sets may also be used.

A process will often be denoted by (X_t), or even by X, together with the indication, when warranted, that (X_t) or X is indexed by $\mathbb{J}$. When no confusion is possible, we will sometimes write "for all t" instead of "for all $t \in \mathbb{J}$." The slight ambiguity that arises from the fact that the letter X is also usually used to denote random variables should not cause any difficulties. In agreement with our conventions, unless otherwise specified, the random variables that make up a process take their values in $\overline{\mathbb{R}}$. A *real-valued process* is a process whose random variables are real valued.

We will say that two processes X and Y are equal, and will write $X = Y$, if they are indexed by the same set $\mathbb{J}$ and if $X_t = Y_t$ for all $t \in \mathbb{J}$. The set of processes indexed by $\mathbb{J}$ will be ordered by the relation $X \le Y$, which means $X_t \le Y_t$ for all $t \in \mathbb{J}$ and reads "X less than or equal to Y" or "Y greater than or equal to X" (or even "Y dominates X"). We can thus speak of nonnegative processes (that is, processes ≥ 0, where 0 denotes the zero process), of the supremum $\sup(X, Y)$ or infimum $\inf(X, Y)$ of two processes X and Y, of the largest or the smallest process (if it exists) in a set of processes, et cetera. Clearly, $\sup(X, Y)$ is the process $(\sup(X_t, Y_t))$ and $\inf(X, Y)$ the process $(\inf(X_t, Y_t))$. The positive [negative] part of a process X is the process $X^+ = \sup(X, 0)$, that is (X_t^+) [$X^- = -\inf(X, 0)$, that is, (X_t^-)]. The absolute value of X is the process $|X| = X^+ + X^-$, that is $(|X_t|)$.

A process X indexed by $\mathbb{J}$ will be termed

a. *adapted* (to the filtration $(\mathscr{F}_t)$) if X_t is $\mathscr{F}_t$-measurable for all $t \in \mathbb{J}$;

b. *nondecreasing* if $X_s \le X_t$ for all couples (s, t) of elements of $\mathbb{J}$ such that $s \le t$;

c. *integrable* if X_t is integrable for all $t \in \mathbb{J}$;

d. *bounded* in L^p if $\sup_{t \in \mathbb{J}} E(|X_t|^p) < \infty$;

e. *uniformly integrable* if the family (X_t) is uniformly integrable.

We will say that $(\tilde{\mathscr{F}}_t, t \in \mathbb{J})$ is the *natural filtration* of a process X indexed by $\mathbb{J}$ if for all $t \in \mathbb{J}$, $\tilde{\mathscr{F}}_t$ is the sigma field generated by the family $(X_s, s \in \mathbb{J}, s \le t)$. Recall that according to our conventions, $\tilde{\mathscr{F}}_t$ contains all null sets of $\mathscr{F}$. Clearly, every process is adapted to its natural filtration, which means that it satisfies condition a above with $\tilde{\mathscr{F}}_t$ instead of $\mathscr{F}_t$.

If X is a process indexed by $\mathbb{J}$ and T is a stopping point with values in $\mathbb{J}$, then X_T will denote the random variable $\omega \mapsto X_{T(\omega)}(\omega)$. If the process X is adapted, then X_T is $\mathscr{F}_T$-measurable.

A Canonical Example

Suppose that d real-valued processes $Y^1 = (Y^1_{t^1}, t^1 \in \mathbb{N}), \dots, Y^d = (Y^d_{t^d}, t^d \in \mathbb{N})$ are given and that $Y_0^1 = \cdots = Y_0^d = 0$. We shall interpret $Y^i_{t^i}$ as the reward obtained from the t^ith play of some game G_i. After n plays, of which t^i are played on game G_i $(t^1 + \cdots + t^d = n)$, the total reward is

$$X_t = \sum_{i=1}^{d} \sum_{s^i=0}^{t^i} Y^i_{s^i}, \qquad \text{where } t = \left(t^1, \dots, t^d\right).$$

If $(\mathscr{F}^i_{t^i}, t^i \in \mathbb{N})$ is the natural filtration of Y^i, then the process $(X_t, t \in \mathbb{I})$ is adapted to the filtration $(\mathscr{F}_t, t \in \mathbb{I})$ defined by

$$\mathscr{F}_t = \mathscr{F}^1_{t^1} \vee \cdots \vee \mathscr{F}^d_{t^d}, \qquad \text{where } t = (t^1, \ldots, t^d).$$

A Generalization

Suppose that a real-valued process $Y = (Y_t, t \in \mathbb{I})$ is given and set

$$X_t = \sum_{s \leq t} Y_s, \qquad \text{for all } t \in \mathbb{I}.$$

Then the process $(X_t, t \in \mathbb{I})$ is adapted to the natural filtration of Y. This example generalizes the previous one, because it reduces to it when $Y_{t^i e_i} = Y^i_{t^i}$ and $Y_t = 0$ if two or more coordinates of t are positive.

Some Notation

From now on, if Γ is an optional or a predictable increasing path and X is a process indexed by $\bar{\mathbb{I}}$ $[\mathbb{I}]$, we will write X^Γ_n instead of X_{Γ_n}, and X^Γ instead of $(X^\Gamma_n, n \in \bar{\mathbb{N}}$ $[\mathbb{N}])$. If X is adapted, then X^Γ_n is $\mathscr{F}^\Gamma_n$-measurable for all $n \in \bar{\mathbb{N}}$ $[\mathbb{N}]$, in other words, X^Γ is adapted to $\mathscr{F}^\Gamma$. Let us stress that X^Γ is a process indexed by $\bar{\mathbb{N}}$ or $\mathbb{N}$ according as X is indexed by $\bar{\mathbb{I}}$ or $\mathbb{I}$.

The subset of $\mathbb{J} \times \Omega$ on which a process X indexed by $\mathbb{J}$ satisfies some set C of conditions will sometimes be denoted by $\{\{C\}\}$, where the double brackets indicate that it is a subset of $\mathbb{I} \times \Omega$ rather than a subset of Ω. For example, $\{\{X > 0\}\}$ denotes the set of (t, ω) such that $X_t(\omega) > 0$. Notice that $\{\{X > 0\}\}$ is adapted (that is, $1_{\{\{X>0\}\}}$ is adapted) if X is adapted. The same will be true for all sets C of conditions that may come up in the sequel.

Processes of Class $D_{\mathbf{A}}$ and $D_{\mathbf{A}_f}$

From now on, we will denote the set of accessible stopping points by $\mathbf{A}$, the set of finite accessible stopping points by $\mathbf{A}_f$ and the set of bounded accessible stopping points by $\mathbf{A}_b$.

We will say that a process X indexed by $\bar{\mathbb{I}}$ $[\mathbb{I}]$ is *of class* $D_{\mathbf{A}}$ $[D_{\mathbf{A}_f}]$ if the family $(X_T, T \in \mathbf{A}$ $[\mathbf{A}_f])$ is uniformly integrable. In dimension 1, the notion of process of class $D_{\mathbf{A}}$ is equivalent to the classical notion of process of class D.

Clearly, all processes of class $D_{\mathbf{A}}$ $[D_{\mathbf{A}_f}]$ are uniformly integrable. On the other hand, even in dimension 1, the uniform integrability of a process X does not in general ensure the integrability of the individual random variables X_T (cf. Exercise 11).

If $E(\sup_t |X_t|) < \infty$, then X is of class $D_{\mathbf{A}}$ $[D_{\mathbf{A}_f}]$, because $|X_T| \leq \sup_t |X_t|$ for all $T \in \mathbf{A}$ $[\mathbf{A}_f]$.

If $X_{\bowtie}$ is integrable when the index set is $\mathbb{I}$, then in order that a process X be of class $D_{\mathbf{A}}$ $[D_{\mathbf{A}_f}]$, it suffices that the family $(X_T, T \in \mathbf{A}_b)$ be uniformly integrable. Indeed, if $T \in \mathbf{A}$ and Γ is a predictable increasing path that passes through T, then $T_n = T \wedge \Gamma_n \in \mathbf{A}_\mathrm{b}$ for all $n \in \mathbb{N}$ and $\lim_{n \to \infty} X_{T_n} 1_{\{T < \bowtie\}} = X_T 1_{\{T < \bowtie\}}$ in L^1, because $T_n = T$ on $\{T < \bowtie\}$ as soon as n is large enough, and the sequence $(X_{T_n}, n \in \mathbb{N})$ is uniformly integrable by hypothesis. Thus, $X_T = X_T 1_{\{T < \bowtie\}} + X_\bowtie 1_{\{T = \bowtie\}}$ belongs to the closure in L^1 of the uniformly integrable set formed by the random variables of the form $X_S 1_F + X_\bowtie 1_{F^c}$, where $S \in \mathbf{A}_\mathrm{b}$ and $F \in \mathscr{F}$. By a property recalled in the subsection "Uniform integrability" in Section 1.2, this closure is uniformly integrable. Omitting the terms involving $X_\bowtie$ handles the case where the index set is $\mathbb{I}$.

Supermartingales and Martingales

A process X indexed by $\mathbb{J}$ is a *supermartingale* [*martingale*] if it is adapted, integrable and

$$\mathrm{E}(X_t | \mathscr{F}_s) \leq X_s \qquad [\mathrm{E}(X_t | \mathscr{F}_s) = X_s], \tag{1.11}$$

for all couples (s, t) of elements of $\mathbb{J}$ such that $s < t$.

Notice that in the case where $\mathbb{J} = \mathbb{I}$, it is equivalent to require only that (1.11) be satisfied by the couples (s, t) such that $t \in \mathbb{D}_s$. Also notice that a process X indexed by $\bar{\mathbb{I}}$ $[\mathbb{I}_t]$ is a martingale if $X_\bowtie$ $[X_t]$ is integrable and $\mathrm{E}(X_\bowtie | \mathscr{F}_s) = X_s$ for all $s \in \bar{\mathbb{I}}$ $[\mathrm{E}(X_t | \mathscr{F}_s) = X_s$ for all $s \in \mathbb{I}_t]$. In dimension 1, the definition above is equivalent to the standard definition of a supermartingale [martingale]. In this case, we say that a supermartingale [martingale] is *closed on the right* to indicate that the index set is $\bar{\mathbb{N}}$.

As in dimension 1, the following assertions follow more or less directly from the definitions (cf. Exercise 12). A supermartingale X is a martingale if and only if the function $t \mapsto \mathrm{E}(X_t)$ is constant. A linear combination of two martingales (with the same index set) is a martingale. A linear combination with nonnegative coefficients of two supermartingales (with the same index set) is a supermartingale. The infimum of two supermartingales (with the same index set) is a supermartingale. Consequently, the negative part X^- of a supermartingale X is a *submartingale* (that is, the opposite of a supermartingale) and the absolute value $|X|$ of a martingale is a submartingale. The composition of $f \circ X$ of a concave and nondecreasing function f with a supermartingale X is a supermartingale, provided that $f(X_t)$ is integrable for all t. Any martingale indexed by $\bar{\mathbb{I}}$ is uniformly integrable. Any supermartingale X indexed by $\bar{\mathbb{I}}$ is bounded in L^1 and its negative part X^- is uniformly integrable. This follows from the relations $|X_t| = X_t + 2X_t^-$ and $X_t^- \leq \mathrm{E}(X_\bowtie^- | \mathscr{F}_t)$.

In the standard theory of martingales, the fact that optional sampling preserves the characteristics of a supermartingale or of a martingale plays an important role. In the same way, Theorem 1 below will be the key to the problems we plan to solve. Corollary 1 will show that this theorem is equivalent to the optional sampling theorem for accessible stopping points. We will see at the end of Section 3.3 that the validity of the optional sampling theorem implies that all stopping points are accessible. For the time being, the reader can easily construct a simple example of an inaccessible stopping point that does not satisfy the conclusion of the optional sampling theorem (cf. Exercise 13). It is also easy to check that without the predictability assumption on the increasing path, the conclusion of Theorem 1 is no longer valid.

Theorem 1 (Composing a supermartingale with a predictable increasing path). *Let X be a supermartingale [martingale] indexed by $\bar{\mathbb{I}}$ or $\mathbb{I}$. For all predictable increasing paths Γ, the process X^Γ is an ordinary supermartingale [martingale] (closed on the right if the index set is $\bar{\mathbb{I}}$) relative to the filtration $\mathscr{F}^\Gamma$.*

Proof. The process X^Γ is adapted to the filtration $\mathscr{F}^\Gamma$ and is integrable, because $|X_n^\Gamma| \leq \sum_{t:\,|t|=n} |X_t|$, for all $n \in \mathbb{N}$, and $X_\infty^\Gamma = X_{\bowtie}$ if the index set is $\bar{\mathbb{I}}$. Assume that X is a supermartingale. Let $n \in \mathbb{N}$, $F \in \mathscr{F}_n^\Gamma$, and consider $t \in \mathbb{I}$ such that $|t| = n + 1$. Clearly,

$$\int_{F \cap \{\Gamma_{n+1} = t\}} X_{n+1}^\Gamma \, dP = \int_{F \cap \{\Gamma_{n+1} = t\}} X_t \, dP = \sum_{s \in \mathbb{P}_t} \int_{F \cap \{\Gamma_{n+1} = t\} \cap \{\Gamma_n = s\}} X_t \, dP,$$

where $\mathbb{P}_t$ denotes the set of direct predecessors of t. But Γ_{n+1} is $\mathscr{F}_n^\Gamma$-measurable; therefore, $F \cap \{\Gamma_{n+1} = t\} \cap \{\Gamma_n = s\}$ belongs to $\mathscr{F}_s$, which implies by (1.11) that the last right-hand side is less than or equal to

$$\sum_{s \in \mathbb{P}_t} \int_{F \cap \{\Gamma_{n+1} = t\} \cap \{\Gamma_n = s\}} X_s \, dP = \int_{F \cap \{\Gamma_{n+1} = t\}} X_n^\Gamma \, dP.$$

By summing over all t such that $|t| = n + 1$, we deduce that

$$\int_F X_{n+1}^\Gamma \, dP \leq \int_F X_n^\Gamma \, dP,$$

therefore, that $E(X_{n+1}^\Gamma | \mathscr{F}_n^\Gamma) \leq X_n^\Gamma$. On the other hand, if the index set is $\bar{\mathbb{I}}$, then by (1.11), it follows that for all $n \in \mathbb{N}$ and all $F \in \mathscr{F}_n^\Gamma$,

$$\int_F X_{\bowtie} \, dP \leq \sum_{t:\,|t|=n} \int_{F \cap \{\Gamma_n = t\}} X_t \, dP = \int_F X_n^\Gamma \, dP,$$

which shows that $E(X_{\bowtie} | \mathscr{F}_n^\Gamma) \leq X_n^\Gamma$ and therefore that X^Γ is a supermartingale. If X is a martingale, then the last three inequalities above become equalities; in other words, X^Γ is a martingale.

Corollary 1 (The optional sampling theorem for accessible stopping points). *Let X be a supermartingale [martingale] indexed by $\bar{\mathbb{I}}$. If S and T are two accessible stopping points such that $S \leq T$, then X_S and X_T are integrable and* $\mathrm{E}(X_T|\mathscr{F}_S) \leq X_S$ $[\mathrm{E}(X_T|\mathscr{F}_S) = X_S]$. *This conclusion also applies if X is a supermartingale [martingale] indexed by $\mathbb{I}$ provided that $(X^-_{\Gamma_n}, n \in \mathbb{N})$ is uniformly integrable for all predictable increasing paths Γ and S and T are finite.*

Proof. By the proposition of Section 1.5, there exists a predictable increasing path Γ that passes through S and T, in other words such that $X^\Gamma_\sigma = X_S$ and $X^\Gamma_\tau = X_T$, where $\sigma = |S|$ and $\tau = |T|$. Therefore the corollary is an immediate consequence of the standard optional sampling theorem in dimension 1 (for supermartingales [martingales] closed on the right).

Corollary 2 (Class $\mathrm{D}_{\mathbf{A}}$ criterion). *All martingales indexed by $\bar{\mathbb{I}}$ are of class $\mathrm{D}_{\mathbf{A}}$. The negative part of a supermartingale indexed by $\bar{\mathbb{I}}$ is of class $\mathrm{D}_{\mathbf{A}}$. A supermartingale indexed by $\bar{\mathbb{I}}$ is therefore of class $\mathrm{D}_{\mathbf{A}}$ if and only if its positive part is.*

Proof. The first assertion follows from Corollary 1 and the second from the inequality $X^-_T \leq \mathrm{E}(X^-_{\bowtie}|\mathscr{F}_T)$, which is true for all accessible stopping points T, by Corollary 1.

Theorem 2 (Closing a supermartingale indexed by $\mathbb{I}$ on the right). *Let X be a supermartingale indexed by $\mathbb{I}$ whose negative part X^- is uniformly integrable. Then X_t converges in probability when $t \twoheadrightarrow \bowtie$ to an integrable random variable $X_{\bowtie}$ and $(X_t, t \in \bar{\mathbb{I}})$ is a supermartingale.*

Proof. Suppose that X_t does not converge in probability when $t \twoheadrightarrow \bowtie$. Then there exists $\varepsilon > 0$ such that for any $t \in \mathbb{I}$, there is $s \gg t$ such that $\mathrm{P}\{|X_s - X_t| > \varepsilon\} > \varepsilon$, and therefore there exists a sequence $(t_n, n \in \mathbb{N})$ of elements of $\mathbb{I}$ such that $t_n \twoheadrightarrow \bowtie$ and X_{t_n} does not converge in probability. Because the hypothesis implies that $\sup_{n \in \mathbb{N}} \mathrm{E}(X^-_{t_n}) < \infty$, and because $(X_{t_n}, n \in \mathbb{N})$ is a supermartingale, a standard result on convergence of ordinary supermartingales implies that X_{t_n} converges in probability to an integrable random variable. The resulting contradiction allows us to conclude that X_t converges in probability when $t \twoheadrightarrow \bowtie$, to a random variable $X_{\bowtie}$ which is necessarily integrable since it is the limit of X_{t_n} for any sequence $(t_n, n \in \mathbb{N})$ of elements of $\mathbb{I}$ such that $t_n \twoheadrightarrow \bowtie$. Because of the uniform integrability of X^-, the convergence of X^-_t to $X^-_{\bowtie}$ also occurs in L^1. By (1.11), if $s < t$ and $F \in \mathscr{F}_s$, then

$$\mathrm{E}(X^+_t; F) - \mathrm{E}(X^-_t; F) \leq \mathrm{E}(X_s; F).$$

It follows that

$$\liminf_{t \twoheadrightarrow \bowtie} \mathrm{E}(X_t^+; F) - \mathrm{E}(X_{\bowtie}^-; F) \leq \mathrm{E}(X_s; F).$$

By considering a sequence $(t_n, n \in \mathbb{N})$ of elements of $\mathbb{I}$ such that $t_n \twoheadrightarrow \bowtie$ and by applying Fatou's lemma, we see that

$$\mathrm{E}(X_{\bowtie}; F) \geq \liminf_{t \twoheadrightarrow \bowtie} \mathrm{E}(X_t^+; F) - \mathrm{E}(X_{\bowtie}^-; F) \leq \mathrm{E}(X_s; F),$$

which proves that $\mathrm{E}(X_{\bowtie} | \mathscr{F}_s) \leq X_s$.

Corollary 3 (Class $\mathrm{D}_{\mathbf{A}_f}$ criterion). *Let X be as in Theorem 2. Then X^- is of class $\mathrm{D}_{\mathbf{A}_f}$. In particular, for all optional increasing paths Γ, $(X_{\Gamma_n}^-, n \in \mathbb{N})$ is uniformly integrable and X_n^Γ converges* (a.s.) *to an integrable random variable.*

Proof. The first two conclusions are a direct consequence of Theorem 2 and Corollary 2. The last statement is a standard result for ordinary supermartingales.

The Increasing Process Associated with a Supermartingale Along a Predictable Increasing Path

Let X be a supermartingale indexed by $\bar{\mathbb{I}}$ and Γ a predictable increasing path. By Theorem 1, X^Γ is a supermartingale (closed on the right), therefore bounded in L^1 and its negative part is uniformly integrable. By Theorem 2, X_n^Γ converges in probability when $n \to \infty$ to an $\mathscr{F}_{\infty-}^\Gamma$-measurable and integrable random variable, which we denote $X_{\infty-}^\Gamma$.

Set

$$A_0^\Gamma = 0,$$

$$A_n^\Gamma = \sum_{m=1}^{n} \left(X_{m-1}^\Gamma - \mathrm{E}\left(X_m^\Gamma | \mathscr{F}_{m-1}^\Gamma\right)\right), \qquad \text{if } n \in \mathbb{N}^*, \tag{1.12}$$

$$A_\infty^\Gamma = A_{\infty-}^\Gamma + X_{\infty-}^\Gamma - \mathrm{E}\left(X_\infty^\Gamma | \mathscr{F}_{\infty-}^\Gamma\right), \qquad \text{where } A_{\infty-}^\Gamma = \sup_{n \in \mathbb{N}} A_n^\Gamma.$$

The process $(A_n^\Gamma, n \in \bar{\mathbb{N}})$ thus defined is nondecreasing, integrable, and *predictable*, this last term meaning that A_n^Γ is $\mathscr{F}_{n-1}^\Gamma$-measurable for all $n \in \mathbb{N}^*$ and A_∞^Γ is $\mathscr{F}_{\infty-}^\Gamma$-measurable. It will be denoted by A^Γ and termed the *increasing process associated with X along Γ*.

In dimension 1, because there is only one increasing path, the upper index is not necessary and A is precisely the ordinary increasing process associated with the supermartingale X.

For all $n \in \bar{\mathbb{N}}$, set

$$M_n^\Gamma = A_n^\Gamma + X_n^\Gamma. \tag{1.13}$$

The following assertions are part of the standard theory of martingales (cf. Exercise 15):

1. $(M_n^\Gamma,\ n \in \mathbb{N})$ is a martingale.
2. $(M_n^\Gamma,\ n \in \overline{\mathbb{N}})$ is a martingale (closed on the right) if and only if the supermartingale X^Γ is uniformly integrable.
3. If $\tilde{A}^\Gamma$ is an increasing process indcxcd by $\overline{\mathbb{N}}$, which is integrable, predictable in the sense just explained above, and such that $\tilde{A}^\Gamma + X^\Gamma$ is a martingale (closed on the right), then $\tilde{A}^\Gamma = A^\Gamma$.

EXERCISES

1. a. Prove the convergence criterion stated in the "Notation and Conventions" section. (*Hint*. To prove that $\limsup_{t \to \bowtie} x_t \leq x$ if $\limsup_{n \to \infty} x_{t_n} \leq x$ for all nondecreasing sequences $(t_n,\ n \in \mathbb{N})$ of elements of $\mathbb{I}$, proceed by contradiction and repeated extraction of d subsequences.)

b. Show that if $\mathbb{I}$ were replaced by $\mathbb{R}_+^d$ in part a, then there would exist families $(x_t,\ t \in \mathbb{R}_+^d)$ of real numbers such that $\limsup_{t \to \bowtie} x_t > \sup \limsup_{n \to \infty} x_{t_n}$, where the supremum on the right-hand side is taken over all nondecreasing sequences $(t_n,\ n \in \mathbb{N})$ of elements of $\mathbb{R}_+^d$.

2. Let $\Omega = [0, 1]$, let $\mathscr{F}$ be the Borel sigma field of Ω and P be Lebesgue measure. For each $x \in [0, 1]$, let Y_x be the random variable defined by

$$Y_x(\omega) = \begin{cases} 1, & \text{if } \omega = x, \\ 0, & \text{if } \omega \neq x. \end{cases}$$

Show that $\sup_{x \in [0,1]} Y_x = 1$, but $\operatorname{esssup}_{x \in [0,1]} Y_x = 0$ (a.s.).

3. Prove property 1 in the Remarks of Section 1.2.

4. Let $(X_l,\ l \in \mathbb{L})$ be a family of random variables. Suppose there exists an integrable random variable Y such that $|X_l| \leq Y$ for all $l \in \mathbb{L}$. Show that $(X_l,\ l \in \mathbb{L})$ is uniformly integrable.

5. Let $(Y_n,\ n \in \mathbb{N})$ be a sequence of independent and identically distributed random variables such that $\mathrm{P}\{Y_n = 1\} = \mathrm{P}\{Y_n = -1\} = \frac{1}{2}$. For $n \in \overline{\mathbb{N}}$, set

$$X_n = \begin{cases} \dfrac{2n}{n+1} \displaystyle\prod_{j=1}^{n} (Y_j + 1), & \text{if } n \in \mathbb{N}^*, \\ 0, & \text{if } n \in \{0, \infty\}. \end{cases}$$

Prove that $(X_n,\ n \in \overline{\mathbb{N}})$ is not uniformly integrable.

6. Prove the assertions in items 1, 3, 4, and 5 of Section 1.3.

7. (An inaccessible stopping point) Suppose $d = 3$ and let $(X_t, t \in \mathbb{I})$ be a family of independent and identically distributed random variables such that $P\{X_0 = 0\} > 0$ and $P\{X_0 = 1\} > 0$. For all $t \in \mathbb{I}$, let $\mathscr{F}_t$ be the sigma field generated by the family $(X_s, s \in \mathbb{I}_t)$. Give an example of an inaccessible stopping point relative to the filtration $(\mathscr{F}_t, t \in \mathbb{I})$. (*Hint*. Proceed along the lines of the example in Section 1.4 that involves independent random variables; see also the first Example in Section 4.3.)

8. Suppose $d = 2$ and let $(Y^1_{t^1}, t^1 \in \mathbb{N})$ and $(Y^2_{t^2}, t^2 \in \mathbb{N})$ be two independent sequences of independent random variables such that $P\{Y^i_{t^i} \geq x\} > 0$ for all $x \in \mathbb{R}_+$, all $t^i \in \mathbb{N}$, and $i = 1, 2$. For all $t = (t^1, t^2) \in \mathbb{I}$, let $\mathscr{F}_t$ denote the sigma field generated by $(Y^1_{s^1}, s^1 \in \mathbb{N}_{t^1})$ and $(Y^2_{s^2}, s^2 \in \mathbb{N}_{t^2})$, and set

$$X_t = \sum_{s^1=0}^{t^1} Y^1_{s^1} + \sum_{s^2=0}^{t^2} Y^2_{s^2}.$$

Equip the probability space with the filtration $(\mathscr{F}_t, t \in \mathbb{I})$, where $\mathscr{F}_t$ is the sigma field just defined, and fix a predictable increasing path Γ. Let τ be the stopping time relative to the filtration $\mathscr{F}^\Gamma$ defined by $\tau = \inf\{n \in \mathbb{N}: X^\Gamma_n \geq 1\}$. Prove that if Δ is any predictable increasing path that passes through $T = \Gamma_\tau$, then $(\Delta_{n \wedge \tau}, n \in \mathbb{N})$ and $(\Gamma_{n \wedge \tau}, n \in \mathbb{N})$ coincide. (*Hint*. Let S be an accessible stopping point such that $S \leq T$. It suffices to prove that Γ passes through S. Assume that $P\{\Gamma_{|S|} \neq S\} > 0$, and show that there are two elements s and t of $\mathbb{I}$ such that $s < t$ and $P(F) > 0$, where $F = \{\Gamma_{|s|} = s, S = t, \Gamma_{|s|+1} \in \mathbb{I} - \mathbb{I}_t\}$. In the case where $s^1 = t^1$ and $s^2 < t^2$, show that because $S \leq T$, F is a subset of $\{Y^1_{s^1+1} < 1 - X_s\}$. On the other hand, by the assumptions on $(Y^1_{t^1}, t^1 \in \mathbb{N})$ and $(Y^2_{t^2}, t^2 \in \mathbb{N})$, $P(F \cap \{Y^1_{s^1+1} > 1 - X_s\}) > 0$. From this contradiction, conclude that $\Gamma_{|S|} = S$.)

9. (Section Theorem in dimension $d = 1$) Suppose $d = 1$ and let H be an adapted subset of $\mathbb{I} \times \Omega$. Show that the debut T of H (along the only predictable increasing path) is a stopping time τ such that $\{\omega: (\tau(\omega), \omega) \in H\} = \{\omega: H_\omega \neq \varnothing\}$.

10. (Absence of a Section Theorem in dimension $d = 2$) Suppose $d = 2$, let $\Omega_1 = \Omega_2 = \{0, 1\}$, and for $i = 1, 2$, set $\mathscr{F}^i_{t^i} = \{\varnothing, \Omega_i\}$ when $t^i = 0$ and $\mathscr{F}^i_{t^i} = \mathscr{F}_i = \mathscr{P}(\Omega_i)$ when $t^i \in \mathbb{N}^*$. Let $\Omega = \Omega_1 \times \Omega_2$, $\mathscr{F} = \mathscr{F}_1 \times \mathscr{F}_2$ and $\mathscr{F}_t = \mathscr{F}^1_{t^1} \times \mathscr{F}^2_{t^2}$ for all $t = (t^1, t^2) \in \mathbb{I}$. Finally, let P be any probability measure on $\Omega \times \mathscr{F}$ such that all nonempty sets have positive probability.

If H is the subset of $\mathbb{I} \times \Omega$ defined by $H = \{((0,1), \omega_2), ((0,1), \omega_3), ((1,0), \omega_1), ((1,0), \omega_3)\}$, where $\omega_1 = (1,0)$, $\omega_2 = (0,1)$, and $\omega_3 = (1,1)$, show that H is adapted but that there is no stopping point T such that $\{\omega\colon (T(\omega), \omega) \in H\} = \{\omega\colon H_\omega \neq \varnothing\}$.

11. Let $(X_n, n \in \mathbb{N})$ be a sequence of independent and identically distributed symmetric random variables such that $\mathrm{E}(X_0^2) < \infty$ and X_0^+ is not bounded.

a. Prove that $(X_n, n \in \mathbb{N})$ is uniformly integrable.

b. Give an example of a finite stopping time τ relative to the natural filtration of $(X_n, n \in \mathbb{N})$ such that $\mathrm{E}(X_\tau) = \infty$. (*Hint*. For $n \in \mathbb{N}^*$, set $\tau_n = \inf\{m \in \mathbb{N}\colon X_m > 2^n\}$ and for an appropriate choice of sets $F_1, F_2, \ldots,$ define τ by

$$\tau = \begin{cases} \tau_1 \text{ on } F_1, \\ \tau_n \text{ on } F_1^c \cap \cdots \cap F_{n-1}^c \cap F_n \qquad \text{if } n > 1.) \end{cases}$$

12. Prove the various properties of supermartingales stated between (1.11) and Theorem 1 of Section 1.6.

13. (An inaccessible stopping point that does not satisfy the conclusion of the optional sampling theorem) Consider the filtration $(\mathscr{F}_t)$ introduced in item 2 of Section 1.3 and set $T = (1,0)\,1_F + (0,1)\,1_{F^c}$. Let (X_t) be defined by $X_{0,0} = 1$, $X_{1,0} = (1/\mathrm{P}(F))\,1_F$, $X_{0,1} = (1/\mathrm{P}(F^c))\,1_{F^c}$ and $X_t = 0$ if $|t| > 1$. Show that (X_t) is a supermartingale but $\mathrm{E}(X_T | \mathscr{F}_{0,0}) > X_{0,0}$.

14. Let X be an integrable process indexed by $\mathbb{N}$ and adapted to a filtration $(\mathscr{F}_n, n \in \mathbb{N})$. Let τ be a stopping time relative to this filtration. Assume that for all $n \in \mathbb{N}$, $\mathrm{E}(X_{n+1} | \mathscr{F}_n) = X_n$ on $\{\tau > n\}$. Prove that $\mathrm{E}(X_{n \wedge \tau}) = \mathrm{E}(X_0)$ for all $n \in \mathbb{N}$.

15. In the context of Section 1.6, prove the assertions 1, 2, and 3 stated at the end of that section. (*Hint*. Assertion 1 is a straightforward consequence of (1.13) and (1.12). Regarding assertion 2, note that

$$\mathrm{E}\left(A_n^\Gamma\right) = \mathrm{E}\left(X_0^\Gamma\right) - \mathrm{E}\left(X_n^\Gamma\right), \qquad \text{for all } n \in \mathbb{N}. \tag{1.14}$$

In particular, $\mathrm{E}(A_\infty^\Gamma) < \infty$. Consequently, $(X_n^\Gamma, n \in \overline{\mathbb{N}})$ is uniformly integrable if and only if $(M_n^\Gamma, n \in \overline{\mathbb{N}})$ is. If X^Γ is uniformly integrable, then X_n^Γ converges to $X_{\infty-}^\Gamma$ in L^1. Conclude using (1.13) and (1.12) that in this case, $\mathrm{E}(M_\infty^\Gamma | \mathscr{F}_n^\Gamma) = M_n^\Gamma$ for all $n \in \mathbb{N}$. In order to verify assertion 3, let

$\tilde{A}^\Gamma$ be as in this assertion and let $\tilde{M}^\Gamma = \tilde{A}^\Gamma + X^\Gamma$. Observe that $M^\Gamma - \tilde{M}^\Gamma = A^\Gamma - \tilde{A}^\Gamma$; therefore $M^\Gamma - \tilde{M}^\Gamma$ is predictable. Conclude that for all $n \in \mathbb{N}$, $M_n^\Gamma - M_{n-1}^\Gamma = \tilde{M}_n^\Gamma - \tilde{M}_{n-1}^\Gamma$; therefore $M^\Gamma = \tilde{M}^\Gamma$ and $A^\Gamma = \tilde{A}^\Gamma$, since $M_0^\Gamma = \tilde{M}_0^\Gamma = X_0^\Gamma$.)

16. a. Let $X = (X_n,\ n \in \overline{\mathbb{N}})$ be a uniformly integrable supermartingale. Prove that X is of class D. (*Hint*. Use assertion 2 of Section 1.6 and (1.14) to conclude that X is the sum of two processes of class D.)

b. Let X be a uniformly integrable supermartingale indexed by $\bar{\mathbb{I}}$. When $d > 1$, does this imply that X is of class $\mathrm{D}_{\mathbf{A}}$? (*Note*. This is an open question. It is not difficult to see that the answer is yes if X is also assumed to be negative.)

HISTORICAL NOTES

The first study of a general class of sequential control problems involving processes indexed by (partially) ordered sets seems to be due to Haggstrom (1966). This author considered a particular class of ordered sets that contains the set described in Section 1.1, and therefore in principle, his framework is sufficient to formalize all the sequential stochastic optimization problems considered in this book. However, in many applications, such as certain statistical applications considered in Robbins and Siegmund (1974), it is convenient to use other types of (partially) ordered sets, particularly $\mathbb{N}^2$ or $\mathbb{I}$ (see Chapter 6). Such sets were also used in mathematical generalizations of processes indexed by $\mathbb{N}$, beginning with Bochner (1955), who formulated a martingale theory for processes indexed by directed sets. Processes indexed by $\mathbb{I}$ were considered in particular in articles of Cairoli (1970), Gut (1976), and Cairoli and Gabriel (1978).

Krengel and Sucheston (1981) proposed a formal notion suitable for the study of sequential optimization problems described using a countable (partially) ordered set. A mathematically equivalent concept was introduced by Mandelbaum and Vanderbei (1981). The notion of accessible stopping point appears in these two references under different names. These studies were pursued in dimension $d = 2$ by Walsh (1981) and Mazziotto and Szpirglas (1983). There, the primary focus was on stopping points and assumptions were made on the filtration to ensure that all stopping points are accessible (see Chapter 5). The emphasis made on accessible stopping points in the first six chapters of this book is new and leads to a unified theory for all dimensions $d \geq 1$.

The setup of Section 1.1 was considered already in Haggstrom (1966). The results presented in Section 1.2 are classical and can be found in such standard references as Neveu (1975) and Dellacherie and Meyer (1978). The version of Fatou's Lemma presented there is taken from Chow, Robbins, and

Siegmund (1971). Most of the properties of stopping points given in Section 1.3 appear in Mazziotto and Szpirglas (1983). Questions discussed in Section 1.4 concerning unicity of predictable increasing paths passing through a given stopping point arise from an article of Fouque (1983), and the example in the text is due to Dalang (1988b).

The proposition of Section 1.5 is new, though the two corollaries can be found in Walsh (1981) and Mazziotto and Szpirglas (1983).

Optional sampling theorems for processes indexed by (partially) ordered sets were given in Kurtz (1980) and Washburn and Willsky (1981), but the proof of Theorem 1 of Section 1.6 does not rely on these references.

Martingale convergence theorems involving convergence in probability were first obtained by Krickeberg (1956), followed by Helms (1958) and Chow (1960); the deeper theorems concerning almost sure convergence, first proved by Cairoli (1970), will not be needed in this book.

In dimension $d = 1$, the increasing process associated with a supermartingale along a predictable increasing path is just the classical Doob decomposition of a supermartingale indexed by $\mathbb{N}$ (see Doob (1953)).

Exercise 2 is taken from Neveu (1975) and Exercise 3 from Haggstrom (1966). The material for Exercise 5 comes from Haggstrom (1966) and Chow, Robbins, and Siegmund (1971). The example of an inaccessible stopping point referred to in Exercise 7 is due independently to Krengel and Sucheston (1981) and to Mandelbaum and Vanderbei (1981), and the example in Exercise 8 is adapted from Dalang (1988b). The statement in Exercise 14 will be used in subsequent chapters. The conclusions of Exercises 15 and 16 can be found, for instance, in Dellacherie and Meyer (1982).

CHAPTER 2

Sums of Independent Random Variables

This chapter is devoted to the study of maximal inequalities and to the strong law of large numbers for arrays of independent random variables which are not necessarily identically distributed.

2.1 MAXIMAL INEQUALITIES

In order to achieve our objectives, we will need to use random variables whose values are families $(x_t,\ t \in \mathbb{I})$ of real numbers. This leads us to work from the very beginning with random variables taking values in a Banach space. The associated notion of expectation (in the sense of Bochner integration) will, however, not be used. At the same time, the main result can be formulated in the setting of random variables taking values in a Banach space.

Let B be a separable Banach space. We let $\| \ \|$ denote the norm on B and let $\mathscr{B}$ be the Borel sigma field of B, that is, the sigma field generated by the open sets of B. Equipped with this sigma field, B is a measurable space, and a random variable taking values in B is merely a measurable map from Ω into B. As in the real-valued setting, we make no distinction between two random variables with values in B that differ only on a null set. Notice that every linear combination of random variables with values in B is again a random variable (since B is separable). Also observe that if X is a random variable with values in B, then $\|X\|$ is a real-valued random variable, because $\|X\| = f \circ X$, where f is the continuous function $x \mapsto \|x\|$.

A random variable X with values in B is *symmetric* if $\mathrm{P}\{X \in B\} = \mathrm{P}\{-X \in B\}$ for all $B \in \mathscr{B}$. Clearly, the sum of two independent and symmetric random variables is again a symmetric random variable.

In the following, we will often refer to the elementary inequality

$$(x_1 + \cdots + x_n)^p \le n^p\,(x_1^p + \cdots + x_n^p), \tag{2.1}$$

which is valid for all $n \geq 1$, all $p \in \mathbb{R}_+$ and all choices of $x_1, \ldots, x_n \in \mathbb{R}_+$. (The best constant in (2.1) is in fact n^{p-1} if $p \geq 1$, and 1 if $p < 1$.)

Throughout this chapter, we will be working with partial sums of random variables. We will denote these sums by the letter S, even though this letter was previously used for stopping points. However, no ambiguity should arise because in this chapter, stopping points will be denoted by other letters. In the terminology we have adopted, a family of random variables is a stochastic process, but in this chapter, we prefer to use the term *family*.

Lemma. *Let $\mathbb{L}$ be a finite set and $(X_1^l, l \in \mathbb{L}), \ldots, (X_n^l, l \in \mathbb{L})$ be independent families of symmetric (not necessarily independent) random variables with values in* B. *For $m = 1, \ldots, n$ and all $l \in \mathbb{L}$, set $S_m^l = \sum_{k=1}^m X_k^l$, Then*

$$\mathrm{P}\Big\{\sup_{m,l} \|S_m^l\| > x\Big\} \leq 2\,\mathrm{P}\Big\{\sup_{l} \|S_n^l\| > x\Big\} \tag{2.2}$$

for all $x \in \mathbb{R}_+$.

Proof. Equip $\mathbb{L}$ with an arbitrary total order, fix x and define τ and λ on $\{\sup_{m,l} \|S_m^l\| > x\}$ by

$$\tau = \inf\Big\{m\colon \sup_{l} \|S_m^l\| > x\Big\} \quad \text{and} \quad \lambda = \inf\big\{l\colon \|S_\tau^l\| > x\big\}.$$

Clearly,

$$\{\tau = m,\ \lambda = l\} \subset \big\{\|S_m^l\| > x\big\} \tag{2.3}$$

and

$$\bigcup_{m,l} \{\tau = m,\ \lambda = l\} = \Big\{\sup_{m,l} \|S_m^l\| > x\Big\}. \tag{2.4}$$

Set $S_n^{l,m} = S_m^l - (S_n^l - S_m^l)$. Since $S_n^l = S_m^l + (S_n^l - S_m^l)$, it follows that $S_m^l = \frac{1}{2}(S_n^{l,m} + S_n^l)$, and therefore

$$\big\{\|S_m^l\| > x\big\} \subset \big\{\|S_n^{l,m}\| > x\big\} \cup \big\{\|S_n^l\| > x\big\}. \tag{2.5}$$

Moreover, because $S_n^l - S_m^l$ is a symmetric random variable independent of the sigma field generated by the families $(X_1^{l'}, l' \in \mathbb{L}), \ldots, (X_m^{l'}, l' \in \mathbb{L})$, and because $\{\tau \in m,\ \lambda = l\}$ belongs to that sigma field,

$$\mathrm{P}\big\{\|S_n^{l,m}\| > x,\ \tau = m,\ \lambda = l\big\} = \mathrm{P}\big\{\|S_n^l\| > x,\ \tau = m,\ \lambda = l\big\}.$$

Together with this equality, (2.3) and (2.5) imply that

$$\mathrm{P}\{\tau = m, \lambda = l\} \le 2\,\mathrm{P}\big\{\|S_n^l\| > x, \tau = m, \lambda = l\big\}.$$

Taking (2.4) into account, we conclude that

$$\begin{aligned}
\mathrm{P}\Big\{\sup_{m,l} \|S_m^l\| > x\Big\} &= \sum_{m,l} \mathrm{P}\{\tau = m, \lambda = l\} \\
&\le 2 \sum_{m,l} \mathrm{P}\big\{\|S_n^l\| > x, \tau = m, \lambda = l\big\} \\
&\le 2 \sum_{m,l} \mathrm{P}\Big\{\sup_{l'} \|S_n^{l'}\| > x, \tau = m, \lambda = l\Big\} \\
&= 2\,\mathrm{P}\Big\{\sup_{l'} \|S_n^{l'}\| > x, \sup_{m,l} \|S_m^l\| > x\Big\} \\
&= 2\,\mathrm{P}\Big\{\sup_{l'} \|S_n^{l'}\| > x\Big\},
\end{aligned}$$

which establishes the lemma.

Theorem 1 (Lévy's inequality). *Let $(X_t, t \in \mathbb{I}_u)$ be a family of independent and symmetric random variables with values in* B. *For all $t \in \mathbb{I}_u$, set $S_t = \sum_{s \le t} X_s$. Then*

$$\mathrm{P}\Big\{\sup_{t \le u} \|S_t\| > x\Big\} \le 2^d\,\mathrm{P}\{\|S_u\| > x\} \tag{2.6}$$

for all $x \in \mathbb{R}_+$.

Proof. The proof is by induction on d. In the case where $d = 1$, (2.6) follows from the lemma with $\mathbb{L} = \{1\}$. Assume that $d > 1$ and that the conclusion is true in dimension $d - 1$; set

$$\mathbb{L} = \big\{(t^2, \dots, t^d) \in \mathbb{N}^{d-1}: t^2 \le u^2, \dots, t^d \le u^d\big\},$$

and, for $m = t^1 = 0, 1, \dots, u^1$ and $(t^2, \dots, t^d)$ in $\mathbb{L}$, let

$$X_{t^1}^{(t^2, \dots, t^d)} = \sum_{s^2=0}^{t^2} \cdots \sum_{s^d=0}^{t^d} X_{(t^1, s^2, \dots, s^d)}$$

and

$$S_{t^1}^{(t^2, \dots, t^d)} = \sum_{s^1=0}^{t^1} X_{s^1}^{(t^2, \dots, t^d)} = S_{(t^1, \dots, t^d)}.$$

By the lemma,

$$\mathrm{P}\Big\{\sup_{t^1 \le u^1, \ldots, t^d \le u^d} \|S_{(t^1, \ldots, t^d)}\| > x\Big\} \le 2\,\mathrm{P}\Big\{\sup_{t^2 \le u^2, \ldots, t^d \le u^d} \|S_{(u^1, t^2, \ldots, t^d)}\| > x\Big\}.$$

Because (2.6) is assumed to be true for dimension $d - 1$,

$$\mathrm{P}\Big\{\sup_{t^2 \le u^2, \ldots, t^d \le u^d} \|S_{(u^1, t^2, \ldots, t^d)}\| > x\Big\} \le 2^{d-1}\,\mathrm{P}\{\|S_u\| > x\}.$$

Thus,

$$\mathrm{P}\Big\{\sup_{t^1 \le u^1, \ldots, t^d \le u^d} \|S_{(t^1, \ldots, t^d)}\| > x\Big\} \le 2\,2^{d-1}\,\mathrm{P}\{\|S_u\| > x\},$$

which establishes the theorem.

Theorem 2 (A maximal inequality). *Let $(X_t,\ t \in \mathbb{I})$ be a family of independent random variables with values in B and fix $p \in \mathbb{R}_+$. For all $t \in \mathbb{I}$, set $S_t = \sum_{s \le t} X_s$. Then*

$$\mathrm{E}\Big(\sup_t \|S_t\|^p\Big) \le c_p^d \sup_t \mathrm{E}(\|S_t\|^p), \tag{2.7}$$

where c_p^d is a constant that only depends on d and p.

Proof. Clearly, we can assume that $p > 0$. If the random variables X_t are symmetric, then (2.7) is a direct consequence of (2.6). Indeed, it suffices to substitute $x^{1/p}$ for x in (2.6), to integrate with respect to x, and to take the supremum over u. In this case, $c_p^d = 2^d$. We will extend this to the general case by using symmetrization. Let $(\Omega', \mathscr{F}', \mathrm{P}')$ be a copy of $(\Omega, \mathscr{F}, \mathrm{P})$, and let $(X_t',\ t \in \mathbb{I})$ be a copy of $(X_t,\ t \in \mathbb{I})$ defined on Ω'. Put $S_t' = \sum_{s \le t} X_s'$ and consider the random variables X_t, S_t, X_t', and S_t' as being defined on $\Omega \times \Omega'$. Let $X_t^* = X_t - X_t'$ and $S_t^* = S_t - S_t'$. With respect to the probability measure $\mathrm{P}^* = \mathrm{P} \times \mathrm{P}'$, $(X_t^*,\ t \in \mathbb{I})$ is a family of independent and symmetric random variables with values in B. Thus, thanks to the result in the symmetric case.

$$\mathrm{E}^*\Big(\sup_t \|S_t^*\|^p\Big) \le 2^d \sup_t \mathrm{E}^*(\|S_t^*\|^p), \tag{2.8}$$

where E^* denotes the expectation operator associated to P^*. From the definition of S_t^* and (2.1), it follows that

$$\|S_t\|^p \le 2^p\Big(\sup_u \|S_u^*\|^p + \|S_t'\|^p\Big).$$

Taking successively the expectation relative to P′, the supremum over t and the expectation relative to P, we see that

$$\mathrm{E}\Big(\sup_t \|S_t\|^p\Big) \le 2^p\Big(\mathrm{E}^*\Big(\sup_u \|S_u^*\|^p\Big) + \sup_t \mathrm{E}(\|S_t'\|^p)\Big).$$

We now use (2.8) and the inequality

$$\mathrm{E}^*(\|S_u^*\|^p) \le 2^p(\mathrm{E}(\|S_u\|^p) + \mathrm{E}'(\|S_u'\|^p))$$

to conclude that (2.7) is true with $c_p^d = 2^p(2^{d+p+1} + 1)$.

2.2 INTEGRABILITY CRITERIA FOR THE SUPREMUM

The main results of this section are contained in two theorems for which we need two lemmas that we will establish first. We also define a symbol we will use frequently in the sequel: for all $t \in \mathbb{I}$, $\langle t \rangle$ denotes the number card $\mathbb{I}_t$ $(= \prod_{i=1}^d (t^i + 1))$.

Lemma 1. *Let X and Y be two independent random variables with values in* B *and fix $p \in \mathbb{R}_+$. If Y is symmetric, then*

$$\mathrm{E}(\|X\|^p) \le 2\,\mathrm{E}(\|X + Y\|^p).$$

Proof. Because

$$2\|X\| = \|X + Y + X - Y\| \le \|X + Y\| + \|X - Y\|,$$

it follows from (2.1) that

$$\|X\|^p \le \|X + Y\|^p + \|X - Y\|^p.$$

However, the laws of the couples (X, Y) and $(X, -Y)$ are identical; therefore

$$\mathrm{E}(\|X + Y\|^p) = \mathrm{E}(\|X - Y\|^p).$$

The conclusion is now clear.

Lemma 2. *Let $X_1, \ldots, X_n$ be independent symmetric random variables with values in* B *and fix $p \in \mathbb{R}_+$. For $m = 1, \ldots, n$, set $S_m = \sum_{k=1}^m X_k$. If $\mathrm{E}(\|S_n\|^p) < \infty$, then*

$$\Big(3^{-p} - 2\,\mathrm{P}\Big\{\sup_m \|S_m\| > x\Big\}\Big)\,\mathrm{E}(\|S_n\|^p) \le \mathrm{E}\Big(\sup_m \|X_m\|^p\Big) + 2x^p \quad (2.9)$$

for all $x \in \mathbb{R}_+$.

Proof. Fix $x \in \mathbb{R}_+$ and set

$$\tau = \begin{cases} \inf\{m: \|S_m\| > x\} & \text{on } \left\{\sup_m \|S_m\| > x\right\}, \\ n+1 & \text{elsewhere}. \end{cases}$$

Moreover, set $S_0 = 0$. Then

$$\begin{aligned} \mathrm{E}(\|S_n\|^p) &= \mathrm{E}(\|S_n\|^p; \tau \le n) + \mathrm{E}(\|S_n\|^p; \tau = n+1) \\ &\le \sum_{k=1}^n \mathrm{E}\Big((\|S_n - S_k\| + \|S_k\|)^p; \tau = k\Big) + x^p \\ &\le \sum_{k=1}^n \mathrm{E}\left(\Big(\|S_n - S_k\| + \|S_{k-1}\| + \sup_m \|X_m\|\Big)^p; \tau = k\right) + x^p \\ &\le 3^p \sum_{k=1}^n \mathrm{E}\left(\|S_n - S_k\|^p + \|S_{k-1}\|^p + \sup_m \|X_m\|^p; \tau = k\right) + x^p, \end{aligned}$$

where the last step is justified by (2.1). But the set $\{\tau = k\}$ is independent of $\|S_n - S_k\|$, because it belongs to the sigma field generated by $X_1, \ldots, X_k$. Moreover, $\|S_{k-1}\| \le x$ on $\{\tau = k\}$ by definition of τ. Taking Lemma 1 into account (applied to $X = S_n - S_k$ and $Y = S_k$), we conclude that

$$\mathrm{E}(\|S_n\|^p) \le 3^p(2\,\mathrm{E}(\|S_n\|^p) + x^p)\,\mathrm{P}\{\tau \le n\} + 3^p\,\mathrm{E}\left(\sup_m \|X_m\|^p\right) + x^p,$$

from which (2.9) follows easily, since $\{\tau \le n\} = \{\sup_m \|S_m\| > x\}$.

Theorem 1 (Integrability of $\sup_t \|S_t\|^p$). *Let $(X_t,\ t \in \mathbb{I})$ be a family of independent random variables with values in* B. *For all $t \in \mathbb{I}$, set $S_t = \sum_{s \le t} X_s$. If $\sup_t \|S_t\| < \infty$ and $\mathrm{E}(\sup_t \|X_t\|^p) < \infty$ for some $p \in \mathbb{R}_+$, then $\mathrm{E}(\sup_t \|S_t\|^p) < \infty$.*

Proof. We first prove the theorem under the assumption that the random variables X_t are symmetric. Fix $u \in \mathbb{I}$, set $n = \operatorname{card} \mathbb{I}_u$ and let $t(1), \ldots, t(n)$ denote the elements of $\mathbb{I}_u$ arranged in *lexicographical order*: $t(k)$ precedes $t(m)$ if $t^i(k) < t^i(m)$, where i is the smallest index j such that $t^j(k) \ne t^j(m)$. For $m = 1, \ldots, n$, set

$$\tilde{X}_m = X_{t(m)} \quad \text{and} \quad \tilde{S}_m = \sum_{k=1}^m \tilde{X}_k.$$

Clearly,

$$\sup_m \|\tilde{X}_m\| = \sup_{t \le u} \|X_t\|.$$

Moreover, by induction on the dimension d, it is easy to see that each random variable $\tilde{S}_m$ can be expressed as a sum over 2^d or fewer terms of the type $\pm S_t$, where $t \in \mathbb{I}_u$, which implies that

$$\sup_m \|\tilde{S}_m\| \leq 2^d \sup_{t \leq u} \|S_t\|.$$

Inequality (2.9) applied to the random variables $\tilde{X}_1, \ldots, \tilde{X}_n$ now implies that

$$\left(3^{-p} - 2\,\mathrm{P}\left\{\sup_t \|S_t\| > 2^{-d}x\right\}\right) \mathrm{E}(\|S_u\|^p) \leq \mathrm{E}\left(\sup_t \|X_t\|^p\right) + 2x^p.$$

Because $\sup_t \|S_t\| < \infty$, the first factor on the left can be made positive by choosing x large enough. Since u is arbitrary, it follows that $\sup_u \mathrm{E}(\|S_u\|^p) < \infty$. But then $\mathrm{E}(\sup_t \|S_t\|^p) < \infty$ by (2.7).

If the random variables X_t are not symmetric, we use the auxiliary random variables X_t', S_t', X_t^*, and S_t^* introduced in the proof of Theorem 2 of Section 2.1. Using the hypothesis of the theorem, we observe that

$$\sup_t \|S_t^*\| \leq \sup_t \|S_t\| + \sup_t \|S_t'\| < \infty \qquad (\mathrm{P}^*\text{-a.s.}),$$

and, by (2.1), that

$$\mathrm{E}^*\left(\sup_t \|X_t^*\|^p\right) \leq 2^p\left(\mathrm{E}\left(\sup_t \|X_t\|^p\right) + \mathrm{E}'\left(\sup_t \|X_t'\|^p\right)\right) < \infty.$$

Because of the result just established, we conclude that

$$\mathrm{E}'\left(\mathrm{E}\left(\sup_t \|S_t - S_t'\|^p\right)\right) = \mathrm{E}^*\left(\sup_t \|S_t^*\|^p\right) < \infty,$$

and therefore that

$$\mathrm{E}\left(\sup_t \| S_t - S_t'(\omega') \|^p\right) < \infty$$

for P′-almost all ω'. We now choose an ω' for which this inequality is true and $\sup_t \|S_t'(\omega')\| < \infty$. Taking (2.1) into account, it follows that

$$\mathrm{E}\left(\sup_t \|S_t\|^p\right) \leq 2^p\left(\sup_t \| S_t'(\omega') \|^p + \mathrm{E}\left(\sup_t \| S_t - S_t'(\omega') \|^p\right)\right) < \infty,$$

which was to be proved.

Theorem 2 (Integrability of $\sup_t (\|S_t\|/\langle t\rangle)^p$). *Let $(X_t,\ t \in \mathbb{I})$ be a family of independent random variables with values in* B. *For all $t \in \mathbb{I}$, set $S_t = \sum_{s \leq t} X_s$. If $\sup_t (\|S_t\|/\langle t\rangle) < \infty$ and $\mathrm{E}(\sup_t(\|X_t\|/\langle t\rangle)^p) < \infty$ for some $p \in \mathbb{R}_+$, then* $\mathrm{E}(\sup_t (\|S_t\|/\langle t\rangle^p) < \infty$.

Proof. Let $c_0^d(\mathsf{B})$ denote the vector space consisting of all families $x = (x_u,\ u \in \mathbb{I})$ of elements of B such that $\lim_{u \to \bowtie} x_u = 0$. Equipped with the norm $\|x\|_\infty = \sup_u \|x_u\|$, $c_0^d(\mathsf{B})$ is a separable Banach space. For all $t \in \mathbb{I}$, set $\tilde{X}_t = (\tilde{X}_t^u,\ u \in \mathbb{I})$, where

$$\tilde{X}_t^u = \begin{cases} 0, & \text{if } u \not\geq t, \\ \dfrac{X_t}{\langle u\rangle}, & \text{if } u \geq t. \end{cases}$$

Clearly, $(\tilde{X}_t,\ t \in \mathbb{I})$ is a family of independent random variables with values in $c_0^d(\mathsf{B})$. Set $\tilde{S}_t = \sum_{s \leq t} \tilde{X}_s$ and observe that $\tilde{S}_t = (\tilde{S}_t^u,\ u \in \mathbb{I})$, where

$$\tilde{S}_t^u = \begin{cases} \dfrac{S_{t \wedge u}}{\langle u\rangle}, & \text{if } u \not\geq t, \\ \dfrac{S_t}{\langle u\rangle}, & \text{if } u \geq t. \end{cases}$$

It is now clear that

$$\|\tilde{S}_t\|_\infty = \sup_{u \leq t} \frac{\|S_u\|}{\langle u\rangle},$$

which leads to

$$\sup_t \|\tilde{S}_t\|_\infty = \sup_t \frac{\|S_t\|}{\langle t\rangle} < \infty.$$

On the other hand,

$$\|\tilde{X}_t\|_\infty = \frac{\|X_t\|}{\langle t\rangle},$$

which implies that

$$\mathrm{E}\left(\sup_t \|\tilde{X}_t\|_\infty^p\right) = \mathrm{E}\left(\sup_t \left(\frac{\|X_t\|}{\langle t\rangle}\right)^p\right) < \infty.$$

It follows that the family $(\tilde{X}_t,\ t \in \mathbb{I})$ satisfies the hypotheses of Theorem 1, from which we conclude that

$$\mathrm{E}\left(\sup_t \|\tilde{S}_t\|_\infty^p\right) < \infty,$$

and, in particular, that

$$\mathrm{E}\left(\sup_t \left(\frac{\|S_t\|}{\langle t\rangle}\right)^p\right) < \infty,$$

which was to be proved.

2.3 THE STRONG LAW OF LARGE NUMBERS

In this section, we present the strong law of large numbers for real-valued random variables. These variables are not assumed to be identically distributed.

We will say that a (formal) series $\sum_t x_t$ of real numbers *converges* if the limit $\lim_{t \to \bowtie} \sum_{s \leq t} x_s$ exists (in $\mathbb{R}$). In this case, this limit will again be denoted by $\sum_t x_t$. If the x_t are nonnegative, we write $\sum_t x_t < \infty$ if the series converges and $\sum_t x_t = \infty$ if it diverges. We will say that a series with random terms *converges* if it converges a.s. Any other type of convergence will be indicated explicitly.

We will begin by proving two lemmas to be applied to families of random variables. The first is a technical lemma. It compensates partially for the absence of a version for random variables of the criterion mentioned in the "Notation and Conventions" section preceding Chapter 1 (see the Remark in Section 3.8). The second is the extension to dimensions $d > 1$ of Kronecker's well-known lemma.

Lemma 1. *Let $(x_t,\ t \in \mathbb{I})$ be a family of real numbers and x a real number. Suppose that for all nonempty subsets D of $\{1, \ldots, d\}$,*

$$\limsup_{t^i \to \infty,\ i \in D} x_t \leq x \qquad \left[\lim_{t^i \to \infty,\ i \in D} x_t = x\right] \tag{2.10}$$

for all choices of (fixed) coordinates t^i, $i \in D^c$ (when $D^c = \{1, \ldots, d\} - D$ is not empty). Then

$$\limsup_{t \to \bowtie} x_t \leq x \qquad \left[\lim_{t \to \bowtie} x_t = x\right]. \tag{2.11}$$

Proof. It is sufficient to prove the assertion concerning lim sup. Consider first the case where x is finite and assume that (2.11) is not true. By the criterion formulated in the "Notation and Conventions" section preceding Chapter 1, there exists $\varepsilon > 0$ and a nondecreasing sequence $(t_n, n \in \mathbb{N})$ of elements of $\mathbb{I}$ such that $t_n \to \bowtie$ and $x_{t_n} \geq x + \varepsilon$ for all $n \in \mathbb{N}$. Set $D = \{i: \lim_{n\to\infty} t_n^i = \infty\}$. For all $i \in D^c$, the sequence $(t_n^i, n \in \mathbb{N})$ is bounded; therefore there exists $n_0 \in \mathbb{N}$ such that $t_n^i = t_{n_0}^i$ for all $n \geq n_0$ and all $i \in D^c$. Consider now the left-hand side of (2.10) with the set D just defined and assume that the coordinates of t that remain fixed are $t^i = t_{n_0}^i$ with $i \in D^c$. Then

$$\limsup_{t^i\to\infty,\, i\in D} x_t \geq \limsup_{n\to\infty} x_{t_n} \geq x + \varepsilon,$$

which shows that (2.10) is not true and proves the lemma in the case where x is finite. The case where $x = -\infty$ is solved by a similar argument. Because there is nothing to prove in the case where $x = \infty$, the lemma is established.

Lemma 2. *Let $(x_t, t \in \mathbb{I})$ be a family of real numbers. If the series $\Sigma_t (x_t/\langle t\rangle)$ converges and $\sup_t |\Sigma_{s\leq t} x_s/\langle s\rangle| < \infty$, then*

$$\lim_{t\twoheadrightarrow\infty} \frac{1}{\langle t\rangle} \sum_{s\leq t} x_s = 0.$$

Proof. For reasons of simplicity, we will limit ourselves to the case where $d = 2$ (the proof in higher dimensions (cf. Exercise 2) can be carried out by using some auxiliary results that require an induction argument on d). For all $t \in \mathbb{I}$, set $y_t = \Sigma_{s\leq t} x_s/\langle s\rangle$. Moreover, let $y = \lim_{t\twoheadrightarrow\bowtie} y_t$. It is not difficult to check that

$$\sum_{s\leq t} x_s = \sum_{s\ll t} y_s - (t^2+1)\sum_{s^1<t^1} y_{(s^1,t^2)} - (t^1+1)\sum_{s^2<t^2} y_{(t^1,s^2)} + \langle t\rangle y_t,$$

and therefore that

$$\frac{1}{\langle t\rangle}\sum_{s\leq t} x_s = \frac{1}{\langle t\rangle}\sum_{s\ll t} y_s - \frac{1}{t^1+1}\sum_{s^1<t^1} y_{(s^1,t^2)} - \frac{1}{t^2+1}\sum_{s^2<t^2} y_{(t^1,s^2)} + y_t.$$

Since $\sup_t |y_t| < \infty$ by hypothesis, the three normalized sums on the right-hand side converge to y when $t \twoheadrightarrow \bowtie$. The last term also converges to y when $t \twoheadrightarrow \bowtie$. Therefore, the left-hand side converges to 0 when $t \twoheadrightarrow \bowtie$.

Theorem 1 (The strong law of large numbers). *Let* $(X_t,\ t \in \mathbb{I})$ *be a family of independent and real-valued random variables with mean* 0. *For all* $t \in \mathbb{I}$, *set* $S_t = \Sigma_{s \le t} X_s$. *If* $\sup_t \mathrm{E}(|X_t|^p) < \infty$ *for some* $p > 1$, *then*

$$\lim_{t \to \bowtie} \frac{S_t}{\langle t \rangle} = 0.$$

Proof. The conclusion will follow from Lemma 1 if we show that

$$\lim_{t^i \to \infty,\ i \in D} \frac{S_t}{\langle t \rangle} = 0 \tag{2.12}$$

for all nonempty subsets D of $\{1, \dots, d\}$ and all choices of coordinates t^i, $i \in D^c$ (when D^c is not empty). In order to show this, it is sufficient to consider the case where $D = \{1, \dots, d\}$, because the case where $D \neq \{1, \dots, d\}$ reduces to this one by considering the family $(\Sigma_{s:\ s^i \le t^i,\ i \in D^c} X_s)$ indexed by $(s^i,\ i \in D) \in \mathbb{N}^D$. Indeed, this family satisfies the assumptions of the theorem. In particular, the last assumption is satisfied, because by (2.1),

$$\sup_{s^i \in \mathbb{N},\ i \in D} \mathrm{E}\left(\left| \sum_{s:\ s^i < t^i,\ i \in D^c} X_s \right|^p\right) \le \left(\prod_{i \in D^c} (t^i + 1)\right)^{p+1} \sup_t \mathrm{E}(|X_t|^p) < \infty. \tag{2.13}$$

We therefore restrict ourselves to the case where $D = \{1, \dots, d\}$. Clearly, we can assume that $p < 2$. For all $t \in \mathbb{I}$, set $Y_t = X_t\ 1_{\{|X_t| \le \langle t \rangle\}}$ and observe that

$$\begin{aligned} \sum_t \mathrm{E}\left(\left(\frac{Y_t - E(Y_t)}{\langle t \rangle}\right)^2\right) &\le \sum_t \mathrm{E}\left(\left(\frac{Y_t}{\langle t \rangle}\right)^2\right) \\ &\le \sum_t \mathrm{E}\left(\left(\frac{|X_t|}{\langle t \rangle}\right)^p;\ |X_t| \le \langle t \rangle\right) \\ &\le \sup_t \mathrm{E}(|X_t|^p) \sum_t \frac{1}{\langle t \rangle^p} \\ &< \infty. \end{aligned} \tag{2.14}$$

Fix temporarily $u, v \in \mathbb{I}$ such that $u < v$, set $\mathbb{I}_v^u = \mathbb{I}_v - \mathbb{I}_u$ and

$$Z_t = \begin{cases} \dfrac{Y_t - \mathrm{E}(Y_t)}{\langle t \rangle}, & \text{if } t \in \mathbb{I}_v^u, \\ 0, & \text{if } t \notin \mathbb{I}_v^u. \end{cases}$$

Using Chebyshev's inequality and inequality (2.7) applied to the family $(Z_t,\ t \in \mathbb{I})$, we see that for all $\varepsilon > 0$,

$$\begin{aligned}
\mathrm{P}\left\{\sup_t \left|\sum_{s \le t} Z_s\right| > \varepsilon\right\} &\le \frac{1}{\varepsilon^2}\mathrm{E}\left(\sup_t\left(\sum_{s\le t} Z_s\right)^2\right) \\
&\le \frac{c_2^d}{\varepsilon^2}\sup_t \mathrm{E}\left(\left(\sum_{s<t} Z_s\right)^2\right) \\
&= \frac{c_2^d}{\varepsilon^2}\sum_{t\in \mathbb{I}_v^u} \mathrm{E}\left(\left(\frac{Y_t - \mathrm{E}(Y_t)}{\langle t\rangle}\right)^2\right).
\end{aligned}$$

By (2.14), the last right-hand side converges to 0 when $u \twoheadrightarrow \bowtie$, and therefore

$$\lim_{u \twoheadrightarrow \bowtie} \mathrm{P}\left\{\sup_{t>u}\left|\sum_{s\in\mathbb{I}_t^u}\frac{Y_s - \mathrm{E}(Y_s)}{\langle s\rangle}\right| > \varepsilon\right\} = 0.$$

As ε is arbitrary, we conclude that the series $\Sigma_t (Y_t - \mathrm{E}(Y_t))/\langle t\rangle$ converges. But because $\mathrm{E}(X_t) = 0$,

$$\mathrm{E}(X_t;\ |X_t| \le \langle t\rangle) = -\mathrm{E}(X_t;\ |X_t| > \langle t\rangle),$$

therefore

$$\begin{aligned}
\sum_t \frac{|\mathrm{E}(Y_t)|}{\langle t\rangle} = \sum_t \frac{1}{\langle t\rangle}|\mathrm{E}(X_t;\ |X_t| > \langle t\rangle)| &\le \sum_t \frac{\mathrm{E}(|X_t|^p)}{\langle t\rangle^p} \\
&\le \sup_t \mathrm{E}(|X_t|^p)\sum_t \frac{1}{\langle t\rangle^p} < \infty.
\end{aligned}$$

It follows that the series $\Sigma_t Y_t/\langle t\rangle$ converges. On the other hand,

$$\mathrm{E}\left(\sum_t 1_{\{X_t \ne Y_t\}}\right) = \sum_t \mathrm{P}\{X_t \ne Y_t\} = \sum_t \mathrm{P}\{|X_t| > \langle t\rangle\} = \sum_t \mathrm{P}\{|X_t|^p > \langle t\rangle^p\},$$

and the right-hand side is less than or equal to

$$\sup_t \mathrm{E}(|X_t|^p)\sum_t \frac{1}{\langle t\rangle^p} < \infty;$$

therefore

$$P\{\omega\colon \mathrm{card}\{t \in \mathbb{I}\colon X_t(\omega) \ne Y_t(\omega)\} < \infty\} = \mathrm{P}\left\{\sum_t 1_{\{X_t \ne Y_t\}} < \infty\right\} = 1.$$

Consequently, the series $\Sigma_t (X_t/\langle t\rangle)$ converges. In order to conclude, using Lemma 2, that assertion (2.12) is true in the case where $D = \{1, \dots, d\}$, we still have to check that $\sup_t |\Sigma_{s \le t} X_s/\langle s\rangle| < \infty$, or equivalently, that $\sup_t |\Sigma_{s \le t} (Y_s - \mathrm{E}(Y_s))/\langle s\rangle| < \infty$. For all $n \in \mathbb{N}^*$, set

$$F_n = \left\{ \sup_t \left| \sum_{s \le t} \frac{Y_s - \mathrm{E}(Y_s)}{\langle s\rangle} \right| > n \right\}.$$

By applying Chebyshev's inequality, (2.7) and (2.14), we see that

$$\begin{aligned}
\mathrm{P}(F_n) &\le \frac{1}{n^2}\mathrm{E}\left(\sup_t \left(\sum_{s \le t} \frac{Y_s - \mathrm{E}(Y_s)}{\langle s\rangle} \right)^2 \right) \\
&\le \frac{c_2^d}{n^2} \sup_t \mathrm{E}\left(\left(\sum_{s \le t} \frac{Y_s - \mathrm{E}(Y_s)}{\langle s\rangle} \right)^2 \right) \\
&= \frac{c_2^d}{n^2} \sup_t \sum_{s \le t} \mathrm{E}\left(\left(\frac{Y_s - \mathrm{E}(Y_s)}{\langle s\rangle} \right)^2 \right) \\
&= \frac{c_2^d}{n^2} \sum_t \mathrm{E}\left(\left(\frac{Y_t - \mathrm{E}(Y_t)}{\langle t\rangle} \right)^2 \right) \\
&\le \frac{c_2^d}{n^2} \sup_t \mathrm{E}(|X_t|^p) \sum_t \frac{1}{\langle t\rangle^p}.
\end{aligned}$$

From this, we conclude that $\Sigma_n \mathrm{P}(F_n) < \infty$, therefore, by the Borel–Cantelli lemma, that $\sup_t |\Sigma_{s \le t} (Y_s - \mathrm{E}(Y_s))/\langle s\rangle| < \infty$, which was to be proved.

The Integrability of the Supremum, Revisited

In light of the theorem just established, we will now return briefly to the question addressed in Section 2.2

Theorem 2 (Integrability of $\sup_t (|S_t|/\langle t\rangle)^q$). *With the hypotheses and notation of Theorem* 1, *if* $\sup_t \mathrm{E}(|X_t|^p) < \infty$ *for some* $p > 1$, *then* $\mathrm{E}(\sup_t (|S_t|/\langle t\rangle)^q) < \infty$ *for all* q *such that* $0 \le q < p$.

Proof. Theorem 1 ensures that $\sup_t (|S_t|/\langle t\rangle) < \infty$. By Theorem 2 of Section 2.2, it is sufficient to prove that $\mathrm{E}(\sup_t (|X_t|/\langle t\rangle)^q) < \infty$. Clearly, we can assume that $q \ge 1$. Choose $\varepsilon > 0$ sufficiently small, in such a way that $q(1 + \varepsilon) \le p$. The conclusion follows from the following inequalities, the third of which uses Chebyshev's inequality:

$$\begin{aligned}\mathrm{E}\left(\sup_t\left(\frac{|X_t|}{\langle t\rangle}\right)^q\right) &\le 1+\int_1^\infty \mathrm{P}\left\{\sup_t\left(\frac{|X_t|}{\langle t\rangle}\right)^q > x\right\}dx\\ &\le 1+\int_1^\infty \sum_t \mathrm{P}\left\{\left(\frac{|X_t|}{\langle t\rangle}\right)^q > x\right\}dx\\ &\le 1+\sup_t \mathrm{E}\left(|X_t|^{q(1+\varepsilon)}\right)\sum_t \frac{1}{\langle t\rangle^{q(1+\varepsilon)}}\int_1^\infty \frac{1}{x^{1+\varepsilon}}\,dx\\ &< \infty.\end{aligned}$$

2.4 CASE WHERE THE RANDOM VARIABLES ARE IDENTICALLY DISTRIBUTED

For identically distributed random variables, the theorems of the previous sections can be refined. We use the convention $0^0 = 1$ and set

$$\log^+ x = \begin{cases}\log x, & \text{if } x \ge 1,\\ 0, & \text{if } 0 \le x < 1.\end{cases}$$

Lemma 1. *Fix $n \ge 1$ and let f be the function $x \mapsto x(\log^+ x)^n$. For all $k \ge 1$ and all choices of $x_1, \dots, x_k \in \mathbb{R}_+$,*

$$f(x_1+\cdots+x_k) \le 2^n(f(x_1)+\cdots+f(x_k))+f(k^2). \tag{2.15}$$

Proof. Because the function f is convex,

$$f\left(\frac{1}{k}(x_1+\cdots+x_k)\right) \le \frac{1}{k}(f(x_1)+\cdots+f(x_k));$$

therefore

$$f(x_1+\cdots+x_k) \le \frac{1}{k}(f(kx_1)+\cdots+f(kx_k)).$$

On the other hand, if $0 \le x \le y$, then

$$f(xy) \le 2^n x f(y); \tag{2.16}$$

therefore

$$f(kx) \le 2^n k(f(x)+f(k)).$$

Consequently,

$$\begin{aligned} f(x_1 + \cdots + x_k) &\leq 2^n(f(x_1) + \cdots + f(x_k) + kf(k)) \\ &= 2^n(f(x_1) + \cdots + f(x_k)) + f(k^2), \end{aligned}$$

which was to be proved.

Lemma 2. *For all $n \geq 1$,*

(a) $$\int_1^\infty \cdots \int_1^\infty \frac{1}{x_1^2 \cdots x_n^2} 1_{\{x \leq x_1 \cdots x_n\}}\, dx_1 \cdots dx_n \leq \frac{n}{x}\left((\log^+ x)^{n-1} + 1\right) \qquad (x \in \mathbb{R}_+^*);$$

(b) $$\int_1^\infty \cdots \int_1^\infty \frac{1}{x_1 \cdots x_n} 1_{\{x > x_1 \cdots x_n\}}\, dx_1 \cdots dx_n = \frac{1}{n!}(\log^+ x)^n \qquad (x \in \mathbb{R}_+);$$

(c) $$\int_1^\infty \cdots \int_1^\infty 1_{\{x > x_1 \cdots x_n\}}\, dx_1 \cdots dx_n \sim \frac{1}{n!} x (\log x)^{n-1} \qquad (x \to \infty),$$

that is, the ratio of the left-hand side to the right-hand side converges to 1 *when* $x \to \infty$.

Proof. We can clearly restrict ourselves to the case where $x \geq 1$. The inequality in (a) is clearly true for $n = 1$. Suppose that it is true for some $n \geq 1$ and let $I_n(x)$ denote the integral on the left-hand side. Since $I_n(y) = 1$ if $y \leq 1$, we see that

$$\begin{aligned} I_{n+1}(x) &= \int_1^x \frac{1}{x_{n+1}^2} I_n\left(\frac{x}{x_{n+1}}\right) dx_{n+1} + \int_x^\infty \frac{1}{x_{n+1}^2}\, dx_{n+1} \\ &\leq n \int_1^x \frac{1}{x_{n+1}^2} \frac{x_{n+1}}{x}\left(\left(\log \frac{x}{x_{n+1}}\right)^{n-1} + 1\right) dx_{n+1} + \frac{1}{x} \\ &= \frac{1}{x}(\log x)^n + \frac{n}{x}\log x + \frac{1}{x} \\ &\leq \frac{n+1}{x}\left((\log x)^n + 1\right). \end{aligned}$$

The inequality in (a) can thus be established by induction on n. Assertions (b) and (c) are proved by induction on n in the same way. The value of the integral in (c) is

$$\frac{1}{n!}x(\log x)^{n-1} - \frac{1}{(n-1)!}x(\log x)^{n-2} + \cdots + (-1)^{n-1}x + (-1)^n.$$

Lemma 3. *For all $n \geq 1$ and all real-valued random variables $X \geq 0$,*

$$\sum_{k_1=1}^{\infty} \cdots \sum_{k_n=1}^{\infty} \mathrm{P}\{X > k_1 \cdots k_n\} < \infty$$

if and only if

$$\mathrm{E}\big(X(\log^+ X)^{n-1}\big) < \infty.$$

Proof. Set

$$I_n(x) = \int_1^{\infty} \cdots \int_1^{\infty} 1_{\{x > x_1 \cdots x_n\}}\, dx_1 \cdots dx_n.$$

By (c) of Lemma 2, $\mathrm{E}(I_n(X)) < \infty$ if and only if $\mathrm{E}(X(\log^+ X)^{n-1}) < \infty$. On the other hand,

$$\mathrm{E}(I_n(X)) = \int_1^{\infty} \cdots \int_1^{\infty} \mathrm{P}\{X > x_1 \cdots x_n\}\, dx_1 \cdots dx_n;$$

therefore

$$\sum_{k_1=2}^{\infty} \cdots \sum_{k_n=2}^{\infty} \mathrm{P}\{X > k_1 \cdots k_n\} \leq \mathrm{E}(I_n(X))$$

$$\leq \sum_{k_1=1}^{\infty} \cdots \sum_{k_n=1}^{\infty} \mathrm{P}\{X > k_1 \cdots k_n\}.$$

It is thus clear that the condition of the lemma is necessary. In order to show that it is sufficient, we proceed by induction on n, using the inequalities

$$\sum_{k_1=1}^{\infty} \cdots \sum_{k_n=1}^{\infty} \mathrm{P}\{X > k_1 \cdots k_n\}$$

$$\leq n \sum_{k_1=1}^{\infty} \cdots \sum_{k_{n-1}=1}^{\infty} \mathrm{P}\{X > k_1 \cdots k_{n-1}\} + \sum_{k_1=2}^{\infty} \cdots \sum_{k_n=2}^{\infty} \mathrm{P}\{X > k_1 \cdots k_n\}$$

$$\leq n \sum_{k_1=1}^{\infty} \cdots \sum_{k_{n-1}=1}^{\infty} \mathrm{P}\{X > k_1 \cdots k_{n-1}\} + \mathrm{E}(I_n(X)).$$

These reduce to

$$\sum_{k_1=1}^{\infty} \mathrm{P}\{X > k_1\} \leq \mathrm{P}\{X > 1\} + \mathrm{E}(I_1(X))$$

in the case where $n = 1$.

Lemma 4. *Let $(x_t,\ t \in \mathbb{I})$ be a family of real numbers. Set $y_t = \sum_{s \leq t} x_s$ if $t \in \mathbb{I}$ and $y_t = 0$ if $t \in \mathbb{Z}^d - \mathbb{I}$. Then*

$$x_t = \sum_{e \in \{0,1\}^d} (-1)^{|e|} y_{t-e}.$$

Proof. The assertion is clearly true in dimension 1. Assume that it is true in dimension $d-1$ $(d > 1)$ and show that it is also true in dimension d. We distinguish the elements of $\mathbb{N}^{d-1}$ from those of $\mathbb{I}$ by topping them with the symbol $\sim$ and set $y_{\tilde{t}}^{t^d} = \sum_{\tilde{s} \leq \tilde{t}} x_{(\tilde{s}, t^d)}$ if $\tilde{t} \in \mathbb{N}^{d-1}$, $y_{\tilde{t}}^{t^d} = 0$ if $\tilde{t} \in \mathbb{Z}^{d-1} - \mathbb{N}^{d-1}$. Then

$$\begin{aligned} x_t &= x_{(\tilde{t}, t^d)} \\ &= \sum_{\tilde{e} \in \{0,1\}^{d-1}} (-1)^{|\tilde{e}|}\, y_{\tilde{t}-\tilde{e}}^{t^d} \\ &= \sum_{\tilde{e} \in \{0,1\}^{d-1}} (-1)^{|\tilde{e}|} \left(y_{(\tilde{t}, t^d)-(\tilde{e},0)} - y_{(\tilde{t}, t^d)-(\tilde{e},1)} \right) \\ &= \sum_{e \in \{0,1\}^d} (-1)^{|e|}\, y_{t-e}. \end{aligned}$$

Theorem 1 (The strong law of large numbers). *Let $(X_t,\ t \in \mathbb{I})$ be a family of independent and identically distributed real-valued random variables. For all $t \in \mathbb{I}$, set $S_t = \sum_{s \leq t} X_s$. If $\mathrm{E}(|X_0|(\log^+|X_0|)^{d-1}) < \infty$, then*

$$\lim_{t \to \bowtie} \frac{S_t}{\langle t \rangle} = \mathrm{E}(X_0).$$

Conversely, if

$$\limsup_{t \twoheadrightarrow \bowtie} \frac{|S_t|}{\langle t \rangle} < \infty,$$

then $\mathrm{E}(|X_0|(\log^+|X_0|)^{d-1}) < \infty$.

Proof. The proof of the first statement is similar to that of Theorem 1 of Section 2.3, and therefore we only indicate the steps that need to be changed. In the initial remark where the problem is reduced to the case where $D = \{1, \ldots, d\}$, the right-hand side of (2.12) becomes $\mathrm{E}(X_0)$ and an inequality similar to (2.13) is obtained using (2.15). After that, the inequalities in (2.14) must be replaced by the following:

$$\begin{aligned}
&\sum_t \mathrm{E}\left(\left(\frac{Y_t - \mathrm{E}(Y_t)}{\langle t\rangle}\right)^2\right)\\
&\quad\le \sum_t \mathrm{E}\left(\left(\frac{Y_t}{\langle t\rangle}\right)^2\right)\\
&\quad= \sum_{t^1=1}^{\infty} \cdots \sum_{t^d=1}^{\infty} \frac{1}{\left(t^1 \cdots t^d\right)^2} \mathrm{E}\left(X_0^2; |X_0| \le t^1 \cdots t^d\right)\\
&\quad\le 4^d \int_1^{\infty} \cdots \int_1^{\infty} \frac{1}{x_1^2 \cdots x_d^2} \mathrm{E}\left(X_0^2; |X_0| \le x_1 \cdots x_d\right) dx_1 \cdots dx_d\\
&\quad= 4^d\, \mathrm{E}\left(X_0^2 \int_1^{\infty} \cdots \int_1^{\infty} \frac{1}{x_1^2 \cdots x_d^2} 1_{\{|X_0| \le x_1 \cdots x_d\}}\, dx_1 \cdots dx_d\right)\\
&\quad\le d\, 4^d\, \mathrm{E}\left(|X_0|\left(\left(\log^+|X_0|\right)^{d-1} + 1\right)\right)\\
&\quad< \infty,
\end{aligned}$$

where the factor 4^d comes from the inequality $1/t^2 \le 4/(t+1)^2$ and the next-to-the-last step is justified by (a) of Lemma 2. The remainder of the proof is unchanged: Because the series $\Sigma_t\, (Y_t - \mathrm{E}(Y_t))/\langle t\rangle$ converges and $\sup_t |\Sigma_{s \le t}\, \mathrm{E}((Y_s - \mathrm{E}(Y_s))/\langle s\rangle)| < \infty$,

$$\lim_{t \twoheadrightarrow \bowtie} \frac{1}{\langle t\rangle} \sum_{s \le t} (Y_s - \mathrm{E}(Y_s)) = 0$$

by Lemma 2 of Section 2.3; therefore

$$\lim_{t \twoheadrightarrow \bowtie} \frac{1}{\langle t\rangle} \sum_{s \le t} Y_s = \mathrm{E}(X_0),$$

since

$$\lim_{t \to \bowtie} \mathrm{E}(Y_t) = \lim_{t \to \bowtie} \mathrm{E}(X_0; |X_0| \le \langle t\rangle) = \mathrm{E}(X_0).$$

However, by Lemma 3,

$$\mathrm{E}\left(\sum_t 1_{\{X_t \neq Y_t\}}\right) = \sum_t \mathrm{P}\{X_t \neq Y_t\} = \sum_t \mathrm{P}\{|X_0| > \langle t \rangle\} < \infty;$$

therefore

$$\mathrm{P}\{\omega \colon \mathrm{card}\{t \in \mathbb{I} \colon X_t(\omega) \neq Y_t(\omega)\} < \infty\} = \mathrm{P}\left\{\sum_t 1_{\{X_t \neq Y_t\}} < \infty\right\} = 1,$$

and consequently

$$\lim_{t \twoheadrightarrow \bowtie} \frac{S_t}{\langle t \rangle} = \mathrm{E}(X_0).$$

We now turn to the proof of the second part of the theorem. Suppose that $\mathrm{E}(|X_0|(\log^+|X_0|)^{d-1}) = \infty$. If $n \geq 1$ and $u \in \mathbb{I}$, we see by (2.16) (written for y/x instead of y) that

$$\mathrm{E}\left(\frac{|X_0|}{n\langle u \rangle}\left(\log^+ \frac{|X_0|}{n\langle u \rangle}\right)^{d-1}\right) = \infty.$$

It follows from Lemma 3 that

$$\sum_t \mathrm{P}\{|X_0| > n\langle t \rangle \langle u \rangle\} = \infty.$$

But if $u \geq (2, \ldots, 2)$, then

$$\mathrm{P}\{|X_0| > n\langle t \rangle\langle u \rangle\} \leq \mathrm{P}\{|X_0| > n\langle t + u \rangle\} = \mathrm{P}\{|X_{t+u}| > n\langle t + u \rangle\},$$

and therefore

$$\sum_{t \geq u} \mathrm{P}\{|X_t| > n\langle t \rangle\} = \infty.$$

By the Borel–Cantelli lemma,

$$\sum_{t \geq u} 1_{\{|X_t| > n\langle t \rangle\}} = \infty,$$

from which it follows that

$$\sum_{t \geq u} 1_{\{|S_t| > n\langle t \rangle / 2^d\}} = \infty.$$

Indeed, if $|X_t| > n\langle t\rangle$, where $t \geq (1, \ldots, 1)$, then $|S_{t-e}| > n\langle t\rangle/2^d$ for at least one $e \in \{0, 1\}^d$, since by Lemma 4,

$$X_t = \sum_{e \in \{0,1\}^d} (-1)^{|e|} S_{t-e}.$$

As n and u are arbitrary, we conclude that

$$\limsup_{t \twoheadrightarrow \bowtie} \frac{|S_t|}{\langle t\rangle} = \infty.$$

Integrability of the Supremum

With the help of the previous theorem and of Theorem 2 of Section 2.2, we can now give necessary and sufficient conditions for the integrability of $\sup_t |S_t|/\langle t\rangle$. Notice that in (a) of Theorem 2 below, the exponent over the logarithm is d rather than $d - 1$ as in the previous theorem.

Theorem 2 (Integrability of $\sup_t (|S_t|/\langle t\rangle)$). Let $(X_t,\ t \in \mathbb{I})$ be a family of independent and identically distributed real-valued random variables. For all $t \in \mathbb{I}$, set $S_t = \sum_{s \leq t} X_s$. The following assertions are equivalent:

(a) $$\mathrm{E}\Big(|X_0|(\log^+ |X_0|)^d\Big) < \infty;$$

(b) $$\mathrm{E}\left(\sup_t \frac{|X_t|}{\langle t\rangle}\right) < \infty;$$

(c) $$\mathrm{E}\left(\sup_t \frac{|S_t|}{\langle t\rangle}\right) < \infty.$$

Proof. By Lemma 4,

$$\mathrm{E}\left(\sup_t \frac{|X_t|}{\langle t\rangle}\right) < 2^d\, \mathrm{E}\left(\sup_t \frac{|S_t|}{\langle t\rangle}\right),$$

therefore (c) implies (b). On the other hand, if (a) holds, then $\sup_t (|S_t|/\langle t\rangle) < \infty$ by Theorem 1, therefore (a) and (b) (together) imply (c), by Theorem 2 of Section 2.2. It now remains to establish the equivalence of (a) and (b).

Clearly,

$$
\begin{aligned}
\mathrm{E}\left(\sup_t \frac{|X_t|}{\langle t\rangle}\right) &\le 1 + \sum_{n=1}^{\infty} \mathrm{P}\left\{\sup_t \frac{|X_t|}{\langle t\rangle} > n\right\} \\
&\le 1 + \sum_{n=1}^{\infty} \sum_{t^1=0}^{\infty} \cdots \sum_{t^d=0}^{\infty} \mathrm{P}\{|X_0| > n\, t^1 \cdots t^d\};
\end{aligned}
$$

therefore (a) implies (b) by Lemma 3. In order to prove the converse implication, equip ▯ with some total order $\prec$ and set

$$
T = \begin{cases} \inf\left\{t\colon \dfrac{|X_t|}{\langle t\rangle} > 1\right\} & \text{on } \left\{\sup_t \dfrac{|X_t|}{\langle t\rangle} > 1\right\}, \\ \bowtie & \text{elsewhere}. \end{cases}
$$

Clearly, $\{t \prec T\} = \bigcap_{s \prec t,\, s \neq t}\{|X_s| \le \langle s\rangle\}$; therefore $\{t \prec T\}$ is independent of X_t. Moreover,

$$
\begin{aligned}
\mathrm{E}\left(\sup_t \frac{|X_t|}{\langle t\rangle}\right) &\ge \sum_t \mathrm{E}\left(\frac{|X_t|}{\langle t\rangle}; T = t\right) \\
&= \sum_t \frac{1}{\langle t\rangle} \mathrm{E}(|X_t|; |X_t| > \langle t\rangle, t \prec T) \\
&\ge \mathrm{P}\{T = \bowtie\} \sum_t \frac{1}{\langle t\rangle} \mathrm{E}(|X_0|; |X_0| > \langle t\rangle) \\
&\ge \mathrm{P}\{T = \bowtie\}\, \mathrm{E}\left(|X_0| \int_1^{\infty} \cdots \int_1^{\infty} \frac{1}{x_1 \cdots x_d} \right. \\
&\qquad\qquad \left. \times 1_{\{|X_0| > x_1 \cdots x_d\}}\, dx_1 \cdots dx_d\right);
\end{aligned}
$$

therefore, by (b) of Lemma 2,

$$
\mathrm{E}\left(\sup_t \frac{|X_t|}{\langle t\rangle}\right) \ge \frac{1}{d!} \mathrm{P}\{T = \bowtie\}\, \mathrm{E}\left(|X_0|(\log^+|X_0|)^d\right).
$$

On the other hand,

$$
\begin{aligned}
\mathrm{P}\{T = \bowtie\} &= \mathrm{P}\left(\bigcap_t \{|X_t| \le \langle t\rangle\}\right) \\
&= \prod_t \mathrm{P}\{|X_t| \le \langle t\rangle\} = \prod_t (1 - \mathrm{P}\{|X_t| > \langle t\rangle\}).
\end{aligned}
$$

But this infinite product is positive if $\sum_t \mathrm{P}\{|X_t| > \langle t\rangle\} < \infty$; therefore, by

Lemma 3, it is positive if $E(|X_0|(\log^+|X_0|)^{d-1}) < \infty$. It is thus proved that with this condition, (b) implies (a). The proof can now be completed by applying the above result to the subfamilies $(X_t,\ t \in \mathbb{N}^j \times \{0\}^{d-j} \times \{0\}^{d-j})$ $(1 \leq j \leq d-1)$. By hypothesis,

$$E\left(\sup_{t \in \mathbb{N}^j \times \{0\}^{d-j}} \frac{|X_t|}{\langle t \rangle} \right) < \infty.$$

Therefore, assuming that $E(|X_0|(\log^+ |X_0|)^{j-1}) < \infty$, we conclude that $E(|X_0|(\log^+|X_0|^j) < \infty$.

Case Where the Dimension Is 1

In the case where $d = 1$, Theorem 2 can be extended to apply to stopping times as follows. A generalization to $d > 1$ will be given in Section 4.2.

Theorem 3 (Additional equivalent conditions). *Let $(X_n,\ n \in \mathbb{N})$ be a sequence of independent and identically distributed real-valued random variables. Set $S_n = \sum_{m=0}^{n} X_m$ and $X_\infty = S_\infty = 0$. The following assertions are equivalent:*

(a) $$E(|X_0|(\log^+|X_0|)) < \infty;$$

(b) $$E\left(\frac{|X_\tau|}{\tau + 1};\ \tau < \infty \right) < \infty, \qquad \textit{for all stopping times } \tau;$$

(c) $$E\left(\frac{|S_\tau|}{\tau + 1};\ \tau < \infty \right) < \infty, \qquad \textit{for all stopping times } \tau.$$

Proof. Of course, the notion of stopping time is to be taken relative to the natural filtration. This being said, notice that (a) implies (b) and (c) by Theorem 2, because

$$\frac{|X_\tau|}{\tau + 1} 1_{\{\tau < \infty\}} \leq \sup_n \frac{|X_n|}{n + 1} \quad \text{and} \quad \frac{|S_\tau|}{\tau + 1} 1_{\{\tau < \infty\}} \leq \sup_n \frac{|S_n|}{n + 1}.$$

On the other hand, letting $\prec$ be the natural order on $\mathbb{N}$ and τ be the stopping time defined in the same way as T in the proof of Theorem 2, we see that if $E(|X_\tau|/(\tau + 1);\ \tau < \infty) < \infty$, then (a) is true. It remains therefore to prove that if (c) holds, then $E(|X_\tau|/(\tau + 1);\ \tau < \infty) < \infty$ for this particular stopping time. Replacing if necessary the random variables X_n by aX_n, where a is an appropriate positive constant, we can assume that $\mathrm{P}\{|X_0| \leq 1\} > 0$. Clearly,

$$\mathrm{E}\left(\frac{|X_\tau|}{\tau + 1}; \tau < \infty\right) \leq \mathrm{E}\left(\frac{|S_\tau|}{\tau + 1}; \tau < \infty\right) + \mathrm{E}\left(\frac{1}{\tau + 1}(|X_0| + \cdots + |X_{\tau-1}|); 0 < \tau < \infty\right);$$

therefore it is sufficient to show that the last expectation is finite. But this expectation can be written

$$\sum_{n=1}^{\infty} \frac{1}{n+1} \mathrm{E}(|X_0| + \cdots + |X_{n-1}|; \tau = n) = \sum_{n=1}^{\infty} \frac{1}{n+1} \mathrm{P}\{\tau = n\} \sum_{m=0}^{n-1} \mathrm{E}(|X_m| \,|\tau = n).$$

On the other hand, using the fact that the X_n are independent and identically distributed, we see that if $\mathrm{P}\{\tau = n\} > 0$, then

$$\begin{aligned} \mathrm{E}(|X_m| \,|\tau = n) &= \frac{\mathrm{E}(|X_m|; |X_0| \leq 1, \ldots, |X_{n-1}| \leq n, |X_n| > n+1)}{\mathrm{P}\{|X_0| \leq 1, \ldots, |X_{n-1}| \leq n, |X_n| > n+1\}} \\ &= \mathrm{E}(|X_m| \,\big|\, |X_m| \leq m+1) \\ &= \mathrm{E}(|X_0| \,\big|\, |X_0| \leq m+1) \\ &\leq \frac{\mathrm{E}(|X_0|)}{\mathrm{P}\{|X_0| \leq 1\}}. \end{aligned}$$

We conclude that

$$\mathrm{E}\left(\frac{1}{\tau+1}(|X_0| + \cdots + |X_{\tau-1}|); 0 < \tau < \infty\right) \leq \frac{\mathrm{P}\{0 < \tau < \infty\}}{\mathrm{P}\{|X_0| \leq 1\}} \mathrm{E}(|X_0|) < \infty,$$

because $\mathrm{E}(|X_0|) = \mathrm{E}(|S_0|) < \infty$.

EXERCISES

1. (A sharpening of Theorem 2 of Section 2.1) Recall that a random variable X with values in B is (Bochner) *integrable* if the real-valued random variable $\|X\|$ is integrable, and the (Bochner) integral $\mathrm{E}(X)$ of X has the property that $\|\mathrm{E}(X)\| \leq \mathrm{E}(\|X\|)$.

a. Show that if X and X' are independent and integrable random variables with values in B and with mean 0, then for any $p \geq 1$,

$$\mathrm{E}(\|X\|^p) \leq \mathrm{E}(\|X - X'\|^p) \leq 2^p\,\mathrm{E}(\|X\|^p).$$

(*Hint*. Observe that $\|x\| = \|\mathrm{E}(x - X')\| \leq \mathrm{E}(\|x - X'\|)$ for all $x \in \mathsf{B}$.)

b. Fix $u \in \mathbb{I}$ and let $(X_t, t \in \mathbb{I})$ be a family of independent and integrable random variables with values in B and with mean 0. For $t \in \mathbb{I}_u$, let $S_t = \Sigma_{s \leq t} X_s$. Show that there exists a constant c_p^d, that depends only on d and p, such that

$$\mathrm{E}\left(\sup_{t \leq u} \|S_t\|^p\right) \leq c_p^d\,\mathrm{E}(\|S_u\|^p).$$

(*Hint*. If the X_t are assumed to be symmetric, then this is an immediate consequence of Theorem 1 of Section 2.1. If not, then let $\mathsf{B}^{\mathbb{I}_u}$ be the vector space of all families $x = (x_t, t \in \mathbb{I}_u)$ of elements of B. Equipped with the norm $\|x\|_\infty = \sup_{t \leq u} \|x_t\|$, $\mathsf{B}^{\mathbb{I}_u}$ is a separable Banach space. For $s \in \mathbb{I}_u$, let $\tilde{X}_s = (\tilde{X}_s^t, t \in \mathbb{I}_u)$ be the random variable with values in $\mathsf{B}^{\mathbb{I}_u}$ defined by

$$\tilde{X}_s^t = \begin{cases} 0, & \text{if } t \not\geq s, \\ X_s, & \text{if } t \geq s. \end{cases}$$

Check that

$$\left\| \sum_{s \leq u} \tilde{X}_s \right\|_\infty^p = \sup_{t \leq u} \|S_t\|^p.$$

Now let $(X_t', t \in \mathbb{I}_u)$ be an independent copy of $(X_t, t \in \mathbb{I}_u)$ as defined in the proof of Theorem 2 of Section 2.1, and for $s \in \mathbb{I}_u$, let $\tilde{X}_s'$ be defined in the same way as $\tilde{X}_s$ but with X_s replaced by X_s'. With the notation used in that proof, conclude from part a that

$$\mathrm{E}\left(\left\| \sum_{s \leq u} \tilde{X}_s \right\|_\infty^p\right) \leq \mathrm{E}\left(\left\| \sum_{s \leq u} (\tilde{X}_s - \tilde{X}_s') \right\|_\infty^p\right).$$

Now apply the result from the symmetric case together with part a.)

c. Show that the conclusion in part b is no longer valid if the variables X_t are not required to have mean 0.

2. (Proof of Kronecker's lemma) The purpose of this exercise is to give a proof of Lemma 2 of Section 2.3 that is valid for all $d \geq 1$. Using the notation of that lemma, follow the steps below.

a. For $n \in \{0, \ldots, d\}$, let $\mathscr{D}_n$ be the family of all subsets of $\{1, \ldots, d\}$ with exactly n elements (in particular, $\mathscr{D}_0 = \{\emptyset\}$ and $\mathscr{D}_d = \{\{1, \ldots, d\}\}$). For $t \in \mathbb{I}$, set $\langle t \rangle_\emptyset = 1$ and for $n \in \{1, \ldots, d\}$ and $D \in \mathscr{D}_n$, let

$$\langle t \rangle_D = \prod_{i \in D} (t^i + 1).$$

In addition, put

$$\mathbb{I}_t(D) = \{s \in \mathbb{I}: s^i = t^i \text{ for } i \in D, \text{ and } s^i < t^i \text{ for } i \in D^c\}.$$

Show that for $d \geq 1$,

$$\sum_{s \leq t} x_s = \sum_{n=0}^{d} (-1)^{d-n} \sum_{D \in \mathscr{D}_n} \langle t \rangle_D \sum_{s \in \mathbb{I}_t(D)} y_s. \tag{2.17}$$

(*Hint*. This formula can be established by induction on d, but another way of understanding it is to observe that for $s \leq t$, the term x_s appears exactly once in the sum on the left-hand side, whereas the term $x_s / \langle s \rangle$ appears

$$\sum_{n=0}^{d} (-1)^{d-n} \sum_{D \in \mathscr{D}_n} \langle t \rangle_D \prod_{i \in D^c} (t^i - s^i)$$

times on the right-hand side (by convention, the product on the right-hand side equals 1 when $D^c = \emptyset$). The fact that this expression is equal to $\langle s \rangle$ can be explained geometrically by observing that the term that corresponds to $n = d$ is the volume $\langle t \rangle$ of the "rectangle" $\mathscr{R}_t = \{(z^1, \ldots, z^d) \in \mathbb{R}^d: 0 \leq z^i \leq t^i \text{ for } i = 1, \ldots, d\}$; the terms that correspond to $n = d - 1$ are the volumes of d subrectangles of $\mathscr{R}_t$, which are subtracted from $\langle t \rangle$; because these d subrectangles are not disjoint, the formula proceeds according to the inclusion-exclusion principle to end up with the volume of the rectangle $\mathscr{R}_s$. In dimension $d = 2$, drawing a picture is helpful.)

b. Prove the lemma. (*Hint*. Divide both sides of (2.17) by $\langle t \rangle$ and observe that each term on the right-hand side can be written

$$\frac{1}{\langle t \rangle_{D^c}} \sum_{s \in \mathbb{I}_t(D)} y_s.$$

Using the assumption that $y = \lim_{t \twoheadrightarrow \bowtie} y_t$ exists and that $\sup_t |y_t| < \infty$, and observing that $\lim_{t \twoheadrightarrow \bowtie} (\langle t \rangle_{D^c} / \mathrm{card}\, \mathbb{I}_t(D)) = 1$, conclude that

$$\lim_{t \twoheadrightarrow \bowtie} \frac{1}{\langle t \rangle_{D^c}} \sum_{s \in \mathbb{I}_t(D)} y_s = y.$$

Finally, notice that according to the binomial formula,

$$\sum_{n=0}^{d} (-1)^{d-n} \,\mathrm{card}\, \mathscr{D}_n = (1-1)^d = 0,$$

which proves the lemma.)

3. Show that the equivalence of (a), (b), and (c) of Theorem 3 of Section 2.4 remains valid if, instead of taking into account all stopping times, only finite stopping times are considered.

HISTORICAL NOTES

The study of sums of independent random variables forms the core of probability theory. Nevertheless, the theorems proved here for processes indexed by $\mathbb{N}^d$ had never yet appeared in a book.

The two theorems of Section 2.1 are due to Gabriel (1975), who extended in particular Lévy's classical inequality to dimensions $d > 1$. Some refinements can be found in Etemadi (1991). The two theorems of Section 2.2 are due to Hoffman-Jorgensen (1974) in the case where $d = 1$, and the extension of Theorems 1 and 2 of that section to $d > 1$ seem to be new.

Lemma 2 of Section 2.3 is well known, but there seem to be no easily accessible proofs in the literature for $d > 2$. The proof in Exercise 2 is taken from Bolle (1993); formula (2.17) is a special case of an extension to multiple sums of a result, known as Abel's lemma, quoted by Hardy (see Hardy (1974)) and by Moore (1973). Theorem 1 of that section is due to Smythe (1973), and Theorem 2 is an extension to $d > 1$ of results of Hoffman-Jorgensen (1974) in the case $d = 1$. Lemma 3 of Section 2.4 is taken from the article of Gut (1978). Theorem 1 of the same section is due to Smythe (1973), Theorem 2 to Gabriel (1977) (see also Gut (1979)) and Theorem 3 to McCabe and Shepp (1970) and Davis (1971).

Exercise 1 uses ideas that can be found in Hoffman-Jorgensen (1974).

CHAPTER 3

Optimal Stopping

The optimization problem considered here arises in the context of a game of chance that pays a reward X_T if the gambler quits the game at stage T. At each stage, the gambler must decide whether to quit the game or to continue. If the decision is to continue, the gambler must also choose at which stage the next observation will be made. It is possible that the gambler never decides to quit, in which case the reward is $X_{\bowtie}$. The problem is to determine T so as to maximize the gambler's expected reward. Because the decision of terminating the game at stage T is preceded by a sequence of decisions leading the player to T, it is clear that the only stages to be considered are those that correspond to accessible stopping points. The essentially theoretical case of optimization over inaccessible stopping points will not be addressed in this book.

3.1 STATING THE PROBLEM

Let X be an adapted integrable process indexed by $\bar{\mathbb{I}}$ whose positive part X^+ is of class $\mathrm{D}_{\mathbf{A}}$. This process is given and fixed, and called a *reward process*. The upper bound

$$\sup_{T \in \mathbf{A}} \mathrm{E}(X_T) \tag{3.1}$$

is termed the *value of the game* (associated with X). Notice that the random variables X_T are not necessarily integrable, because the value of $\mathrm{E}(X_T)$ can be $-\infty$ for some accessible stopping points T. These stopping points will not play any role, as we will see as our analysis progresses. Let us notice to begin with that the value of the game is finite, because

$$-\infty < \mathrm{E}(X_0) \leq \sup_{T \in \mathbf{A}} \mathrm{E}(X_T) \leq \sup_{T \in \mathbf{A}} \mathrm{E}(X_T^+) < \infty.$$

An accessible stopping point T^o will be termed *optimal* (in $\mathbf{A}$) if

$$\mathrm{E}(X_{T^o}) = \sup_{T \in \mathbf{A}} \mathrm{E}(X_T). \tag{3.2}$$

The first problem is to establish the existence of an optimal accessible stopping point. However, without some restriction on the behavior of X at infinity, no optimal stopping point will exist. This situation is illustrated by the following simple deterministic example in dimension $d = 1$. Set $X_0 = 0$, $X_t = 1 - 1/t$ if $t \in \mathbb{N}^*$ and $X_{\bowtie} = 0$; the value of the game is 1 but $\mathrm{E}(X_T) < 1$ for all stopping points T, and thus there are no optimal stopping points. However, optimal stopping points will exist if X is upper semicontinuous at infinity, and the main objective of this chapter is to construct an optimal accessible stopping point using an extension of Snell's method to dimensions greater than 1.

Behavior of the Reward Process at Infinity

The following condition will be used at a specific point in the resolution of the problem and thus will only be assumed when needed. It expresses that X is upper semicontinuous at infinity:

$$\limsup_{t \to \bowtie} X_t \le X_{\bowtie}. \tag{3.3}$$

According to this definition, X is upper semicontinuous at infinity if and only if it is upper semicontinuous according to the topology on $\bar{\mathbb{I}}$ having for (countable) base the singletons $\{t\}$, with $t \in \mathbb{I}$, and the sets $\bar{\mathbb{I}} - \mathbb{I}_t$, also with $t \in \mathbb{I}$. Notice that $\bar{\mathbb{I}}$ endowed with this topology is a compact set.

This condition cannot be replaced with the weaker condition, involving convergence of all coordinates to infinity, which stipulates that

$$\limsup_{t \twoheadrightarrow \bowtie} X_t \le X_{\bowtie}. \tag{3.4}$$

In order to see this, suppose $d = 2$ and consider the deterministic process X defined by

$$X_t = \begin{cases} 1 - \dfrac{1}{t^1}, & \text{if } t^1 \in \mathbb{N}^* \text{ and } t^2 = 0, \\ 0, & \text{for all other values of } t, \text{ including } t = \bowtie. \end{cases} \tag{3.5}$$

This process satisfies (3.4), but not (3.3). The associated value of the game is 1, but $\mathrm{E}(X_T) < 1$ for all stopping points T, which shows that there is no optimal stopping point.

Similarly, the hypothesis

$$\limsup_{n\to\infty} X_n^\gamma \leq X_{\bowtie} \text{ for all deterministic increasing paths } \gamma, \qquad (3.6)$$

which is also weaker than (3.3), does not imply the existence of an optimal stopping point, as we will see in Section 3.8 in a nontrivial example.

Remark

If the values of the random variables X_t approach 0 or become negative in a neighborhood of infinity (that is, when $|t|$ is large enough), then condition (3.3) will, for instance, be satisfied if $X_{\bowtie} = 0$.

Criterion for Upper Semicontinuity at Infinity

Condition (3.3) is generally difficult to verify. However, it may happen that we will have at hand information concerning the behavior of X_t when some of the coordinates of t become large and the others are fixed. It is then often possible to establish that for all nonempty subsets D of $\{1,\ldots,d\}$ and all fixed choices of coordinates of t whose indices do not belong to D,

$$\limsup_{t^i\to\infty,\, i\in D} X_t \leq X_{\bowtie}\,. \qquad (3.7)$$

By Lemma 1 of Section 2.3, condition (3.3) is then satisfied.

Finite Optimal Stopping Points

It will often happen that the value of the game stays the same when $\mathbf{A}$ is replaced by the set of finite accessible stopping points (this is true, for instance, when X is of class $\mathrm{D}_{\mathbf{A}}$ and $\lim_{t\to\bowtie} X_t$ exists and is equal to $X_{\bowtie}$). In this case, we can ask whether finite accessible optimal stopping points exist. The deterministic example above shows that the answer is no, in general. Indeed, the value of the game computed using only finite stopping points remains 1, but $\mathrm{E}(X_T) < 1$ for all finite stopping points T. In certain cases and under various appropriate conditions, it is possible to establish that there are indeed finite optimal stopping points. A nontrivial example will be considered in Section 3.7.

3.2 SNELL'S ENVELOPE

Using the reward process X, we define a new process indexed by $\bar{\mathbb{I}}$, called *Snell's envelope of* X, which will play a key role in the construction of an optimal accessible stopping point. This process is denoted by Z and is defined by

$$Z_t = \operatorname*{esssup}_{U \in \mathbf{A}^t} \mathrm{E}(X_U | \mathscr{F}_t), \tag{3.8}$$

where $\mathbf{A}^t$ denotes the set of accessible stopping points $U \geq t$. In the interpretation given at the beginning of the chapter, Z_t represents the maximal expected reward for the player when the game is stopped at some stage $U \geq t$, under the assumption that the player uses only the information available at stage t.

Clearly, $Z_{\bowtie} = X_{\bowtie}$ and $Z_t \geq X_t$ for all $t \in \mathbb{I}$, therefore Snell's envelope dominates the reward process.

Equality (3.8) extends to all accessible stopping points; more precisely, if $T \in \mathbf{A}$, then

$$Z_T = \operatorname*{esssup}_{U \in \mathbf{A}^T} \mathrm{E}(X_U | \mathscr{F}_T), \tag{3.9}$$

where $\mathbf{A}^T$ denotes the set of accessible stopping points $U \geq T$. Indeed, $\mathrm{E}(X_U | \mathscr{F}_t)$ and $\mathrm{E}(X_U | \mathscr{F}_T)$ both coincide with $\mathrm{E}(X_U 1_{\{T=t\}} | \mathscr{F}_t)$ on $\{T = t\}$, and the essential upper bound of the last conditional expectation is the same regardless of whether U runs over $\mathbf{A}^t$ or $\mathbf{A}^T$, because if U belongs to $\mathbf{A}^t$ [$\mathbf{A}^T$], then $U' = U \vee T$ [$U' = U \vee t$] belongs to $\mathbf{A}^T$ [$\mathbf{A}^t$] by the proposition of Section 1.5, and $U' = U$ on $\{T = t\}$.

The Upwards-Directed Property of Conditional Rewards

We are going to prove that for all $T \in \mathbf{A}$, the family of random variables $(\mathrm{E}(X_U | \mathscr{F}_T), U \in \mathbf{A}^T)$ is upwards directed. Let U_1 and U_2 be two elements of $\mathbf{A}^T$. Let F denote the set $\{\mathrm{E}(X_{U_1} | \mathscr{F}_T) \geq \mathrm{E}(X_{U_2} | \mathscr{F}_T)\}$ and let

$$U = \begin{cases} U_1 & \text{on } F, \\ U_2 & \text{on } F^c. \end{cases}$$

By Corollary 1 of Section 1.5, U belongs to $\mathbf{A}^T$. Moreover,

$$\mathrm{E}(X_U | \mathscr{F}_T) = \mathrm{E}\big(X_{U_1} \big| \mathscr{F}_T\big) 1_F + \mathrm{E}\big(X_{U_2} \big| \mathscr{F}_T\big) 1_{F^c} = \sup\Big(\mathrm{E}\big(X_{U_1} \big| \mathscr{F}_T\big), \mathrm{E}\big(X_{U_2} \big| \mathscr{F}_T\big)\Big),$$

which proves the assertion.

Taking into account the fact that $U = \bowtie$ belongs to $\mathbf{A}^T$ and that X_U is integrable, we use the theorem of Section 1.2 to deduce the following lemma.

Lemma. *For each $T \in \mathbf{A}$, there exists a sequence $(U_n, n \in \mathbb{N})$ of elements of $\mathbf{A}^T$ such that X_{U_n} is integrable for all $n \in \mathbb{N}$ and $\mathrm{E}(X_{U_n} | \mathscr{F}_T)$ increases* a.s. *to Z_T.*

From this lemma, it follows immediately that

$$E(Z_T) = \sup_{U \in \mathbf{A}^T} E(X_U), \tag{3.10}$$

and, in particular, that

$$E(Z_0) = \sup_{U \in \mathbf{A}} E(X_U). \tag{3.11}$$

Therefore, Z_T is an integrable random variable for all $T \in \mathbf{A}$. We will show below that Z is in fact of class $D_{\mathbf{A}}$.

Properties of Snell's Envelope

The main properties of Snell's envelope are presented in the following propositions.

Proposition 1 (Smallest supermartingale dominating X and recursivity property). *Z is a supermartingale, more precisely, the smallest supermartingale that dominates the reward process X. Moreover, for all $t \in \mathbb{I}$,*

$$Z_t = \sup\Big(X_t, \sup_{u \in \mathbb{D}_t} E(Z_u | \mathscr{F}_t)\Big). \tag{3.12}$$

Proof. It has already been observed that Z is an integrable process. Moreover, it is clear that this process is adapted. Fix $t \in \bar{\mathbb{I}}$ and let $(U_n,\ n \in \mathbb{N})$ be a sequence of elements of $\mathbf{A}^t$ such that X_{U_n} is integrable for all $n \in \mathbb{N}$ and $E(X_{U_n} | \mathscr{F}_t)$ increases a.s. to Z_t. If $s < t$, by conditioning with respect to $\mathscr{F}_s$ and taking the inclusion $\mathbf{A}^t \subset \mathbf{A}^s$ into account, we see that

$$E(Z_t | \mathscr{F}_s) = \lim_{n \to \infty} E\big(X_{U_n} | \mathscr{F}_s\big) \le \operatorname*{esssup}_{U \in \mathbf{A}^t} E(X_U | \mathscr{F}_s) \le Z_s,$$

which proves that Z is a supermartingale. We already know that Z dominates X. On the other hand, if Z' is a supermartingale indexed by $\bar{\mathbb{I}}$ that dominates X and $t \in \bar{\mathbb{I}}$, then $E(Z'_U | \mathscr{F}_t) \le Z'_t$ for all $U \in \mathbf{A}^t$, by Corollary 1 of Section 1.6, and therefore $E(X_U | \mathscr{F}_t) \le Z'_t$ for the same U. It follows that $Z \le Z'$, establishing the first part of the proposition. As for (3.12), notice that the inequality $\ge$ between the two sides is an immediate consequence of the fact that Z is supermartingale that dominates X. It is now sufficient to prove the converse inequality. For $t \in \mathbb{I}$, let $\tilde{Z}_t$ denote the right-hand side of (3.12) and set $\tilde{Z}_{\bowtie} = X_{\bowtie}$. Because $\tilde{Z}_t \ge X_t$ for all $t \in \bar{\mathbb{I}}$, we only need to prove that the process $(\tilde{Z}_t,\ t \in \bar{\mathbb{I}})$ is a supermartingale. This process is clearly adapted and integrable. Fix $s \in \mathbb{I}$ and $t \in \bar{\mathbb{I}}$ such that $s < t$, and let $u \in \mathbb{D}_s$ be such that $u \leq t$. Since $\tilde{Z}_t \leq Z_t$,

$$E(\tilde{Z}_t | \mathscr{F}_s) \leq E(Z_t | \mathscr{F}_s) = E(E(Z_t | \mathscr{F}_u) | \mathscr{F}_s) \leq E(Z_u | \mathscr{F}_s) \leq \tilde{Z}_s.$$

The proof is complete.

Proposition 2 (Class $D_{\mathbf{A}}$ property). *Z is of class $D_{\mathbf{A}}$.*

Proof. By Corollary 2 of Section 1.6, it is sufficient to prove that Z^+ is of class $D_{\mathbf{A}}$. Because X^+ is of class $D_{\mathbf{A}}$, there exists a test function for uniform integrability f such that $\sup_{T \in \mathbf{A}} E(f(X_T^+)) < \infty$ (see the subsection "Uniform Integrability" in Section 1.2). We are going to prove that $\sup_{T \in \mathbf{A}} E(f(Z_T^+)) < \infty$, which will establish the proposition. Let T be some element of $\mathbf{A}$. By the lemma, there exists a sequence $(U_n, n \in \mathbb{N})$ of elements of $\mathbf{A}^T$ such that $E(X_{U_n} | \mathscr{F}_T)$ increases a.s. to Z_T. It follows that $E(X_{U_n} | \mathscr{F}_T)^+$ increases a.s. to Z_T^+. Using the continuity, the monotonicity, and the convexity of f, which implies in particular that $x \mapsto f(x^+)$ is convex, we see that

$$\begin{aligned} E(f(Z_T^+)) &= E\left(f\left(\lim_{n \to \infty} E(X_{U_n} | \mathscr{F}_T)^+\right)\right) \\ &= E\left(\lim_{n \to \infty} f\left(E(X_{U_n} | \mathscr{F}_T)^+\right)\right) \\ &\leq \liminf_{n \to \infty} E\left(f\left(E(X_{U_n}^+ | \mathscr{F}_T)\right)\right) \\ &\leq \liminf_{n \to \infty} E\left(E\left(f(X_{U_n}^+) | \mathscr{F}_T\right)\right) \\ &\leq \sup_{U \in \mathbf{A}} E(f(X_U^+)). \end{aligned}$$

This completes the proof.

Proposition 3 (Behavior at infinity). *Suppose that X satisfies condition* (3.3). *If $(T_n, n \in \mathbb{N})$ is an increasing sequence of accessible stopping points and $\lim_{n \to \infty} T_n = T$, then $\lim_{n \to \infty} E(Z_{T_n}) = E(Z_T)$. In particular, if Γ is a predictable increasing path, then $E(Z_{\infty-}^\Gamma) = E(Z_\infty^\Gamma)$. In fact, $Z_{\infty-}^\Gamma = E(Z_\infty^\Gamma | \mathscr{F}_{\infty-}^\Gamma)$, where $Z_{\infty-}^\Gamma = \lim_{n \to \infty} Z_n^\Gamma$.*

Proof. By the theorem of Section 1.6 and Propositions 1 and 2, Z^Γ is a uniformly integrable supermartingale (which is closed on the right). Consequently, Z_n^Γ converges a.s. and in L^1 to an integrable random variable $Z_{\infty-}^\Gamma$ such that $Z_{\infty-}^\Gamma \geq E(Z_\infty^\Gamma | \mathscr{F}_{\infty-}^\Gamma)$. On the other hand, given $\varepsilon > 0$ and for all $n \in \mathbb{N}$, there exists by (3.10) stopping points $U_n \in \mathbf{A}^{\Gamma_n}$ such that $E(Z_n^\Gamma) - \varepsilon \leq E(X_{U_n})$. Now $U_n \geq \Gamma_n$; therefore $U_n \to \bowtie$, and taking (3.3) into account, this implies

$$\limsup_{n\to\infty} X_{U_n} \le \sup\Big(\limsup_{t\to\bowtie} X_t, X_{\bowtie}\Big) \le X_{\bowtie}.$$

Consequently, $\limsup_{n\to\infty} X_{U_n}$ is a semi-integrable random variable. Since X^+ is of class $\mathrm{D}_{\mathbf{A}}$, Fatou's lemma applies and allows us to write

$$\begin{aligned} \mathrm{E}\big(Z^{\Gamma}_{\infty-}\big) - \varepsilon &= \lim_{n\to\infty} \mathrm{E}\big(Z^{\Gamma}_{n}\big) - \varepsilon \\ &\le \limsup_{n\to\infty} \mathrm{E}\big(X_{U_n}\big) \\ &\le \mathrm{E}\Big(\limsup_{n\to\infty} X_{U_n}\Big) \\ &\le \mathrm{E}(X_{\bowtie}). \end{aligned}$$

Because ε is arbitrary and $X_{\bowtie} = Z^{\Gamma}_{\infty}$, it follows that $\mathrm{E}(Z^{\Gamma}_{\infty-}) \le \mathrm{E}(Z^{\Gamma}_{\infty})$ and therefore that $Z^{\Gamma}_{\infty-} = \mathrm{E}(Z^{\Gamma}_{\infty} | \mathscr{F}^{\Gamma}_{\infty-})$, which establishes the second assertion of the proposition. In order to prove the first, consider a predictable increasing path Γ that passes through all the T_n and therefore also through T. Set $\tau_n = |T_n|$ and $\tau = |T|$. Clearly, τ_n increases a.s. to τ; therefore $Z_{T_n} = Z^{\Gamma}_{\tau_n}$ converges a.s. and in L^1 to the random variable

$$Z^{\Gamma}_{\tau-} = Z^{\Gamma}_{\tau} 1_{\{\tau<\infty\}} + Z^{\Gamma}_{\infty-} 1_{\{\tau=\infty\}}.$$

But $\mathrm{E}(Z^{\Gamma}_{\infty-} 1_{\{\tau=\infty\}}) = \mathrm{E}(Z^{\Gamma}_{\infty} 1_{\{\tau=\infty\}})$, because $\{\tau = \infty\} = \{\tau < \infty\}^c \in \mathscr{F}^{\Gamma}_{\infty-}$ and $Z^{\Gamma}_{\infty-} = \mathrm{E}(Z^{\Gamma}_{\infty} | \mathscr{F}^{\Gamma}_{\infty-})$. Consequently, $\mathrm{E}(Z^{\Gamma}_{\tau-}) = \mathrm{E}(Z^{\Gamma}_{\tau})$, which proves the first assertion.

Proposition 4 (Snell's envelope by successive approximations). *Set $Z^{(0)} = X$ and define $Z^{(n+1)}$ by induction using the formula*

$$Z^{(n+1)}_t = \begin{cases} \sup\Big(X_t, \mathrm{E}(X_{\bowtie} | \mathscr{F}_t), \sup_{u\in\mathbb{D}_t} \mathrm{E}\big(Z^{(n)}_u \big| \mathscr{F}_t\big)\Big), & \text{if } t \in \mathbb{I}, \\ X_{\bowtie}, & \text{if } t = \bowtie. \end{cases}$$

Then the sequence of processes $(Z^{(n)}, n \in \mathbb{N})$ is nondecreasing and

$$\lim_{n\to\infty} Z^{(n)} = Z \qquad \Big(\text{that is, } \lim_{n\to\infty} Z^{(n)}_t = Z_t \text{ for all } t \in \bar{\mathbb{I}}\Big).$$

Proof. By induction, it is clear that $Z^{(n)} \le Z^{(n+1)}$ for all $n \in \mathbb{N}$. Moreover, since Z is a supermartingale that dominates X, we observe by induction that $Z^{(n)} \le Z$ for all $n \in \mathbb{N}$. If $Z^{(\infty)}$ denotes the process $\lim_{n\to\infty} Z^{(n)}$, it is clear that $X \le Z^{(\infty)} \le Z$. Passing to the limit in the formula that defines $Z^{(n+1)}_t$, we see that for all $t \in \mathbb{I}$ and all $u \in \mathbb{D}_t$,

$$\mathrm{E}\big(Z_u^{(\infty)}|\mathcal{F}_t\big) \le Z_t^{(\infty)} \quad \text{and} \quad \mathrm{E}\big(Z_{\bowtie}^{(\infty)}|\mathcal{F}_t\big) = \mathrm{E}(X_{\bowtie}|\mathcal{F}_t) \le Z_t^{(\infty)},$$

which proves that $Z^{(\infty)}$ is a supermartingale. Using Proposition 1, we conclude that $Z^{(\infty)} = Z$.

3.3 SOLVING THE PROBLEM

Snell's envelope Z plays a key role in the solution of the problem at hand.

Proposition 1 (Optimality criterion). *An accessible stopping point T is optimal if and only if*

(a) $$Z_T = X_T, \quad \text{and}$$

(b) $$\mathrm{E}(Z_T) = \mathrm{E}(Z_0).$$

Proof. By (3.2) and (3.11), T is optimal if and only if $\mathrm{E}(Z_0) = \mathrm{E}(X_T)$. It follows that if conditions (a) and (b) are satisfied, then T is optimal. Conversely, if T is optimal, then $\mathrm{E}(Z_0) = \mathrm{E}(X_T)$; therefore $\mathrm{E}(Z_T) \le \mathrm{E}(X_T)$, because $\mathrm{E}(Z_T) \le \mathrm{E}(Z_0)$ by Corollary 1 of Section 1.6. The validity of (a) is now clear, because $X_T \le Z_T$, and (b) follows immediately.

Proposition 2 (Optimality criterion). *Let T be an optimal accessible stopping point and S an accessible stopping point such that $S \le T$. In order that S be optimal, it is necessary and sufficient that $Z_S = X_S$. In particular, the debut of $\{\{Z = X\}\}$ along each predictable increasing path that passes through any given optimal accessible stopping point is also optimal.*

Proof. By (b) of Proposition 1, $\mathrm{E}(Z_T) = \mathrm{E}(Z_0)$; therefore $\mathrm{E}(Z_S) = \mathrm{E}(Z_0)$, because $\mathrm{E}(Z_T) \le \mathrm{E}(Z_S) \le \mathrm{E}(Z_0)$ by Corollary 1 of Section 1.6. The conclusion now follows from Proposition 1.

Optimizing Increasing Paths

Intuitively, the best strategy is the one that consists of choosing at each stage one of the immediately following stages where the expected future reward is highest possible. This suggests the following construction.

For all $s \in \mathbb{I}$, equip $\mathbb{D}_s$ with some total order. This order formalizes the idea that certain direct successors of s have priority over others. Let Γ denote the sequence $(\Gamma_n, n \in \mathbb{N})$ of random variables with values in $\mathbb{I}$ defined by induction as follows. Set $\Gamma_0 = 0$ and, for all $n \in \mathbb{N}$ and all $s \in \mathbb{I}$ such that $\{\Gamma_n = s\} \neq \varnothing$, set

$$\Gamma_{n+1} = \inf\Big\{t \in \mathbb{D}_s\colon \mathrm{E}(Z_t|\mathcal{F}_s) = \sup_{u \in \mathbb{D}_s} \mathrm{E}(Z_u|\mathcal{F}_s)\Big\} \qquad \text{on } \{\Gamma_n = s\}. \quad (3.13)$$

Clearly, $\Gamma_{n+1} \in \mathbb{D}_{\Gamma_n}$. Moreover, under the assumption that Γ_n is a stopping point, we see that $\{\Gamma_n = s\} \cap \{\Gamma_{n+1} = t\} \in \mathscr{F}_s$ for all $s \in \mathbb{I}$ and all $t \in \mathbb{D}_s$, which proves that Γ_{n+1} is an $\mathscr{F}_n^\Gamma$-measurable stopping point. Therefore Γ is a predictable increasing path.

Notice that for all $n \in \mathbb{N}$, this path satisfies

$$E\left(Z_{n+1}^\Gamma \middle| \mathscr{F}_n^\Gamma\right) = \sup_{u \in \mathbb{D}_{\Gamma_n}} E\left(Z_u \middle| \mathscr{F}_n^\Gamma\right). \tag{3.14}$$

Indeed, on $\{\Gamma_n = s\} \cap \{\Gamma_{n+1} = t\}$, this equality reduces to

$$E(Z_t | \mathscr{F}_s) = \sup_{u \in \mathbb{D}_s} E(Z_u | \mathscr{F}_s),$$

which is a consequence of (3.13).

From now on, any predictable increasing path Γ that satisfies (3.14) for all $n \in \mathbb{N}$ will be termed an *optimizing increasing path*.

Optimization

Under our assumptions, all optimizing increasing paths, and in particular the path defined by (3.14), do in fact lead to an optimal stopping point.

Theorem (Smallest and largest optimal accessible stopping point on an optimizing increasing path). *Let X be an adapted integrable process indexed by $\bar{\mathbb{I}}$ whose positive part X^+ is of class* $\mathrm{D}_{\mathbf{A}}$. *Suppose that condition* (3.3) *is satisfied. Let Z be Snell's envelope of X, Γ an optimizing increasing path, S the debut of $\{\{Z = X\}\}$ along Γ, and let $T = \Gamma_\tau$, where τ is the first time Z^Γ stops being a martingale; that is*

$$\tau = \inf\left\{n \in \mathbb{N}: Z_n^\Gamma - E\left(Z_{n+1}^\Gamma \middle| \mathscr{F}_n^\Gamma\right) > 0\right\}. \tag{3.15}$$

Then S and T are optimal stopping points, more precisely the smallest and largest of the optimal stopping points located on Γ.

Proof. By Proposition 2, it is sufficient to prove the assertion concerning T. Clearly, τ is a stopping time relative to the filtration $\mathscr{F}^\Gamma$; therefore T is an accessible stopping point (located on Γ). Moreover, for all $n \in \mathbb{N}$, $Z_n^\Gamma > E(Z_{n+1}^\Gamma | \mathscr{F}_n^\Gamma)$ on $\{\tau = n\}$ by (3.15); therefore $Z_n^\Gamma = X_n^\Gamma$ on this set, by (3.14) and (3.12). Because $Z_\infty^\Gamma = X_\infty^\Gamma$, it follows that $Z_T = X_T$. We now prove that $E(Z_T) = E(Z_0)$. By (3.15), $Z_n^\Gamma = E(Z_{n+1}^\Gamma | \mathscr{F}_n^\Gamma)$ on $\{\tau > n\}$, and therefore $E(Z_{n \wedge \tau}^\Gamma) = E(Z_0)$. Since Z^Γ is a supermartingale by Proposition 1 of Section 3.2, it follows that $(Z_{n \wedge \tau}^\Gamma, n \in \mathbb{N})$ is in fact a martingale, which is uniformly integrable by Proposition 2. By Corollary 3 of Section 1.6, $Z_{n \wedge \tau}^\Gamma$ converges a.s. and in L^1 to an integrable random variable Y. Taking Proposition 3 of

Section 3.2 into account,we now observe that

$$Y = \begin{cases} Z^{\Gamma}_{\tau} & \text{on } \{\tau < \infty\}, \\ Z^{\Gamma}_{\infty -} & \text{on } \{\tau = \infty\}, \end{cases}$$

and therefore that

$$\mathrm{E}(Z_0) = \lim_{n \to \infty} \mathrm{E}(Z^{\Gamma}_{n \wedge \tau}) = \mathrm{E}\big(Z^{\Gamma}_{\tau} 1_{\{\tau < \infty\}} + Z^{\Gamma}_{\infty -} 1_{\{\tau = \infty\}}\big).$$

Since $Z^{\Gamma}_{\infty -} = \mathrm{E}(Z^{\Gamma}_{\infty} | \mathscr{F}^{\Gamma}_{\infty -})$ by Proposition 3 of Section 3.2, we can replace $Z^{\Gamma}_{\infty -}$ by Z^{Γ}_{∞}, which equals Z^{Γ}_{τ} on $\{\tau = \infty\}$, to get

$$\mathrm{E}(Z_0) = \mathrm{E}(Z^{\Gamma}_{\tau}) = \mathrm{E}(Z_T).$$

Therefore conditions (a) and (b) of Proposition 1 are satisfied; hence T is optimal. Let T' be an optimal stopping point located on Γ and set $\tau' = |T'|$. Then $\mathrm{E}(Z^{\Gamma}_{\tau'}) = \mathrm{E}(Z_0)$ by (b) of Proposition 1; therefore $(Z^{\Gamma}_{n \wedge \tau'}, n \in \overline{\mathbb{N}})$ is a martingale, or equivalently, $\mathrm{E}(Z^{\Gamma}_{n+1} | \mathscr{F}^{\Gamma}_{n}) = Z^{\Gamma}_{n}$ on $\{\tau' > n\}$. By (3.15), it follows that $\tau \geq \tau'$, and thus T is the largest optimal accessible stopping point located on Γ.

Case Where the Dimension Is 1

In dimension 1, the conclusions of the theorem are expressed as follows: The debut of $\{\{Z = X\}\}$ and the first time Z^{Γ} stops being a martingale are, respectively, the smallest and largest optimal stopping time.

Remarks

The following remarks can be verified on simple deterministic examples.

1. Let Γ, S, and T be as in the theorem. A stopping point located on Γ between S and T is not necessarily optimal.
2. In general, the debut of $\{\{Z = X\}\}$ along an arbitrary predictable increasing path is not optimal.
3. If the debut of $\{\{Z = X\}\}$ along some predictable increasing path Γ is optimal, then the stopping point T defined as in the theorem, but with the help of that path Γ, is not optimal in general.
4. Unless the dimension is 1, the existence of finite accessible optimal stopping points does not guarantee that the optimal stopping points defined in the theorem are finite.

Optimization Beyond a Given Stage

Let X be as in the theorem and fix $u \in \mathbb{I}$. It is natural to ask whether there exists an accessible stopping point $T^o \geq u$ such that

$$\mathrm{E}(X_{T^o}) = \sup_{T \in \mathbf{A}^u} \mathrm{E}(X_T).$$

Such a point will be termed *optimal in* $\mathbf{A}^u$. Notice that by the upwards-directed property of conditional rewards and the theorem of Section 1.2, T^o is optimal in $\mathbf{A}^u$ if and only if

$$\mathrm{E}(X_{T^o} | \mathscr{F}_u) = \operatorname*{esssup}_{T \in \mathbf{A}^u} \mathrm{E}(X_T | \mathscr{F}_u).$$

We are now going to show how to reduce this problem to the one that has just been solved. Let $\tilde{X}$ denote the process $(\tilde{X}_t,\ t \in \bar{\mathbb{I}})$, where $\tilde{X}_t = X_{t+u}$ for all $t \in \mathbb{I}$ and $\tilde{X}_{\bowtie} = X_{\bowtie}$. Set $\tilde{\mathscr{F}}_t = \mathscr{F}_{t+u}$ for all $t \in \mathbb{I}$ and $\tilde{\mathscr{F}}_{\bowtie} = \mathscr{F}_{\bowtie}$. Clearly, $(\tilde{\mathscr{F}}_t)$ is a filtration and $\tilde{X}$ is an integrable process adapted to this filtration. Let $\tilde{\mathbf{A}}$ denote the set of accessible stopping points relative to the filtration $(\tilde{\mathscr{F}}_t)$ and let $\tilde{\mathbf{A}}^t$ be the set of elements $\tilde{T}$ of $\tilde{\mathbf{A}}$ such that $\tilde{T} \geq t$. It is easy to check that $\tilde{T}$ belongs to $\tilde{\mathbf{A}}^t$ if and only if $\tilde{T}$ is of the form $T - u$, where T belongs to $\mathbf{A}^{t+u}$. It follows that $\tilde{X}^+$ is of class $\mathrm{D}_{\mathbf{A}}$. In addition, if Z and $\tilde{Z}$ are Snell's envelopes of X and $\tilde{X}$, respectively, then for all $t \in \mathbb{I}$,

$$\begin{aligned}
\tilde{Z}_t &= \operatorname*{esssup}_{\tilde{T} \in \tilde{\mathbf{A}}^t} \mathrm{E}\left(\tilde{X}_{\tilde{T}} | \tilde{\mathscr{F}}_t\right) \\
&= \operatorname*{esssup}_{\tilde{T} \in \tilde{\mathbf{A}}^t} \mathrm{E}\left(X_{\tilde{T}+u} | \mathscr{F}_{t+u}\right) \\
&= \operatorname*{esssup}_{T \in \mathbf{A}^{t+u}} \mathrm{E}\left(X_T | \mathscr{F}_{t+u}\right) \\
&= Z_{t+u}.
\end{aligned}$$

Moreover, $\tilde{Z}_{\bowtie} = Z_{\bowtie}$. Using $\tilde{X}$ and $\tilde{Z}$, the following corollary can easily be deduced from the theorem.

Corollary . (Optimal stopping points in $\mathbf{A}^u$). *Let X and Z be as in the theorem and fix $u \in \mathbb{I}$. Let $(\Gamma_n,\ n \geq |u|)$ be the sequence of stopping points defined inductively by letting $\Gamma_{|u|} = u$ and then defining Γ_{n+1} using formula* (3.13). *Assume that condition* (3.3) *is satisfied. Then the debut of $\{\{Z = X\}\}$ along $(\Gamma_n,\ n \geq |u|)$ and the stopping point $T = \Gamma_\tau$, where τ is defined as in* (3.15) *but taking the lower bound over $n \geq |u|$, are, respectively, the smallest and the largest of the optimal stopping points in $\mathbf{A}^u$ located on $(\Gamma_n,\ n \geq |u|)$.*

A Condition Ensuring that a Given Stopping Point Is Accessible

The proof of the following proposition relies on the theorem above. This proposition shows in particular that the validity of the conclusion of the optional sampling theorem implies that all stopping points are accessible.

Proposition 3 (Accessibility condition). *If T is a stopping point such that $\mathrm{E}(Y_T) \le \mathrm{E}(Y_0)$ for all bounded supermartingales Y indexed by $\bar{\mathbb{I}}$, then T is accessible.*

Proof. The process X defined by $X_t = 1_{\{T=t\}}$ $(t \in \bar{\mathbb{I}})$ is adapted, bounded, and satisfies condition (3.3). Let us take it as reward process and denote its Snell's envelope by Z. Clearly, $0 \le Z \le 1$ and $1 = X_T \le Z_T \le 1$, which shows that $Z_T = 1$. On the other hand, under the assumption of the proposition, $\mathrm{E}(Z_T) \le \mathrm{E}(Z_0) \le 1$; therefore $\mathrm{E}(Z_0) = 1$. By the theorem, there exists an optimal accessible stopping point T^o. Because $\mathrm{E}(X_{T^o}) = \mathrm{E}(Z_0)$ by Proposition 1 and $\mathrm{E}(X_{T^o}) = \mathrm{P}\{T = T^o\}$, we conclude that $T = T^o$, and therefore T is accessible.

3.4 A RELATED PROBLEM

A variation on the optimization problem formulated at the beginning of Section 3.1 is obtained by starting with a reward process indexed by $\mathbb{I}$ instead of $\bar{\mathbb{I}}$. That is, the reward process is an adapted integrable process X indexed by $\mathbb{I}$ whose positive part X^+ is of class $\mathrm{D}_{\mathbf{A}_f}$, where $\mathbf{A}_f$ denotes the set of finite accessible stopping points. The notion of value of the game and that of optimal accessible stopping point are defined as in the beginning of Section 3.1, but only taking into account finite accessible stopping points.

At first, we could be tempted to incorporate into X the random variable $X_{\bowtie} = \limsup_{t \to \bowtie} X_t$ and to treat the modified process as in the previous sections. Unfortunately, this approach is ineffective, because it may increase the value of the game and of Snell's envelope (cf. Section 3.8), and in fact, the problem considered here is quite different from the one formulated in Section 3.1. This should be no surprise, since even in dimension $d = 1$, the optimization problem for stochastic processes indexed by $\bar{\mathbb{I}}$ is the random version of maximizing an upper semicontinuous function defined on the compact set $\overline{\mathbb{N}}$, whereas the optimization problem for processes indexed by $\mathbb{I}$ extends maximizing a function defined on the (half) open set $\mathbb{N}$. In the first case, existence of a global maximum is established, whereas in the second, it is only possible to provide conditions under which certain candidates for a maximum are indeed optimal.

In order to determine whether an optimal accessible stopping point exists, we are going to review step by step the three previous sections and point out which modifications are necessary. We replace $\bar{\mathbb{I}}$ by $\mathbb{I}$ everywhere from now on, and for $T \in \mathbf{A}_f$, we let $\mathbf{A}_f^T$ denote the set of $U \in \mathbf{A}_f$ such that $U \geq T$ and $\mathbf{A}_X^T$ the set of $U \in \mathbf{A}_f^T$ such that $E(X_U^-) < \infty$. Instead of $\mathbf{A}_X^0$, we will usually write $\mathbf{A}_X$. Observe that $\mathbf{A}_X$ contains all accessible stopping points that take only a finite number of finite values, and in particular stopping points of the form Γ_n or $\Gamma_{n \wedge \tau}$, where Γ is a predictable increasing path and τ is a stopping time relative to the filtration $\mathscr{F}^\Gamma$. However, $\mathbf{A}_X^T$ may be empty (cf. Exercise 3).

Stating the Problem

In (3.1) and (3.2), $\mathbf{A}$ should be replaced by $\mathbf{A}_f$. The value of the game is finite due to the inequalities

$$-\infty < E(X_0) \leq \sup_{T \in \mathbf{A}_f} E(X_T) \leq \sup_{T \in \mathbf{A}_f} E(X_T^+) < \infty.$$

A stopping point $T^o \in \mathbf{A}_f$ is optimal (in $\mathbf{A}_f$) if it satisfies (3.2). It is then necessarily an element of $\mathbf{A}_X$. The remainder of Section 3.1 is to be omitted, except for the example in (3.5) (without $X_{\bowtie}$), which shows that there is not necessarily an optimal stopping point (in $\mathbf{A}_f$).

Snell's Envelope

The first part of Section 3.2 up to and including (3.11) remains valid, provided $\mathbf{A}$, $\mathbf{A}^t$, and $\mathbf{A}^T$ are replaced, respectively, by $\mathbf{A}_f$, $\mathbf{A}_f^t$, and $\mathbf{A}_f^T$; the conclusion concerning the integrability of the X_{U_n} in the lemma is removed if $\mathbf{A}_X^T$ is empty; and the stopping point T in (3.10) is assumed to belong to $\mathbf{A}_X$. Therefore, Z is an integrable process. Moreover, if $\mathbf{A}_X^T$ is not empty, then Z_T is integrable and

$$Z_T = \operatorname*{esssup}_{U \in \mathbf{A}_X^T} E(X_U | \mathscr{F}_T) \quad \text{and} \quad E(Z_T) = \sup_{U \in \mathbf{A}_X^T} E(X_U).$$

Some modifications to Propositions 1, 2, and 3 are necessary. For the convenience of the reader, we will present their new statements followed by a remark concerning Proposition 4. But first we need a new notion.

Regular Supermartingales

We say that a stochastic process Y indexed by $\mathbb{I}$ is a *regular supermartingale* (relative to $\mathbf{A}_X$) if Y is adapted, Y_T is integrable for all $T \in \mathbf{A}_X$ and

$$E(Y_T | \mathscr{F}_s) \leq Y_s, \qquad \text{for all } s \in \mathbb{I} \text{ and all } T \in \mathbf{A}_X^s. \tag{3.16}$$

A regular supermartingale is clearly a supermartingale. Conversely, by Theorem 2 of Section 1.6, any supermartingale Y indexed by $\mathbb{I}$ such that Y^-

is uniformly integrable is a regular supermartingale. In particular, all nonnegative supermartingales indexed by $\mathbb{I}$ are regular supermartingales.

We can now state the modifications of Propositions 1, 2, and 3 of Section 3.2.

Proposition 1 (Smallest regular supermartingale dominating X and recursion property). *Z is a regular supermartingale, more precisely, the smallest regular supermartingale that dominates the reward process X. Moreover, for all $t \in \mathbb{I}$,* (3.12) *is satisfied.*

Proof. It is sufficient to modify a few points in the proof of Proposition 1 of Section 3.2. More precisely, in order to establish that Z is a regular supermartingale, it suffices to replace t by $T \in \mathbf{A}_X^s$ and $\mathbf{A}^t$ by $\mathbf{A}_X^T$. In the part concerning unicity, Z' should denote a regular supermartingale and U should be taken in $\mathbf{A}_X^t$ (the reference to Corollary 1 of Section 1.6 is no longer needed). In the proof of (3.12), the inequality $\geq$ between the two sides is again an immediate consequence of the fact that Z is a supermartingale dominating X. The proof of the converse inequality should be replaced by the following. Fix $t \in \mathbb{I}$ and consider some element U of $\mathbf{A}_X^t$ and a predictable increasing path Γ passing though t and U. For all $u \in \mathbb{D}_t$, set $F_u = \{\Gamma_{|t|+1} = u\} \cap \{U \neq t\}$. Note that $\bigcup_{u \in \mathbb{D}_t} F_u = \{U \neq t\}$, because $\Gamma_{|t|+1}$ is a direct successor of t. Moreover, F_u belongs to $\mathscr{F}_t$, since $\Gamma_{|t|+1}$ is $\mathscr{F}_{|t|}^{\Gamma}$-measurable and $\Gamma_{|t|} = t$. For all $u \in \mathbb{D}_t$, put

$$T_u = \begin{cases} U & \text{on } F_u, \\ u & \text{on } F_u^c. \end{cases}$$

Clearly, $T_u \geq u$ and $\{T_u = s\} = (\{U = s\} \cap F_u) \cup (\{u = s\} \cap F_u^c) \in \mathscr{F}_s$ for all $s \geq u$, which shows that T_u is a stopping point. On the other hand, putting $T_1 = U$, $T_2 = u$, $S = t$ and $F = F_u$ in Corollary 1 of Section 1.5, we see that T_u is accessible. Write

$$X_U = X_t \, 1_{\{U=t\}} + \sum_{u \in \mathbb{D}_t} X_{T_u} 1_{F_u}$$

and take the conditional expectation of both sides with respect to $\mathscr{F}_t$. Since Z dominates X and $\mathrm{E}(Z_{T_u} | \mathscr{F}_t) = \mathrm{E}(\mathrm{E}(Z_{T_u} | \mathscr{F}_u) | \mathscr{F}_t) \leq \mathrm{E}(Z_u | \mathscr{F}_t)$, we see that

$$\begin{aligned} \mathrm{E}(X_U | \mathscr{F}_t) &\leq X_t \, 1_{\{U=t\}} + \sup_{u \in \mathbb{D}_t} \mathrm{E}(Z_u | \mathscr{F}_t) \, 1_{\{U \neq t\}} \\ &\leq \sup\Big(X_t, \sup_{u \in \mathbb{D}_t} \mathrm{E}(Z_u | \mathscr{F}_t) \Big). \end{aligned}$$

Because U was arbitrary in $\mathbf{A}_X^t$, it follows that

$$Z_t = \operatorname*{esssup}_{U \in \mathbf{A}_X^t} \mathrm{E}(X_U | \mathscr{F}_t) \leq \sup\Big(X_t, \sup_{u \in \mathbb{D}_t} \mathrm{E}(Z_u | \mathscr{F}_t) \Big),$$

which establishes (3.12).

Proposition 2 (Class $\mathrm{D}_{\mathbf{A}_\mathrm{f}}$ property). *Z^+ is of class $\mathrm{D}_{\mathbf{A}_\mathrm{f}}$.*

Proof. It suffices to replace $\mathbf{A}$ and $\mathbf{A}^T$, respectively, by $\mathbf{A}_\mathrm{f}$ and $\mathbf{A}_\mathrm{f}^T$ in the proof of Proposition 2 of Section 3.2.

Proposition 3 (Behavior at infinity). *Suppose* $\mathrm{E}(\sup_{t \in \mathbb{I}} X_t^+) < \infty$. *Let Γ be a predictable increasing path such that*

$$\lim_{n \to \infty} \sup_{t \in \mathbb{I}:\, t \geq \Gamma_n} X_t = \limsup_{n \to \infty} X_n^\Gamma. \tag{3.17}$$

(*This equality holds in particular if* $\lim_{t \to \bowtie} X_t$ *exists* (*in* $\overline{\mathbb{R}}$) *or if the dimension is* 1.) *Then*

$$\limsup_{n \to \infty} Z_n^\Gamma = \limsup_{n \to \infty} X_n^\Gamma \quad \Big(= \lim_{t \to \bowtie} X_t \textit{ if this limit exists} \Big).$$

Proof. By (3.8) (with $\mathbf{A}_\mathrm{f}^t$ instead of $\mathbf{A}^t$), if $n \geq m$, then

$$X_n^\Gamma \leq Z_n^\Gamma \leq \mathrm{E}\Big(\sup_{t \in \mathbb{I}:\, t \geq \Gamma_m} X_t | \mathscr{F}_n^\Gamma \Big).$$

Therefore, since $\sup_{t \in \mathbb{I}:\, t \geq \Gamma_m} X_t$ is an integrable random variable,

$$\limsup_{n \to \infty} X_n^\Gamma \leq \limsup_{n \to \infty} Z_n^\Gamma \leq \mathrm{E}\Big(\sup_{t \in \mathbb{I}:\, t \geq \Gamma_m} X_t \Big| \mathscr{F}_{\infty-}^\Gamma \Big).$$

Letting m converge to infinity, we see that

$$\limsup_{n \to \infty} X_n^\Gamma \leq \limsup_{n \to \infty} Z_n^\Gamma \leq \mathrm{E}\Big(\lim_{m \to \infty} \sup_{t \in \mathbb{I}:\, t \geq \Gamma_m} X_t \Big| \mathscr{F}_{\infty-}^\Gamma \Big).$$

Assume that (3.17) holds. Then the conditional expectation on the right-hand side is equal to $\limsup_{n \to \infty} X_n^\Gamma$, and the proof is complete.

Case Where X^- is Uniformly Integrable

Let $(Z^{(n)}, n \in \mathbb{N})$ be the nondecreasing sequence of processes indexed by $\mathbb{I}$ and defined by induction as follows:

$$Z_t^{(0)} = X_t \quad \text{and} \quad Z_t^{(n+1)} = \sup\Big(X_t, \sup_{u \in \mathbb{D}_t} \mathrm{E}(Z_u^{(n)} | \mathscr{F}_t) \Big). \tag{3.18}$$

Then $Z^{(n)}$ converges to the smallest supermartingale indexed by $\mathbb{I}$ that dominates X, which is not necessarily Snell's envelope Z of X (cf. Exercise 7). The proof is similar to that of Proposition 4 of Section 3.2. In the case where X^- is uniformly integrable, any supermartingale indexed by $\mathbb{I}$ that dominates X is a regular supermartingale, by Theorem 2 and Corollary 2 of Section 1.6. In this case, Snell's envelope Z of X is the smallest supermartingale indexed by $\mathbb{I}$ that dominates X and is also the limit of the sequence $(Z^{(n)},\ n \in \mathbb{N})$ defined above. In the proposition of Section 3.6, we shall characterize this limit without the assumption on X^-.

Optimization

Once it is assumed that the stopping points are finite, the only modification necessary in Proposition 1 of Section 3.3 and its proof consists of removing the reference to Corollary 1 of Section 1.6 and pointing out instead that Z is a regular supermartingale. Proposition 2 will not be used, but remains valid (cf. Exercises 4 and 5). The construction of the optimizing increasing path remains unchanged. The theorem reads as follows.

Theorem 1 (Existence of an optimal accessible stopping point). *Let X be an adapted integrable process indexed by $\mathbb{I}$ whose positive part X^+ is of class $D_{\mathbf{A}_f}$. Let Z be Snell's envelope of X and let Γ be an optimizing increasing path. If the debut S of $\{\{Z = X\}\}$ along Γ is finite, then it is optimal (in $\mathbf{A}_f$); more precisely, S is the smallest of the optimal stopping points (in $\mathbf{A}_f$) located on Γ. In particular, S is finite if $\mathrm{E}(\sup_{t \in \mathbb{I}} X_t^+) < \infty$ and $\limsup_{t \to \bowtie} X_t = -\infty$. If the stopping time τ defined in (3.15) is finite, then $T = \Gamma_\tau$ is optimal (in $\mathbf{A}_f$) and is larger than any other optimal stopping point (in $\mathbf{A}_f$) located on Γ.*

Proof. If S is finite, then $Z_S = X_S$ by definition of S. Condition (a) of Proposition 1 of Section 3.3 is therefore satisfied. In order to conclude using that proposition, it remains to verify its condition (b). Put $\sigma = |S|$. Even if S is not necessarily finite, it follows from the definition of S that $Z_n^\Gamma > X_n^\Gamma$ on $\{n < \sigma\}$; therefore, by (3.12) and (3.14), $\mathrm{E}(Z_{n+1}^\Gamma \mid \mathscr{F}_n^\Gamma) = Z_n^\Gamma$ on $\{n < \sigma\}$. Thus $(Z_{n \wedge \sigma}^\Gamma,\ n \in \mathbb{N})$ is a martingale (relative to the filtration $\mathscr{F}^\Gamma$). By Proposition 2, the positive part of this martingale is uniformly integrable, therefore Fatou's lemma applies and shows, under the assumption that S is finite, that

$$\infty > \mathrm{E}(Z_S) = \mathrm{E}(Z_\sigma^\Gamma) = \mathrm{E}\Big(\lim_{n \to \infty} Z_{n \wedge \sigma}^\Gamma\Big) \geq \lim_{n \to \infty} \mathrm{E}(Z_{n \wedge \sigma}^\Gamma) = \mathrm{E}(Z_0).$$

Since $Z_S = X_S$, $\mathrm{E}(Z_S) = \mathrm{E}(X_S) \leq \mathrm{E}(Z_0)$ by (3.11) (with $\mathbf{A}$ replaced by $\mathbf{A}_f$); therefore $\mathrm{E}(Z_S) = \mathrm{E}(Z_0)$, which establishes the validity of (b). By Proposition 1 of Section 3.3, S is clearly the smallest of the optimal stopping points located on Γ. Assume now that $\mathrm{E}(\sup_{t \in \mathbb{I}} X_t^+) < \infty$ and $\limsup_{t \to \infty} X_t = -\infty$. Then $\limsup_{n \to \infty} Z_n^\Gamma = -\infty$, by Proposition 3. But $(Z_{n \wedge \sigma}^\Gamma,\ n \in \mathbb{N})$ is a martingale whose positive part is uniformly integrable; therefore $\lim_{n\to\infty} Z_n^\Gamma$ exists and is finite on $\{S = \bowtie\}$. It follows that $\{S = \bowtie\}$ is a null set; therefore S is finite. The proof of the assertion concerning T is as follows. Under the assumption that τ is finite, we first conclude as in the proof of the theorem of Section 3.3 that $Z_T = X_T$. Therefore $\mathrm{E}(Z_T) = \mathrm{E}(X_T) \le \mathrm{E}(Z_0)$ by (3.11), and by (b) of Proposition 1 of Section 3.3, we only need to show that $\mathrm{E}(Z_T) \ge \mathrm{E}(Z_0)$. Since $Z_n^\Gamma = \mathrm{E}(Z_{n+1}^\Gamma | \mathscr{F}_n^\Gamma)$ on $\{\tau > n\}$ by (3.15), (3.14), and (3.12), we deduce that $\mathrm{E}(Z_{n\wedge\tau}^\Gamma) = \mathrm{E}(Z_0)$. Because τ is finite, $Z_\tau^\Gamma = \limsup_{n\to\infty} Z_{n\wedge\tau}^\Gamma$, and because Z^+ is uniformly integrable by Proposition 2, Fatou's lemma applies and lets us conclude that

$$\mathrm{E}(Z_T) = \mathrm{E}(Z_\tau^\Gamma) \ge \limsup_{n\to\infty} \mathrm{E}(Z_{n\wedge\tau}^\Gamma) = \mathrm{E}(Z_0).$$

The proof of the last assertion of the theorem is similar to that of the corresponding assertion of the theorem of Section 3.3.

Optimization to Within ε

Given $\varepsilon > 0$, there exists $T^\varepsilon \in \mathbf{A}_\mathrm{f}$ such that

$$\mathrm{E}(X_{T^\varepsilon}) \ge \sup_{T \in \mathbf{A}_\mathrm{f}} \mathrm{E}(X_T) - \varepsilon.$$

T^ε is said to be an *ε-optimal* stopping point (in $\mathbf{A}_\mathrm{f}$). Because optimal stopping points do not generally exist (in $\mathbf{A}_\mathrm{f}$), it is natural to ask whether it is possible to determine an ε-optimal stopping point in a constructive manner. We will see that under the assumptions of Proposition 3, the optimizing increasing path Γ defined by (3.14) can be used for this purpose. Another possibility will be mentioned at the end of Section 3.6.

Theorem 2 (Constructing an ε-optimal stopping point). *Let X be an adapted integrable process indexed by $\mathbb{I}$ such that* $\mathrm{E}(\sup_{t\in\mathbb{I}} X_t^+) < \infty$. *Let Z be Snell's envelope of X and Γ an optimizing increasing path. If condition* (3.17) *is satisfied, then the debut T^ε of $\{\{Z \le X + \varepsilon\}\}$ along Γ is ε-optimal (in $\mathbf{A}_\mathrm{f}$).*

Proof. On $\{T^\varepsilon = \bowtie\}$, $Z_n^\Gamma > X_n^\Gamma + \varepsilon$ for all $n \in \mathbb{N}$; therefore $\limsup_{n\to\infty} Z_n^\Gamma > \limsup_{n\to\infty} X_n^\Gamma$. By Proposition 3, this is only possible if $\{T^\varepsilon = \bowtie\}$ is a null set, that is, if T^ε is finite. But then $Z_{T^\varepsilon} \le X_{T^\varepsilon} + \varepsilon$; therefore $\mathrm{E}(Z_{T^\varepsilon}) \le \mathrm{E}(X_{T^\varepsilon}) + \varepsilon$. On the other hand, the argument used to establish that $\mathrm{E}(Z_S) \ge \mathrm{E}(Z_0)$ in the proof of Theorem 1 also shows that $\mathrm{E}(Z_{T^\varepsilon}) \ge \mathrm{E}(Z_0)$. Thus, $\mathrm{E}(Z_0) \le \mathrm{E}(X_{T^\varepsilon}) + \varepsilon$, and using (3.11) (with $\mathbf{A}$ replaced by $\mathbf{A}_\mathrm{f}$), the conclusion follows.

3.5 MAXIMAL ACCESSIBLE STOPPING POINTS

Returning to the reward process X considered at the beginning of Section 3.1, we denote once again Snell's envelope of X by Z and consider an optimizing increasing path Γ as well as the optimal accessible stopping point T defined in the theorem of Section 3.3. Moreover, we denote by $\tilde{\mathbf{A}}^T$ the set of elements of $\mathbf{A}^T$ distinct from T. By the definition of T, $\mathrm{E}(Z_U) < \mathrm{E}(Z_T)$ for all elements U of $\tilde{\mathbf{A}}^T$ located on Γ. It is reasonable to expect that the inequality between these two expectations still holds when U belongs to $\tilde{\mathbf{A}}^T$ but is no longer located on Γ. We are going to prove that this is indeed the case. In order to do this and also to study the converse problem, we will term *maximal accessible stopping point* any maximal element of the (partially) ordered set $\mathbf{M} = \{S \in \mathbf{A}\colon \mathrm{E}(Z_S) = \mathrm{E}(Z_0)\}$. Notice that all optimal accessible stopping points belong to $\mathbf{M}$ (by (b) of Proposition 1 of Section 3.3).

Proposition (Maximality of T). *The optimal accessible stopping point T defined in the theorem of Section* 3.3 *is a maximal element of* $\mathbf{M}$.

Proof. Assume that U belongs to $\tilde{\mathbf{A}}^T$ and let F denote the set $\{T < U\}$. Notice that F is not a null set and that it belongs to $\mathscr{F}_T$ (by property 4 of Section 1.3). We are going to show that

$$\int_F Z_U \, \mathrm{dP} < \int_F Z_T \, \mathrm{dP}.$$

By Corollary 1 of Section 1.6, this will imply that $\mathrm{E}(Z_U) < \mathrm{E}(Z_T)$, therefore that U is not an element of $\mathbf{M}$ (because $\mathrm{E}(Z_T) = \mathrm{E}(Z_0)$). Let Γ be the optimizing increasing path used to define T. Using the proposition of Section 1.5, we fix a predictable increasing path Γ' passing through T and U such that $(\Gamma'_{n\wedge\tau}) = (\Gamma_{n\wedge\tau})$, where τ denotes $|T|$ (in agreement with the notation of the theorem of Section 3.3). By the classical optional sampling theorem (applied to $Z^{\Gamma'}$), and since $\Gamma'_{|U|} = U$ and $\Gamma'_{|T|} = \Gamma_{|T|} = T$, it follows that

$$\mathrm{E}(Z_U|\mathscr{F}_T) \le \mathrm{E}\big(Z^{\Gamma'}_{\tau+1}\big|\mathscr{F}_T\big) \text{ on } F \quad \text{and} \quad \mathrm{E}\big(Z^{\Gamma'}_{\tau+1}\big|\mathscr{F}_T\big) \le Z_T \text{ on } \{\tau < \infty\}.$$

Since $F \subset \{\tau < \infty\}$, it suffices to show that for each $n \in \mathbb{N}$,

$$\mathrm{E}\big(Z^{\Gamma'}_{n+1}\big|\mathscr{F}_T\big) < Z_T \qquad \text{on } F \cap \{\tau = n\}.$$

Now

$$\mathrm{E}\big(Z^{\Gamma'}_{n+1}\big|\mathscr{F}_T\big) \le \sup_{u \in \mathbb{D}_T} \mathrm{E}(Z_u|\mathscr{F}_T) \qquad \text{on } F \cap \{\tau = n\},$$

because on $G = F \cap \{\tau = n\} \cap \{\Gamma'_n = t\} \cap \{\Gamma'_{n+1} = v\}$ (which belongs to $\mathscr{F}_t$), this inequality becomes

$$E(Z_v|\mathscr{F}_t) \le \sup_{u \in \mathbb{D}_t} E(Z_u|\mathscr{F}_t)$$

and is thus obviously satisfied (if v does not belong to $\mathbb{D}_t$, then G is a null set). By (3.14), the right-hand side of this inequality is equal to $E(Z^{\Gamma}_{n+1}|\mathscr{F}_t)$ on G. We conclude using (3.15) that

$$E\left(Z^{\Gamma'}_{n+1}\middle|\mathscr{F}_T\right) \le E\left(Z^{\Gamma}_{n+1}\middle|\mathscr{F}_T\right) < Z^{\Gamma}_n = Z_T \qquad \text{on } F \cap \{\tau = n\}.$$

Studying the Converse

An optimal accessible stopping point is not necessarily maximal (for instance, this is generally the case for the stopping point S in the theorem of Section 3.3). On the other hand, the converse is true. This is not surprising, because $E(Z_T)$ represents the largest amount that can be won on average by a player who decides to quit the game after stage T.

Theorem (Optimality of maximal accessible stopping points). *All maximal elements of* $\mathbf{M}$ *are optimal.*

Proof. Let T be a maximal element of $\mathbf{M}$. By the lemma of Section 3.2, there exists a sequence $(U_n, n \in \mathbb{N})$ of elements of $\mathbf{A}^T$ such that X_{U_n} is integrable for all $n \in \mathbb{N}$ and $E(X_{U_n}|\mathscr{F}_T)$ increases a.s. to Z_T. It follows that $E(Z_{U_n}|\mathscr{F}_T)$ converges a.s. to Z_T, because $X \le Z$ and $E(Z_{U_n}|\mathscr{F}_T) \le Z_T$, by Corollary 1 of Section 1.6. For all $n \in \mathbb{N}$, consider a predictable increasing path ${}^n\Gamma$ passing through T and U_n, and set

$$\tau_n = \begin{cases} |U_n| \wedge (|T| + 1) & \text{on } \{T < \bowtie\}, \\ \infty & \text{on } \{T = \bowtie\}, \end{cases} \qquad T_n = {}^n\Gamma_{\tau_n}.$$

Again by Corollary 1 of Section 1.6, $E(Z_{U_n}|\mathscr{F}_T) \le E(Z_{T_n}|\mathscr{F}_T) \le Z_T$; therefore $E(Z_{T_n}|\mathscr{F}_T)$ converges a.s. to Z_T. Set $Y = \sup_{1 \le i \le d} E(Z_{T+e_i}|\mathscr{F}_T)$, agreeing by convention that $T + e_i = \bowtie$ if $T = \bowtie$. Notice that $Y \le Z_T$, by the corollary just mentioned. Also notice that $\{T_n = T + e_i\} \cap \{T_n > T\}$ belongs to $\mathscr{F}_T$, because

$$\begin{aligned} &\{T_n = T + e_i\} \cap \{T_n > T\} \cap \{T = t\} \\ &\quad = \left\{{}^n\Gamma_{|t|+1} = t + e_i\right\} \cap \{\tau_n > |t|\} \cap \{T = t\}, \end{aligned}$$

and this set belongs to $\mathscr{F}_t$, by (1.8), for all $t \in \mathbb{I}$. Since $\{T_n = T\}$ also belongs to $\mathscr{F}_T$, it follows that

$$\begin{aligned} E\left(Z_{T_n} \middle| \mathscr{F}_T\right) &= Z_T \, 1_{\{T_n = T\}} + \sum_{1 \le i \le d} E\left(Z_{T+e_i} \middle| \mathscr{F}_T\right) 1_{\{T_n = T+e_i\} \cap \{T_n > T\}} \\ &\le Z_T \, 1_{\{T_n = T\}} + Y \, 1_{\{T_n > T\}} \\ &\le Z_T, \end{aligned}$$

and, passing to the limit, that

$$\lim_{n \to \infty} \left(Z_T \, 1_{\{T_n = T\}} + Y \, 1_{\{T_n > T\}}\right) = Z_T.$$

This last is equivalent to

$$\lim_{n \to \infty} (Z_T - Y) \, 1_{\{T_n > T\}} = 0. \tag{3.19}$$

But for all i,

$$E\left(Z_{T+e_i} \middle| \mathscr{F}_T\right) < Z_T \qquad \text{on } \{T < \bowtie\},$$

because otherwise, if F denotes the set $\{E(Z_{T+e_i} | \mathscr{F}_T) = Z_T\} \cap \{T < \bowtie\}$, then $P(F)$ would be positive and

$$T' = \begin{cases} T + e_i & \text{on } F, \\ T & \text{on } F^c \end{cases}$$

would be an element of $\mathbf{A}^T$ distinct from T such that $E(Z_T) = E(Z_{T'})$, contradicting the assumption that T is maximal. Consequently, $Y < Z_T$ on $\{T < \bowtie\}$. Therefore, by (3.19),

$$\lim_{n \to \infty} 1_{\{T_n > T\}} = 0.$$

Since $\{T_n > T\} = \{U_n > T\}$, we conclude that U_n converges a.s. stationarily to T. It follows that $E(Z_T) = E(X_T)$, because $E(Z_T) = \lim_{n \to \infty} E(X_{U_n})$ by the choice of U_n. Therefore, condition (a) of Proposition 1 of Section 3.3 is satisfied. Given that condition (b) of that proposition is also satisfied because T belongs to $\mathbf{M}$, that proposition implies that T is optimal.

Another Proof of the Existence of Optimal Accessible Stopping Points

By appealing to Zorn's lemma, the theorem just established provides an alternative but nonconstructive proof of the existence of optimal accessible stopping points. To see this, it is necessary to check that any family $(T_l, l \in \mathbb{L})$ of totally ordered elements of $\mathbf{M}$ has an upper bound in $\mathbf{M}$. By the theorem of Section 1.2 (applied to the family $(|T_l|, l \in \mathbb{L})$), it is possible to extract an increasing sequence $(T_{l_n}, n \in \mathbb{N})$ from $(T_l, l \in \mathbb{L})$ whose limit T dominates all the T_l. By Corollary 2 of Section 1.5, T is an accessible stopping point. Moreover, if condition (3.3) is satisfied, then T belongs to $\mathbf{M}$ by Proposition 3 of Section 3.2. It follows that the hypothesis of Zorn's lemma is satisfied, so the conclusion is also: $\mathbf{M}$ admits a maximal element. By the above theorem, this element is optimal.

3.6 CASE WHERE THE INDEX SET IS FINITE

The central role Snell's envelope plays in the resolution of the optimization problem stated at the beginning of the chapter is evident from the developments in the preceding sections. We will now see that in the case where the index set is finite, this envelope can be constructed step by step, together with a family of accessible stopping points that lead to an optimal accessible stopping point.

In this section, the index set $\bar{\mathbb{I}}$ will be replaced with $\mathbb{I}_u$, where u is some nonzero element of $\mathbb{I}$. Because of this, the meaning of all objects related to the index set will be modified. In particular, the basic filtration will be a family of sigma fields $(\mathscr{F}_t)$ indexed by $\mathbb{I}_u$ satisfying conditions a and b of Section 1.1; stopping points will take their values in $\mathbb{I}_u$; predictable increasing paths will be finite sequences $(\Gamma_n, n \in \mathbb{N}_{|u|})$ of stopping points satisfying (1.7) for all $n \leq |u|$ and (1.8) for all $n < |u|$; accessible stopping points will be stopping points located on predictable increasing paths. We denote the set of accessible stopping points by $\mathbf{A}_u$ and the set of elements U of $\mathbf{A}_u$ such that $U \geq t$ by $\mathbf{A}_u^t$. Moreover, $\mathbb{D}_t$ stands for the set of direct successors of t that belong to $\mathbb{I}_u$.

Let X be an adapted and integrable process indexed by $\mathbb{I}_u$. As mentioned earlier, we are going to present a direct construction of an optimal accessible stopping point (in $\mathbf{A}_u$).

Let Z denote the process indexed by $\mathbb{I}_u$, defined by (backwards) induction as follows:

a. $Z_u = X_u$;
b. if $s < u$ and Z_t has been defined for all $t \in \mathbb{D}_s$, then

$$Z_s = \sup\left(X_s, \sup_{t \in \mathbb{D}_s} \mathrm{E}(Z_t | \mathscr{F}_s)\right). \tag{3.20}$$

For all $s < u$, we equip $\mathbb{D}_s$ with some arbitrary total order and set

$$D(s) = \inf\left\{t \in \mathbb{D}_s \colon \mathrm{E}(Z_t | \mathscr{F}_s) = \sup_{r \in \mathbb{D}_s} \mathrm{E}(Z_r | \mathscr{F}_s)\right\}. \tag{3.21}$$

Denote by (T_t) the family of stopping points indexed by $\mathbb{I}_u$ and defined by (backwards) induction as follows:

a. $T_u = u$;
b. if $s < u$ and T_t has been defined for all $t \in \mathbb{D}_s$, then

$$T_s = \begin{cases} s & \text{on } \{Z_s = X_s\}, \\ T_{D(s)} & \text{on } \{Z_s > X_s\}. \end{cases} \tag{3.22}$$

By (3.20), Z is a supermartingale that dominates X. On the other hand, if Z' is also a supermartingale indexed by $\mathbb{I}_u$ that dominates X, we see by (backwards) induction using (3.20) that $Z' \geq Z$. In other words, Z is the smallest supermartingale indexed by $\mathbb{I}_u$ that dominates X. In view of Proposition 1 of Section 3.2, it is natural to expect that Z is Snell's envelope of X as defined by (3.8), but with $\mathbf{A}^t$ replaced by $\mathbf{A}^t_u$. The theorem that follows will confirm this and will show in addition that T_0 is optimal.

Theorem (Optimal stopping over a finite index set). *For all $t \in \mathbb{I}_u$, T_t belongs to $\mathbf{A}^t_u$ and*

$$Z_t = \mathrm{E}\left(X_{T_t} \middle| \mathscr{F}_t\right) \geq \mathrm{E}(X_U | \mathscr{F}_t) \tag{3.23}$$

for all $U \in \mathbf{A}^t_u$. In particular, Z is Snell's envelope of X and

$$\mathrm{E}(Z_0) = \mathrm{E}(X_{T_0}) \geq \mathrm{E}(X_U)$$

for all $U \in \mathbf{A}_u$, which shows that T_0 is an optimal accessible stopping point (in $\mathbf{A}_u$).

Proof. The second part of the theorem is an immediate consequence of the first, which we now establish by (backwards) induction. If $t = u$, the assertions are clearly true. The induction hypothesis is now that they are true for all $t \in \mathbb{D}_s$ $(s < u)$ and we will show their validity for s.

We begin with the fact that T_s belongs to $\mathbf{A}^s_u$. By (3.22) and the induction hypothesis, $T_s \geq s$ and $\{T_s = r\} \in \mathscr{F}_r$ for all $r \geq s$. Again by the induction hypothesis, for all $t \in \mathbb{D}_s$, there exists a predictable increasing path ${}^t\Gamma$ passing through t and T_t. Let γ denote a deterministic increasing path that passes through s and define Γ by

$$\Gamma_n = \begin{cases} \gamma_n, & \text{if } n \leq |s|, \\ \gamma_n & \text{on } \{Z_s = X_s\}, \text{ if } n > |s|, \\ {}^t\Gamma_n & \text{on } \{Z_s > X_s\} \cap \{D(s) = t\}, \text{ if } n > |s| \text{ and } t \in \mathbb{D}_s. \end{cases}$$

Clearly, Γ is a predictable increasing path passing through T_s; therefore T_s is an accessible stopping point.

We now turn to the equality in (3.23). By (3.22), $Z_s = X_{T_s}$ on $\{Z_s = X_s\}$. Moreover, by (3.20) and (3.21), $Z_s = \mathrm{E}(Z_t|\mathscr{F}_s)$ on $\{Z_s > X_s\} \cap \{D(s) = t\}$. On the other hand, by the induction hypothesis, $Z_t = \mathrm{E}(X_{T_t}|\mathscr{F}_t)$ for all $t \in \mathbb{D}_s$. Taking (3.22) into account, it follows that $Z_s = \mathrm{E}(X_{T_s}|\mathscr{F}_s)$ on $\{Z_s > X_s\} \cap \{D(s) = t\}$. Therefore $Z_s = \mathrm{E}(X_{T_s}|\mathscr{F}_s)$.

It remains to establish the inequality in (3.23). Let $T \in \mathbf{A}_u^s$ and Γ be a predictable increasing path passing through s and T. Set $m = |s| + 1$. By (1.8), $\{\Gamma_m = t\} \in \mathscr{F}_s$ for all $t \in \mathbb{D}_s$. For each $t \in \mathbb{D}_s$, set

$$U_t = \begin{cases} T & \text{on } \{\Gamma_m = t\}, \\ t & \text{on } \{\Gamma_m \neq t\}. \end{cases}$$

Clearly, $U_t \geq t$ and $\{U_t = r\} \in \mathscr{F}_r$ for all $r \geq t$. Furthermore, letting γ denote a deterministic increasing path that passes through t and replacing Γ_n by γ_n on $\{\Gamma_m \neq t\}$ if $n \geq m$, we define a predictable increasing path passing through U_t, which shows that $U_t \in \mathbf{A}_u^t$. Using successively the inequality $Z \geq X$, the fact that Z is a supermartingale, and the induction hypothesis applied to U_t, we obtain, for all $F \in \mathscr{F}_s$,

$$\begin{aligned} \int_F Z_s \, \mathrm{dP} &= \int_{F \cap \{T=s\}} Z_s \, \mathrm{dP} + \sum_{t \in \mathbb{D}_s} \int_{F \cap \{T>s\} \cap \{\Gamma_m = t\}} Z_s \, \mathrm{dP} \\ &\geq \int_{F \cap \{T=s\}} X_s \, \mathrm{dP} + \sum_{t \in \mathbb{D}_s} \int_{F \cap \{T>s\} \cap \{\Gamma_m = t\}} Z_t \, \mathrm{dP} \\ &\geq \int_{F \cap \{T=s\}} X_T \, \mathrm{dP} + \sum_{t \in \mathbb{D}_s} \int_{F \cap \{T>s\} \cap \{\Gamma_m = t\}} X_{U_t} \, \mathrm{dP} \\ &= \int_F X_T \, \mathrm{dP}, \end{aligned}$$

which shows that $Z_s \geq \mathrm{E}(X_T|\mathscr{F}_s)$.

Comparing the Results

If the orders with which the sets $\mathbb{D}_s$ were equipped in (3.13) and (3.21) are the same, then the optimal stopping point T_0 constructed by induction is nothing but the optimal accessible stopping point S defined in the theorem of Section 3.3 (adapted to the present setting). To see this, let Γ denote the predictable increasing path defined by (3.13). Set

$$F_n = \begin{cases} \{Z_0^\Gamma = X_0^\Gamma\}, & \text{if } n = 0, \\ \{Z_0^\Gamma > X_0^\Gamma, \ldots, Z_{n-1}^\Gamma > X_{n-1}^\Gamma, Z_n^\Gamma = X_n^\Gamma\}, & \text{if } 0 < n \leq |u|. \end{cases}$$

By definition of S, $S = \Gamma_n$ on F_n. On the other hand, again on F_n, $T_{\Gamma_n} = \Gamma_n$ and $T_{\Gamma_m} = T_{D(\Gamma_m)}$ if $0 \leq m < n$, by (3.22). But by (3.13) and (3.21) $D(\Gamma_m) = \Gamma_{m+1}$; therefore $T_0 = T_{\Gamma_0} = T_{\Gamma_n} = \Gamma_n$ on F_n, from which we conclude that $T_0 = S$.

Returning to the Index Set $\mathbb{I}$

We again place ourselves in the setting of Section 3.4 and denote by Z Snell's envelope of the reward process X, by X^u the restriction of X to $\mathbb{I}_u$, by Z^u Snell's envelope of X^u, and by T^u the optimal accessible stopping point (in $\mathbf{A}_u$) defined by the theorem. Since Z^u and T^u can be computed recursively, it is interesting to determine whether they converge, respectively, to Z and to an optimal stopping point when $u \twoheadrightarrow \bowtie$. Unfortunately, Z^u does not necessarily converge to Z and in general, nothing can be said about the behavior of T^u (cf. Exercise 7). On the other hand, it is also interesting to compare the limit of Z^u with the limit of the nondecreasing sequence $(Z^{(n)},\ n \in \mathbb{N})$ defined in (3.18) (this comparison will be useful in Chapter 6 in particular). For $t \in \mathbb{I}$, we denote the set of bounded accessible stopping points U such that $U \geq t$ by $\mathbf{A}_{\mathrm{b}}^t$.

Proposition (Approximating Snell's envelope of a process indexed by $\mathbb{I}$). *Let X be an adapted integrable process indexed by $\mathbb{I}$ whose positive part X^+ is of class $\mathrm{D}_{\mathbf{A}_{\mathrm{f}}}$. For $t \in \mathbb{I}$, set*

$$Z_t^{\mathrm{b}} = \operatorname*{esssup}_{T \in \mathbf{A}_{\mathrm{b}}^t} \mathrm{E}(X_T | \mathscr{F}_t). \tag{3.24}$$

Then the process $Z^{\mathrm{b}} = (Z_t^{\mathrm{b}},\ t \in \mathbb{I})$ is the smallest supermartingale that dominates X and

$$Z_t^{\mathrm{b}} = \lim_{u \twoheadrightarrow \bowtie} Z_t^u = \lim_{n \to \infty} Z_t^{(n)}, \qquad \text{for all } t \in \mathbb{I}.$$

Moreover, if X^- is uniformly integrable, then Z^{b} is Snell's envelope of X.

Proof. Since Z^u is the smallest supermartingale indexed by $\mathbb{I}_u$ that dominates X^u, $X_t^u \leq Z_t^u \leq Z_t^v \leq Z_t^{\mathrm{b}}$ if $t \leq u \leq v$. It follows that $X \leq Z^{\bowtie} \leq Z^{\mathrm{b}}$, where $Z^{\bowtie} = (Z_t^{\bowtie},\ t \in \mathbb{I})$ and $Z_t^{\bowtie} = \lim_{u \twoheadrightarrow \bowtie} Z_t^u$. In addition, for any $T \in \mathbf{A}_{\mathrm{b}}^t$ and $v \geq t$ such that $T \leq v$, $\mathrm{E}(X_T | \mathscr{F}_t) \leq Z_t^v \leq Z_t^{\bowtie}$; therefore $Z_t^b \leq Z_t^{\bowtie}$ and so Z^{b} is equal to $Z^{\bowtie}$. It is clear that $Z^{\bowtie}$ is a supermartingale that dominates X. On the other hand, since the restriction to $\mathbb{I}_u$ of any supermartingale that dominates X dominates X^u, therefore Z^u, for all $u \in \mathbb{I}$, $Z^{\bowtie}$ is the smallest supermartingale that dominates X. In order to complete the proof of the first statement of the proposition, for all $t \in \mathbb{I}$, we let $\mathbf{A}_n^t$ denote the set of elements T of $\mathbf{A}^t$ such that $|T - t| \leq n$. We claim that

$$Z_t^{(n)} = \operatorname*{esssup}_{T \in \mathbf{A}_n^t} \mathrm{E}(X_T | \mathcal{F}_t), \qquad \text{for all } t \in \mathbb{I}. \tag{3.25}$$

Assuming this equality for the moment and letting e denote the element $(1, \dots, 1)$ of $\mathbb{I}$, we deduce that whenever $t \leq u$,

$$Z_t^u \leq Z_t^{(|u-t|)} \leq Z_t^{t+|u-t|e},$$

because $\mathbf{A}_u^t$ is a subset of $\mathbf{A}_{|u-t|}^t$, which is itself a subset of $\mathbf{A}_{t+|u-t|e}^t$. We conclude therefore that $\lim_{u \to \infty} Z_t^u = \lim_{n \to \infty} Z_t^{(n)}$, which was to be proved. The proof of (3.25) follows the lines of the proof of (3.23). We therefore leave it to the reader as an exercise (cf. Exercise 6).

The last statement of the proposition follows from the fact that the sequence $(Z^{(n)}, n \in \mathbb{N})$ converges to Snell's envelope of X if X^- is uniformly integrable, by what was proved after (3.18).

Optimization to Within ε

Again in the setting of Section 3.4 and under the assumption that X is of class $\mathrm{D}_{\mathbf{A}_f}$, the result that has just been proved can be used to construct an ε-optimal stopping point (in $\mathbf{A}_f$). Indeed, given $\varepsilon > 0$, we can first choose $u \in \mathbb{I}$ such that $\mathrm{E}(Z_0^u) \geq \mathrm{E}(Z_0) - \varepsilon$ and then set $T^\varepsilon = T_0$, where T_0 is the stopping point of the theorem (which is optimal in $\mathbf{A}_u$). Then

$$\mathrm{E}(X_{T^\varepsilon}) = \mathrm{E}(Z_0^u) \geq \mathrm{E}(Z_0) - \varepsilon = \sup_{T \in \mathbf{A}_f} \mathrm{E}(X_T) - \varepsilon,$$

which proves that T^ε is ε-optimal (in $\mathbf{A}_f$).

3.7 AN APPLICATION TO NORMALIZED PARTIAL SUMS

Let $(Y_t, t \in \mathbb{I})$ be a family of independent real-valued random variables with mean zero such that $\sup_t \mathrm{E}(|Y_t|^p) < \infty$ for some $p > 1$. For all $t \in \mathbb{I}$, set $S_t = \sum_{s \leq t} Y_s$ and $\langle t \rangle = \operatorname{card} \mathbb{I}_t$. By Theorems 2 and 1 of Section 2.3,

$$\mathrm{E}\left(\sup_t \frac{|S_t|}{\langle t \rangle} \right) < \infty \quad \text{and} \quad \lim_{t \to \infty} \frac{S_t}{\langle t \rangle} = 0.$$

Set

$$X_t = \begin{cases} \dfrac{S_t}{\langle t \rangle}, & \text{if } t \in \mathbb{I}, \\ 0, & \text{if } t = \infty. \end{cases} \tag{3.26}$$

The process X thus defined satisfies the properties required of a reward process as well as condition (3.3). The underlying filtration is here the natural filtration of X (which is also the natural filtration of the family $(Y_t, t \in \mathbb{I})$). Using the theorem of Section 3.3, we conclude that an optimal stopping point exists.

A Condition Under which an Optimal Accessible Stopping Point Is Finite

In general, it is extremely difficult to provide natural conditions that guarantee that a finite accessible stopping point exists. In the case where the reward process is defined by (3.26) using a family $(Y_t, t \in \mathbb{I})$ of independent real-valued random variables with mean zero and variance 1, it is possible to exhibit a sufficient condition (see (3.27)) that seems to be natural but is nevertheless difficult to verify. The reader will observe that in dimension 1, this condition follows from Lindeberg's condition (by the central limit theorem). It is natural to expect that the same thing occurs in dimensions greater than 1, but no existing result confirms this conjecture.

The remainder of this section will be devoted to the proof of the following theorem.

Theorem (Condition under which an optimal accessible stopping point is finite). *Let $(Y_t, t \in \mathbb{I})$ be a family of independent real-valued random variables with mean zero and variance* 1. *Let X be the reward process defined by* (3.26) *and let T be an optimal accessible stopping point. If there exists a predictable increasing path Γ passing through T such that*

$$P\left\{\limsup_{n \to \infty} \frac{S_n^\Gamma}{\langle \Gamma_n \rangle^{1/2}} > 0\right\} = 1, \tag{3.27}$$

then T is finite.

The main point of this theorem is that the variables Y_t are *not* assumed to be identically distributed. Under this additional assumption, we will see in Chapter 4 that the conclusion of the theorem holds.

The proof relies on two lemmas.

Lemma 1. *If Γ is a predictable increasing path, then for all $n \in \mathbb{N}$ and all p such that $1 \le p < 2$,*

$$\mathrm{E}\left(\sup_{m>n}\left|\frac{S_m^\Gamma - S_n^\Gamma}{\langle \Gamma_m \rangle}\right|^p \,\middle|\, \mathscr{F}_n^\Gamma\right) \le \frac{2d}{2-p}\,\frac{1}{\langle \Gamma_n \rangle^{p/2}}.$$

Proof. Let $t \in \mathbb{I}$ and $F \in \mathscr{F}_t$ be such that $F \subset \{\Gamma_n = t\}$. Clearly,

$$\mathrm{E}\left(\sup_{m>n}\left|\frac{S_m^\Gamma - S_n^\Gamma}{\langle \Gamma_m \rangle}\right|^p; F\right)$$

$$= \mathrm{E}\left(\sup_{m>n}\left|\frac{S_m^\Gamma - S_t}{\langle \Gamma_m \rangle}\right|^p; F\right)$$

$$= \int_0^\infty px^{p-1}\,\mathrm{P}\left\{1_F \sup_{m>n}\left|\frac{S_m^\Gamma - S_t}{\langle \Gamma_m \rangle}\right| > x\right\} dx$$

$$\le \mathrm{P}(F)\int_0^{\langle t\rangle^{-1/2}} px^{p-1}\,dx + \int_{\langle t\rangle^{-1/2}}^\infty px^{p-1}\,\mathrm{P}\left\{1_F \sup_{m>n}\left|\frac{S_m^\Gamma - S_t}{\langle \Gamma_m \rangle}\right|^2 > x^2\right\} dx.$$

By the Hajek–Renyi inequality (applied to the transform of the submartingale $(1_F(S_m^\Gamma - S_t)^2, \mathscr{F}_m^\Gamma, m > n)$ by the predictable sequence $(1/\langle \Gamma_m \rangle^2, m > n)$), the second term on the last right-hand side is dominated by

$$\int_{\langle t\rangle^{-1/2}}^\infty px^{p-3}\,dx \sum_{m>n} \mathrm{E}\left(\frac{(S_m^\Gamma - S_t)^2 - (S_{m-1}^\Gamma - S_t)^2}{\langle \Gamma_m \rangle^2}; F\right)$$

$$= \frac{p}{2-p}\langle t\rangle^{1-p/2}\,\mathrm{E}\left(\sum_{m>n}\frac{1}{\langle \Gamma_m\rangle^2}\mathrm{E}\left((S_m^\Gamma)^2 - (S_{m-1}^\Gamma)^2 \middle| \mathscr{F}_{m-1}^\Gamma\right); F\right).$$

Now

$$\begin{aligned}\mathrm{E}\left((S_m^\Gamma)^2 - (S_{m-1}^\Gamma)^2 \middle| \mathscr{F}_{m-1}^\Gamma\right) &= \sum_{u:\,|u|=m-1}\ \sum_{v\in\mathbb{D}_u} 1_{\{\Gamma_m = v,\, \Gamma_{m-1}=u\}}\,\mathrm{E}\left(S_v^2 - S_u^2 \middle| \mathscr{F}_u\right)\\ &= \sum_{u:\,|u|=m-1}\ \sum_{v\in\mathbb{D}_u} 1_{\{\Gamma_m = v,\, \Gamma_{m-1}=u\}}(\langle v\rangle - \langle u\rangle)\\ &= \langle \Gamma_m\rangle - \langle \Gamma_{m-1}\rangle,\end{aligned}$$

and moreover,

$$\begin{aligned}&\sum_{m>n}\frac{\langle \Gamma_m\rangle - \langle \Gamma_{m-1}\rangle}{\langle \Gamma_m\rangle^2}\\ &\quad= \sum_{i=1}^d\sum_{m>n}\frac{\langle \Gamma_m\rangle - \langle \Gamma_{m-1}\rangle}{\langle \Gamma_m\rangle^2}1_{\{\Gamma_m^i = \Gamma_{m-1}^i + 1\}}\\ &\quad= \sum_{i=1}^d\sum_{m>n}\frac{\prod_{j=1}^d(\Gamma_m^j + 1) - \prod_{j=1}^d(\Gamma_{m-1}^j + 1)}{\prod_{j=1}^d(\Gamma_m^j + 1)^2}1_{\{\Gamma_m^i = \Gamma_{m-1}^i + 1\}}\\ &\quad\le \sum_{i=1}^d\frac{1}{\prod_{j\ne i}(\Gamma_n^j + 1)}\sum_{m>n}\frac{1}{(\Gamma_m^i + 1)^2}.\end{aligned}$$

Using the elementary inequality $\Sigma_{m>n} 1/m^2 \leq 1/n$, we see that the last right-hand side is less than or equal to

$$\sum_{i=1}^{d} \frac{1}{\langle \Gamma_n \rangle} = \frac{d}{\langle \Gamma_n \rangle}.$$

We conclude that the expectation written at the beginning of the proof is dominated by

$$\frac{1}{\langle t \rangle^{p/2}} \mathrm{P}(F) + \frac{pd}{2-p} \frac{1}{\langle t \rangle^{p/2}} \mathrm{P}(F) \leq \frac{2d}{2-p} \frac{1}{\langle t \rangle^{p/2}} \mathrm{P}(F).$$

Because t and F can be chosen arbitrarily in $\mathbb{I}$ and in $\mathscr{F}_t$, the inequality of the lemma follows.

Lemma 2. *Let Γ be a predictable increasing path and let τ be a stopping time relative to the filtration $\mathscr{F}^\Gamma$. Then for all $n \in \mathbb{N}$, all $\varepsilon > 0$, and all p such that $1 < p < 2$,*

$$\mathrm{E}\left(X_\tau^\Gamma \middle| \mathscr{F}_n^\Gamma\right) < X_n^\Gamma \qquad \text{on } F_n^\varepsilon(\tau),$$

where

$$F_n^\varepsilon(\tau) = \left\{ \frac{S_n^\Gamma}{\langle \Gamma_n \rangle^{1/2}} > \varepsilon, \mathrm{P}\left(\tau = \infty \middle| \mathscr{F}_n^\Gamma\right) \geq \frac{1}{2}, \right.$$
$$\left. \mathrm{P}\left(\tau < \infty \middle| \mathscr{F}_n^\Gamma\right) \leq \left(\frac{2-p}{2d}\right)^{1/(p-1)} \left(\frac{\varepsilon}{2}\right)^{p/(p-1)} \right\}.$$

Proof. We begin by observing that

$$\mathrm{E}\left(X_\tau^\Gamma \middle| \mathscr{F}_n^\Gamma\right) = \mathrm{E}\left(\frac{S_\tau^\Gamma}{\langle \Gamma_\tau \rangle} 1_{\{\tau \leq n\}} + \left(\frac{S_n^\Gamma}{\langle \Gamma_\tau \rangle} + \frac{S_\tau^\Gamma - S_n^\Gamma}{\langle \Gamma_\tau \rangle} \right) 1_{\{n < \tau < \infty\}} \middle| \mathscr{F}_n^\Gamma \right).$$

Notice moreover that $F_n^\varepsilon(\tau) \subset \{\tau > n\}$, because

$$\mathrm{P}\left\{\tau = \infty \middle| \mathscr{F}_n^\Gamma\right\} = 0 \qquad \text{on } \{\tau \leq n\}.$$

Indeed,

$$\int_{\{\tau \leq n\}} \mathrm{P}\left\{\tau = \infty \middle| \mathscr{F}_n^\Gamma\right\} d\mathrm{P} = \int_{\{\tau \leq n\}} 1_{\{\tau = \infty\}} \, d\mathrm{P} = 0.$$

Thus, we see that on $F_n^\varepsilon(\tau)$,

$$\mathrm{E}\left(X_\tau^\Gamma \middle| \mathscr{F}_n^\Gamma\right) \le \mathrm{E}\left(\frac{S_n^\Gamma}{\langle \Gamma_\tau \rangle} 1_{\{\tau<\infty\}} \middle| \mathscr{F}_n^\Gamma\right) + \mathrm{E}\left(\sup_{m>n}\left|\frac{S_m^\Gamma - S_n^\Gamma}{\langle \Gamma_m\rangle}\right| 1_{\{n<\tau<\infty\}} \middle| \mathscr{F}_n^\Gamma\right).$$

On $F_n^\varepsilon(\tau)$, the first term on the right-hand side is less than or equal to

$$\frac{S_n^\Gamma}{\langle \Gamma_n \rangle} - \frac{S_n^\Gamma}{\langle \Gamma_n \rangle} \mathrm{P}\{\tau = \infty | \mathscr{F}_n^\Gamma\} < X_n^\Gamma - \frac{\varepsilon}{2\langle \Gamma_n \rangle^{1/2}}.$$

By applying Hölder's inequality to the last conditional expectation above and using Lemma 1, we conclude that on $F_n^\varepsilon(\tau)$,

$$\begin{aligned} \mathrm{E}\left(X_\tau^\Gamma | \mathscr{F}_n^\Gamma\right) &< X_n^\Gamma - \frac{\varepsilon}{2\langle \Gamma_n \rangle^{1/2}} \\ &\quad + \mathrm{E}\left(\sup_{m>n}\left|\frac{S_m^\Gamma - S_n^\Gamma}{\langle \Gamma_m \rangle}\right|^p \middle| \mathscr{F}_n^\Gamma\right)^{1/p} \mathrm{P}\left(n<\tau<\infty | \mathscr{F}_n^\Gamma\right)^{1/q} \\ &\le X_n^\Gamma - \frac{\varepsilon}{2\langle \Gamma_n \rangle^{1/2}} \\ &\quad + \left(\frac{2d}{2-p}\right)^{1/p} \frac{1}{\langle \Gamma_n \rangle^{1/2}} \left(\frac{2-p}{2d}\right)^{1/(p-1)q} \left(\frac{\varepsilon}{2}\right)^{p/(p-1)q} \\ &= X_n^\Gamma, \end{aligned}$$

where the final equality is a consequence of the relation

$$\frac{1}{p} + \frac{1}{q} = 1, \text{ which implies that } \frac{1}{(p-1)q} = \frac{1}{p}.$$

The lemma is established.

Proof of the Theorem

Let Γ be a predictable increasing path passing through T and satisfying (3.27). Set $\tau = |T|$ and assume that $\mathrm{P}\{\tau = \infty\} > 0$. Moreover, let

$$F = \left\{\limsup_{n\to\infty} \frac{S_n^\Gamma}{\langle \Gamma_n \rangle^{1/2}} > \varepsilon\right\} \cap \{\tau = \infty\},$$

where ε is chosen small enough so that $\mathrm{P}(F) > 0$. Define $F_n^\varepsilon(\tau)$ as in Lemma 2 (with some arbitrary p such that $1 < p < 2$), and let σ be the stopping time relative to the filtration $\mathscr{F}^{\Gamma}$ defined by

$$\sigma(\omega) = \inf\{n \in \mathbb{N}: \omega \in F_n^{\varepsilon}(\tau)\}.$$

Since

$$\lim_{n\to\infty} \mathrm{P}\left(\tau = \infty \middle| \mathscr{F}_n^{\Gamma}\right) = \mathrm{P}\left(\tau = \infty \middle| \mathscr{F}_{\infty-}^{\Gamma}\right) = 1_{\{\tau=\infty\}} = 1 \qquad \text{on } F,$$

and therefore

$$\lim_{n\to\infty} \mathrm{P}\left(\tau < \infty \middle| \mathscr{F}_n^{\Gamma}\right) = 0 \qquad \text{on } F,$$

it is clear that σ is finite on F. It follows that $\mathrm{P}\{\sigma < \tau\} > 0$, therefore that $\mathrm{P}\{\sigma = n,\ \sigma < \tau\} > 0$ for at least one $n \in \mathbb{N}$. Since $\{\sigma = n,\ \sigma < \tau\}$ belongs to $\mathscr{F}_n^{\Gamma}$,

$$\int_{\{\sigma<\tau\}} X_\tau^{\Gamma}\, \mathrm{dP} = \sum_{n=0}^{\infty} \int_{\{\sigma=n,\ \sigma<\tau\}} \mathrm{E}\left(X_\tau^{\Gamma} \middle| \mathscr{F}_n^{\Gamma}\right) \mathrm{dP},$$

and since $\{\sigma = n,\ \sigma < \tau\}$ is contained in $F_n^{\varepsilon}(\tau)$ for all $n \in \mathbb{N}$, we see using Lemma 2 that the sum is strictly less than

$$\sum_{n=0}^{\infty} \int_{\{\sigma=n,\ \sigma<\tau\}} X_n^{\Gamma}\, \mathrm{dP} = \int_{\{\sigma<\tau\}} X_\sigma^{\Gamma}\, \mathrm{dP} = \int_{\{\sigma<\tau\}} X_{\sigma\wedge\tau}^{\Gamma}\, \mathrm{dP},$$

which implies that

$$\begin{aligned}
\mathrm{E}(X_T) &= \mathrm{E}\left(X_\tau^{\Gamma} 1_{\{\sigma<\tau\}}\right) + \mathrm{E}\left(X_\tau^{\Gamma} 1_{\{\sigma\geq\tau\}}\right) \\
&< \mathrm{E}\left(X_{\sigma\wedge\tau}^{\Gamma} 1_{\{\sigma<\tau\}}\right) + \mathrm{E}\left(X_{\sigma\wedge\tau}^{\Gamma} 1_{\{\sigma\geq\tau\}}\right) = \mathrm{E}\left(X_{\sigma\wedge\tau}^{\Gamma}\right) = \mathrm{E}(X_{T'}),
\end{aligned}$$

where T' denotes the accessible stopping point $\Gamma_{\sigma\wedge\tau}$. It is thus apparent that T is not optimal, which contradicts the hypothesis. Consequently, T must be finite.

3.8 COMPLEMENTS

This section contains some complements concerning several questions mentioned earlier.

About Condition (3.6)

With the help of an example, we are going to show that condition (3.6) does not ensure the existence of an optimal accessible stopping point.

For each $n \in \mathbb{N}$, let f_n be a Borel function with the following property: If Y is a uniformly distributed random variable on $[0,1]$, then the random variable $f_n(Y)$ is uniformly distributed on the set $\mathbb{N}_n$. Consider a sequence $(Y_n,\ n \in \mathbb{N})$ of independent and uniformly distributed random variables on $[0,1]$. Set $d = 2$, and for all $n \in \mathbb{N}^*$, let

$$k_n = 2^{2n^2} \quad \text{and} \quad \mathbb{J}_n = \left\{t \in \mathbb{I}\colon |t| = k_n,\ |t^1 - t^2| \leq 2n^2\right\}.$$

Notice that if $t \in \mathbb{J}_n$ and $u \in \mathbb{J}_{n+1}$, then $t \leq u$. Let X denote the process indexed by $\bar{\mathbb{I}}$ defined by

$$X_t = \begin{cases} \left(1 - \dfrac{Y_n}{n}\right) 1_{\{f_{2n^2}(Y_{n-1}) = t^1 - \frac{1}{2}k_n + n^2\}}, & \text{if } t \in \mathbb{J}_n, \\ 0, & \text{if } t \notin \cup_{n \in \mathbb{N}^*} \mathbb{J}_n. \end{cases}$$

This process is bounded and adapted to the filtration defined by

$$\mathscr{F}_t = \mathscr{F}_n, \qquad \text{if } k_n \leq |t| < k_{n+1},$$

where $\mathscr{F}_n$ denotes the sigma field generated by $(Y_0, \ldots, Y_n)$. Let us denote by Γ any predictable increasing path such that

$$\Gamma^1_{k_n} = \tfrac{1}{2}k_n - n^2 + m \qquad \text{on } \left\{f_{2n^2}(Y_{n-1}) = m\right\} \qquad \left(n > 0,\ m = 0, \ldots, 2n^2\right).$$

Such a path is easily constructed by induction. By definition of Γ, $X^\Gamma_{k_n} = 1 - Y_n/n$ for all $n > 0$. Because $|Y_n| \leq 1$, it follows that

$$\limsup_{n \to \infty} X^\Gamma_{k_n} = \limsup_{n \to \infty} X^\Gamma_n = 1.$$

Setting $\tau_\varepsilon = \inf\{n\colon X^\Gamma_n \geq 1 - \varepsilon\}$, we see that $\mathrm{E}(X^\Gamma_{\tau_\varepsilon}) \geq 1 - \varepsilon$ for all $\varepsilon > 0$, which implies that the value of the game (associated with X) is 1. But since $\mathrm{P}\{X_t < 1\} = 1$ for all $t \in \bar{\mathbb{I}}$, $\mathrm{E}(X_T) < 1$ for all accessible stopping points T, therefore there is no optimal stopping point. However, condition (3.6) is satisfied, because if γ denotes a deterministic increasing path, then

$$\mathrm{P}\left\{X^\gamma_{k_n} \neq 0\right\} = \mathrm{P}\left\{f_{2n^2}(Y_{n-1}) = \gamma^1_{k_n} - \tfrac{1}{2}k_n + n^2\right\} = \frac{1}{2n^2},$$

by definition of X and of f_{2n^2}. Therefore $\limsup_{n \to \infty} X^\gamma_n = 0\ (= X_\bowtie)$, by the Borel–Cantelli lemma.

Remark

The example we have just presented also shows that it is possible to have $\lim_{t \to \bowtie} X_t = 1$ while $\lim_{n \to \infty} X_n^{\gamma} = 0$ for all deterministic increasing paths γ.

Closure on the Right of a Reward Process

In Section 3.4, we mentioned that incorporating the random variable $X_{\bowtie} = \limsup_{t \to \bowtie} X_t$ into the reward process leads in general to an increase of Snell's envelope. We are going to present two simple examples that illustrate this situation. The reader who refers to Section 5.3 concerning the assumption of conditional independence will notice that this assumption is not satisfied in the first example but is satisfied in the second.

Example 1. Let $\Omega = \{\omega_1, \omega_2\}$, $\mathscr{F} = \mathscr{P}(\Omega)$, $P(\{\omega_1\}) = \frac{1}{2}$ and $\mathbb{I} = \mathbb{N}^2$. Set $\mathscr{F}_0 = \{\varnothing, \Omega\}$ and $\mathscr{F}_t = \mathscr{F}$ if $t \in \mathbb{I}^*$. Define X by

$$X_t = \begin{cases} 0, & \text{if } t = 0, \\ 1_{\{\omega_1\}}, & \text{if } t \in \mathbb{N}^* \times \{0\}, \\ 1_{\{\omega_2\}}, & \text{if } t \in \{0\} \times \mathbb{N}^*, \\ 0, & \text{if } t \gg 0. \end{cases}$$

Clearly, Snell's envelope Z of X is defined by $Z_0 = \frac{1}{2}$ and $Z_t = X_t$ if $t \in \mathbb{I}^*$. On the other hand, it is obvious that $\limsup_{t \to \bowtie} X_t = 1$, which implies that Snell's envelope of the process X closed on the right by $X_{\bowtie} = 1$ is the constant process equal to 1. Notice that the value of the game associated with X is $\frac{1}{2}$, whereas the one associated with X closed on the right is 1. Evidently, $T = (1, 0)\, 1_{\{\omega_1\}}$ is optimal in the first case and $T = \bowtie$ is optimal in the second case. It is interesting to notice that the value of the game associated with X goes from $\frac{1}{2}$ to 1 if inaccessible stopping points are taken into account as well. In this case, $T = (1, 0)\, 1_{\{\omega_1\}} + (0, 1)\, 1_{\{\omega_2\}}$ is an optimal inaccessible stopping point.

Example 2. Let Ω, $\mathscr{F}$, P, and $\mathbb{I}$ be as in Example 1. Set $\mathscr{F}_t = \{\varnothing, \Omega\}$ if $t \in \mathbb{N} \times \{0\}$ and $\mathscr{F}_t = \mathscr{F}$ if $t \in \mathbb{I} - (\mathbb{N} \times \{0\})$. Define X by

$$X_t = \begin{cases} 0, & \text{if } t = 0, \\ \frac{1}{3}, & \text{if } t \in \mathbb{N}^* \times \{0\}, \\ 1_{\{\omega_1\}}, & \text{if } t \in \{0\} \times \mathbb{N}^*, \\ 0, & \text{if } t \gg 0. \end{cases}$$

Clearly, Snell's envelope Z of X is defined by $Z_0 = \frac{1}{2}$ and $Z_t = X_t$ if $t \in \mathbb{I}^*$. On the other hand, it is obvious that $\limsup_{t \to \bowtie} X_t = 1_{\{\omega_1\}} + \frac{1}{3}\, 1_{\{\omega_2\}}$, which means that Snell's envelope of the reward process X closed on the right by $X_{\bowtie} = 1_{\{\omega_1\}} + \frac{1}{3}1_{\{\omega_2\}}$ is defined by

$$Z_t = \begin{cases} \frac{2}{3}, & \text{if } t \in \mathbb{N} \times \{0\}, \\ 1_{\{\omega_1\}} + \frac{1}{3}1_{\{\omega_2\}}, & \text{if } t \in \bar{\mathbb{I}} - (\mathbb{N} \times \{0\}). \end{cases}$$

Notice that the value of the game associated with X is $\frac{1}{2}$, whereas that associated with X closed on the right by $X_{\bowtie}$ is $\frac{2}{3}$. In this example, every stopping point is accessible.

EXERCISES

1. Consider the sequences $(Y_n, n \in \mathbb{N})$ and $(X_n, n \in \overline{\mathbb{N}})$ defined in Exercise 5 of Chapter 1.

a. Let $(\mathcal{F}_n, n \in \mathbb{N})$ be the natural filtration of $(Y_n, n \in \mathbb{N})$ and $\mathbf{T}$ the set of stopping times (relative to this filtration). Show that $\sup_{\tau \in \mathbf{T}} \mathrm{E}(X_\tau) = \sup_{n \in \mathbb{N}} \mathrm{E}(X_n) = 2$, but that there exists no optimal stopping time. Prove that $\mathrm{E}(X_{n+1} | X_n) > X_n$ on $\{X_n > 0\}$. Show also that from the point of view of maximizing expected rewards, it is never optimal to stop as long as $X_n > 0$.

b. Suppose now that for each $n \in \mathbb{N}$, $\mathcal{F}_n$ is the sigma field generated by $(Y_1, \ldots, Y_n, Y_{n+1})$. Evaluate $\sup_{\tau \in \mathbf{T}} \mathrm{E}(X_\tau)$, where $\mathbf{T}$ now denotes the set of stopping times relative to the new filtration $(\mathcal{F}_n, n \in \mathbb{N})$.

2. Let $(Y_n, n \in \mathbb{N})$ be a sequence of independent and identically distributed random variables such that $\mathrm{P}\{Y_0 = 1\} = \mathrm{P}\{Y_0 = -1\} = \frac{1}{2}$. Let Z be a random variable uniformly distributed on $[0, 1]$. Suppose that for each $n \in \mathbb{N}$, $\mathcal{F}_n$ is the sigma field generated by $(Y_1, \ldots, Y_n, Z)$. Set $X_0 = 0$ and, for $n \in \mathbb{N}^*$, let $X_n = \inf(1, Y_1 + \cdots + Y_n)$.

a. Give an example of a finite stopping time τ relative to the filtration $(\mathcal{F}_n, n \in \mathbb{N})$ such that $\mathrm{E}(X_\tau) = -\infty$. (*Hint*. Set $\tau_n = \inf\{m \in \mathbb{N}: X_m = -2^n\}$ and define $\tau = \tau_n$ on $\{Z \in B_n\}$, where B_n is an appropriate Borel subset of $[0, 1]$.)

b. Give an example of a finite stopping time τ relative to the natural filtration of $(Y_n, n \in \mathbb{N})$ such that $\mathrm{E}(X_\tau) = -\infty$.

c. Let $\mathbf{T}_{\mathrm{f}}$ be the set of finite stopping times relative to the natural filtration of $(Y_n, n \in \mathbb{N})$. For all $\tau \in \mathbf{T}_{\mathrm{f}}$, let $\mathbf{T}_X^\tau$ be the set of $\sigma \in \mathbf{T}_{\mathrm{f}}$ such that $\sigma \geq \tau$ and $\mathrm{E}(X_\sigma^-) < \infty$. Show that $\mathbf{T}_X^\tau \neq \varnothing$ for all $\tau \in \mathbf{T}_{\mathrm{f}}$.

3. Fix $p \in]0, \frac{1}{2}[$, and let $(Y_n, n \in \mathbb{N})$ be a sequence of independent and identically distributed random variables such that $\mathrm{P}\{Y_0 = 1\} = p$ and

$P\{Y_0 = -1\} = 1 - p$. Let $(\mathcal{F}_n,\ n \in \mathbb{N})$ be the natural filtration of $(Y_n,\ n \in \mathbb{N})$. Set $X_0 = 0$ and, for $n \in \mathbb{N}^*$, let $X_n = \inf(1, Y_1 + \cdots + Y_n)$.

a. Give an example of a finite stopping time τ relative to the natural filtration of $(Y_n,\ n \in \mathbb{N})$ such that $E(X_\tau) = -\infty$.

b. Prove that $E(\sup_{n \in \mathbb{N}}(Y_1 + \cdots + Y_n)) < \infty$. (*Hint*. Recall that $P\{\tau_k < \infty\} = (p/(1-p))^k$ for all $k \in \mathbb{N}^*$, where $\tau_k = \inf\{n \in \mathbb{N}^*: Y_1 + \cdots + Y_n = k\}$.)

c. Show that if $\mathbf{T}_X^\tau$ is defined as in part c of Exercise 2, then $\mathbf{T}_X^\tau = \varnothing$ when τ is the stopping time constructed in part a.

4. In the setting of Section 3.4, show that Snell's envelope Z is not only a regular supermartingale but it also satisfies the inequality $E(Z_T | \mathcal{F}_S) \leq Z_S$ for all couples (S, T) of accessible stopping points such that $S \leq T$ and $T \in \mathbf{A}_X$.

5. Deduce from Exercise 4 that Proposition 2 of Section 3.3 remains valid in the setting of Section 3.4.

6. Prove the equality in (3.25) by following the approach used to prove (3.23). (*Hint*. Formula (3.25) is trivial for $n = 0$, then proceed by induction.)

7. Let $(Y_n,\ n \in \mathbb{N})$ be a sequence of independent and identically distributed random variables such that $P\{Y_0 = 1\} = P\{Y_0 = -1\} = \frac{1}{2}$. Let X be the process defined by

$$X_n = \begin{cases} 0, & \text{if } n = 0, \\ \inf(1, Y_1 + \cdots + Y_n) - n/(n+1), & \text{if } n \in \mathbb{N}^*, \end{cases}$$

and assume that the underlying filtration is the natural filtration of $(Y_n,\ n \in \mathbb{N})$.

a. Show that X is a supermartingale.

b. Deduce from part a that $X = Z^{\mathrm{b}}$, where $Z^{\mathrm{b}} = (Z_n^{\mathrm{b}},\ n \in \mathbb{N})$ and Z_n^{b} is defined in (3.24).

c. Show that for all $k \in \mathbb{N}$, 0 is the unique optimal stopping time for the restriction of X to $\mathbb{N}_k$.

d. Deduce that $Z^{\mathrm{b}} \neq Z$, where Z is Snell's envelope of X, and therefore that X is not a regular supermartingale.

e. Conclude that

$$\sup_{T \in \mathbf{T}_{\mathrm{b}}} E(X_T) < \sup_{T \in \mathbf{T}_{\mathrm{f}}} E(X_T),$$

where $\mathbf{T}_{\mathrm{b}}$ $[\mathbf{T}_{\mathrm{f}}]$ denotes the set of bounded [finite] accessible stopping times.

Open Problem

(1) Suppose $d > 1$ and let $(Y_t, t \in \mathbb{I})$ be a family of independent real-valued random variables with mean zero and variance 1. Let X be the reward process defined by (3.26) and let T be an optimal accessible stopping point.

a. Does there exist a predictable increasing path Γ passing through T for which condition (3.27) holds?

b. Is T necessarily finite?

HISTORICAL NOTES

The problem of optimal stopping for processes indexed by a directed set has a long history, going back to Haggstrom (1966), though similar problems were studied using a different formulation in the classical book of Dubins and Savage (1965), and in articles from the econometrics and statistics literature, such as Arrow, Blackwell, and Girshick (1949) and Whittle (1965).

Optimal stopping problems involving stopping points in the plane were specifically considered by Cairoli and Gabriel (1978) for reward processes involving sums of independent random variables. Many further results were given in Krengel and Sucheston (1981). Mandelbaum and Vanderbei (1981) studied a more general class of processes but considered only finite accessible stopping points. Mazziotto and Szpirglas (1983) solved the optimal stopping problem over all stopping points for general processes indexed by $\overline{\mathbb{N}}^2$. The existence proof in this last reference relies on Zorn's lemma as outlined in Section 3.5, and is carried out under assumptions that guarantee that all stopping points in the plane are accessible (see Chapter 5).

Focusing the optimal stopping problem on accessible stopping points as presented in this chapter is new and generalizes the results of the authors mentioned above. It also provides a unified way of solving the optimal stopping problem in both the one-dimensional and higher-dimensional cases, without putting restrictive assumptions on the underlying filtration. The one-dimensional problem has been considered in classical books such as Chow, Robbins, and Siegmund (1971), Dynkin and Yushkevitch (1969), Neveu (1975) and Shiryayev (1978).

The importance of the process introduced in Section 3.2 and called Snell's envelope was first observed by Snell (1952) in his study of processes in discrete time and in dimension $d = 1$. Our initial proof of Proposition 1 of Section 3.2 was simplified by J. B. Walsh. The resolution in Section 3.3 is new for $d > 1$. The solution of the related problem considered in Section 3.4 is due to Mandelbaum and Vanderbei (1981), though Proposition 1 and Theorem 1 of this section are extensions of their results. They also listed the conclusions of Proposition 3 of Section 3.3 and Theorem 2 of Section 3.4 as open problems. The study of maximal accessible stopping points in dimension

$d = 2$ is due to Mazziotto and Szpirglas (1983). A reference to Zorn's lemma can be found in Hewitt and Stromberg (1965, Chapter 1).

Optimal stopping over finite (partially) ordered index sets as considered in Section 3.6 was considered in detail by Krengel and Sucheston (1981). The backwards induction method used there is sometimes referred to as the dynamic programming method, on which there is an extensive literature (see e.g. Whittle (1982)). A result related to that of the proposition of Section 3.6 was proved by Haggstrom (1966).

The study of normalized partial sums carried out in Section 3.7 generalizes results of Chow and Robbins (1965) and Dvoretzky (1967). The extension to $d > 1$ of the fact that existence of finite optimal stopping points is implied by condition (3.27) is new. Except in the case where the random variables are identically distributed (which we consider in Section 2 of Chapter 4), it is an open problem to determine whether condition (3.27) (or the conclusion of the theorem of Section 3.7) is satisfied in dimension $d > 1$. The Hajek–Renyi inequality used in this section can be found in Chow and Teicher (1988), as can the statement of Lindeberg's conditions. The first example in Section 3.8 is due to T. Mountford, though the presentation given here was simplified by J.B. Walsh.

Extensions of the optimal stopping problem in dimension $d = 2$ to continuous-time reward processes can be found in Dalang (1989), in Nualart (1992), and in the references therein.

Exercises 1 and 7 are taken from Chow, Robbins, and Siegmund (1971).

CHAPTER 4

Reduction to a Single Dimension

In this chapter, we show how in some contexts, if X is a reward process and T is an accessible stopping point, then the law of X_T is identical to the law of a particular process indexed by $\overline{\mathbb{N}}$ and stopped at a stopping time. This is notably useful to compute the expectation of X_T. For this reduction from d dimensions to one dimension to succeed, it will be necessary that the process X be defined in a well-determined manner from a family of exchangeable random variables. Typical examples are partial sums or averages of independent and identically distributed random variables. The first section is devoted to the proof of the main result. In the second section, this result will be applied to the study of the problem of optimal stopping. The third section is devoted to considerations concerning inaccessible stopping points.

4.1 LINEAR REPRESENTATION OF ACCESSIBLE STOPPING POINTS

The term linear used in the title is to be taken in a geometric rather than algebraic sense, for it refers to a one-dimensional representation of stopping points.

Stating the Problem

Suppose that a family $(Y_t,\ t \in \mathbb{I})$ of exchangeable random variables is given. Recall that such a family is characterized by the following property: If p is a permutation of $\mathbb{I}$ that leaves invariant all except a finite number of elements of $\mathbb{I}$, then the laws of $(Y_t,\ t \in \mathbb{I})$ and $(Y_{p(t)},\ t \in \mathbb{I})$ coincide. Let $(f_n,\ n \in \mathbb{N}^*)$ be a sequence of Borel functions defined on $\overline{\mathbb{R}}^n$ such that

$$f_n(y_1, \ldots, y_{n-1}, y_n) = f_n(y_{\pi(1)}, \ldots, y_{\pi(n-1)}, y_n) \tag{4.1}$$

for all $n > 1$ and all permutations π of $\{1, \ldots, n-1\}$. Because the order of the first $n-1$ variables is irrelevant, we can write without ambiguity $f_n(y, y_n)$

every time that y is a family of $n-1$ variables. In order to avoid treating the case $n = 1$ separately, we agree that the symbol $f_n(y, y_n)$ represents $f_1(y_1)$ when $n = 1$.

Let $(\mathscr{F}_t,\ t \in \mathbb{I})$ be the natural filtration of $(Y_t, t \in \mathbb{I})$, and set

$$X_t = \begin{cases} f_{\langle t\rangle}((Y_s,\ s \in \mathbb{I}_t - \{t\}), Y_t), & \text{if } t \in \mathbb{I}, \\ Y_{\bowtie}, & \text{if } t = \bowtie, \end{cases} \tag{4.2}$$

where $\langle t\rangle$ denotes card $\mathbb{I}_t$ and $Y_{\bowtie}$ denotes a random variable that is either measurable with respect to the tail sigma field $\mathscr{G}$, or conditionally independent of $\mathscr{F}_{\bowtie-}$ given $\mathscr{G}$. Recall that $\mathscr{G} = \bigcap_{t \in \mathbb{I}} \mathscr{G}_t$, where $\mathscr{G}_t$ is the sigma field generated by $(Y_u,\ u \in \mathbb{I} - \mathbb{I}_t)$. It is clear that $(Y_s,\ s \in \mathbb{I}_t - \{t\})$ denotes the empty family when $t = 0$.

The process X defined by (4.2) plays the role of the reward process.

Exchangeability in the Presence of $Y_{\bowtie}$

From de Finetti's theorem, the random variables Y_t $(t \in \mathbb{I})$ are conditionally independent and identically distributed given $\mathscr{G}$. Consequently, if $\{t_1, \dots, t_n\}$ and $\{t'_1, \dots, t'_n\}$ are subsets of $\mathbb{I}$ and $B_1, \dots, B_n, B_\infty$ are Borel sets, then

$$\begin{aligned} P\big(Y_{t_1} \in B_1, \dots, Y_{t_n} \in B_n, Y_{\bowtie} \in B_\infty | \mathscr{G}\big) &= \mathrm{P}\big(Y_{t_1} \in B_1, \dots, Y_{t_n} \in B_n | \mathscr{G}\big)\, \mathrm{P}\big(Y_{\bowtie} \in B_\infty | \mathscr{G}\big) \\ &= \mathrm{P}\big(Y_{t_1} \in B_1 | \mathscr{G}\big) \cdots \mathrm{P}\big(Y_{t_n} \in B_n | \mathscr{G}\big)\, \mathrm{P}\big(Y_{\bowtie} \in B_\infty | \mathscr{G}\big) \\ &= \mathrm{P}\big(Y_{t'_1} \in B_1 | \mathscr{G}\big) \cdots \mathrm{P}\big(Y_{t'_n} \in B_n | \mathscr{G}\big)\, \mathrm{P}\big(Y_{\bowtie} \in B_\infty | \mathscr{G}\big) \\ &= \mathrm{P}\big(Y_{t'_1} \in B_1, \dots, Y_{t'_n} \in B_n, Y_{\bowtie} \in B_\infty | \mathscr{G}\big), \end{aligned}$$

which shows that the conditional laws given $\mathscr{G}$, hence the laws, of $(Y_{t_1}, \dots, Y_{t_n}, Y_{\bowtie})$ and $(Y_{t'_1}, \dots, Y_{t'_n}, Y_{\bowtie})$ are identical.

Main Result

The theorem that we are about to state and prove achieves the reduction mentioned in the introduction. In Section 4.2, this theorem will be applied to questions discussed in Sections 2.4 and 3.7.

In the sequel, in order to simplify the notation, we will write X_n^1, Y_n^1, and $\mathscr{F}_n^1$ instead of $X_{(n,0,\dots,0)}$, $Y_{(n,0,\dots,0)}$, and $\mathscr{F}_{(n,0,\dots,0)}$ and will set $X_\infty^1 = Y_\infty^1 = Y_{\bowtie}$. Notice that for all $n \in \mathbb{N}$,

$$X_n^1 = f_{n+1}\big(Y_0^1, \dots, Y_n^1\big). \tag{4.3}$$

By convention, we set $\langle T(\omega)\rangle = \langle T(\omega)\rangle - 1 = \infty$ if $T(\omega) = \bowtie$.

Theorem (Linear representation). *To any accessible stopping point T relative to the filtration $(\mathscr{F}_t)$, there corresponds a stopping time τ relative to the filtration $(\mathscr{F}_n^1)$ such that the laws of the couples $(X_T, \langle T\rangle - 1)$ and (X_τ^1, τ) are identical.*

The proof of this theorem relies on three lemmas that require some preliminaries.

To each deterministic increasing path γ, we associate an enumeration $(\lambda(\gamma, n), n \in \mathbb{N})$ of $\mathbb{I}$ defined as follows:

a. $\lambda(\gamma, 0) = 0$;

b. for all $n > 0$, the terms

$$\lambda(\gamma, \langle\gamma_{n-1}\rangle), \lambda(\gamma, \langle\gamma_{n-1}\rangle + 1), \ldots, \lambda(\gamma, \langle\gamma_n\rangle - 1)$$

denote the elements of $\mathbb{I}_{\gamma_n} - \mathbb{I}_{\gamma_{n-1}}$ written down in lexicographical order.

Notice that $(\lambda(\gamma, 0), \ldots, \lambda(\gamma, \langle\gamma_n\rangle - 1))$ is an enumeration of $\mathbb{I}_{\gamma_n}$ such that $\lambda(\gamma, \langle\gamma_n\rangle - 1) = \gamma_n$. Also note that $\lambda(\gamma, n)$ depends on γ only through the γ_m of index $m \le m(\gamma)$, where

$$m(\gamma) = \inf\{k \in \mathbb{N}: \langle\gamma_k\rangle > n\} \tag{4.4}$$

(for notational convenience, the explicit dependence of $m(\gamma)$ on n is omitted).

If Γ is a predictable increasing path (or even only an optional increasing path), then $\lambda(\Gamma, n)$ will denote the random variable $\omega \mapsto \lambda(\Gamma(\omega), n)$, where $\Gamma(\omega) = (\Gamma_m(\omega), m \in \mathbb{N})$. From what was observed above, if γ is a deterministic increasing path, then

$$\lambda(\Gamma(\omega), 0) = \lambda(\gamma, 0), \ldots, \lambda(\Gamma(\omega), n) = \lambda(\gamma, n),$$
$$\text{for all } \omega \text{ such that } \Gamma_0(\omega) = \gamma_0, \ldots, \Gamma_{m(\gamma)}(\omega) = \gamma_{m(\gamma)}, \tag{4.5}$$

where $m(\gamma)$ is defined in (4.4).

Let us choose, once and for all, a predictable increasing path Γ passing through the stopping point T of the theorem. This path will be fixed throughout the proof. For all $n \in \mathbb{N}$ and all ω, we set

$$\tilde{Y}_n(\omega) = Y_{\lambda(\Gamma(\omega), n)}(\omega). \tag{4.6}$$

By (4.5), if γ is a deterministic increasing path, then

$$\tilde{Y}_0 = Y_{\lambda(\gamma, 0)}, \ldots, \tilde{Y}_n = Y_{\lambda(\gamma, n)} \qquad \text{on } \{\Gamma_0 = \gamma_0, \ldots, \Gamma_{m(\gamma)} = \gamma_{m(\gamma)}\}. \tag{4.7}$$

Let $(\tilde{\mathscr{F}}_n)$ be the natural filtration of $(\tilde{Y}_n)$ and set $\tilde{Y}_\infty = Y_{\bowtie}$.

Lemma 1. *The processes* $(\tilde{Y}_n, n \in \overline{\mathbb{N}})$ *and* $(Y_n^1, n \in \overline{\mathbb{N}})$ *have the same law.*

Proof. Fix $n \in \mathbb{N}^*$, and for all deterministic increasing paths γ, set

$$F(\gamma) = \{\Gamma_0 = \gamma_0, \dots, \Gamma_{m(\gamma)} = \gamma_{m(\gamma)}\},$$

where $m(\gamma)$ is defined in (4.4). Since $m(\gamma) \le n$ for all γ (because $\langle \gamma_n \rangle > n$), we can extract from the set of all deterministic increasing paths a finite subset G such that $(F(\gamma), \gamma \in \mathrm{G})$ forms a partition of Ω. Using (4.7), for all choices of Borel sets $B_0, \dots, B_n, B_\infty$, we can write

$$\begin{aligned} &\mathrm{P}\left\{\tilde{Y}_0 \in B_0, \dots, \tilde{Y}_n \in B_n, \tilde{Y}_\infty \in B_\infty\right\} \\ &\quad = \sum_{\gamma \in \mathrm{G}} \mathrm{P}\left(F(\gamma) \cap \left\{Y_{\lambda(\gamma,0)} \in B_0, \dots, Y_{\lambda(\gamma,n)} \in B_n, Y_\infty^1 \in B_\infty\right\}\right). \quad (4.8) \end{aligned}$$

Let us fix momentarily a deterministic increasing path γ. Because Γ is predictable, $F(\gamma)$ is an element of $\mathcal{F}^{\gamma}_{m(\gamma)-1}$. But $\mathcal{F}^{\gamma}_{m(\gamma)-1}$ is the sigma field generated by the family $(Y_t,\ t \in \mathbb{I}_{\gamma_{m(\gamma)-1}})$, thus also by the family $(Y_{\lambda(\gamma,0)}, \dots, Y_{\lambda(\gamma, \langle \gamma_{m(\gamma)-1} \rangle - 1)})$. Moreover, by (4.4), $\langle \gamma_{m(\gamma)-1} \rangle \le n$. Consequently, there is a Borel set B of $\overline{\mathbb{R}}^n$ such that

$$F(\gamma) \cap \{Y_{\lambda(\gamma,0)} \in B_0, \dots, Y_{\lambda(\gamma,n-1)} \in B_{n-1}\} = \left\{\left(Y_{\lambda(\gamma,0)}, \dots, Y_{\lambda(\gamma,n-1)}\right) \in B\right\},$$

which allows us to write

$$\begin{aligned} &\mathrm{P}\left(F(\gamma) \cap \left\{Y_{\lambda(\gamma,0)} \in B_0, \dots, Y_{\lambda(\gamma,n)} \in B_n, Y_\infty^1 \in B_\infty\right\}\right) \\ &\quad = \mathrm{P}\left\{\left(Y_{\lambda(\gamma,0)}, \dots, Y_{\lambda(\gamma,n-1)}\right) \in B, Y_{\lambda(\gamma,n)} \in B_n, Y_\infty^1 \in B_\infty\right\}. \quad (4.9) \end{aligned}$$

Therefore, thanks to the exchangeability hypothesis and because none of the indices $\lambda(\gamma,0), \dots, \lambda(\gamma, n-1)$ are equal to $(n, 0, \dots, 0)$, we can replace $Y_{\lambda(\gamma,n)}$ by Y_n^1 in the right-hand side of (4.9), and thus in the left-hand side, without alterating their common value. By (4.8), we conclude that

$$\begin{aligned} &\mathrm{P}\left\{\tilde{Y}_0 \in B_0, \dots, \tilde{Y}_n \in B_n, \tilde{Y}_\infty \in B_\infty\right\} \\ &\quad = \mathrm{P}\left\{\tilde{Y}_0 \in B_0, \dots, \tilde{Y}_{n-1} \in B_{n-1}, Y_n^1 \in B_n, Y_\infty^1 \in B_\infty\right\}. \quad (4.10) \end{aligned}$$

Unless $n-1$ is already zero, we can pursue the substitution process. Indeed, with the same argument, we observe that the random variables $\tilde{Y}_{n-1}, \dots, \tilde{Y}_1$ can be replaced successively by $Y_{n-1}^1, \dots, Y_1^1$ in the right-hand side of (4.10). Since $\tilde{Y}_0 = Y_0^1$, we conclude that the laws of $(\tilde{Y}_0, \dots, \tilde{Y}_n, \tilde{Y}_\infty)$ and of $(Y_0^1, \dots, Y_n^1, Y_\infty^1)$ are identical, which we had set out to prove.

Lemma 2. *For all deterministic increasing paths* γ *and all* $n \in \mathbb{N}^*$, $\{\Gamma_0 = \gamma_0, \ldots, \Gamma_n = \gamma_n\} \in \tilde{\mathscr{F}}_{\langle \gamma_{n-1} \rangle - 1}$.

Proof. Because Γ is predictable, $\{\Gamma_0 = \gamma_0, \ldots, \Gamma_n = \gamma_n\}$ is an element of $\mathscr{F}^{\gamma}_{n-1}$. But we have already observed that $\mathscr{F}^{\gamma}_{n-1}$ is the sigma field generated by the family $(Y_{\lambda(\gamma, 0)}, \ldots, Y_{\lambda(\gamma, \langle \gamma_{n-1} \rangle - 1)})$. Consequently, there is a Borel subset B_n of $\overline{\mathbb{R}}^{\langle \gamma_{n-1} \rangle}$ such that

$$\{\Gamma_0 = \gamma_0, \ldots, \Gamma_n = \gamma_n\} = \left\{ \left(Y_{\lambda(\gamma, 0)}, \ldots, Y_{\lambda(\gamma, \langle \gamma_{n-1} \rangle - 1)} \right) \in B_n \right\}. \quad (4.11)$$

We are going to show that

$$\{\Gamma_0 = \gamma_0, \ldots, \Gamma_n = \gamma_n\} = \bigcap_{m=1}^{n} \left\{ \left(Y_{\lambda(\Gamma, 0)}, \ldots, Y_{\lambda(\Gamma, \langle \gamma_{m-1} \rangle - 1)} \right) \in B_m \right\}, \quad (4.12)$$

from which the conclusion of the lemma follows, by (4.6). Equality (4.12) is true in the case where $n = 1$, because it then reduces to (4.11), due to the fact that $Y_{\lambda(\gamma, 0)} = Y_{\lambda(\Gamma, 0)}$. Suppose that (4.12) is true in the case where $n - 1$ replaces n. Then

$$\bigcap_{m=1}^{n} \left\{ \left(Y_{\lambda(\Gamma, 0)}, \ldots, Y_{\lambda(\Gamma, \langle \gamma_{m-1} \rangle - 1)} \right) \in B_m \right\}$$
$$= \{\Gamma_0 = \gamma_0, \ldots, \Gamma_{n-1} = \gamma_{n-1}\} \cap \left\{ \left(Y_{\lambda(\Gamma, 0)}, \ldots, Y_{\lambda(\Gamma, \langle \gamma_{n-1} \rangle - 1)} \right) \in B_n \right\}.$$

By (4.7) (written for $\langle \gamma_{n-1} \rangle - 1$ instead of n) and (4.11), the right-hand side is equal to

$$\{\Gamma_0 = \gamma_0, \ldots, \Gamma_{n-1} = \gamma_{n-1}\} \cap \left\{ \left(Y_{\lambda(\gamma, 0)}, \ldots, Y_{\lambda(\gamma, \langle \gamma_{n-1} \rangle - 1)} \right) \in B_n \right\}$$
$$= \{\Gamma_0 = \gamma_0, \ldots, \Gamma_n = \gamma_n\},$$

which establishes (4.12).

Lemma 3. *If* τ *is a stopping time relative to the filtration* $\mathscr{F}^{\Gamma}$, *then* $\tilde{\tau} = \langle \Gamma_\tau \rangle - 1$ *is a stopping time relative to the filtration* $(\tilde{\mathscr{F}}_n)$, *and* $X^{\Gamma}_{\tau} = f_{n+1}(\tilde{Y}_0, \ldots, \tilde{Y}_n)$ *on* $\{\tilde{\tau} = n\}$ *for all* $n \in \mathbb{N}$.

Proof. Fix $n \in \mathbb{N}$ and write the decomposition

$$\{\tilde{\tau} = n\} = \{\langle \Gamma_\tau \rangle = n + 1\} = \bigcup_{t \in \mathbb{I}:\, \langle t \rangle = n+1} \{\Gamma_{|t|} = t,\, \tau = |t|\}.$$

Moreover, observe that

$$\{\Gamma_{|t|} = t,\, \tau = |t|\} = \bigcup_{\gamma \in \mathrm{G}} \{\Gamma_0 = \gamma_0, \ldots, \Gamma_{|t|} = \gamma_{|t|}\} \cap \{\tau = |t|\}, \quad (4.13)$$

where G denotes a suitable finite set of deterministic increasing paths that pass through t. By Lemma 2, if γ is such a path and if $\langle t \rangle = n+1$, then $\{\Gamma_0 = \gamma_0, \ldots, \Gamma_{|t|} = \gamma_{|t|}\}$ belongs to $\tilde{\mathscr{F}}_n$, because $\tilde{\mathscr{F}}_{\langle \gamma_{|t|-1} \rangle} \subset \tilde{\mathscr{F}}_{\langle \gamma_{|t|} \rangle - 1} = \tilde{\mathscr{F}}_n$. On the other hand, in view of the fact that $\{\tau = |t|\}$ belongs to $\mathscr{F}^{\gamma}_{|t|}$, there exists a Borel subset B of $\overline{\mathbb{R}}^{\langle \gamma_{|t|} \rangle}$ such that

$$\{\tau = |t|\} = \Big\{\big(Y_{\lambda(\gamma, 0)}, \ldots, Y_{\lambda(\gamma, \langle \gamma_{|t|} \rangle - 1)}\big) \in B\Big\}.$$

Thus, if $\gamma_{|t|} = t$ and $\langle t \rangle = n+1$, we conclude using (4.7) that

$$\begin{aligned}\{\Gamma_0 = \gamma_0, \ldots, \Gamma_{|t|} = \gamma_{|t|}\} &\cap \{\tau = |t|\} \\ &= \{\Gamma_0 = \gamma_0, \ldots, \Gamma_{|t|} = \gamma_{|t|}\} \cap \Big\{\big(\tilde{Y}_0, \ldots, \tilde{Y}_n\big) \in B\Big\}.\end{aligned}$$

Therefore, every term of the union in (4.13) belongs to $\tilde{\mathscr{F}}_n$, which leads to the conclusion that $\tilde{\tau}$ is a stopping time relative to the filtration $(\tilde{\mathscr{F}}_n)$. Regarding the second statement of the lemma, it suffices to observe that on $\{\tilde{\tau} = n\}$,

$$\begin{aligned} X^{\Gamma}_{\tau} &= f_{\langle \Gamma_\tau \rangle}\Big(\big(Y_s,\ s \in \mathbb{I}_{\Gamma_\tau} - \{\Gamma_\tau\}\big), Y^{\Gamma}_{\tau}\Big) \\ &= f_{\langle \Gamma_\tau \rangle}\big(Y_{\lambda(\Gamma, 0)}, \ldots, Y_{\lambda(\Gamma, \langle \Gamma_\tau \rangle - 1)}\big) \\ &= f_{n+1}\big(\tilde{Y}_0, \ldots, \tilde{Y}_n\big).\end{aligned}$$

Proof of the Theorem

Set $\tilde{\tau} = \langle \Gamma_{|T|} \rangle - 1$. Because $|T|$ is a stopping time relative to the filtration $\mathscr{F}^{\Gamma}$, $\tilde{\tau}$ is a stopping time relative to $(\tilde{\mathscr{F}}_n)$ and

$$X_T = X^{\Gamma}_{|T|} = f_{n+1}\big(\tilde{Y}_0, \ldots, \tilde{Y}_n\big) \qquad \text{on } \{\tilde{\tau} = n\}, \tag{4.14}$$

for all $n \in \mathbb{N}$, by Lemma 3. Again for all $n \in \mathbb{N}$, let B_n denote a Borel set of $\overline{\mathbb{R}}^{n+1}$ such that $\{\tilde{\tau} = n\} = \{(\tilde{Y}_0, \ldots, \tilde{Y}_n) \in B_n\}$ and set $F_n = \{(Y^1_0, \ldots, Y^1_n) \in B_n\}$. By Lemma 1, $\mathrm{P}(F_n) = \mathrm{P}\{\tilde{\tau} = n\}$ and $\mathrm{P}(F_m \cap F_n) = \mathrm{P}\{\tilde{\tau} = m, \tilde{\tau} = n\} = 0$ if $m \neq n$. Consequently, putting

$$\tau(\omega) = \begin{cases} 0, & \text{if } \omega \in F_0, \\ n, & \text{if } \omega \in F_n - \bigcup_{m=0}^{n} F_m \text{ and } n > 0, \\ \infty, & \text{if } \omega \in \Big(\bigcup_{n \in \mathbb{N}} F_n\Big)^c, \end{cases}$$

we define a stopping time τ relative to the filtration $(\mathscr{F}^1_n)$ whose law coincides with that of $\tilde{\tau}$. We now show that the law of $(X_T, \langle T \rangle - 1)$

coincides with the law of (X^1_τ, τ). By (4.14), it is sufficient to prove that if B is a Borel set, then

$$\mathrm{P}\left\{f_{n+1}\left(\tilde{Y}_0, \ldots, \tilde{Y}_n\right) \in B, \tilde{\tau} = n\right\} = \mathrm{P}\left\{f_{n+1}\left(Y^1_0, \ldots, Y^1_n\right) \in B, \tau = n\right\}$$

for all $n \in \mathbb{N}$, and

$$\mathrm{P}\{X_{\bowtie} \in B, \tilde{\tau} = \infty\} = \mathrm{P}\{X^1_\infty \in B, \tau = \infty\}.$$

However, defining B_n as above and taking into account that $X_{\bowtie} = \tilde{Y}_\infty = X^1_\infty = Y^1_\infty$, we see that these two equalities can be written in the form

$$\mathrm{P}\left\{\left(\tilde{Y}_0, \ldots, \tilde{Y}_n\right) \in f^{-1}_{n+1}(B) \cap B_n\right\} = \mathrm{P}\left\{\left(Y^1_0, \ldots, Y^1_n\right) \in f^{-1}_{n+1}(B) \cap B_n\right\}$$

and

$$\mathrm{P}\left(\bigcap_{n \in \mathbb{N}} \left\{\left(\tilde{Y}_0, \ldots, \tilde{Y}_n\right) \in B^c_n, \tilde{Y}_\infty \in B\right\}\right)$$
$$= \mathrm{P}\left(\bigcap_{n \in \mathbb{N}} \left\{\left(Y^1_0, \ldots, Y^1_n\right) \in B^c_n, Y^1_\infty \in B\right\}\right).$$

The validity of these equalities is a direct consequence of Lemma 1.

4.2 APPLICATIONS

When it applies, the theorem of the previous section lets us take advantage of techniques specific to dimension one.

Optimal Stopping

With the data and notation of Section 4.1, the values of the games associated respectively with X and with $X^1 = (X^1_n,\ n \in \overline{\mathbb{N}})$ coincide; moreover, there exists an optimal accessible stopping point for X if and only if there exists an optimal stopping time for X^1 (of course, these assertions make sense only if at least one of the two games has a finite value). Indeed, in this case, they follow directly from the theorem of Section 4.1, for to each accessible stopping point T can be associated a stopping time τ such that $\mathrm{E}(X_T) = \mathrm{E}(X^1_\tau)$ and for each stopping time τ, the random variable

$$T = \begin{cases} (\tau, 0, \ldots, 0), & \text{if } \tau < \infty, \\ \bowtie, & \text{if } \tau = \infty, \end{cases}$$

is an accessible stopping point such that $\mathrm{E}(X^1_\tau) = \mathrm{E}(X_T)$.

In this connection, it may be useful to remember that there is an optimal stopping time for the process X^1 if this process is of class D and $\limsup_{n\to\infty} X_n^1 \leq X_\infty^1$.

Optimal Stopping of Normalized Partial Sums

Let us assume that the random variables Y_t ($t \in \mathbb{I}$) are integrable. For all $n \in \mathbb{N}^*$, set

$$f_n(x_1,\ldots,x_n) = \begin{cases} \dfrac{1}{n}(x_1 + \cdots + x_n), & \text{if } (x_1,\ldots,x_n) \in \mathbb{R}^n, \\ 0, & \text{otherwise}, \end{cases} \tag{4.15}$$

and define X by (4.2) (or by (3.26), but with $X_{\bowtie} = Y_{\bowtie}$). The process $(X_n^1,\, n \in \mathbb{N})$ is then a reversed martingale, which implies that X_n^1 converges a.s. The second of the conditions recalled above is then fulfilled if we put $X_\infty^1\ (= Y_{\bowtie}) = \lim_{n\to\infty} X_n^1$.

Case Where the Random Variables Are Independent and Identically Distributed

When the random variables Y_t ($t \in \mathbb{I}$) are independent, integrable and identically distributed and the functions f_n are defined by (4.15), $\lim_{n\to\infty} X_n^1 = \mathrm{E}(Y_0)$. Moreover, X^1 is of class D if and only if one of the following conditions is fulfilled:

a. $\mathrm{E}(|Y_0|\,|\log^+|Y_0|) < \infty$;
b. $\mathrm{E}(\sup_n |X_n^1|) < \infty$;
c. $\mathrm{E}(|X_\tau^1|) < \infty$, for all stopping times τ;
d. $\sup_\tau \mathrm{E}(|X_\tau^1|) < \infty$;
e. $\sup_{T\in\mathbf{A}} \mathrm{E}(|X_T|) < \infty$.

Indeed, by Theorems 2 and 3 of Section 2.4, conditions a, b, and c are equivalent. Moreover, it is obvious that b implies d and that d implies c. On the other hand, we have already noticed in the theorem of Section 4.1 that the upper bounds in d and in e coincide. It remains to observe that if X^1 is of class D, then condition d is fulfilled, and under condition b, X^1 is of class D.

Two additional conditions equivalent to a–e are

f. $\mathrm{E}(|Y_T|/\langle T\rangle) < \infty$, for all $T \in \mathbf{A}_{\mathrm{f}}$;
g. $\sup_{T\in\mathbf{A}_{\mathrm{f}}} \mathrm{E}(|Y_T|/\langle T\rangle) < \infty$.

(cf. Exercise 1.)

Existence of Finite Optimal Stopping Points

Assume again that the random variables Y_t $(t \in \mathbb{I})$ are independent and identically distributed and that the functions f_n are defined by (4.15). The theorem of Section 3.7 implies that if $\mathrm{E}(Y_0^2) < \infty$, then all optimal stopping times for X^1 are finite. It can be shown (but the proof is not easy) that this conclusion is also true under the hypothesis that $\mathrm{E}(|Y_0|^p) < \infty$ for some p with $1 < p < 2$ (cf. Historical Notes). In either case, by the theorem of Section 4.1, this same conclusion extends to all optimal accessible stopping points for X. Indeed, if T is such a stopping point and τ is the stopping time associated with T as in the theorem just mentioned, then τ is finite because it is optimal, and thus T is finite.

4.3 LINEAR REPRESENTATION IN THE SETTING OF INACCESSIBLE STOPPING POINTS

The reader who has imagined that the only stopping points with a linear representation are those that are accessible will be surprised by the example that we are about to present. A second example will follow that establishes the existence of stopping points admitting no linear representation.

Stopping Points which Admit a Linear Representation

With the data and notation of Section 4.1, we will say that a stopping point T relative to the filtration $(\mathscr{F}_t)$ *admits a linear representation* if there is a stopping time ν relative to the filtration $(\mathscr{F}_n^1)$ satisfying the conclusion of the theorem of Section 4.1 for all choices of the sequence (f_n) used to define X.

By the theorem just mentioned, all accessible stopping points admit a linear representation.

Example: An Inaccessible Stopping Point which Admits a Linear Representation

Consider the case where $d = 3$ and suppose that the random variables Y_t $(t \in \mathbb{I})$ are independent and identically distributed with common law given by $\mathrm{P}\{Y_0 = 0\} = \mathrm{P}\{Y_0 = 1\} = \frac{1}{2}$. Let $(\mathscr{F}_t)$ denote the natural filtration of (Y_t) and define F_1, F_2, and F_3 by

$$F_1 = \{Y_{(0,1,0)} = 1, Y_{(0,0,1)} = 1\},$$

$$F_2 = \{Y_{(1,0,0)} = 1, Y_{(0,0,1)} = 0\},$$

$$F_3 = \{Y_{(1,0,0)} = 0, Y_{(0,1,0)} = 0\}.$$

Clearly, $F_i \cap F_j = \varnothing$ if $i \neq j$. Put

$$T = \begin{cases} (0,1,1) & \text{on } F_1, \\ (1,0,1) & \text{on } F_2, \\ (1,1,0) & \text{on } F_3, \\ (1,1,1) & \text{on } (F_1 \cup F_2 \cup F_3)^c. \end{cases}$$

It is clear that T is a stopping point and it is not difficult to check, for example, with the help of Theorem 1 of Section 5.1, that T is not accessible. Nevertheless, we shall prove that T admits a linear representation. To this end, set $e = (1,1,1)$ and for $i = 1,2,3$, let $\varphi_i: \mathbb{N}_7 \to \mathbb{I}_e$ be any enumeration of $\mathbb{I}_e$ with the properties

$$\varphi_i(i-1) = (0,0,0), \quad \varphi_i(j-1) = e_j,$$
$$\varphi_i(k-1) = e_k, \quad \text{and} \quad \varphi_i(3) = e_j + e_k,$$

if (i,j,k) is a cyclic permuation of $(1,2,3)$, and

$$\varphi_i(7) = e.$$

In addition, let $\psi: \mathbb{N}_7 \to \mathbb{I}_e$ be any enumeration of $\mathbb{I}_e$ such that

$$\psi(j-1) = e_j, \quad \text{for } j \in \{1,2,3\}, \quad \psi(3) = (0,0,0), \quad \text{and} \quad \psi(7) = e.$$

Let $(n,\omega) \mapsto \lambda(n,\omega)$ be the function defined on $\mathbb{N}_7 \times \Omega$ with values in $\mathbb{I}_e$ by

$$\lambda(n,\omega) = \begin{cases} \varphi_i(n), & \text{if } n \in \mathbb{N}_7 \text{ and } \omega \in F_i, \quad i = 1,2,3, \\ \psi(n), & \text{if } n \in \mathbb{N}_7 \text{ and } \omega \in (F_1 \cup F_2 \cup F_3)^c. \end{cases}$$

For all $n \in \mathbb{N}_7$, let $\tilde{Y}_n$ denote the random variable defined by $\tilde{Y}_n(\omega) = Y_{\lambda(n,\omega)}(\omega)$ and let $(\tilde{\mathscr{F}}_n)$ be the natural filtration of $(\tilde{Y}_n, n \in \mathbb{N}_7)$. Again for all $n \in \mathbb{N}_7$, put $\tilde{X}_n = f_{n+1}(\tilde{Y}_0, \ldots, \tilde{Y}_n)$, where (f_n) is the sequence of functions introduced just before (4.1). Moreover, set $\tilde{\tau} = \langle T \rangle - 1$.

Lemma. *The random variables $\tilde{Y}_n$ $(n \in \mathbb{N}_7)$ are independent and have the same distribution as Y_0, $\tilde{\tau}$ is a stopping time relative to the filtration $(\tilde{\mathscr{F}}_n)$, and if X is defined by* (4.2), *then the law of the couple $(X_T, \langle T \rangle - 1)$ is identical to that of the couple $(\tilde{X}_{\tilde{\tau}}, \tilde{\tau})$.*

With the help of this lemma, a linear representation of T is obtained by defining τ as in the proof of the theorem of Section 4.1, that is, by setting

$$\tau = n \text{ on } \{Y_0^1 = i_0, \ldots, Y_n^1 = i_n\}, \qquad \text{if } \tilde{\tau} = n \text{ on } \{\tilde{Y}_0 = i_0, \ldots, \tilde{Y}_n = i_n\},$$

and by observing that the laws of (X_τ^1, τ) and of $(\tilde{X}_{\tilde{\tau}}, \tilde{\tau})$ are then identical.

Proof of the Lemma

In order to determine the distribution of $(\tilde{Y}_n, n \in \mathbb{N}_7)$, let $(x_n, n \in \mathbb{N}_7)$ be a family of elements of $\{0, 1\}$. Observe that

$$\begin{aligned} \mathrm{P}\{\tilde{Y}_n = x_n, n \in \mathbb{N}_7\} &= \mathrm{P}\{Y_{\lambda(n,\cdot)} = x_n, n \in \mathbb{N}_7\} \\ &= \sum_{i=1}^{3} \mathrm{P}\big(\{Y_{\varphi_i(n)} = x_n, n \in \mathbb{N}_7\} \cap F_i\big) \\ &\quad + \mathrm{P}\big(\{Y_{\psi(n)} = x_n, n \in \mathbb{N}_7\} \cap (F_1 \cup F_2 \cup F_3)^c\big). \end{aligned}$$

For $i = 1, 2, 3$, let (i, j, k) be a cyclic permutation of $(1, 2, 3)$. Use the definition of φ_i to see that the term in the summation is equal to

$$\begin{aligned} \mathrm{P}\Big(\Big\{&Y_{(0,0,0)} = x_{i-1}, Y_{e_j} = x_{j-1}, Y_{e_k} = x_{k-1}, Y_{e_j+e_k} = x_3, \\ &\big(Y_s = x_{\varphi_i^{-1}(s)}, s \in \mathbb{I}_e - \mathbb{I}_{e_j+e_k}\big)\Big\} \cap F_i\Big). \end{aligned}$$

Because F_i is measurable with respect to the sigma field generated by $\{Y_{e_j}, Y_{e_k}\}$, and $\psi(j-1) = e_j$ and $\psi(k-1) = e_k$, the independence of the Y_t $(t \in \mathbb{I})$ allows us to permute the random variables Y_s $(s \in \mathbb{I}_e - \{e_j, e_k\})$ to see that this probability is equal to

$$\begin{aligned} \mathrm{P}\Big(\Big\{&Y_{\psi(i-1)} = x_{i-1}, Y_{\psi(j-1)} = x_{j-1}, Y_{\psi(k-1)} = x_{k-1}, Y_{\psi(3)} = x_3, \\ &\big(Y_s = x_{\psi^{-1}(s)}, s \in \mathbb{I}_e - (\mathbb{I}_{e_1} \cup \mathbb{I}_{e_2} \cup \mathbb{I}_{e_3})\big)\Big\} \cap F_i\Big), \end{aligned}$$

and therefore

$$\mathrm{P}\{\tilde{Y}_n = x_n, n \in \mathbb{N}_7\} = \mathrm{P}\{Y_{\psi(n)} = x_n, n \in \mathbb{N}_7\},$$

which implies that the $\tilde{Y}_n$ $(n \in \mathbb{N}_7)$ are independent and have the same distribution as Y_0.

We now show how to express the sets F_1, F_2, and F_3 in terms of the variables $\tilde{Y}_n$ $(n \in \mathbb{N}_7)$. Let

$$\tilde{F}_1 = \{\tilde{Y}_1 = 1, \tilde{Y}_2 = 1\}, \qquad \tilde{F}_2 = \{\tilde{Y}_0 = 1, \tilde{Y}_2 = 0\},$$
$$\tilde{F}_3 = \{\tilde{Y}_0 = 0, \tilde{Y}_1 = 0\}.$$

It is not difficult to check from the definitions that $F_1 \subset \tilde{F}_1$, $F_2 \subset \tilde{F}_2$, $F_3 \subset \tilde{F}_3$, and $(F_1 \cup F_2 \cup F_3)^c \subset (\tilde{F}_1 \cup \tilde{F}_2 \cup \tilde{F}_3)^c$, and since $(F_1, F_2, F_3, (F_1 \cup F_2 \cup F_3)^c)$ is a partition of Ω, these four inclusions must in fact be equalities.

In order to verify that $\tilde{\tau}$ is a stopping time relative to $(\tilde{\mathscr{F}}_n)$, notice that $\tilde{\tau}$ takes values in $\{3, 7\}$, and therefore that $\{\tilde{\tau} = 3\}$ is equal to

$$\{\tilde{\tau} = 7\}^c = \{\langle T \rangle = 8\}^c = (F_1 \cup F_2 \cup F_3)^c = \left(\tilde{F}_1 \cup \tilde{F}_2 \cup \tilde{F}_3\right)^c \in \tilde{\mathscr{F}}_2 \subset \tilde{\mathscr{F}}_3.$$

To prove the final conclusion of the lemma, observe that for any Borel subset B of $\mathbb{R}$,

$$\begin{aligned}
&\mathrm{P}\{X_T \in B, \langle T \rangle - 1 = 7\} \\
&\quad = \mathrm{P}\left(\{f_8((Y_t, t < e), Y_{(1,1,1)}) \in B\} \cap (F_1 \cup F_2 \cup F_3)^c\right) \\
&\quad = \mathrm{P}\left(\left\{f_8\left((\tilde{Y}_n, n \in \mathbb{N}_6), \tilde{Y}_7\right) \in B\right\} \cap (F_1 \cup F_2 \cup F_3)^c\right) \\
&\quad = \mathrm{P}\left\{\tilde{X}_{\tilde{\tau}} \in B, \tilde{\tau} = 7\right\},
\end{aligned}$$

and $\mathrm{P}\{X_T \in B, \langle T \rangle - 1 = 3\}$ is equal to

$$\begin{aligned}
&\mathrm{P}\left(\{f_4(Y_{(0,0,0)}, Y_{(0,1,0)}, Y_{(0,0,1)}, Y_{(0,1,1)}) \in B\} \cap F_1\right) \\
&\quad + \mathrm{P}\left(\{f_4(Y_{(0,0,0)}, Y_{(1,0,0)}, Y_{(0,0,1)}, Y_{(1,0,1)}) \in B\} \cap F_2\right) \\
&\quad + \mathrm{P}\left(\{f_4(Y_{(0,0,0)}, Y_{(1,0,0)}, Y_{(0,1,0)}, Y_{(1,1,0)}) \in B\} \cap F_3\right).
\end{aligned}$$

Using the definition of the variables $\tilde{Y}_n$, this is equal to

$$\begin{aligned}
&\mathrm{P}\left(\left\{f_4\left(\tilde{Y}_0, \tilde{Y}_1, \tilde{Y}_2, \tilde{Y}_3\right) \in B\right\} \cap F_1\right) + \mathrm{P}\left(\left\{f_4\left(\tilde{Y}_1, \tilde{Y}_0, \tilde{Y}_2, \tilde{Y}_3\right) \in B\right\} \cap F_2\right) \\
&\qquad + \mathrm{P}\left(\left\{f_4\left(\tilde{Y}_2, \tilde{Y}_0, \tilde{Y}_1, \tilde{Y}_3\right) \in B\right\} \cap F_3\right).
\end{aligned}$$

Due to the invariance of $f_4(\cdot, \cdot, \cdot, \cdot)$ under permutations of its first three arguments, this is equal to

$$\mathrm{P}\left(\left\{f_4\left((\tilde{Y}_n, n \in \mathbb{N}_2), \tilde{Y}_3\right) \in B\right\} \cap (F_1 \cup F_2 \cup F_3)\right) = \mathrm{P}\left\{\tilde{X}_{\tilde{\tau}} \in B, \tilde{\tau} = 3\right\},$$

and the proof is complete.

Example: A Stopping Point that Admits No Linear Representation

Assume again that $d = 3$ and that the random variables Y_t $(t \in \mathbb{I})$ are independent and identically distributed with common law defined by $\mathrm{P}\{Y_0 = 0\} = \mathrm{P}\{Y_0 = 1\} = \frac{1}{2}$. Let $(\mathscr{F}_t)$ be the natural filtration of (Y_t). Define the process X by (4.2) with $f_n(x_1, \ldots, x_n) = x_1 + \cdots + x_n$, that is, by setting

$X_t = S_t$ for all $t \in \mathbb{I}$, where $S_t = \Sigma_{s \le t} Y_s$. Put

$$T = \begin{cases} (1,1,1) & \text{on } \{X_{(1,1,1)} > 0\}, \\ (1,2,2) & \text{on } \{X_{(1,1,2)} = 0,\ X_{(1,2,1)} = 1\}, \\ (2,1,2) & \text{on } \{X_{(1,1,2)} = 1,\ X_{(2,1,1)} = 0\}, \\ (2,2,1) & \text{on } \{X_{(1,2,1)} = 0,\ X_{(2,1,1)} = 1\}, \\ (2,2,2) & \text{elsewhere.} \end{cases}$$

Clearly, T is a stopping point relative to the filtration $(\mathscr{F}_t)$ and

$$\begin{aligned} P\{X_T = 1, \langle T\rangle - 1 = 17\} &= P\{X_{(1,2,2)} = 1, T = (1,2,2)\} \\ &\quad + P\{X_{(2,1,2)} = 1, T = (2,1,2)\} \\ &\quad + P\{X_{(2,2,1)} = 1, T = (2,2,1)\} \\ &= P\{X_{(1,2,2)} = 1,\ X_{(1,1,2)} = 0,\ X_{(1,2,1)} = 1\} \\ &\quad + P\{X_{(2,1,2)} = 1,\ X_{(1,1,2)} = 1,\ X_{(2,1,1)} = 0\} \\ &\quad + P\{X_{(2,2,1)} = 1,\ X_{(1,2,1)} = 0,\ X_{(2,1,1)} = 1\} \\ &= 3\frac{1}{2^{12}}\frac{4}{2^4}\frac{1}{2^2} \\ &= \frac{12}{2^{18}}. \end{aligned}$$

In addition

$$\begin{aligned} P\{X_T > 0, \langle T\rangle - 1 = 7\} &= P\{X_T > 0, T = (1,1,1)\} \\ &= P\{X_{(1,1,1)} > 0,\ X_{(1,1,1)} > 0\} \\ &= P\{X_{(1,1,1)} > 0\}. \end{aligned}$$

Assume that the law of the couple $(X_T, \langle T\rangle - 1)$ is identical to that of the couple (X_τ^1, τ), where τ is a stopping time relative to the filtration $(\mathscr{F}_n^1)$. Then

$$P\{X_\tau^1 = 1, \tau = 17\} = P(X_{17}^1 = 1, \tau = 17\} = \frac{12}{2^{18}}. \tag{4.16}$$

Moreover,

$$P\{X_\tau^1 > 0, \tau = 7\} = P\{X_7^1 > 0, \tau = 7\} = P\{X_{(1,1,1)} > 0\} = P\{X_7^1 > 0\},$$

which leads to $\{X_7^1 > 0\} \subset \{\tau = 7\}$ or, equivalently, $\{\tau \neq 7\} \subset \{X_7^1 = 0\}$. But then $\{\tau = 17\} \subset \{X_7^1 = 0\}$, and so

$$\begin{aligned} P\{X_{17}^1 = 1, \tau = 17\} &= P\{X_{17}^1 = 1, X_7^1 = 0, \tau = 17\} \\ &\leq P\{X_{17}^1 = 1, X_7^1 = 0\} \\ &= \frac{1}{2^8} \frac{10}{2^{10}} \\ &= \frac{10}{2^{18}}. \end{aligned}$$

This contradicts (4.16).

EXERCISES

1. In the setting of Section 4.2, prove that conditions f and g are equivalent to condition a. (*Hint*. Regarding condition f, use the theorem of Section 4.1 together with the functions $f_n(y_1, \ldots, y_{n-1}, y_n) = y_n/n$ and Theorem 3 of Section 2.4. To check that a implies g, let $Y_n^1 = Y_{(n, 0, \ldots, 0)}$ and $Y_\infty^1 = 0$, and check that $\lim_{n \to \infty} Y_n^1/n = 0$. Using Theorem 2 of Section 2.4, show that there exists an optimal stopping time for the process $(Y_n^1, n \in \overline{\mathbb{N}})$. Then use f and the theorem of Section 4.1.)

2. (Another inaccessible stopping point which admits a linear representation). Assume that $d = 3$ and suppose that the random variables Y_t $(t \in \mathbb{I})$ are independent and identically distributed with common law given by $P\{Y_0 = 0\} = P\{Y_0 = 1\} = 1/2$. Let $(\mathscr{F}_t)$ denote the natural filtration of (Y_t) and define F_1, F_2, and F_3 by

$$\begin{aligned} F_1 &= \{S_{(1,3,1)} - S_{(1,1,1)} > 0, S_{(1,1,3)} - S_{(1,1,1)} = 0\}, \\ F_2 &= \{S_{(3,1,1)} - S_{(1,1,1)} = 0, S_{(1,1,3)} - S_{(1,1,1)} > 0\}, \\ F_3 &= \{S_{(3,1,1)} - S_{(1,1,1)} > 0, S_{(1,3,1)} - S_{(1,1,1)} = 0\}, \end{aligned}$$

where for $t \in \mathbb{I}$, $S_t = \Sigma_{s \leq t} Y_s$. Clearly, $F_i \cap F_j = \varnothing$ if $i \neq j$. Put

$$T = \begin{cases} (1,3,3) & \text{on } F_1, \\ (3,1,3) & \text{on } F_2, \\ (3,3,1) & \text{on } F_3, \\ (3,3,3) & \text{on } (F_1 \cup F_2 \cup F_3)^c. \end{cases}$$

a. Verify that T is a stopping point, but that T is not accessible. (*Hint*. This can be done directly, or for example with the help of Theorem 1 of Section 5.1.)

b. Set $e = (1, 1, 1)$ and choose arbitrary one-to-one maps

$$
\begin{aligned}
&\varphi_0 \colon \mathbb{N}_7 \to \mathbb{I}_e, && \\
&\varphi_i \colon \mathbb{N}_7 \to \mathbb{I}_{e+2e_i} - \mathbb{I}_e && (i \in \{1,2,3\}), \\
&\varphi_{ij} \colon \mathbb{N}_7 \to \mathbb{I}_{e+2e_i+2e_j} - \left(\mathbb{I}_{e+2e_i} \cup \mathbb{I}_{e+2e_j}\right) && (\{i,j\} \subset \{1,2,3\}), \\
&\psi_{ij} \colon \mathbb{N}_{31} \to \mathbb{I}_{3e} - \mathbb{I}_{e+2e_i+2e_j} && (\{i,j\} \subset \{1,2,3\}), \\
&\psi \colon \mathbb{N}_{31} \to \mathbb{I}_{3e} - \left(\mathbb{I}_{e+2e_1} \cup \mathbb{I}_{e+2e_2} \cup \mathbb{I}_{e+2e_3}\right), &&
\end{aligned}
$$

such that $\varphi_{ij}(7) = e + 2e_i + 2e_j$ and $\psi_{ij}(31) = \psi(31) = 3e$. Set $I_0 = \mathbb{N}_7$, $I_1 = \mathbb{N}_{15} - \mathbb{N}_7$, $I_2 = \mathbb{N}_{23} - \mathbb{N}_{15}$, $I_3 = \mathbb{N}_{31} - \mathbb{N}_{23}$, $I_4 = \mathbb{N}_{63} - \mathbb{N}_{31}$, and let $n_0 = 0$, $n_1 = 8$, $n_2 = 16$, $n_3 = 24$, $n_4 = 32$. For $i = 1, 2, 3$, let $\lambda_i \colon \mathbb{N}_{63} \to \mathbb{I}_{3e}$ be the enumeration of $\mathbb{I}_{3e}$ defined as follows: if (i, j, k) is a cyclic permutation of $(1, 2, 3)$, then

$$
\lambda_i(n) = \begin{cases}
\varphi_0(n - n_{i-1}), & \text{if } n \in I_{i-1}, \\
\varphi_j(n - n_{j-1}), & \text{if } n \in I_{j-1}, \\
\varphi_k(n - n_{k-1}), & \text{if } n \in I_{k-1}, \\
\varphi_{jk}(n - n_3), & \text{if } n \in I_3, \\
\psi_{jk}(n - n_4), & \text{if } n \in I_4.
\end{cases}
$$

In addition, let $\lambda_0 \colon \mathbb{N}_{63} \to \mathbb{I}_{3e}$ be the enumeration of $\mathbb{I}_{3e}$ defined by

$$
\lambda_0(n) = \begin{cases}
\varphi_1(n - n_0), & \text{if } n \in I_0, \\
\varphi_2(n - n_1), & \text{if } n \in I_1, \\
\varphi_3(n - n_2), & \text{if } n \in I_2, \\
\varphi_0(n - n_3), & \text{if } n \in I_3, \\
\psi(n - n_4), & \text{if } n \in I_4.
\end{cases}
$$

Let $(n, \omega) \mapsto \lambda(n, \omega)$ be the function defined on $\mathbb{N}_{63} \times \Omega$ with values in $\mathbb{I}_{3e}$ by

$$
\lambda(n, \omega) = \begin{cases}
\lambda_i(n), & \text{if } n \in \mathbb{N}_{63} \text{ and } \omega \in F_i, \quad i = 1,2,3, \\
\lambda_0(n), & \text{if } n \in \mathbb{N}_{63} \text{ and } \omega \in (F_1 \cup F_2 \cup F_3)^c.
\end{cases}
$$

For all $n \in \mathbb{N}_{63}$, let $\tilde{Y}_n$ denote the random variable defined by $\tilde{Y}_n(\omega) = Y_{\lambda(n,\omega)}(\omega)$ and let $(\tilde{\mathcal{F}}_n)$ be the natural filtration of $(\tilde{Y}_n, n \in \mathbb{N}_{63})$. Again for all $n \in \mathbb{N}_{63}$, put $\tilde{X}_n = f_{n+1}(\tilde{Y}_0, \ldots, \tilde{Y}_n)$, where (f_n) is the sequence of functions introduced just before (4.1). Moreover, set $\tilde{\tau} = \langle T \rangle - 1$.

b1. Prove that the random variables $\tilde{Y}_n$ $(n \in \mathbb{N}_{63})$ are independent and have the same distribution as Y_0. (*Hint*. Let $(x_n, n \in \mathbb{N}_{63})$ be a family of elements of $\{0, 1\}$. Let (i, j, k) be a cyclic permutation of $(1, 2, 3)$ and observe that $\mathrm{P}(\{\tilde{Y}_n = x_n, n \in \mathbb{N}_{63}\} \cap F_i)$ is equal to

$$\begin{aligned}
\mathrm{P}\Big(F_i \cap \{Y_s &= x_{n_{i-1}+\varphi_0^{-1}(s)}, s \in \mathbb{I}_e\} \\
&\cap \{Y_s = x_{n_{j-1}+\varphi_j^{-1}(s)}, s \in \mathbb{I}_{e+2e_j} - \mathbb{I}_e\} \\
&\cap \{Y_s = x_{n_{k-1}+\varphi_k^{-1}(s)}, s \in \mathbb{I}_{e+2e_k} - \mathbb{I}_e\} \\
&\cap \Big\{Y_s = x_{n_3+\varphi_{jk}^{-1}(s)}, s \in \mathbb{I}_{e+2e_j+2e_k} - \big(\mathbb{I}_{e+2e_j} \cup \mathbb{I}_{e+2e_k}\big)\Big\} \\
&\cap \{Y_s = x_{n_4+\psi_{jk}^{-1}(s)}, s \in \mathbb{I}_{3e} - \mathbb{I}_{e+2e_j+2e_k}\}\Big).
\end{aligned}$$

Because F_i is measurable with respect to the sigma field generated by Y_s $(s \in (\mathbb{I}_{e+2e_j} - \mathbb{I}_e) \cup (\mathbb{I}_{e+2e_k} - \mathbb{I}_e))$, it is possible to permute the random variables Y_s $(s \in \mathbb{I}_{3e} - ((\mathbb{I}_{e+2e_j} - \mathbb{I}_e) \cup (\mathbb{I}_{e+2e_k} - \mathbb{I}_e)))$ to see that

$$\mathrm{P}\big(\{\tilde{Y}_n = x_n, n \in \mathbb{N}_{63}\} \cap F_i\big) = \mathrm{P}\big(\{Y_s = x_{\lambda_0^{-1}(s)}, s \in \mathbb{I}_{3e}\} \cap F_i\big).)$$

b2. Show that $\tilde{\tau}$ is a stopping time relative to the filtration $(\tilde{\mathcal{F}}_n)$. (*Hint*. For $n \in \mathbb{N}_{63}$, let $\tilde{S}_n = \Sigma_{m \leq n} \tilde{Y}_m$ and set

$$\begin{aligned}
\tilde{F}_1 &= \{\tilde{S}_{15} - \tilde{S}_7 > 0, \tilde{S}_{23} - \tilde{S}_{15} = 0\}, \\
\tilde{F}_2 &= \{\tilde{S}_7 = 0, \tilde{S}_{23} - \tilde{S}_{15} > 0\}, \\
\tilde{F}_3 &= \{\tilde{S}_7 > 0, \tilde{S}_{15} - \tilde{S}_7 = 0\},
\end{aligned}$$

and prove, as in the corresponding part of the proof of the lemma of Section 4.3, that $F_i = \tilde{F}_i$ for $i = 1, 2, 3$.)

b3. Show that if X is defined by (4.2), then the law of the couple $(X_T, \langle T \rangle - 1)$ is identical to that of the couple $(\tilde{X}_{\tilde{\tau}}, \tilde{\tau})$. (*Hint*. Proceed as in the corresponding part of the proof of the lemma of Section 4.3.)

b4. Prove that T admits a linear representation. (*Hint*. Proceed as explained just before the proof of the lemma of Section 4.3.)

Open Problems

1. In the setting of Section 4.3, characterize those inaccessible stopping points which admit a linear representation.
2. In the setting of Section 4.3, if a stopping point has no linear embedding, what is the best approximation of the law of (X_T, T) by laws of (X^1_τ, τ), where τ is a stopping time?
3. Let $d \geq 3$. For the reward process defined in (3.26), is the supremum of $E(X_T)$ over all stopping points T strictly greater than the supremum over $T \in \mathbf{A}$?

HISTORICAL NOTES

In response to a question posed by Cairoli and Gabriel (1978) concerning the degree of integrability required to ensure that condition e of Section 4.2 be satisfied, Krengel and Sucheston (1981) proposed the method of linear embedding of accessible stopping points. They proved in particular the theorem of Section 4.1 in the case of independent and identically distributed random variables. The extension to exchangeable random variables given in Section 4.1 is new, and the proof, simpler than the one in the article just mentioned, handles measure-theoretic details with care. Other extensions were considered, for example, in Grillenberger and Krengel (1981).

The statement and proof of de Finetti's theorem can be found in Chapter 7 of Chow and Teicher (1988).

The results of Section 4.2 extend those of McCabe and Shepp (1970) and Davis (1971) presented in Theorem 3 of Section 2.4. They are due to Cairoli and Gabriel (1978) and Krengel and Sucheston (1981). Both of these references contain additional related results. As mentioned at the end of Section 4.2, all stopping times for the process X^1 of that section are finite if $E(|Y_0|^p) < \infty$ for some p with $p > 1$. This result is due to Davis (1973).

The stopping point considered in the first example of Section 4.3 was the example used by Krengel and Sucheston (1981) and Mandelbaum and Vanderbei (1981) to show the existence of inaccessible stopping points. The notion of linear embedding of inaccessible stopping points is new, as is the observation that this particular stopping point admits a linear embedding. Also new is the second example in that section. From these examples arise the question of classification of stopping points that admit a linear embedding and the other open problems mentioned above.

The result of Exercise 1 is contained in Cairoli and Gabriel (1978) and Krengel and Sucheston (1981). The example in Exercise 2 is new.

CHAPTER 5

Accessibility and Filtration Structure

The notion of accessibility plays an important role in dimension $d > 1$, so it is natural to ask if it is ever true that all stopping points are accessible. This depends of course on the filtration. In the first part of this chapter, we will see that it is true for several interesting classes of two-dimensional filtrations, but that in higher dimensions, it is true only in degenerate cases. In the second part of the chapter, we will study the structure of a finite filtration and show that in dimension 2 and if all stopping points are accessible, then the problem of optimal stopping can be reduced to a problem of linear programming.

Unless mentioned otherwise, we will assume that $d > 1$. In Sections 5.1–5.4, we will be working with a general probability space and filtration, while in Sections 5.5–5.7, the sigma field $\mathscr{F}$ of the probability space will be finite.

5.1 CONDITIONS FOR ACCESSIBILITY

By Corollary 1 of Section 1.6 and Proposition 3 of Section 3.3, in order that all stopping points be accessible, it is necessary and sufficient that the conclusion of the optional sampling theorem be valid. Although interesting, this result does not explain the mechanisms that determine whether or not a stopping point is accessible. In this section, we intend to establish two criteria for accessibility. The first concerns accessibility of a given stopping point T. The criterion relates to the conditional probability that T belong to certain hyperplanes through t, for each $t \in \mathbb{I}$. The second criterion will furnish a condition indicating the constraints the filtration must satisfy in order that *all* stopping points be accessible.

Theorem 1 (Criterion for the accessibility of a given stopping point). *In order that a stopping point T be accessible, it is necessary and sufficient that*

$$\bigcap_{i=1}^{d} \{P(T > t, T^i = t^i | \mathscr{F}_t) > 0\} = \varnothing \tag{5.1}$$

for all $t \in \mathbb{I}$.

Proof. Assume that (5.1) is satisfied for all $t \in \mathbb{I}$ and define $\Gamma = (\Gamma_n, n \in \mathbb{N})$ by setting $\Gamma_0 = 0$ and, by induction,

$$\Gamma_{n+1} = t + e_i \quad \text{on } \{\Gamma_n = t\} \cap \{\inf\{j: P(T > t, T^j = t^j | \mathscr{F}_t) = 0\} = i\}. \tag{5.2}$$

It is necessary to point out that this definition of Γ is coherent, because $\{j: P(T > t, T^j = t^j | \mathscr{F}_t) = 0\} \neq \varnothing$ by (5.1). Evidently, $\Gamma_{n+1} \in \mathbb{D}_{\Gamma_n}$ and $\{\Gamma_{n+1} = u\} \cap \{\Gamma_n = t\} \in \mathscr{F}_t$ for all $t \in \mathbb{I}$ and all $u \in \mathbb{D}_t$, which shows that Γ is a predictable increasing path. Assume that $P\{\Gamma_{|T|} \neq T\} > 0$ and set

$$\tau = \begin{cases} \sup\{n \in \mathbb{N}: \Gamma_n \leq T\} & \text{on } \{T < \bowtie\}, \\ \infty & \text{on } \{T = \bowtie\}. \end{cases}$$

Then

$$T > \Gamma_\tau \quad \text{and} \quad T \ngeq \Gamma_{\tau+1} \qquad \text{on } \{\Gamma_{|T|} \neq T\},$$

therefore there exist $t \in \mathbb{I}$, i and n such that

$$P\{T > t, T \ngeq t + e_i, \Gamma_n = t, \Gamma_{n+1} = t + e_i\} > 0.$$

By (5.2),

$$\{\Gamma_n = t, \Gamma_{n+1} = t + e_i\} \subset \{P(T > t, T^i = t^i | \mathscr{F}_t) = 0\}.$$

On the other hand,

$$\{T > t, T \ngeq t + e_i\} = \{T > t, T^i = t^i\}.$$

Consequently,

$$P\{T > t, T^i = t^i, P(T > t, T^i = t^i | \mathscr{F}_t) = 0\} > 0,$$

which is impossible because $F \subset \{P(F | \mathscr{F}_t) > 0\}$ for all $F \in \mathscr{F}$. Thus $P\{\Gamma_{|T|} \neq T\} = 0$; in other words, Γ passes through T.

Assume now that there exists a predictable increasing path Γ passing through T and show that (5.1) is satisfied for all $t \in \mathbb{I}$. Assume the contrary, that is, there exists $t \in \mathbb{I}$ such that $\mathrm{P}(F_t) > 0$, where F_t denotes the left-hand side of (5.1). Since

$$\bigcup_{n \le |t|} \{\Gamma_n \le t, \Gamma_{n+1} \not\le t\} = \Omega,$$

we can choose $n \le |t|$ such that

$$\mathrm{P}(F_t \cap \{\Gamma_n \le t, \Gamma_{n+1} \not\le t\}) > 0.$$

We thus see that there exist $s \in \mathbb{I}$ and i such that

$$s \le t, \qquad s + e_i \not\le t \tag{5.3}$$

and

$$\mathrm{P}(F_t \cap \{\Gamma_n = s, \Gamma_{n+1} = s + e_i\}) > 0.$$

Because

$$F_t \subset \{\mathrm{P}(T > t, T^i = t^i | \mathscr{F}_t) > 0\},$$

it follows that

$$\mathrm{P}\{\mathrm{P}(T > t, T^i = t^i | \mathscr{F}_t) > 0, \Gamma_n = s, \Gamma_{n+1} = s + e_i\} > 0.$$

But

$$\{\Gamma_n = s, \Gamma_{n+1} = s + e_i\} \in \mathscr{F}_s \subset \mathscr{F}_t;$$

therefore

$$\mathrm{P}\{T > t, T^i = t^i, \Gamma_{n+1} = s + e_i\} > 0. \tag{5.4}$$

However, $t^i = s^i$ by (5.3); therefore

$$\{T > t, T^i = t^i\} \subset \{T > t, T \not\ge s + e_i\}.$$

Using (5.4), it now follows that

$$\mathrm{P}\{T > t, T \not\ge \Gamma_{n+1}\} > 0,$$

which is impossible since $n \le |t|$ and $\Gamma_{|T|} = T$. Condition (5.1) is therefore satisfied for all $t \in \mathbb{I}$.

Theorem 2 (Criterion for the accessibility of all stopping points). *In order that all stopping points be accessible, it is necessary and sufficient that the following condition be satisfied*:

A. *For all j such that $1 < j \leq d$, all choices $t_1, \ldots, t_j$ of elements of $\mathbb{I}$ (or even only all choices of pairwise incomparable elements $t_1, \ldots, t_j$ of $\mathbb{I}$) and all choices of pairwise disjoint sets $F_1 \in \mathcal{F}_{t_1}, \ldots, F_j \in \mathcal{F}_{t_j}$,*

$$\bigcap_{i=1}^{j} \left\{ \mathrm{P}\left(F_i | \mathcal{F}_{t_1 \wedge \ldots \wedge t_j}\right) > 0 \right\} = \varnothing. \tag{5.5}$$

Proof. Assume that the condition is satisfied and consider an arbitrary stopping point T. Fix $t \in \mathbb{I}$ and observe that

$$\begin{aligned}\bigcap_{i=1}^{d} \{\mathrm{P}(T > t, T^i = t^i | \mathcal{F}_t) > 0\} &= \bigcap_{i=1}^{d} \bigcup_{u \in \mathbb{I}:\, u > t,\, u^i = t^i} \{\mathrm{P}(T = u | \mathcal{F}_t) > 0\} \\ &= \bigcup_{\substack{(u_1, \ldots, u_d) \in \mathbb{I}^d \\ u_1 > t, \ldots, u_d > t \\ u_1^1 = t^1, \ldots, u_d^d = t^d}} \bigcap_{i=1}^{d} \{\mathrm{P}(T = u_i | \mathcal{F}_t) > 0\}.\end{aligned}$$

Choose an arbitrary d-tuple $(u_1, \ldots, u_d)$ that satisfies the conditions in this union and let $t_1, \ldots, t_j$ denote the (necessarily pairwise incomparable) minimal elements of the set of values taken by $u_1, \ldots, u_d$. Because $u_1 \wedge \ldots \wedge u_d = t$, it follows that $t_1 \wedge \ldots \wedge t_j = t$. Since $u_i > t$ for $i = 1, \ldots, d$, j is necessarily > 1. By (5.5), because the sets $\{T = t_1\}, \ldots, \{T = t_j\}$ are pairwise disjoint,

$$\bigcap_{i=1}^{j} \{\mathrm{P}(T = t_i | \mathcal{F}_t) > 0\} = \varnothing.$$

But since

$$\bigcap_{i=1}^{d} \{\mathrm{P}(T = u_i | \mathcal{F}_t) > 0\} \subset \bigcap_{i=1}^{j} \{\mathrm{P}(T = t_i | \mathcal{F}_t) > 0\}$$

and $(u_1, \ldots, u_d)$ has been chosen arbitrarily, it follows that

$$\bigcap_{i=1}^{d} \{\mathrm{P}(T > t, T^i = t^i | \mathcal{F}_t) > 0\} = \varnothing.$$

This is true for all t, so Theorem 1 implies that T is accessible.

Assume now that Condition A is not satisfied; that is, there exists a set $\{t_1, \ldots, t_j\}$ $(j > 1)$ of elements of $\mathbb{I}$ and pairwise disjoint sets $F_1 \in \mathcal{F}_{t_1}, \ldots, F_j \in \mathcal{F}_{t_j}$ such that

$$\bigcap_{i=1}^{j} \{P(F_i | \mathcal{F}_t) > 0\} \neq \varnothing,$$

where $t = t_1 \wedge \ldots \wedge t_j$. Observe that $t_i > t$ for $i = 1, \ldots, j$, because if there were an index i such that $t_i = t$, then the intersection above would be empty. Indeed, if $k \neq i$, then

$$\begin{aligned}\{P(F_i | \mathcal{F}_t) > 0\} \cap \{P(F_k | \mathcal{F}_t) > 0\} &= F_i \cap \{P(F_k | \mathcal{F}_t) > 0\} \\ &= \{P(F_i \cap F_k | \mathcal{F}_t) > 0\} = \varnothing.\end{aligned}$$

Set

$$T = \begin{cases} t_i & \text{on } F_i, \\ t_1 \vee \cdots \vee t_j & \text{on } \left(\bigcup_{i=1}^{j} F_i\right)^c. \end{cases}$$

Clearly, T is a stopping point and $P\{T > t\} = 1$. We are going to show that T is not accessible. Because $t^i = \inf(t_1^i, \ldots, t_j^i)$ for $i = 1, \ldots, j$, there exists an index $m(i)$ such that $t^i = t_{m(i)}^i$. Thus,

$$\begin{aligned}\bigcap_{i=1}^{d} \{P(T > t, T^i = t^i | \mathcal{F}_t) > 0\} &\supset \bigcap_{i=1}^{d} \{P(T^i = t_{m(i)}^i | \mathcal{F}_t) > 0\} \\ &\supset \bigcap_{i=1}^{d} \{P(F_{m(i)} | \mathcal{F}_t) > 0\} \\ &\supset \bigcap_{i=1}^{j} \{P(F_i | \mathcal{F}_t) > 0\}.\end{aligned}$$

The intersection on the last right-hand side is nonempty; hence the intersection on the left-hand side is also nonempty. It remains to apply Theorem 1 to conclude that T is not accessible.

5.2 CONSEQUENCES FOR THE STRUCTURE OF THE FILTRATION

It turns out that in dimension $d > 2$, Condition A is so restrictive that it is almost never satisfied. Even in the case where $(\mathcal{F}_t)$ is the natural filtration of a process of independent random variables, this condition is generally unfulfilled, as we saw in the example of Section 1.4 where the existence of inaccessible stopping points was pointed out. As we will see in the next section, the situation is different in dimension $d = 2$, where Condition A seems to be more natural and easier to work with. In this section, we intend to analyze in some depth the consequences of imposing Condition A in dimension $d > 2$. In fact, in order to avoid complicated explanations, we will limit ourselves to the case where $d = 3$, but the general case could be treated using similar arguments.

Examples of Filtrations Satisfying Condition A

Let us first give two examples of tridimensional filtrations that satisfy Condition A.

Example 1. Let $(\mathscr{G}_u)$ be a bidimensional filtration for which Condition A is satisfied (with $d = 2$). Moreover, let φ be a map from $\mathbb{N}^3$ into $\mathbb{N}^2$ such that $\varphi(s \wedge t) = \varphi(s) \wedge \varphi(t)$ for all couples (s, t) of elements of $\mathbb{N}^3$ (for example, $\varphi(t^1, t^2, t^3) = (t^1 \wedge t^2, t^3)$). For all $t \in \mathbb{N}^3$, set $\mathscr{F}_t = \mathscr{G}_{\varphi(t)}$. It is easy to see that the tridimensional filtration $(\mathscr{F}_t)$ thus defined satisfies Condition A (with $d = 3$). With the help of the results given in Section 5.4, it can be shown that from a combinatorial point of view, this filtration is indistinguishable from a bidimensional filtration.

Example 2. Set $\Omega = \{\omega_1, \omega_2, \omega_3, \omega_4\}$, $\mathscr{F} = \mathscr{P}(\Omega)$, $\mathrm{P}(\{\omega_1\}) = \mathrm{P}(\{\omega_2\}) = \mathrm{P}(\{\omega_3\}) = \mathrm{P}(\{\omega_4\}) = \frac{1}{4}$ and

$$\mathscr{F}_t = \begin{cases} \{\varnothing, \Omega\}, & \text{if } t = ne_i, \quad n \in \mathbb{N}, \quad i = 1,2,3, \\ \{\varnothing, \{\omega_1, \omega_2\}, \{\omega_3, \omega_4\}, \Omega\}, & \text{if } t^1, t^3 > 0 \quad \text{and} \quad t^2 = 0, \\ \{\varnothing, \{\omega_1, \omega_3\}, \{\omega_2, \omega_4\}, \Omega\}, & \text{if } t^1, t^2 > 0 \quad \text{and} \quad t^3 = 0, \\ \{\varnothing, \{\omega_1, \omega_4\}, \{\omega_2, \omega_3\}, \Omega\}, & \text{if } t^2, t^3 > 0 \quad \text{and} \quad t^1 = 0, \\ \mathscr{P}(\Omega), & \text{if } t \gg 0. \end{cases}$$

It is easy to check that the filtration $(\mathscr{F}_t)$ thus defined satisfies Condition A (with $d = 3$). It is also possible to show that this filtration cannot be obtained by a transformation of indices from a bidimensional filtration satisfying Condition A (cf. Exercise 4).

Effect of Condition A on the Growth of the Sigma Fields in the Filtration

Looking at the examples above, we see that the sigma fields of the filtration never increase simultaneously in the three directions defined by e_1, e_2, e_3. The theorem we are about to establish confirms this observation and shows that if Condition A is fulfilled, then the filtration is essentially bidimensional.

If F belongs to $\mathscr{F}$ and $\mathscr{G}$ is a sub-sigma field of $\mathscr{F}$, we denote by $\mathscr{G} \cap F$ the trace of $\mathscr{G}$ on F, that is, the sigma field generated by the sets of the form $G \cap F$, where G belongs to $\mathscr{G}$.

Theorem (Local bidimensional increase under Condition A). *Let $d = 3$ and assume that the filtration $(\mathscr{F}_t)$ satisfies Condition* A. *Then for all $t \in \mathbb{I}$, there exists a partition (F_1, F_2, F_3) of Ω consisting of elements of $\mathscr{F}_t$ such that*

$$\mathscr{F}_{t+ne_i} \cap F_i = \mathscr{F}_t \cap F_i \tag{5.6}$$

for all $n \in \mathbb{N}$ and $i = 1, 2, 3$.

This theorem states in particular that on F_i, the filtration does not increase in direction i, but can only increase in the other two directions. Of course, the directions of increase depend on $\omega \in \Omega$ and $t \in \mathbb{I}$. The proof will be preceded by four lemmas that present the principal properties of an operator that quantifies, so to speak, the measurability gap between a set in $\mathscr{F}$ and a given sub-sigma field of $\mathscr{F}$. If $\mathscr{G}$ is this sub-sigma field, the operator in question is the map V from $\mathscr{F}$ into $\mathscr{G}$ defined by

$$\begin{aligned} V(F) &= \{\mathrm{P}(F|\mathscr{G}) > 0\} \cap \{\mathrm{P}(F^c|\mathscr{G}) > 0\} \\ &= \{\mathrm{P}(\{\mathrm{P}(F|\mathscr{G}) > 0\} \cap F^c|\mathscr{G}) > 0\}. \end{aligned} \tag{5.7}$$

Lemma 1.

(a) $\mathrm{P}(V(F)) = 0$ *if and only if* $F \in \mathscr{G}$.

(b) $V(F) = \{\mathrm{P}(F \cap V(F)|\mathscr{G}) > 0\}$.

(c) $V(F \cap G) = V(F) \cap G, \qquad$ *for all* $G \in \mathscr{G}$.

(d) $V(F) = V(F \cap V(F))$.

(e) $V(F) = V(\{\mathrm{P}(F|\mathscr{G}) > 0\} \cap F^c)$.

Proof. Clearly, $\mathrm{P}(V(F)) = 0$ if $F \in \mathscr{G}$. Conversely, if $\mathrm{P}(V(F)) = 0$, then $\{\mathrm{P}(F|\mathscr{G}) > 0\} \subset F$ by (5.7), and therefore $\{\mathrm{P}(F|\mathscr{G}) > 0\} = F$ because the converse inclusion always holds. It follows that $F \in \mathscr{G}$, which establishes assertion (a). Concerning assertion (b), observe that $V(F) \in \mathscr{G}$ and therefore

$$\{\mathrm{P}(F \cap V(F)|\mathscr{G}) > 0\} = \{\mathrm{P}(F|\mathscr{G}) > 0\} \cap V(F) = V(F).$$

Assertion (c) follows from the equalities

$$\begin{aligned} V(F \cap G) &= \{P(F \cap G|\mathscr{G}) > 0\} \cap \{P(F^c \cup G^c|\mathscr{G}) > 0\} \\ &= \{P(F|\mathscr{G}) > 0\} \cap G \cap (\{P(F^c|\mathscr{G}) > 0\} \cup G^c) \\ &= \{P(F|\mathscr{G}) > 0\} \cap \{P(F^c|\mathscr{G}) > 0\} \cap G \\ &= V(F) \cap G. \end{aligned}$$

As to the equalities in (d) and (e), they follow immediately from the assertions just established.

Lemma 2. *If $(F_n,\ n \in \mathbb{N})$ is a sequence of elements of $\mathscr{F}$ such that $\{P(F_m|\mathscr{G}) > 0\} \cap F_n = \varnothing$ (or, equivalently, such that $\{P(F_m|\mathscr{G}) > 0\} \cap \{P(F_n|\mathscr{G}) > 0\} = \varnothing$) for all couples (m, n) with $m \neq n$, then*

$$V\Big(\bigcup_n F_n\Big) = \bigcup_n V(F_n).$$

Proof. The conclusion results from the equalities

$$\begin{aligned} V\Big(\bigcup_n F_n\Big) &= \left\{P\left(\left\{P\Big(\bigcup_n F_n\Big|\mathscr{G}\Big) > 0\right\} \cap \Big(\bigcup_n F_n\Big)^c \Big|\mathscr{G}\right) > 0\right\} \\ &= \left\{P\left(\Big(\bigcup_n \{P(F_n|\mathscr{G}) > 0\}\Big) \cap \Big(\bigcap_n F_n^c\Big)\Big|\mathscr{G}\right) > 0\right\} \\ &= \left\{P\left(\bigcup_n (\{P(F_n|\mathscr{G}) > 0\} \cap F_n^c)\Big|\mathscr{G}\right) > 0\right\} \\ &= \bigcup_n \{P(\{P(F_n|\mathscr{G}) > 0\} \cap F_n^c|\mathscr{G}) > 0\} \\ &= \bigcup_n V(F_n), \end{aligned}$$

where the third step is justified by the condition in the lemma.

Lemma 3. *Let $H = F_2 \cap V(F_1)^c$, where F_1 and F_2 are sets in $\mathscr{F}$. Then*

$$V(F_1) \cup V(F_2) = V(F_1) \cup V(H) = V\big((F_1 \cap V(F_1)) \cup \big(F_2 \cap V(F_1)^c\big)\big).$$

Proof. By (c) of Lemma 1, $V(H) = V(F_2) \cap V(F_1)^c$, from which the first equality in the conclusion follows. Concerning the second equality, observe that

$$\{P(F_1 \cap V(F_1)|\mathscr{G}) > 0\} \cap \{P(F_2 \cap V(F_1)^c|\mathscr{G}) > 0\} \subset V(F_1) \cap V(F_1)^c$$
$$= \varnothing,$$

which allows us to conclude, thanks to Lemma 2 and to (d) of Lemma 1, that

$$V((F_1 \cap V(F_1)) \cup (F_2 \cap V(F_1)^c) = V(F_1) \cup V(F_2 \cap V(F_1)^c)$$
$$= V(F_1) \cup V(H).$$

Lemma 4. *Set* $G = \operatorname{esssup}_{F \in \mathscr{F}} V(F)$. *Then* $\mathscr{F} \cap G^c = \mathscr{G} \cap G^c$ *and there exists a set* $F^* \in \mathscr{F}$ *such that* $V(F^*) = G$.

This lemma shows in particular that G^c is the maximal subset of Ω on which $\mathscr{F}$ and $\mathscr{G}$ coincide.

Proof of Lemma 4

The first assertion is true because the existence of an $F \subset G^c$ such that $F \in \mathscr{F} - \mathscr{G}$ would contradict the definition of G. Indeed, this would imply

$$V(F) \subset \{P(F|\mathscr{G}) > 0\} \subset G^c$$

and $P(V(F)) > 0$, by (a) of Lemma 1.

As for the second statement, observe by Lemma 3 that the family $(V(F), F \in \mathscr{F})$ is upwards-directed. Therefore, by the theorem of Section 1.2, there exists a sequence $(F_n, n \in \mathbb{N})$ of elements of $\mathscr{F}$ such that $V(F_n) \subset V(F_{n+1})$ for all n and

$$\bigcup_n V(F_n) = G.$$

Set $\tilde{F}_n = F_n \cap V(F_n)$ and

$$H_n = \begin{cases} \varnothing, & \text{if } n = 0, \\ \tilde{F}_n \cap V(\tilde{F}_{n-1})^c, & \text{if } n > 0. \end{cases}$$

By (d) of Lemma 1, $V(\tilde{F}_n) = V(F_n)$, and by (b) of that lemma,

$$\{P(H_n|\mathscr{G}) > 0\} \subset \{P(\tilde{F}_n|\mathscr{G}) > 0\} = V(F_n).$$

Consequently, if $m < n$, then

$$\{\mathrm{P}(H_m|\mathscr{G}) > 0\} \subset V(\tilde{F}_{n-1}),$$

which implies that

$$\{\mathrm{P}(H_m|\mathscr{G}) > 0\} \cap \{\mathrm{P}(H_n|\mathscr{G}) > 0\} \subset V(\tilde{F}_{n-1}) \cap V(\tilde{F}_{n-1})^c = \varnothing.$$

It follows by Lemma 2 that

$$V\left(\bigcup_n H_n\right) = \bigcup_n V(H_n).$$

On the other hand, by (c) of Lemma 1, if $n > 0$, then

$$V(H_n) = V(\tilde{F}_n) \cap V(\tilde{F}_{n-1})^c = V(F_n) \cap V(F_{n-1})^c.$$

Consequently,

$$V\left(\bigcup_n H_n\right) = \bigcup_n V(F_n).$$

It is therefore sufficient to set $F^* = \bigcup_n H_n$.

Proof of the Theorem

Fix $t \in \mathbb{I}$ and define $V(F)$ using (5.7), but with $\mathscr{G}$ replaced by $\mathscr{F}_t$. For all $n \in \mathbb{N}$ and $i = 1, 2, 3$, define $F_{i,n}$ by

$$F_{i,n}^c = \operatorname{esssup}_{F \in \mathscr{F}_{t+ne_i}} V(F).$$

Clearly, $F_{i,n} \in \mathscr{F}_t$ and $F_{i,n} \supset F_{i,n+1}$. Moreover, by Lemma 4,

$$\mathscr{F}_{t+ne_i} \cap F_{i,n} = \mathscr{F}_t \cap F_{i,n}.$$

For $i = 1, 2, 3$, set $F_i = \bigcap_n F_{i,n}$. Clearly, (5.6) occurs on F_i. It remains therefore to prove that $\mathrm{P}(F_1 \cup F_2 \cup F_3) = 1$, since it is easy to make F_1, F_2, and F_3 pairwise disjoint. Suppose that $\mathrm{P}(F_1 \cup F_2 \cup F_3) < 1$. Then for sufficiently large n, $\mathrm{P}(F_{1,n} \cup F_{2,n} \cup F_{3,n}) < 1$, or equivalently $\mathrm{P}(F_{1,n}^c \cap F_{2,n}^c \cap F_{3,n}^c) > 0$. Fix such an n. By Lemma 4 (applied to the case where $\mathscr{F}_{t+ne_i}$ replaces $\mathscr{F}$) and (e) of Lemma 1, there exists $F_i^* \in \mathscr{F}_{t+ne_i}$ such that

$$F_{i,n}^c = V(F_i^*) = V(\{\mathrm{P}(F_i^* | \mathscr{F}_t) > 0\} \cap (F_i^*)^c).$$

For $i = 1, 2, 3$, set

$$H_1 = F_2^* \cap \{\mathrm{P}(F_3^* | \mathscr{F}_t) > 0\} \cap (F_3^*)^c,$$
$$H_2 = F_3^* \cap \{\mathrm{P}(F_1^* | \mathscr{F}_t) > 0\} \cap (F_1^*)^c,$$
$$H_3 = F_1^* \cap \{\mathrm{P}(F_2^* | \mathscr{F}_t) > 0\} \cap (F_2^*)^c.$$

Clearly H_1, H_2, and H_3 are pairwise disjoint and $H_1 \in \mathscr{F}_{t+ne_2+ne_3}$, $H_2 \in \mathscr{F}_{t+ne_1+ne_3}$, $H_3 \in \mathscr{F}_{t+ne_1+ne_2}$. On the other hand, by Condition A and taking into account the theorem and Lemma 1 of Section 5.3 below as well as (5.7), if (i, j, k) is a cyclic permutation of $(1, 2, 3)$, then

$$\begin{aligned}\{\mathrm{P}(H_i | \mathscr{F}_t) > 0\} &= \{\mathrm{P}(F_j^* | \mathscr{F}_t) > 0\} \cap \{\mathrm{P}(\{\mathrm{P}(F_k^* | \mathscr{F}_t) > 0\} \cap (F_k^*)^c | \mathscr{F}_t) > 0\} \\ &\supset V(F_j^*) \cap V(F_k^*) \\ &= F_{j,n}^c \cap F_{k,n}^c.\end{aligned}$$

Therefore,

$$\bigcap_{i=1}^{3} \{\mathrm{P}(H_i | \mathscr{F}_t) > 0\} \supset F_{1,n}^c \cap F_{2,n}^c \cap F_{3,n}^c.$$

Because the set on the right-hand side has positive probability, this inclusion contradicts (5.5). It is thus proved that $\mathrm{P}(F_1 \cup F_2 \cup F_3) = 1$.

Remark

If the filtration $(\mathscr{F}_t)$ satisfies Condition A, then it is clear that for all $t \in \mathbb{I}$ and all choices of $\{i, j\} \subset \{1, 2, 3\}$, the subfiltration $(\mathscr{F}_{t+me_i+ne_j}, (m, n) \in \mathbb{N}^2)$ satisfies Condition A with $d = 2$. Conversely, it can be proved that if this conclusion and that of the theorem are valid, then $(\mathscr{F}_t)$ satisfies Condition A (cf. Exercise 1).

5.3 THE BIDIMENSIONAL CASE

In this section, we shall assume that $d = 2$. We are going to show that Condition A stipulates a sort of conditional independence that is satisfied in a variety of contexts.

Conditional Qualitative Independence

We say that two sub-sigma fields $\mathscr{F}_1$ and $\mathscr{F}_2$ of $\mathscr{F}$ are *conditionally qualitatively independent* with respect to a sub-sigma field $\mathscr{G}$ of $\mathscr{F}$ if

$$\{\mathrm{P}(F_1 | \mathscr{G}) > 0\} \cap \{\mathrm{P}(F_2 | \mathscr{G}) > 0\} = \{\mathrm{P}(F_1 \cap F_2 | \mathscr{G}) > 0\} \tag{5.8}$$

for all choices of sets $F_1 \in \mathscr{F}_1$ and $F_2 \in \mathscr{F}_2$.

Clearly, conditional independence of $\mathscr{F}_1$ and $\mathscr{F}_2$ with respect to $\mathscr{G}$ implies conditional qualitative independence of these sigma fields with respect to $\mathscr{G}$.

We say that the filtration $(\mathscr{F}_t)$ satisfies the *conditional qualitative independence* [*conditional independence*] *property* if for all couples (t_1, t_2) of elements of $\mathbb{I}$ (or even only for all couples of incomparable elements of $\mathbb{I}$), the sigma fields $\mathscr{F}_{t_1}$ and $\mathscr{F}_{t_2}$ are conditionally qualitatively independent [conditionally independent] with respect to the sigma field $\mathscr{F}_{t_1 \wedge t_2}$. This property will be denoted by the symbol CQI [CI].

Clearly, CI implies CQI. However, property CQI is unaffected if P is replaced by some other probability measure with the same null sets as P, whereas such a substitution may affect property CI.

It is obvious that if CQI [CI] is satisfied, then for all couples (t_1, t_2) of elements of $\mathbb{I}$, the sigma fields $\mathscr{F}_{t_1}$ and $\mathscr{F}_{t_2}$ are conditionally qualitatively independent [conditionally independent] with respect to $\mathscr{F}_t$, for any $t \in \mathbb{I}$ such that $t \geq t_1 \wedge t_2$. The two typical situations where CI holds are the following.

1. Let $(\mathscr{G}_t)$ be a family of independent sub-sigma-fields of $\mathscr{F}$. If $\mathscr{F}_t$ is generated by $\bigcup_{s \leq t} \mathscr{G}_s$ for all t, then the filtration $(\mathscr{F}_t)$ satisfies CI.
2. Suppose that $(\Omega, \mathscr{F}, \mathrm{P})$ is the product of two probability spaces $(\Omega_i, \mathscr{F}_i, \mathrm{P}_i)$, $i = 1, 2$. Suppose, moreover, that each of these two spaces is equipped with a filtration $(\mathscr{F}^i_{t^i})$, $i = 1, 2$. If $\mathscr{F}_t = \mathscr{F}^1_{t^1} \times \mathscr{F}^2_{t^2}$ for all t, then the filtration $(\mathscr{F}_t)$ satisfies CI. This type of filtration is termed a *product filtration*.

The main motivation for CQI is contained in the following theorem.

Theorem (Equivalence of A and CQI in dimension 2). *The filtration* $(\mathscr{F}_t)$ *satisfies Condition* A *if and only if* CQI *is satisfied. In other words, all stopping points are accessible if and only if* CQI *is satisfied.*

Proof. Bringing together Condition A and CQI, we see that the theorem is a direct consequence of the following lemma.

Lemma 1. *Let* $\mathscr{F}_1$, $\mathscr{F}_2$, *and* $\mathscr{G}$ *be sub-sigma fields of* $\mathscr{F}$ *such that* $\mathscr{G} \subset \mathscr{F}_1 \cap \mathscr{F}_2$. *In order that* $\mathscr{F}_1$ *and* $\mathscr{F}_2$ *be conditionally qualitatively independent with respect to* $\mathscr{G}$, *it is necessary and sufficient that*

$$\{\mathrm{P}(F_1|\mathscr{G}) > 0\} \cap \{\mathrm{P}(F_2|\mathscr{G}) > 0\} = \varnothing$$

for all choices of disjoint sets $F_1 \in \mathscr{F}_1$ *and* $F_2 \in \mathscr{F}_2$.

Proof. The condition of this lemma is clearly necessary. In order to show that it is sufficient, consider two arbitrary sets $F_1 \in \mathscr{F}_1$ and $F_2 \in \mathscr{F}_2$. Set

$$H_1 = F_1 \cap \{P(F_1 \cap F_2|\mathscr{G}) = 0\} \quad \text{and} \quad H_2 = F_2 \cap \{P(F_1 \cap F_2|\mathscr{G}) = 0\}.$$

Since $\mathscr{G} \subset \mathscr{F}_1 \cap \mathscr{F}_2$, it is clear that $H_1 \in \mathscr{F}_1$ and $H_2 \in \mathscr{F}_2$. Moreover,

$$H_1 \cap H_2 = F_1 \cap F_2 \cap \{P(F_1 \cap F_2|\mathscr{G}) = 0\} = \varnothing,$$

since $F_1 \cap F_2 \subset \{P(F_1 \cap F_2|\mathscr{G}) > 0\}$. Consequently,

$$\{P(H_1|\mathscr{G}) > 0\} \cap \{P(H_2|\mathscr{G}) > 0\} = \varnothing,$$

which implies

$$\{P(F_1|\mathscr{G}) > 0\} \cap \{P(F_2|\mathscr{G}) > 0\} \cap \{P(F_1 \cap F_2|\mathscr{G}) = 0\} = \varnothing,$$

and therefore

$$\{P(F_1|\mathscr{G}) > 0\} \cap \{P(F_2|\mathscr{G}) > 0\} \subset \{P(F_1 \cap F_2|\mathscr{G}) > 0\}.$$

Because the converse inclusion is obvious, (5.8) follows.

Stability with Respect to Intersection

The filtration $(\mathscr{F}_t)$ is *stable with respect to intersection* provided

$$\mathscr{F}_{t_1} \cap \mathscr{F}_{t_2} = \mathscr{F}_{t_1 \wedge t_2} \tag{5.9}$$

for all couples (t_1, t_2) of elements of $\mathbb{I}$.

Proposition 1 (Stability under CQI in dimension 2). *If the filtration* $(\mathscr{F}_t)$ *satisfies* CQI, *then it is stable with respect to intersection.*

Proof. Fix t_1 and t_2 and observe that the inclusion $\mathscr{F}_{t_1} \cap \mathscr{F}_{t_2} \supset \mathscr{F}_{t_1 \wedge t_2}$ is always true. The converse inclusion follows directly from the following lemma.

Lemma 2. *Let $\mathscr{F}_1$, $\mathscr{F}_2$, and $\mathscr{G}$ be sub-sigma fields of $\mathscr{F}$. If $\mathscr{F}_1$ and $\mathscr{F}_2$ are conditionally qualitatively independent with respect to $\mathscr{G}$, then $\mathscr{F}_1 \cap \mathscr{F}_2 \subset \mathscr{G}$.*

Proof. Let $F \in \mathscr{F}_1 \cap \mathscr{F}_2$. Then $F \in \mathscr{F}_1$ and $F^c \in \mathscr{F}_2$; therefore, using the hypothesis, we see that

$$\{P(F|\mathscr{G}) > 0\} \cap \{P(F^c|\mathscr{G}) > 0\} = \{P(F \cap F^c|\mathscr{G}) > 0\} = \varnothing.$$

It follows that

$$\{P(F|\mathscr{G}) > 0\} \subset \{P(F^c|\mathscr{G}) = 0\}.$$

Because $F \subset \{P(F|\mathscr{G}) > 0\}$ and $F^c \subset \{P(F^c|\mathscr{G}) > 0\}$, both sides of the above inclusion contain F and are contained in F. It follows that $F = \{P(F|\mathscr{G}) > 0\}$, and therefore $F \in \mathscr{G}$.

CQI and the Stability Property Restricted to Direct Successors

Property CQI and the property of stability with respect to intersection can be formulated using only the couples (t_1, t_2) for which t_1 and t_2 are direct successors of $t_1 \wedge t_2$.

Proposition 2 (Criteria for CQI and for stability with respect to intersection in dimension 2). *If for all $t \in \mathbb{I}$, the sigma fields $\mathscr{F}_{t+e_1}$ and $\mathscr{F}_{t+e_2}$ are conditionally qualitatively independent with respect to the sigma field $\mathscr{F}_t$, then the filtration $(\mathscr{F}_t)$ satisfies* CQI. *If for all $t \in \mathbb{I}$, $\mathscr{F}_{t+e_1} \cap \mathscr{F}_{t+e_2} = \mathscr{F}_t$, then this filtration is stable with respect to intersection.*

Proof. We will only establish the second assertion, since the assertion concerning CQI follows from the same kind of argument (cf. Exercise 2). We must show that $\mathscr{F}_{t+me_1} \cap \mathscr{F}_{t+ne_2} = \mathscr{F}_t$ for all couples (m, n) of elements of $\mathbb{N}$. This is trivially true if $m = 0$ and is also true by hypothesis if $m = 1$ and $n = 1$. Moreover, by induction on m, we see that this equality is satisfied for all $m > 1$ if $n = 1$, because

$$\begin{aligned}\mathscr{F}_{t+me_1} \cap \mathscr{F}_{t+e_2} &= \mathscr{F}_{t+me_1} \cap \mathscr{F}_{t+(m-1)e_1+e_2} \cap \mathscr{F}_{t+e_2} \\ &= \mathscr{F}_{t+(m-1)e_1} \cap \mathscr{F}_{t+e_2} \\ &= \mathscr{F}_t,\end{aligned}$$

where the second step is justified by assumption and the last one by the induction hypothesis. Finally, by induction on n, we see that the relation to be established is true for all $m \geq 1$ and all $n \geq 1$, because

$$\begin{aligned}\mathscr{F}_{t+me_1} \cap \mathscr{F}_{t+ne_2} &= \mathscr{F}_{t+me_1} \cap \mathscr{F}_{t+me_1+(n-1)e_2} \cap \mathscr{F}_{t+ne_2} \\ &= \mathscr{F}_{t+me_1} \cap \mathscr{F}_{t+(n-1)e_2} \\ &= \mathscr{F}_t,\end{aligned}$$

where the second step relies on the particular case already established (with $t + (n - 1)e_2$ instead of t) and the last one on the induction hypothesis.

Two Examples

We give here an example which shows that CQI can occur when CI fails and another example that demonstrates that stability with respect to intersection can occur when CQI fails.

Example 1. Set $\Omega_1 = \Omega_2 = \{0, 1\}$, $\Omega = \Omega_1 \times \Omega_2$, $\mathcal{F} = \mathcal{P}(\Omega)$,

$$\mathrm{P}(\{\omega\}) = \begin{cases} \frac{1}{5}, & \text{if } \omega \neq (1,1), \\ \frac{2}{5}, & \text{if } \omega = (1,1), \end{cases}$$

and

$$\mathcal{F}_t = \begin{cases} \{\varnothing, \Omega\}, & \text{if } t = 0, \\ \mathcal{P}(\{0,1\}) \times \{\varnothing, \Omega_2\}, & \text{if } t \in \mathbb{N}^* \times \{0\}, \\ \{\varnothing, \Omega_1\} \times \mathcal{P}(\{0,1\}), & \text{if } t \in \{0\} \times \mathbb{N}^*, \\ \mathcal{F}, & \text{if } t \gg 0. \end{cases}$$

It is easy to check that CQI is satisfied but CI is not.

Example 2. Set $\Omega = \{0, 1, 2\}$, $\mathcal{F} = \mathcal{P}(\Omega)$, $\mathrm{P}(\{0\}) = \mathrm{P}(\{1\}) = \mathrm{P}(\{2\}) = \frac{1}{3}$ and

$$\mathcal{F}_t = \begin{cases} \{\varnothing, \Omega\}, & \text{if } t = 0, \\ \{\varnothing, \{0\}, \{1,2\}, \Omega\}, & \text{if } t \in \mathbb{N}^* \times \{0\}, \\ \{\varnothing, \{1\}, \{0,2\}, \Omega\}, & \text{if } t \in \{0\} \times \mathbb{N}^*, \\ \mathcal{F}, & \text{if } t \gg 0. \end{cases}$$

The property of intersection is clearly satisfied, whereas CQI fails because

$$\{\mathrm{P}(\{0\} \mid \mathcal{F}_0) > 0\} \cap \{\mathrm{P}(\{1\} \mid \mathcal{F}_0) > 0\} \neq \varnothing.$$

CQI in Higher Dimensions

The notion of conditional qualitative independence can be extended in a natural way to the case of $n > 2$ sub-sigma fields of $\mathcal{F}$. Thus extended, this notion can be used as a basis for an extension of CQI to dimensions $d > 2$. However, such an extension is of little interest since it gives rise to an even more restrictive property than Condition A. This can be seen from Example 2 of Section 5.2, where Condition A is satisfied but this tridimensional extension of CQI is not, because

$$\begin{aligned} \varnothing &= \{\mathrm{P}(\{\omega_1, \omega_2\} \cap \{\omega_2, \omega_4\} \cap \{\omega_1, \omega_4\} \mid \mathcal{F}_0) > 0\} \\ &\neq \{\mathrm{P}(\{\omega_1, \omega_2\} \mid \mathcal{F}_0) > 0\} \cap \{\mathrm{P}(\{\omega_2, \omega_4\} \mid \mathcal{F}_0) > 0\} \cap \{\mathrm{P}(\{\omega_1, \omega_4\} \mid \mathcal{F}_0) > 0\} \\ &= \Omega. \end{aligned}$$

On the other hand, the following extension of CQI is of interest and will be adopted in the following: The filtration $(\mathscr{F}_t)$ satisfies the *conditional qualitative independence property* (or satisfies CQI) if for all $t \in \mathbb{I}$ and all choices of distinct elements i and j of $\{1, \dots, d\}$, the sigma fields $\mathscr{F}_{t+e_i}$ and $\mathscr{F}_{t+e_j}$ are conditionally qualitatively independent relative to the sigma field $\mathscr{F}_t$. By Proposition 2, this property is equivalent in the bidimensional case to the one we have just formulated. However, for $d > 2$, it is clearly weaker than Condition A.

5.4 PREDICTABILITY OF OPTIONAL INCREASING PATHS

In this section, we are going to present various conditions that ensure that all optional increasing paths are predictable. These conditions will be used in particular in Chapter 8.

Proposition 1 (Necessary and sufficient condition). *In order that all optional increasing paths be predictable, it is necessary and sufficient that the filtration $(\mathscr{F}_t)$ satisfy the following condition: for all $t \in \mathbb{I}$ and all partitions $(F_1, \dots, F_d)$ of Ω such that $F_i \in \mathscr{F}_{t+e_i}$ for $i = 1, \dots, d$, the relation $F_i \in \mathscr{F}_t$ holds for $i = 1, \dots, d$.*

Proof. Assume that this condition does not hold; that is, there exists $t \in \mathbb{I}$ and a partition $(F_1, \dots, F_d)$ of Ω such that $F_i \in \mathscr{F}_{t+e_i}$ for $i = 1, \dots, d$, but at least one of the F_i, say F_{i_0}, does not belong to $\mathscr{F}_t$. Let γ be a deterministic increasing path passing through t and let $(\Gamma_n,\ n \in \mathbb{N})$ be the sequence defined by

$$\Gamma_n = \begin{cases} \gamma_n, & \text{if } n \leq |t|, \\ t + e_i & \text{on } F_i, \text{ if } n = |t| + 1, \\ \Gamma_{n-1} + e_1, & \text{if } n > |t| + 1. \end{cases}$$

It is easy to check that (Γ_n) is an optional increasing path. However, this path is not predictable, because

$$\{\Gamma_{|t|} = t,\ \Gamma_{|t|+1} = t + e_{i_0}\} = F_{i_0} \notin \mathscr{F}_t.$$

Assume now that the condition of the proposition holds and consider an optional increasing path Γ. In the case $d = 1$, there is nothing to prove, so we assume $d > 1$ and show that Γ is predictable. Let $t \in \mathbb{I}$ and $n \in \mathbb{N}$. Set $F_1 = \{\Gamma_n = t,\ \Gamma_{n+1} = t + e_1\} \cup \{\Gamma_n \neq t\}$, and for $i = 2, \dots, d$, set

$$F_i = \{\Gamma_n = t,\ \Gamma_{n+1} = t + e_i\}.$$

It is clear that $(F_1, \ldots, F_d)$ is a partition of Ω such that $F_i \in \mathscr{F}_{t+e_i}$ for $i = 1, \ldots, d$. By the condition of the proposition, $F_i \in \mathscr{F}_t$ for $i = 1, \ldots, d$. However, $\{\Gamma_n \neq t\}$ and $\{\Gamma_n = t, \Gamma_{n+1} = t + e_1\}$ are disjoint and $\{\Gamma_n \neq t\} \in \mathscr{F}_t$; therefore $\{\Gamma_n = t, \Gamma_{n+1} = t + e_i\} \in \mathscr{F}_t$ for $i = 1, \ldots, d$, which proves that Γ is predictable.

Proposition 2 (Sufficient condition). *If the filtration* $(\mathscr{F}_t)$ *satisfies* CQI, *then all optional increasing paths are predictable.*

Proof. It is sufficient to check that the condition of Proposition 1 holds. Let $t \in \mathbb{I}$ and $(F_1, \ldots, F_d)$ be a partition of Ω such that $F_i \in \mathscr{F}_{t+e_i}$ for $i = 1, \ldots, d$. For $i = 1, \ldots, d$, set

$$G_i = \{P(F_i | \mathscr{F}_t) > 0\}.$$

By the conditional qualitative independence assumption, if $i \neq j$, then

$$G_i \cap G_j = \{P(F_i \cap F_j | \mathscr{F}_t) > 0\} = \varnothing.$$

But $F_i \subset G_i$; therefore $\Omega = \bigcup_{i=1}^d F_i \subset \bigcup_{i=1}^d G_i = \Omega$. It follows that $F_i = G_i$ and therefore $F_i \in \mathscr{F}_t$ for $i = 1, \ldots, d$, which shows that the condition of Proposition 1 holds.

Proposition 3 (Predictability criterion in dimension 2). *If* $d = 2$, *then in order that all optional increasing paths be predictable, it is necessary and sufficient that the filtration* $(\mathscr{F}_t)$ *be stable with respect to intersection.*

Proof. Assume that $(\mathscr{F}_t)$ is not stable with respect to intersection. By Proposition 2 of Section 5.3, there exist $t \in \mathbb{I}$ and a set $F \in \mathscr{F}_{t+e_1} \cap \mathscr{F}_{t+e_2}$ such that $F \notin \mathscr{F}_t$. It suffices to set $F_1 = F$ and $F_2 = \Omega - F$ to see that the condition of Proposition 1 does not hold. By this proposition, it is not true that all optional increasing paths are predictable.

Now assume that $(\mathscr{F}_t)$ is stable with respect to intersection and show that the condition of Proposition 1 holds. Let $t \in \mathbb{I}$ and (F_1, F_2) be a partition of Ω such that $F_1 \in \mathscr{F}_{t+e_1}$ and $F_2 \in \mathscr{F}_{t+e_2}$. Then $F_1 = \Omega - F_2$ and $F_2 = \Omega - F_1$; therefore $F_1 \in \mathscr{F}_{t+e_2}$ and $F_2 \in \mathscr{F}_{t+e_1}$. Consequently, F_1 and F_2 belong to $\mathscr{F}_{t+e_1} \cap \mathscr{F}_{t+e_2} = \mathscr{F}_t$, which proves that the condition of Proposition 1 holds. By this proposition, all optional increasing paths are predictable.

5.5 THE COMBINATORIAL STRUCTURE OF A FILTRATION

In this and in the next two sections, we will study the combinatorial structure of a filtration under the assumption that the basic sigma field $\mathscr{F}$ contains only a finite number of sets with positive probability. Throughout the study, the null sets and the probability measure P only play a secondary role. We will therefore use as basic probability space a probability space $(\Omega, \mathscr{F}, \mathrm{P})$, where $\mathscr{F}$ is a finite algebra and P is such that the empty set is the only set with probability 0. The filtration will consist of subalgebras of $\mathscr{F}$, and because the set of these subalgebras is finite, there is no loss of generality in assuming that the index set of the filtrations under consideration is $\mathbb{I}_u$, where u denotes a nonzero element of $\mathbb{I}$.

Let $(\mathscr{F}_l,\ l \in \mathbb{L})$ be a finite family of subalgebras of $\mathscr{F}$. For all $l \in \mathbb{L}$, $\mathscr{F}_l$ is generated by a unique finite partition $\mathscr{P}_l = (F_1^l, \ldots, F_{n_l}^l)$ of Ω consisting of elements of $\mathscr{F}$. We will say that $\mathscr{P}_l$ is the *generating partition* of $\mathscr{F}_l$ and that $(\mathscr{P}_l,\ l \in \mathbb{L})$ is the *family of generating partitions associated with* $(\mathscr{F}_l,\ l \in \mathbb{L})$. From the point of view of combinatorial structure, the basic objects are the elements of $\bigcup_{l \in \mathbb{L}} \mathscr{P}_l$. These elements are related to each other by their mutual intersections, which will be studied using the notion of intersection graph that we will introduce following some definitions.

Graphs

A *graph*, which we denote G or $(V, \mathscr{E})$, is any couple consisting of a nonempty finite set V and a (possibly empty) set $\mathscr{E}$ of pairs of distinct elements of V (in the specialized literature, this is more precisely termed a finite graph without loops and without directed or multiple edges). Elements of V are termed *vertices* and elements of $\mathscr{E}$ *edges*. Two vertices v and w of a graph $(V, \mathscr{E})$ are *neighbors* if $\{v, w\} \in \mathscr{E}$. For $v \in V$, $N_{\mathsf{G}}(v)$ denotes the set of neighbors of v. A vertex v of a graph G is a *center* of G if $N_{\mathsf{G}}(v) = V - \{v\}$. If G has only one vertex, then this vertex is a center of G.

An *n-path* ($n > 1$) is any graph $(V, \mathscr{E})$ whose vertices can be ordered in a sequence $v_1, \ldots, v_n$ such that $\mathscr{E} = \{\{v_1, v_2\}, \ldots, \{v_{n-1}, v_n\}\}$. The vertices v_1 and v_n are the *extremities* of the n-path. An *n-cycle* ($n > 1$) is any graph $(V, \mathscr{E})$ whose vertices can be ordered in a sequence $v_1, \ldots, v_n$ such that $\mathscr{E} = \{\{v_1, v_2\}, \ldots, \{v_{n-1}, v_n\}, \{v_n, v_1\}\}$. When the value of n is not relevant, such a graph will be referred to simply as a *path* or a *cycle*.

A graph $(V', \mathscr{E}')$ is a *subgraph* of a graph $(V, \mathscr{E})$ if $V' \subset V$ and $\mathscr{E}' \subset \mathscr{E}$. If in addition, $\{v, w\} \in \mathscr{E}'$ for all couples (v, w) of distinct element of V' such that $\{v, w\} \in \mathscr{E}$, then $(V', \mathscr{E}')$ is an *induced subgraph* of $(V, \mathscr{E})$. In this case, we also say that $(V, \mathscr{E})$ *contains* $(V', \mathscr{E}')$. Obviously, any subset V' of V is the set of vertices of exactly one induced subgraph of $(V, \mathscr{E})$. This subgraph is called the *subgraph induced by* V'. A *clique* is a subgraph $(V', \mathscr{E}')$ of $(V, \mathscr{E})$ such that $\{v, w\} \in \mathscr{E}$ for all couples (v, w) of distinct elements of V'. In particular, a subgraph with a single vertex is a clique. A clique of $(V, \mathscr{E})$ is termed *maximal* if all cliques of $(V, \mathscr{E})$ that contain it are equal to it. Note that each vertex of a graph is a vertex of a maximal clique of this graph.

Two graphs $(V, \mathscr{E})$ and $(V', \mathscr{E}')$ are *isomorphic* if there exists a one-to-one map f from V onto V' such that $\{f(v), f(w)\} \in \mathscr{E}'$ if and only if $\{v, w\} \in \mathscr{E}$. In a graph G, the relation "$v = w$ or $v \neq w$ and G contains a path with extremities v and w" is an equivalence relation between vertices. The subgraphs induced by the equivalence classes of this relation are the *connected components* of the graph G. A graph is *connected* if it has only one connected component. It is clear that paths, cycles, and cliques are connected. A set of vertices of a graph (or its induced subgraph) is *totally disconnected* if each of its vertices is a connected component of its induced subgraph, in other words, does not have any neighbors in this subgraph.

The Intersection Graph of a Family of Algebras

Let $(\mathscr{F}_l,\ l \in \mathbb{L})$ be a finite family of subalgebras of $\mathscr{F}$ and $(\mathscr{P}_l,\ l \in \mathbb{L})$ the associated family of generating partitions. The *intersection graph* of $(\mathscr{F}_l,\ l \in \mathbb{L})$ is the graph whose vertices are the couples (l, F) with $l \in \mathbb{L}$ and $F \in \mathscr{P}_l$ and whose edges are the pairs $\{(l_1, F_1), (l_2, F_2)\}$ such that $F_1 \cap F_2 \neq \emptyset$.

The Intersection Graph of a Filtration

The combinatorial structure of a filtration can be conveniently studied with the help of the intersection graph of this filtration. For example, the subgraphs of intersection graphs of filtrations in dimension 1 and in dimension $d > 1$ can be characterized. It is also possible to use properties of the intersection graph to show how general bidimensional filtrations differ from bidimensional filtrations that satisfy CQI. We will return to this last question in the next section. The two theorems below show the difference between the combinatorial structure of filtrations in dimension 1 and in dimension $d > 1$.

Theorem 1 (Combinatorial structure of a filtration in dimension 1). *In order that a graph be isomorphic to an induced subgraph of the intersection graph of a one-dimensional filtration, it is necessary and sufficient that it contain no 4-path or 4-cycle.*

Theorem 2 (Combinatorial structure of a filtration in dimension 2). *Every graph is isomorphic to an induced subgraph of the intersection graph of a bidimensional filtration.*

Theorem 1 shows in particular that the combinatorial structure of a filtration in dimension 1 is quite simple: The intersection graph of such a filtration cannot contain graphs as simple as 4-paths or 4-cycles. Because all bidimensional filtrations can be interpreted as a filtration in dimension $d > 2$, Theorem 2 shows that from dimension 2 on, the combinatorial structure of a filtration is generally very complex. In fact, from a combinatorial point of view, filtrations in dimension 2 are as complex as filtrations in dimension $d > 2$.

Proof of Theorem 1

Necessity. We shall show that if $(V, \mathscr{E})$ is an induced subgraph of the intersection graph of a one-dimensional filtration $(\mathscr{F}_k, k \in \mathbb{N}_n)$, then $(V, \mathscr{E})$ is neither a 4-path nor a 4-cycle. Suppose the contrary. Then card $V = 4$ and the four vertices can be numbered in such a way that $\mathscr{E} = \{\{v_1, v_2\}, \{v_2, v_3\}, \{v_3, v_4\}\}$ or $\mathscr{E} = \{\{v_1, v_2\}, \{v_2, v_3\}, \{v_3, v_4\}, \{v_4, v_1\}\}$. Let $v_j = (k_j, F_j)$, $j = 1, \dots, 4$. Because $F_2 \cap F_3 \neq \emptyset$, necessarily $k_2 \neq k_3$. If $k_2 < k_3$, then $F_3 \subset F_2$; therefore $F_2 \cap F_4 \neq \emptyset$ because $F_3 \cap F_4 \neq \emptyset$. But then $\{v_2, v_4\} \in \mathscr{E}$, which contradicts the hypothesis. With the same argument, we can check that the inequality $k_3 < k_2$ leads to a contradiction.

Sufficiency. The proof relies on the following assertion, which we will establish first: *if a connected graph contains no 4-path or 4-cycle, then it has a center*.

Let $\mathsf{G} = (V, \mathscr{E})$ be such a graph. If card $V < 4$, either G has only one vertex, or G is a 2-path or a 3-path or a 3-cycle. In all four cases, the graph has a center. Assume now that card $V \geq 4$ and choose v_0 in such a way that card $N_{\mathsf{G}}(v_0) \geq$ card $N_{\mathsf{G}}(v)$ for all $v \in V$. Since G is connected, card $N_{\mathsf{G}}(v_0) \geq 2$. Assume that there exists a vertex $v_2 \neq v_0$ that is not a neighbor of v_0. Again by connexity, G contains a path with extremities v_0 and v_2. This path must be a 3-path, because G does not contain any 4-path. Consequently, there exists a vertex $v_1 \in N_{\mathsf{G}}(v_0)$ such that $\{v_1, v_2\} \in \mathscr{E}$. It follows that for all vertices $v \in N_{\mathsf{G}}(v_0) - \{v_1\}$, $\{\{v, v_0\}, \{v_0, v_1\}, \{v_1, v_2\}\} \subset \mathscr{E}$. Since $\{v_0, v_2\} \notin \mathscr{E}$ and G contains neither 4-paths nor 4-cycles, it is necessary that $\{v, v_1\} \in \mathscr{E}$. Consequently $N_{\mathsf{G}}(v_1) \supset \{v_0\} \cup (N_{\mathsf{G}}(v_0) - \{v_1\}) \cup \{v_2\}$; therefore card $N_{\mathsf{G}}(v_1) \geq$ card $N_{\mathsf{G}}(v_0) + 1$, which contradicts the maximal character of card $N_{\mathsf{G}}(v_0)$. It follows that v_0 is a center of G and the assertion is thus proved.

Now let G be a graph without any 4-paths or 4-cycles. All induced subgraphs of G are then also without 4-paths and 4-cycles, which means that if they are connected, then they have a center, by the assertion just established. This property will be used below without any further justification. By induction on k, we shall build connected induced subgraphs $\mathsf{G}_1^k, \dots, \mathsf{G}_{j_k}^k$ of G, no two of which have any vertices in common. At each step of the construction, we will choose $v_1^k, \dots, v_{j_k}^k$ such that v_j^k is a center of G_j^k for $j = 1, \dots, j_k$. For $k = 0$, the subgraphs in question are simply the connected components of G. The step from $k - 1$ to k is accomplished as follows. Let J be the set of all indices j such that G_j^{k-1} is reduced to its center. If $J = \{1, \dots, j_{k-1}\}$, then the construction stops. If $J \neq \{1, \dots, j_{k-1}\}$ and $j \in \{1, \dots, j_{k-1}\} - J$, we consider the subgraph of G_j^{k-1} induced by the vertices of G_j^{k-1} distinct from v_j^{k-1}. We then list the connected components of this subgraph, getting as many lists as there are elements of $\{1, \dots, j_{k-1}\} - J$. The subgraphs that make up these lists, to which we add the subgraphs (with a single vertex) G_j^{k-1} with index $j \in J$, form the finite sequence $\mathsf{G}_1^k, \dots, \mathsf{G}_{j_k}^k$. The construction we have just described stops after a finite number of steps, more precisely at the step where for the first time all the subgraphs $\mathsf{G}_1^k, \ldots, \mathsf{G}_{j_k}^k$ are reduced to their center. Assume that this occurs at stage n.

Set $\Omega = \{v_1^n, \ldots, v_{j_n}^n\}$ and $\mathscr{F} = \mathscr{P}(\Omega)$. For all $k \in \mathbb{N}_n$ and $j = 1, \ldots, j_k$, let $F_j^k = V_j^k \cap \Omega$, where V_j^k denotes the set of vertices of G_j^k. It is not difficult to check that $\mathscr{P}_k = (F_1^k, \ldots, F_{j_k}^k)$ is a partition of Ω and that if $k > 0$, then to each index j corresponds an index j' such that $F_j^k \subset F_{j'}^{k-1}$. Denoting by $\mathscr{F}_k$ the algebra generated by $\mathscr{P}_k$, we obtain therefore a one-dimensional filtration $(\mathscr{F}_k,\ k \in \mathbb{N}_n)$. Let V be the set of vertices of G and V' the set of vertices of the intersection graph of $(\mathscr{F}_k)$. By construction, if v belongs to V, then among the couples (j, k) such that $v = v_j^k$, there is exactly one whose second component k is smaller than all the others. Denote this couple by $\varphi(v)$ and consider the map f from V into V' defined by

$$f(v) = \left(k, F_j^k\right), \qquad \text{if } \varphi(v) = (j, k).$$

Because of the way the filtration has been constructed, f is a one-to-one map and $\{v, w\}$ is an edge of G if and only if $\{f(v), f(w)\}$ is an edge of the intersection graph of $(\mathscr{F}_k)$. In other words, G is isomorphic to an induced subgraph of this graph.

Proof of Theorem 2

The theorem is trivial in the case where the graph has only one vertex. Consider therefore a graph G with at least two vertices. Let $V = \{v_1, \ldots, v_n\}$ the set of its vertices and $K_1, \ldots, K_m$ be the subsets of V that induce the maximal cliques of G. Starting with these objects, we shall construct for all couples of indices (j, k) such that $1 \le j \le m$ and $1 \le k \le n$, a graph G_j^k and two sets D_j^k and K_j^k of vertices of G_j^k such that the first is totally disconnected and the second induces a maximal clique of G_j^k. The construction is done by induction. First we set $\mathsf{G}_0^1 = \mathsf{G}$ and $D_0^1 = \{v_1\}$. Then we define G_j^1, D_j^1, and K_j^1 from G_{j-1}^1, D_{j-1}^1, and K_j in the following manner. If $v_1 \in K_j$, then $\mathsf{G}_j^1 = \mathsf{G}_{j-1}^1$, $D_j^1 = D_{j-1}^1$, and $K_j^1 = K_j$. If $v_1 \notin K_j$, then a vertex v_j^1 and edges are added to G_{j-1}^1 in order to obtain a graph G_j^1 such that $N_{\mathsf{G}_j^1}(v_j^1) = K_j$; we then set $D_j^1 = D_{j-1}^1 \cup \{v_j^1\}$ and $K_j^1 = K_j \cup \{v_j^1\}$. In the case of an index $k > 1$, we proceed in a similar way by first setting $\mathsf{G}_0^k = \mathsf{G}_m^{k-1}$, $D_0^k = \{v_k\}$ and then defining G_j^k, D_j^k, and K_j^k from G_{j-1}^k, D_{j-1}^k, and K_j^{k-1} according to the following rule. If $v_k \in K_j$, then $\mathsf{G}_j^k = \mathsf{G}_{j-1}^k$, $D_j^k = D_{j-1}^k$, $K_j^k = K_j^{k-1}$. If $v_k \notin K_j$, then a vertex v_j^k and edges are added to G_{j-1}^k in order to obtain a graph G_j^k such that $N_{\mathsf{G}_j^k}(v_j^k) = K_j^{k-1}$; we then set $D_j^k = D_{j-1}^k \cup \{v_j^k\}$ and $K_j^k = K_j^{k-1} \cup \{v_j^k\}$.

Now that the construction is completed, it is important to point out certain conclusions that follow from it. The graph G is an induced subgraph of G_m^n. The family $(D_m^1, \ldots, D_m^n)$ is a partition of the set of vertices of G_m^n into totally disconnected subsets. The sets $K_1^n, \ldots, K_m^n$ each induce a maximal clique of G_m^n and all maximal cliques of G_m^n are obtained in this manner. For $j =$

$1,\ldots,m$ and $k = 1,\ldots,n$, K_j^n and D_m^k have exactly one vertex in common, namely, v_j^k.

Our next goal is to define a finite measurable space $(\Omega, \mathscr{F})$ and a family $(\mathscr{F}_k,\ k \in \mathbb{N}_n)$ of subalgebras of $\mathscr{F}$ whose intersection graph contains an induced subgraph isomorphic to G. Set $\Omega = \{K_1^n, \ldots, K_m^n\}$. For all vertices v of G_m^n, let F_v denote the set of elements K_j^n of Ω such that $v \in K_j^n$. For $k = 1,\ldots,n$, set $\mathscr{P}_k = (F_v,\ v \in D_m^k)$. Clearly, $\mathscr{P}_k$ is a partition of Ω. Set $\mathscr{F} = \mathscr{P}(\Omega)$ and let $\mathscr{F}_k$ denote the subalgebra of $\mathscr{F}$ generated by $\mathscr{P}_k$. The family $(\mathscr{F}_k,\ k \in \mathbb{N}_n)$ has the required property. Indeed, the map f from $\{v_1, \ldots, v_n\}$ into the set of vertices of the intersection graph of $(\mathscr{F}_k)$ defined by

$$f(v_k) = \left(k, F_{v_k}\right), \qquad k = 1,\ldots,n,$$

is clearly one to one. Moreover, $F_{v_i} \cap F_{v_k} \neq \varnothing$ if and only if there exists j such that $\{v_i, v_k\} \subset K_j^n$ (or, equivalently, K_j), in other words, if and only if $\{v_i, v_k\}$ is an edge of G. Therefore, G is isomorphic to a subgraph of the intersection graph of $(\mathscr{F}_k)$.

It remains to establish that this graph is isomorphic to a subgraph of the intersection graph of a bidimensional filtration. But this is immediate, because this is true with the filtration $(\tilde{\mathscr{F}}_t,\ t \in \mathbb{I}_{(n,n-1)})$ defined by

$$\tilde{\mathscr{F}}_t = \begin{cases} \{\varnothing, \Omega\}, & \text{if } |t| < n, \\ \mathscr{F}_{t^1}, & \text{if } |t| = n, \\ \mathscr{P}(\Omega), & \text{if } |t| > n. \end{cases}$$

5.6 THE COMBINATORIAL STRUCTURE OF A FILTRATION SATISFYING CQI

In this section, which proceeds in the same spirit as the previous one, we will study the consequences of hypothesis CQI on the combinatorial structure of a bidimensional filtration. We will see that under this hypothesis, Helly's intersection property is satisfied and the intersection graph of the filtration is perfect. Considerable further insight into the combinatorial structure of a filtration can be obtained by describing the graphs that can or cannot appear as subgraphs of its intersection graph. It will be established that even in the presence of CQI, this structure is far more complex than that of one-dimensional filtrations.

Expressing CQI Without the Probability Measure and Helly's Intersection Property

Throughout this section, we assume that $d = 2$. Moreover, if s and t are elements of $\mathbb{I}$, we write $s \angle t$ to express that $s^1 < t^1$ and $s^2 > t^2$. Clearly, "$s \angle t$ or $t \angle s$" means that s and t are incomparable.

Recall that $\mathcal{F}$ is a finite algebra and that the filtrations that we will consider are formed by subalgebras of $\mathcal{F}$. If $(V, \mathcal{E})$ is the intersection graph of a filtration, we will denote by I and term *index function* the function defined on V by

$$I(v) = t, \qquad \text{if } v = (t, F).$$

Recall that CQI remains valid when the probability measure is replaced by an equivalent probability measure, and therefore a filtration $(\mathcal{F}_t,\ t \in \mathbb{I}_u)$ satisfies CQI if and only if it satisfies that hypothesis relative to some arbitrary probability measure P on $\mathcal{F}$ such that $\mathrm{P}(F) > 0$ if $F \neq \varnothing$.

Proposition 1 (Graph-theoretic interpretation of CQI). *Let $(V, \mathcal{E})$ be the intersection graph of a bidimensional filtration $(\mathcal{F}_t,\ t \in \mathbb{I}_u)$ and let I be the index function of this graph. In order that $(\mathcal{F}_t)$ satisfy* CQI, *it is necessary and sufficient that the following condition on $(V, \mathcal{E})$ hold: For all choices of elements v_1, v_2, and v_3 of V such that $I(v_1) \angle I(v_2) \angle I(v_3)$, if $\{v_1, v_2\} \in \mathcal{E}$ and $\{v_2, v_3\} \in \mathcal{E}$, then $\{v_1, v_3\} \in \mathcal{E}$.*

Proof. Let $(\mathcal{P}_t,\ t \in \mathbb{I}_u)$ be the family of generating partitions associated with the filtration. Suppose that CQI is satisfied and consider three vertices $v_1 = (t_1, F_1)$, $v_2 = (t_2, F_2)$, and $v_3 = (t_3, F_3)$ such that $t_1 \angle t_2 \angle t_3$, $\{v_1, v_2\} \in \mathcal{E}$ and $\{v_2, v_3\} \in \mathcal{E}$. Because F_2 belongs to $\mathcal{P}_{t_2}$, the fact that $F_1 \cap F_2 \neq \varnothing$ and $F_2 \cap F_3 \neq \varnothing$ implies that

$$F_2 \subset \left\{\mathrm{P}\left(F_1 | \mathcal{F}_{t_2}\right) > 0\right\} \cap \left\{\mathrm{P}\left(F_3 | \mathcal{F}_{t_2}\right) > 0\right\}.$$

By CQI, the right-hand side is equal to $\{\mathrm{P}(F_1 \cap F_3 | \mathcal{F}_{t_2}) > 0\}$. Thus $F_1 \cap F_3 \neq \varnothing$; in other words, $\{v_1, v_3\} \in \mathcal{E}$, which shows that the condition of the proposition is satisfied. Conversely, assume that the condition holds, consider two elements t_1 and t_2 of $\mathbb{I}_u$ such that $t_1 \angle t_2$ and set $t = t_1 \wedge t_2$. Let $F_1 \in \mathcal{P}_{t_1}$ and $F_2 \in \mathcal{P}_{t_2}$ be such that

$$\left\{\mathrm{P}\left(F_1 | \mathcal{F}_t\right) > 0\right\} \cap \left\{\mathrm{P}\left(F_2 | \mathcal{F}_t\right) > 0\right\} \neq \varnothing.$$

The two terms of the intersection clearly belong to $\mathcal{P}_t$; therefore they are identical. Denote them both by F. Clearly, $F_1 \subset F$ and $F_2 \subset F$, which means that $F_1 \cap F = F_1 \neq \varnothing$ and $F \cap F_2 = F_2 \neq \varnothing$. The condition of the proposition applied to the case where $v_1 = (t_1, F_1)$, $v_2 = (t, F)$, and $v_3 = (t_2, F_2)$ implies therefore that $F_1 \cap F_2 \neq \varnothing$. But then $\{\mathrm{P}(F_1 \cap F_2 | \mathcal{F}_t) > 0\}$ is nonempty and belongs to $\mathcal{P}_t$. On the other hand, it is clear that $\{\mathrm{P}(F_1 \cap F_2 | \mathcal{F}_t) > 0\} \subset F$. Consequently, the two sides of this inclusion are identical, in other words,

$$\left\{\mathrm{P}\left(F_1 | \mathcal{F}_t\right) > 0\right\} \cap \left\{\mathrm{P}\left(F_2 | \mathcal{F}_t\right) > 0\right\} = \left\{\mathrm{P}\left(F_1 \cap F_2 | \mathcal{F}_t\right) > 0\right\}. \quad (5.10)$$

This equality also holds in the case where the intersection on the left-hand side is empty, because the trivial inclusions $F_1 \subset \{P(F_1 | \mathscr{F}_t) > 0\}$ and $F_2 \subset \{P(F_2 | \mathscr{F}_t) > 0\}$ imply in this case that the right-hand side is also empty.

It remains to extend the validity of (5.10) to all $F_1 \in \mathscr{F}_{t_1}$ and $F_2 \in \mathscr{F}_{t_2}$. This is immediate if F_1 or F_2 are empty. Otherwise, $F_1 = \bigcup_j F_1^j$ and $F_2 = \bigcup_k F_2^k$, where $F_1^j \in \mathscr{P}_{t_1}$ and $F_2^k \in \mathscr{P}_{t_2}$. Consequently,

$$\begin{aligned}
&\{P(F_1 | \mathscr{F}_t) > 0\} \cap \{P(F_2 | \mathscr{F}_t) > 0\} \\
&= \left\{P\left(\bigcup_j F_1^j \middle| \mathscr{F}_t\right) > 0\right\} \cap \left\{P\left(\bigcup_k F_2^k \middle| \mathscr{F}_t\right) > 0\right\} \\
&= \bigcup_{j,k} \{P(F_1^j | \mathscr{F}_t) > 0\} \cap \{P(F_2^k | \mathscr{F}_t) > 0\} \\
&= \bigcup_{j,k} \{P(F_1^j \cap F_2^k | \mathscr{F}_t) > 0\} \\
&= \left\{P\left(\bigcup_{j,k} (F_1^j \cap F_2^k) \middle| \mathscr{F}_t\right) > 0\right\} \\
&= \{P(F_1 \cap F_2 | \mathscr{F}_t) > 0\}.
\end{aligned}$$

Proposition 2 (Helly's intersection property). *The family* $(\mathscr{P}_t,\ t \in \mathbb{I}_u)$ *of generating partitions associated with a bidimensional filtration satisfying* CQI *has Helly's intersection property: For any* $n > 2$, *if* $(F_1, \ldots, F_n)$ *is a family of elements of* $\bigcup_{t \in \mathbb{I}_u} \mathscr{P}_t$ *such that* $F_j \cap F_k \neq \varnothing$ *for* $j = 1, \ldots, n$ *and* $k = 1, \ldots, n$, *then* $\bigcap_{j=1}^n F_j \neq \varnothing$.

Proof. Let $t_1, \ldots, t_n$ be such that $F_j \in \mathscr{P}_{t_j}$ for $j = 1, \ldots, n$. Observe that if $t_j \leq t_k$ for some couple (j, k), then either $F_j \cap F_k = \varnothing$ or $F_k \subset F_j$. There is therefore no loss of generality in assuming that $t_1 \angle \cdots \angle t_n$. Set $\tilde{F}_k = \bigcap_{j=1}^k F_j$. By hypothesis, $\tilde{F}_2 \neq \varnothing$. Assume that $\tilde{F}_k \neq \varnothing$ for some k such that $2 \leq k < n$ and show that $\tilde{F}_{k+1} \neq \varnothing$. Let F be an element of $\mathscr{P}_{t_1 \vee t_k}$ such that $F \subset \tilde{F}_k$. Since $F \cap F_k = F \neq \varnothing$ and $F_k \cap F_{k+1} \neq \varnothing$, by applying Proposition 1 to the situation where $v_1 = (t_1 \vee t_k, F)$, $v_2 = (t_k, F_k)$, and $v_3 = (t_{k+1}, F_{k+1})$, we see that $F \cap F_{k+1} \neq \varnothing$. Consequently, $\tilde{F}_k \cap F_{k+1} \neq \varnothing$, which was to be proved, because $\tilde{F}_{k+1} = \tilde{F}_k \cap F_{k+1}$.

Perfection of the Intersection Graph Under CQI

If G is a graph, then $\kappa(\mathsf{G})$ will denote the cardinality of the set of vertices of the largest clique of G, and $\chi(\mathsf{G})$ will denote the minimal number of colors needed to color the vertices of G in such a way that no two neighbors receive the same color. A graph G is *perfect* if $\kappa(\tilde{\mathsf{G}}) = \chi(\tilde{\mathsf{G}})$ for all induced subgraphs $\tilde{\mathsf{G}}$ of G. G is *strongly perfect* if all induced subgraphs $\tilde{\mathsf{G}}$ of G admit a set of totally disconnected vertices that has at least one vertex in common with each maximal clique of $\tilde{\mathsf{G}}$.

We shall show that the intersection graph of a bidimensional filtration satisfying CQI is perfect. The proof relies on two lemmas that use the notion of *strong order*. This is any total order $\prec$ between vertices of a graph such that $v_2 \prec v_1$ or $v_3 \prec v_4$ each time that $\{v_1, v_2\}$, $\{v_2, v_3\}$, and $\{v_3, v_4\}$ are edges of a 4-path of this graph.

Lemma 1. *Let* G *be a graph and let* K *be the set of vertices of a clique of this graph. Suppose that each* $v \in K$ *has a neighbor* $f(v) \notin K$ *and that* $f(v)$ *is not a neighbor of* $f(w)$ *if* (v, w) *is a couple of distinct elements of* K*. Suppose, moreover, that* G *can be equipped with a strong order* $\prec$ *such that* $f(w) \prec w$ *for all* $w \in K$*. Then there exists* $v \in K$ *such that* $f(v)$ *is a neighbor of each element of* K*.*

Proof. The conclusion is trivially true if K only has one element. Suppose it is true for all cliques with n vertices satisfying the hypotheses of the lemma and consider a clique with $n + 1$ vertices satisfying the same hypotheses. Let K be the set of vertices of this clique. For all $v \in K$, $K - \{v\}$ induces a clique with n vertices that again satisfies the hypothesis. Therefore there exists an element $\tilde{v}$ of $K - \{v\}$ such that $f(\tilde{v})$ is a neighbor of all elements of $K - \{v\}$. Let us show that there exists $v \in K$ such that $f(\tilde{v})$ is also a neighbor of v. Assume the contrary. Then the map $v \mapsto \tilde{v}$ from K into K is one to one, therefore onto. Moreover, $f(\tilde{v}) \neq f(\tilde{w})$ for all couples (v, w) of distinct elements of K. Let v be the smallest element of K (for $\prec$) and v_2, v_3 the elements of K defined by $\tilde{v}_2 = v$ and $\tilde{v}_3 = v_2$. Set $v_1 = f(\tilde{v}_3)$ and $v_4 = f(\tilde{v}_2)$. By construction, v_1 is a neighbor of v_2, v_2 of v_3, and v_3 of v_4. On the other hand, v_1 is neither a neighbor of v_3 nor of v_4, and v_2 is not a neighbor of v_4. In other words, $\{v_1, v_2, v_3, v_4\}$ induces a 4-path whose edges are $\{v_1, v_2\}$, $\{v_2, v_3\}$, and $\{v_3, v_4\}$. But $v_4 = f(\tilde{v}_2) = f(v) \prec v \prec v_3$ and $v_1 = f(\tilde{v}_3) \prec \tilde{v}_3 = v_2$, which contradicts the fact that $\prec$ is a strong order. An element v of K such that $f(v)$ is a neighbor of all elements of K must therefore exist.

Lemma 2. *A graph equipped with a strong order is strongly perfect. A strongly perfect graph is perfect.*

Proof. In order to establish the first part of the lemma, it is sufficient to show that a graph G equipped with a strong order $\prec$ admits a totally disconnected set of vertices D that has at least one vertex in common with each maximal clique of G. Let $v_1, \ldots, v_n$ be the vertices of G enumerated in increasing order. Set $D_0 = \varnothing$ and define $D_1, \ldots, D_n$ by induction using the formula

$$D_j = \begin{cases} D_{j-1} \cup \{v_j\}, & \text{if } N_{\mathsf{G}}(v_j) \cap D_{j-1} = \varnothing, \\ D_{j-1}, & \text{if } N_{\mathsf{G}}(v_j) \cap D_{j-1} \neq \varnothing. \end{cases}$$

Set $D = D_n$. It is clear that D is totally disconnected. On the other hand, if K is the set of vertices of a clique such that $D \cap K = \varnothing$, then each $v \in K$ has a neighbor $f(v) \in D$ such that $f(v) \prec v$; therefore, there exists by Lemma 1 an element $w \notin K$ that is a neighbor of each $v \in K$. This shows that the clique is not maximal.

Concerning the second part of the lemma, we proceed by induction on the value of $\kappa(\mathsf{G})$. If $\kappa(\mathsf{G}) = 1$, then the set of vertices of G is totally disconnected; therefore $\kappa(\tilde{\mathsf{G}}) = \chi(\tilde{\mathsf{G}}) = 1$ for all subgraphs $\tilde{\mathsf{G}}$ of G, which implies that G is perfect. Suppose that the assertion of the lemma is true for all strongly perfect graphs G such that $\kappa(\mathsf{G}) \leq n$ and consider a strongly perfect graph G such that $\kappa(\mathsf{G}) = n + 1$. Let $\tilde{\mathsf{G}}$ be an induced subgraph of G and let $\tilde{V}$ be the set of its vertices. Because $\tilde{\mathsf{G}}$ is also strongly perfect, if $\kappa(\tilde{\mathsf{G}}) \leq n$, then the induction hypothesis implies that $\tilde{\mathsf{G}}$ is perfect. Suppose that $\kappa(\tilde{\mathsf{G}}) = n + 1$ and consider a totally disconnected set of vertices D that has one vertex in common with each maximal clique of $\tilde{\mathsf{G}}$. Let $\hat{\mathsf{G}}$ be the subgraph of $\tilde{\mathsf{G}}$ induced by $\tilde{V} - D$. This subgraph is strongly perfect and $\kappa(\hat{\mathsf{G}}) = n$. Again by the induction hypothesis, $\kappa(\hat{\mathsf{G}}) = \chi(\hat{\mathsf{G}}) = n$. But because one additional color suffices to color the vertices of D, it follows that $\chi(\tilde{\mathsf{G}}) = n + 1$, therefore that $\kappa(\tilde{\mathsf{G}}) = \chi(\tilde{\mathsf{G}})$. We have thus established that G is a perfect graph.

Theorem 1 (Perfection of the intersection graph). *If a bidimensional filtration satisfies* CQI, *then its intersection graph is perfect.*

Proof. Let $\mathsf{G} = (V, \mathscr{E})$ be the intersection graph of a bidimensional filtration $(\mathscr{F}_t,\ t \in \mathbb{I}_u)$ that satisfies CQI and let I be the index function of this graph. By Lemma 2, it is sufficient to prove that G can be equipped with a strong order. Let $\prec$ be the total order on V obtained by arbitrarily ordering the vertices v such that $I(v) = t$, for each $t \in \mathbb{I}_u$, and by setting $v \prec w$ if $I(v)$ precedes $I(w)$ in the lexicographical order on $\mathbb{I}_u$. We are going to show that $\prec$ is a strong order. Suppose the contrary; that is, G contains a 4-path with edges $\{v_1, v_2\}$, $\{v_2, v_3\}$, and $\{v_3, v_4\}$ such that $v_1 \prec v_2$ and $v_4 \prec v_3$. It is then impossible that $I(v_2) \leq I(v_3)$, because this would imply that $\{v_2, v_4\}$ is an edge. Similarly, it is not possible that $I(v_3) \leq I(v_2)$, because this would imply that $\{v_1, v_3\}$ is an edge. $I(v_2)$ and $I(v_3)$ must therefore be incomparable. Assume that $I(v_2) \angle I(v_3)$, as the case where $I(v_3) \angle I(v_2)$ can be treated in a similar manner. Since $v_1 \prec v_2$, either $I(v_1) \leq I(v_2)$ or $I(v_1) \angle I(v_2)$. But the first case is excluded because it implies that $\{v_1, v_3\}$ is an edge. The second is also excluded because it leads to the same conclusion by Proposition 1. The resulting contradiction shows that $\prec$ is a strong order.

Forbidden Subgraphs

We will term *forbidden subgraphs* a graph that is not isomorphic to any induced subgraph of a bidimensional filtration that satisfies CQI.

Let G be the intersection graph of a bidimensional filtration. Imposing CQI on this filtration prevents certain graphs from being isomorphic to a subgraph of G. From a structure point of view, a bidimensional filtration is therefore less complex in the presence of CQI than in the absence of this hypothesis. In order to make this statement more precise, we will exhibit a class of forbidden subgraphs that are *minimal* in the sense that any subgraph of a graph of this class *is* isomorphic to a subgraph of this intersection graph of a bidimensional filtration satisfying CQI. Among the graphs of the class just mentioned are some that contain arbitrarily large cliques, which means that from a structural point of view, a bidimensional filtration satisfying CQI is still generally far more complex than a one-dimensional filtration.

By Theorem 1, if the filtration satisfies CQI, then G is perfect. Consequently, it does not contain any n-cycle with n odd and > 3, because if C is such a cycle, then $\kappa(\mathsf{C}) = 2$ and $\chi(\mathsf{C}) = 3$. In fact, G also does not contain any n-cycle with n even and > 4 (cf. Exercise 6). Except for 2-, 3-, and 4-cycles (cf. Exercise 7), all other n-cycles are therefore forbidden. Moreover, they are minimal in the sense explained above (cf. Exercise 8). These facts already illustrate the effect of CQI. Now here is the new type of graph announced above.

Clans

Let $n \geq 3$. We will term *clan* all graphs with $4n - 3$ vertices

$$v_1, \ldots, v_n, \quad v'_1, \ldots, v'_{n-1}, \quad v''_1, v''_{n-1}, \quad w_1, \ldots, w_{n-2}, \quad w'_1, \ldots, w'_{n-2},$$

whose edges are the pairs of vertices determined by conditions a, b, and c below and, if $n > 3$, the additional pairs of vertices $\{v_2, v'_2\}, \ldots, \{v_{n-2}, v'_{n-2}\}$ and $\{v_i, w_j\}$ for all (i, j) such that $1 \leq i < j \leq n - 2$ or $2 \leq j + 1 < i \leq n - 1$. The conditions are

a. $\{v_1, \ldots, v_n\}$ and $\{v_n, w_1, \ldots, w_{n-2}\}$ each induce a clique;

b. for $j = 1, \ldots, n - 2$ $\{v_n, w_j, w'_j\}$ induces a path (with extremities v_n, w'_j);

c. $\{v_1, v'_1, v''_1\}$ and $\{v_{n-1}, v'_{n-1}, v''_{n-1}\}$ each induce a path (with respective extremities v_1, v''_1, and v_{n-1}, v''_{n-1}).

We will say that three vertices v, v_1, and v_2 of a graph have the property $p(v; v_1, v_2)$ if $\{v, v_1, v_2\}$ induces a clique and if there exist two other vertices w and w' of this graph such that $\{v, w, w'\}$ induces a path (with extremities v, w') and $\{v_1, w\}$ are not edges.

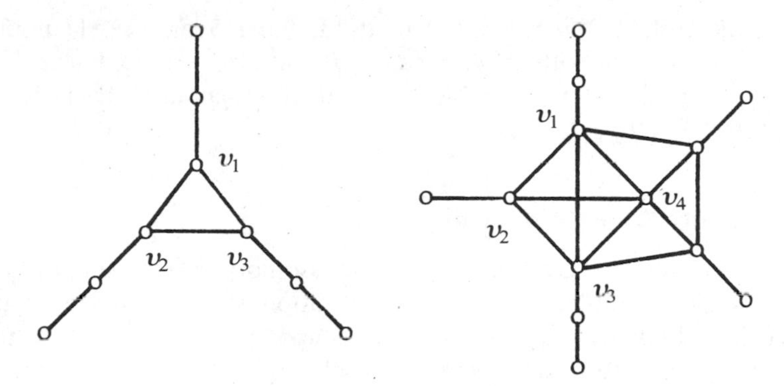

Figure 1. Clans with $n = 3$ (left) and $n = 4$ (right).

Lemma 3. *Let* G *be the intersection graph of a bidimensional filtration satisfying* CQI *and let I be the index function of* G. *Let v, v_1, and v_2 be three vertices of* G *that have the property $p(v; v_1, v_2)$. Then neither $I(v_1) \angle I(v) \angle I(v_2)$ nor $I(v_2) \angle I(v) \angle I(v_1)$ occurs.*

Proof. Because v_1 (or v_2) is neighbor of v but not of w, and because w' is neighbor of w but not of v, it follows that $I(v)$ and $I(w)$ are incomparable. Suppose, for example, that $I(v_1) \angle I(v) \angle I(v_2)$. Then either $I(v_1) \angle I(v) \angle I(w)$ or $I(w) \angle I(v) \angle I(v_2)$. In the first case, Proposition 1 implies that $\{v_1, w\}$ is an edge; in the second case, the same proposition implies that $\{v_2, w\}$ is an edge. This contradicts the definition of $p(v; v_1, v_2)$.

Theorem 2 (Clans are forbidden subgraphs). *The intersection graph of a bidimensional filtration satisfying* CQI *contains no clans.*

Proof. Suppose the contrary; that is, the graph in question does contain a clan, and denote the vertices of this clan as in the definition. We immediately observe that by definition, the property $p(v_i; v_j, v_k)$ holds for all triplets (v_i, v_j, v_k) such that $i \in \{1, n-1\}$, $j, k \in \{1, \ldots, n\}$, $i \neq j \neq k \neq i$, and the property $p(v_n; v_j, v_{j+1})$ holds for $j \in \{1, \ldots, n-2\}$. For $j, k \in \{1, \ldots, n-1\}$, $j \neq k$, v_j' is a neighbor of v_j but not of v_k; therefore $I(v_j)$ and $I(v_k)$ are incomparable. Similarly, using in addition the vertices $w_1, \ldots, w_{n-2}$, we see that for $j \in \{1, \ldots, n-1\}$, $I(v_j)$ and $I(v_n)$ are incomparable. On the other hand, because the properties $p(v_1; v_j, v_k)$ and $p(v_{n-1}; v_j, v_k)$ are satisfied, Lemma 3 implies that for $j \in J \cup \{n\}$, where $J = \emptyset$ if $n = 3$ and $J = \{2, \ldots, n-2\}$ if $n > 3$, either $I(v_1) \angle I(v_j) \angle I(v_{n-1})$ or $I(v_{n-1}) \angle I(v_j) \angle I(v_1)$. For reasons of symmetry, we can assume that

$$I(v_1) \angle I(v_j) \angle I(v_{n-1}), \qquad \text{for } j \in J \cup \{n\}. \tag{5.11}$$

Because the property $p(v_n; v_1, v_2)$ holds, Lemma 3 and (5.11) imply that $I(v_2) \angle I(v_n)$. Consequently, if $k = \sup\{j\colon I(v_j) \angle I(v_n)\}$, then $1 < k < n - 1$. Thus $I(v_k) \angle I(v_n) \angle I(v_{k+1})$, which contradicts Lemma 3 because of the property $p(v_n; v_k, v_{k+1})$.

The Minimality Property of Clans

Let $\mathscr{G}$ denote the set of all graphs that are isomorphic to a subgraph of the intersection graph of a bidimensional filtration satisfying CQI. Clans are minimal forbidden subgraphs in the sense used previously; that is, they do not belong to $\mathscr{G}$ but all of their proper subgraphs do. The first assertion is the subject of Theorem 2. The proof of the second assertion is not difficult but is rather lengthy, which is the reason we leave it as an exercise (cf. Exercise 9).

5.7 OPTIMAL STOPPING AND LINEAR OPTIMIZATION

In this section, we again assume that $\mathscr{F}$ is a finite algebra and that a probability measure P is defined on this algebra with the property that $\mathrm{P}(F) > 0$ for all nonempty sets $F \in \mathscr{F}$. Moreover, we denote by $(\mathscr{F}_t,\ t \in \mathbb{I}_u)$ a bidimensional filtration (formed of subalgebras of $\mathscr{F}$). Of course, by *stopping point* we will mean a stopping point relative to this filtration (with values in $\mathbb{I}_u$). The processes we shall consider will be indexed by $\mathbb{I}_u$. It is clear that the coordinate random variables take a finite number of values. In order to avoid trivialities, we assume that these values are finite.

The Matrix of the Filtration

It will be convenient to have at hand a total order on $\mathbb{I}_u$. We assume that such an order (for example, the lexicographical order) has been chosen and fixed and we will refer to it as *the total order in* $\mathbb{I}_u$.

We denote by $(F_1^t, \ldots, F_{n_t}^t)$ and $(F_1, \ldots, F_m)$ the respective generating partitions of $\mathscr{F}_t$ and $\mathscr{F}$. Moreover, we set $n = \sum_{t \in \mathbb{I}_u} n_t$.

We term *matrix of the algebra* $\mathscr{F}_t$ the $m \times n_t$ matrix $\mathbf{M}_t = (m_{i,j}^t)$ defined by

$$m_{i,j}^t = \begin{cases} 1, & \text{if } F_i \subset F_j^t, \\ 0, & \text{otherwise.} \end{cases}$$

The *matrix of the filtration* $(\mathscr{F}_t)$ is the $m \times n$ matrix $\mathbf{M}$ made up of one row of submatrices obtained by arranging the matrices $\mathbf{M}_t$ according to the total order in $\mathbb{I}_u$.

Clearly, $\mathbf{M}_t$ determines the algebra $\mathscr{F}_t$ and $\mathbf{M}$ determines the filtration $(\mathscr{F}_t)$. Observe that in each row of $\mathbf{M}_t$, all the entries are 0 except for one, which is equal to 1. Moreover, in each column of $\mathbf{M}_t$, there is at least one entry equal to 1.

In the sequel, we will write $\mathbf{x}$ (or use another letter in boldface) to denote an element of $\mathbb{R}^n$ that is to be thought of as a (column) vector. Moreover, we will write $\mathbf{1}$ to denote the vector of $\mathbb{R}^m$ whose components are all equal to 1.

Optimal Stopping

Under the above hypotheses, the problem of optimal stopping is equivalent to an optimization problem over integers. The key to this assertion is the representation of stopping points by vectors. We denote by $\mathbb{X}_{\text{int}}$ the set $\{\mathbf{x} \in \{0, 1\}^n: \mathbf{Mx} = \mathbf{1}\}$, where $\mathbf{M}$ is the matrix of $(\mathscr{F}_t)$.

Lemma 1. *The set of stopping points is in one-to-one correspondence with* $\mathbb{X}_{\text{int}}$. *More precisely, to the stopping point T corresponds the vector $\mathbf{x}$ obtained by ordering according to the total order in $\mathbb{I}_u$ the subvectors $(x_1^t, \ldots, x_{n_t}^t)$ defined by*

$$x_j^t = \begin{cases} 1, & \text{if } F_j^t \subset \{T = t\}, \\ 0, & \text{otherwise.} \end{cases} \tag{5.12}$$

Proof. Let T be a stopping point. For all $t \in \mathbb{I}_u$,

$$1_{\{T=t\}} = \sum_{j=1}^{n_t} x_j^t \, 1_{F_j^t}, \tag{5.13}$$

where $x_1^t, \ldots, x_{n_t}^t$ are as defined in (5.12). Thus

$$\sum_{t \in \mathbb{I}_u} \sum_{j=1}^{n_t} x_j^t \, 1_{F_j^t} = 1.$$

But on each set F_i, this equality can be written

$$\sum_{t \in \mathbb{I}_u} \sum_{j=1}^{n_t} m_{i,j}^t \, x_j^t = 1.$$

In other words, if $\mathbf{x}$ is the vector of $\{0, 1\}^n$ obtained by ordering the vectors $(x_1^t, \ldots, x_{n_t}^t)$ according to the total order in $\mathbb{I}_u$, then $\mathbf{Mx} = \mathbf{1}$. Conversely, consider a vector of $\{0, 1\}^n$ such that $\mathbf{Mx} = \mathbf{1}$ and divide it into subvectors $(x_1^t, \ldots, x_{n_t}^t)$ according to the total order in $\mathbb{I}_u$. Then define T by (5.13). Clearly, T is a stopping point and (5.12) holds.

Theorem 1 (Optimal stopping as an integer programming problem). *Solving the optimal stopping problem in the set of all stopping points when X is a (not necessarily adapted) process is equivalent to determining a vector $\mathbf{x}^o \in \mathbb{X}_{\text{int}}$ such that*

$$\mathbf{c} \cdot \mathbf{x}^o = \sup_{\mathbf{x} \in \mathbb{X}_{\text{int}}} \mathbf{c} \cdot \mathbf{x}, \tag{5.14}$$

where $\mathbf{c}$ is the vector of $\mathbb{R}^n$ defined by putting $c_j^t = \mathrm{E}(X_t; F_j^t)$ and ordering the vectors $(c_1^t, \ldots, c_{n_t}^t)$ according to the total order in $\mathbb{I}_u$.

Proof. Let $\mathbf{x}$ be the element of $\mathbb{X}_{\text{int}}$ that corresponds to T. Taking (5.13) into account, observe that

$$\mathrm{E}(X_T) = \sum_{t \in \mathbb{I}_u} \mathrm{E}(X_t; T = t) = \sum_{t \in \mathbb{I}_u} \sum_{j=1}^{n_t} x_j^t \, \mathrm{E}\left(X_t; F_j^t\right) = \sum_{t \in \mathbb{I}_u} \sum_{j=1}^{n_t} c_j^t x_j^t = \mathbf{c} \cdot \mathbf{x}.$$

The conclusion follows immediately.

The Relationship with Linear Programming

At the present time, there are no efficient algorithms for determining a vector $\mathbf{x}^o \in \mathbb{X}_{\text{int}}$ that satisfies (5.14). On the other hand, there are algorithms (e.g., the simplex algorithm, Karmarkar's algorithm) that efficiently determine a vector $\mathbf{x}^o \in \mathbb{X}$ such that

$$\mathbf{c} \cdot \mathbf{x}^o = \sup_{\mathbf{x} \in \mathbb{X}} \mathbf{c} \cdot \mathbf{x}, \tag{5.15}$$

where $\mathbb{X}$ denotes the set $\{\mathbf{x} \in \mathbb{R}_+^n : \mathbf{M}\mathbf{x} = 1\}$ and $\mathbf{M}$ is the matrix of $(\mathscr{F}_t)$.

Since each column of $\mathbf{M}$ has a nonzero entry, it follows that all elements $\mathbf{x}$ of $\mathbb{X}$ belong to $[0, 1]^n$. Thus $\mathbb{X}$ is a closed subset of $[0, 1]^n$ and therefore is compact. Moreover, all convex combinations of elements of $\mathbb{X}$ belong to $\mathbb{X}$, so $\mathbb{X}$ is convex. It follows that the upper bound on the right-hand side of (5.15) is attained at an extremal element $\mathbf{x}^o$ of $\mathbb{X}$. But if this extremal element belongs to $\{0, 1\}^n$ (therefore to $\mathbb{X}_{\text{int}}$), it obviously also satisfies (5.14). It is thus important to know whether extremal elements of $\mathbb{X}$ belong to $\{0, 1\}^n$, or, equivalently, if the convex envelope of $\mathbb{X}_{\text{int}}$ is $\mathbb{X}$. This is generally not the case, but it is when the filtration $(\mathscr{F}_t)$ satisfies CQI. We will devote the remainder of this section to the proof of this assertion.

The Derived Graph of a Matrix and the Clique Matrix of a Graph

Let $\mathbf{A} = (a_{i,j})$ be an $m \times n$ matrix with entries in $\{0, 1\}$ and with no column consisting only of zeros. The *derived graph* of $\mathbf{A}$ is the graph with n vertices $v_1, \ldots, v_n$ whose edges are the pairs $\{v_j, v_k\}$ $(j \neq k)$ such that there exists i

with $a_{i,j} = a_{i,k} = 1$. Observe that each row of $\mathbf{A}$ represents a (not necessarily maximal) clique of the derived graph of $\mathbf{A}$. More precisely, the clique corresponding to the ith row is determined by the vertices v_j such that $a_{i,j} = 1$. Conversely, if G is a graph with n vertices $v_1, \ldots, v_n$ and $\mathsf{K}_1, \ldots, \mathsf{K}_m$ are cliques of this graph, putting $a_{i,j} = 1$ if v_j is a vertex of K_i and $a_{i,j} = 0$ otherwise, we define a matrix $\mathbf{A} = (a_{i,j})$ with entries in $\{0, 1\}$. If all the maximal cliques of G are present among the cliques $\mathsf{K}_1, \ldots, \mathsf{K}_m$, the matrix $\mathbf{A}$ is termed a *clique matrix* of G.

Case Where the Filtration Satisfies CQI

The theorem we are about to establish relies on a result from graph theory that we will not prove. The reader interested in the proof can consult Golumbic's book (which contains a more complete result; see the Historical Notes).

Lemma 2. *Let* $\mathbf{A}$ *be an* $m \times n$ *matrix with entries in* $\{0, 1\}$ *such that* $\{\mathbf{x} \in \mathbb{R}^n_+ : \mathbf{A}\mathbf{x} = \mathbf{1}\} \neq \varnothing$. *Let* G *be the derived graph of* $\mathbf{A}$. *If* $\mathbf{A}$ *is a clique matrix of* G *and* G *is perfect, then all extremal elements of* $\{\mathbf{x} \in \mathbb{R}^n_+ : \mathbf{A}\mathbf{x} = \mathbf{1}\}$ *belong to* $\{0, 1\}^n$.

Theorem 2 (Optimal stopping by reduction to a linear programming problem). *Let* $(\mathscr{F}_t, t \in \mathbb{I}_u)$ *be a bidimensional filtration satisfying* CQI. *Let* X *be a (not necessarily adapted) process and* $\mathbf{c}$ *be the associated vector defined in Theorem* 1. *Let* $\mathbf{M}$ *be the matrix of* $(\mathscr{F}_t)$ *and* $\mathbb{X}$ *the set* $\{\mathbf{x} \in \mathbb{R}^n_+ : \mathbf{M}\mathbf{x} = 1\}$. *In order that a stopping point* T *be optimal for* X, *it is necessary and sufficient that the corresponding vector* $\mathbf{x}$ *defined in Lemma* 1 *satisfy* (5.15).

Proof. As already explained, it is sufficient to establish that the extremal elements of $\mathbb{X}$ belong to $\{0, 1\}^n$. Let G be the derived graph of $\mathbf{M}$. By Lemma 2, it is sufficient to prove that $\mathbf{M}$ is a clique matrix of G and that G is perfect. Given the manner in which $\mathbf{M}$ and the notion of derived graph are defined, it is easy to see that G is isomorphic to the intersection graph of $(\mathscr{F}_t)$. By Theorem 1 of Section 5.6, this last graph is perfect, therefore G is also perfect. Moreover, it has been pointed out that each row of $\mathbf{M}$ corresponds to a clique of G. It remains to prove that each *maximal* clique of G is represented by a row of $\mathbf{M}$. Let $v_{j_1}, \ldots, v_{j_k}$ be the vertices of a maximal clique of G and let $(t_1, F^{t_1}_{j_1}), \ldots, (t_k, F^{t_k}_{j_k})$ be the vertices of the corresponding maximal clique in the intersection graph of $(\mathscr{F}_t)$. Then the sets $F^{t_1}_{j_1}, \ldots, F^{t_k}_{j_k}$ are pairwise intersecting. Because $(\mathscr{F}_t)$ satisfies CQI, Proposition 2 of Section 5.6 implies that their common intersection is also nonempty. Therefore there exists i such that $F_i \subset F^{t_1}_{j_1} \cap \cdots \cap F^{t_k}_{j_k}$. On the other hand, no $F \in \bigcup_{t \in \mathbb{I}_u} \mathscr{P}_t$, which is different from $F^{t_1}_{j_1}, \ldots, F^{t_k}_{j_k}$ contains F_i; otherwise the clique would not be maximal. Consequently, the ith row of $\mathbf{M}$ represents the maximal clique under consideration. This completes the proof.

EXERCISES

1. In the setting of Section 5.2, suppose $d = 3$ and let $(\mathcal{F}_t,\ t \in \mathbb{I})$ be a filtration that satisfies the conclusion of the theorem of that section. Moreover, assume that for all $t \in \mathbb{I}$ and all choices of $\{i, j\} \subset \{1, 2, 3\}$, the subfiltration $(\mathcal{F}_{t+me_i+ne_j}, (m, n) \in \mathbb{N}^2)$ satisfies Condition A with $d = 2$ (or, equivalently, satisfies CQI). Show that $(\mathcal{F}_t,\ t \in \mathbb{I})$ satisfies Condition A. (*Hint.* Fix $t \in \mathbb{I}$; let F_1, F_2, and F_3 be the three sets in $\mathcal{F}_t$ whose existence is affirmed by the theorem. Fix n_1, n_2, and n_3 in $\mathbb{N}$ and set $t_i = t + n_j e_j + n_k e_k$, where (i, j, k) is a cyclic permutation of $(1, 2, 3)$. Let $G_1 \in \mathcal{F}_{t_1}$, $G_2 \in \mathcal{F}_{t_2}$, and $G_3 \in \mathcal{F}_{t_3}$ be pairwise disjoint sets. Check that for $i = 1, 2, 3$,

$$\{\mathrm{P}(G_i | \mathcal{F}_t) > 0\} = \bigcup_{j=1}^{3} \{\mathrm{P}(G_i \cap F_j | \mathcal{F}_t) > 0\},$$

and show that

$$\left\{\mathrm{P}\left(G_1 \cap F_3 | \mathcal{F}_{t+n_3e_3}\right) > 0\right\} \cap \left\{\mathrm{P}\left(G_2 \cap F_3 | \mathcal{F}_{t+n_3e_3}\right) > 0\right\} = \varnothing.$$

Conclude that

$$\bigcap_{i=1}^{3} \{\mathrm{P}(G_i | \mathcal{F}_t) > 0\} = \varnothing,$$

and deduce that Condition A holds.)

2. Prove the first assertion of Proposition 2 of Section 5.3.

3. (Canonical examples of filtrations that satisfy CQI) Let R be a nonempty finite set and fix $u \in \mathbb{I}$. Denote by Ω the set of all functions $t \mapsto \omega(t)$ from $\mathbb{I}_u$ into R, set $\mathcal{F} = \mathcal{P}(\Omega)$, and consider a probability measure P on $\mathcal{F}$ such that all nonempty sets have positive probability. For each $t \in \mathbb{I}_u$, let X_t be the function from Ω into R defined by $X_t(\omega) = \omega(t)$ and let $(\mathcal{F}_t,\ t \in \mathbb{I}_u)$ be the natural filtration of $(X_t,\ t \in \mathbb{I}_u)$. Show that this filtration satisfies CQI.

4. Show that the family of generating partitions associated with the filtration considered in Example 2 of Section 5.2 does not have Helly's intersection property and that this proves the last statement in that example.

5. Assume that $d = 3$ and for $i = 1, 2, 3$, set $\Omega_i = \{0, 1\}$, $\mathcal{F}_0^i = \{\emptyset, \Omega_i\}$ and $\mathcal{F}_{t^i}^i = \mathcal{P}(\Omega_i)$ if $t^i \in \mathbb{N}^*$. Let $\Omega = \Omega_1 \times \Omega_2 \times \Omega_3$, $\mathcal{F} = \mathcal{P}(\Omega)$ and $\mathcal{F}_t = \mathcal{F}_{t^1}^1 \times \mathcal{F}_{t^2}^2 \times \mathcal{F}_{t^3}^3$ for all $t = (t^1, t^2, t^3) \in \mathbb{I}$. Show that the intersection graph of the filtration $(\mathcal{F}_t, t \in \mathbb{I})$ contains a 5-cycle, and therefore is not perfect.

6. In the setting of Section 5.5, let G be the intersection graph of a bidimensional filtration $(\mathcal{F}_t)$ that satisfies CQI. Show that G contains no n-cycle with n even and > 4. (*Hint.* Suppose the contrary; that is, G contains an n-cycle with vertices $v_1, \dots, v_n$ and edges $\{v_1, v_2\}, \dots,$ $\{v_{n-1}, v_n\}, \{v_n, v_1\}$. For $j = 1, \dots, n$, let $v_j = (t_j, F_j)$, where $F_j \in \mathcal{F}_{t_j}$. Without loss of generality, assume that $t_1^1 = \inf(t_1^1, t_2^1, \dots, t_n^1)$.

a. Show that for $j = 1, \dots, n-1$, t_j and t_{j+1}, as well as t_n and t_1 are not comparable.

b. Use Proposition 1 of Section 5.6 to show that $t_j \angle t_{j+1}$ if and only if $t_{j+2} \angle t_{j+1}$, and therefore, by the assumption on t_1^1, that $t_{2j+1} \angle t_{2j}$ when $2 \leq 2j \leq n-1$.

c. Prove that $t_{2j-1}^1 \leq t_{2j+1}^1$ implies $t_{2j}^2 < t_{2j+2}^2$ and that $t_{2j}^2 \leq t_{2j+2}^2$ implies $t_{2j+1}^1 < t_{2j+3}^1$. For the first implication, use properties of the sets $\{\mathrm{P}(F_{t_{2j}} \cap F_{t_{2j+1}} | \mathcal{F}_{(t_{2j+1}^1, t_{2j}^2)}) > 0\}$ and $\{\mathrm{P}(F_{t_{2j-1}} \cap F_{t_{2j}} | \mathcal{F}_{(t_{2j-1}^1, t_{2j}^2)}) > 0\}$.

d. Conclude from c that $t_2^2 < t_4^2$, $t_3^1 < t_5^1$, $t_4^2 < t_6^2, \dots,$ and therefore that $t_1^1 \leq t_3^1 < t_5^1 < \cdots < t_n^1 < t_1^1$, which is a contradiction to the assumption that G contains an n-cycle with n even and > 4.)

7. For $n = 2, 3, 4$, use Exercise 3 to construct a bidimensional filtration $(\mathcal{F}_t,$ $t \in \mathbb{I}_{(1,1)})$ that satisfies CQI and that contains an n-cycle.

8. For $n > 1$, use Exercise 3 to construct a bidimensional filtration satisfying CQI whose intersection graph contain an $(n-1)$-path. Conclude that n-cycles with $n > 4$ are minimal forbidden induced subgraphs in the sense explained in Section 5.6.

9. Prove that a clan with $n = 3$ is a minimal forbidden induced subgraph in the sense explained in Section 5.6. Do the same for $n = 4$ and then give a general proof for $n > 4$.

10. Let G be the graph with six vertices v_1, v_2, v_3 and w_1, w_2, w_3 whose edges are determined by the condition that $\{v_1, v_2, v_3\}$ induces a clique and $\{w_i, v_j, v_k\}$ induces a clique for each cyclic permutation (i, j, k) of $(1, 2, 3)$.

a. Prove that G is not isomorphic to an induced subgraph of a product filtration $(\mathcal{F}_t)$ (as defined in Section 5.3) on a probability space $(\Omega, \mathcal{F}, \mathrm{P})$ whose sigma field $\mathcal{F}$ is finite and contains only sets with positive probability (except for the empty set).

b. Use Exercise 3 to construct a bidimensional filtration satisfying CQI whose intersection graph contains G. Conclude that the structure of bidimensional filtrations satisfying CQI is more complex than that of product filtrations.

11. In the setting of Section 5.6, let $\mathsf{G} = (V, \mathscr{E})$ be the intersection graph of a bidimensional filtration $(\mathscr{F}_t,\ t \in \mathbb{I}_u)$ on a probability space $(\Omega, \mathscr{F}, \mathrm{P})$ that satisfies CQI. Show that there is a finite set R and a canonical filtration $(\tilde{\mathscr{F}}_t,\ t \in \mathbb{I}_u)$ as described in Exercise 3 whose intersection graph $\tilde{\mathsf{G}} = (\tilde{V}, \tilde{\mathscr{E}})$ contains G. (*Hint*. Assume that the generating partition of $\mathscr{F}$ has n elements, and for $t \in \mathbb{I}_u$, let $\mathscr{P}_t = (F_1^t, \dots, F_{n_t}^t)$ be the generating partition of $\mathscr{F}_t$. Using the notation of Exercise 3, let $R = \{1, \dots, n\}$ and let $\tilde{\mathscr{P}}_t$ be the generating partition of $\tilde{\mathscr{F}}_t$. For $t \in \mathbb{I}_u$, let g_t be the map from $\mathscr{P}_t$ into R defined by $g_t(F_j^t) = j$, and let f be the map from V into $\tilde{V}$ defined by $f(t, F) = (t, f_t(F))$ for all $t \in \mathbb{I}_u$ and $F \in \mathscr{P}_t$, where $f_t(F) = \{\omega \in \tilde{\Omega}\colon \omega(s) = g_s(\{\mathrm{P}(F \mid \mathscr{F}_s) > 0\})$, for all $s \in \mathbb{I}_t\}$. Prove that f is a one-to-one map from V into $\tilde{V}$ and that G is isomorphic to the subgraph of $\tilde{\mathsf{G}}$ induced by $f(V)$.

12. For $i = 1, 2$, let $\Omega_i = [0, 1]$, $\mathscr{F}_i$ be the algebra generated by the partition $([0, \frac{1}{3}[, [\frac{1}{3}, \frac{2}{3}[, [\frac{2}{3}, 1])$, $\mathscr{F}_0^i = \{\varnothing, \Omega_i\}$, $\mathscr{F}_2^i = \mathscr{F}_2$ and let $\mathscr{F}_1^1$ be the algebra generated by the partition $([0, \frac{2}{3}[, [\frac{2}{3}, 1])$. Set $\Omega = \Omega_1 \times \Omega_2$, $\mathscr{F} = \mathscr{F}_1 \times \mathscr{F}_2$ and, for $t \in \mathbb{I}_{(2,2)}$, $\mathscr{F}_t = \mathscr{F}_{t_1}^1 \times \mathscr{F}_{t_2}^2$. Write out explicitly the matrix of the filtration $(\mathscr{F}_t,\ t \in \mathbb{I}_{(2,2)})$ (which has nine rows and 36 columns).

13. (An example of a filtration that satisfies the conclusion of Theorem 1 of Section 5.6, but does not satisfy CQI) Set $\Omega = \mathbb{N}_6$, $\mathscr{F} = \mathscr{P}(\Omega)$ and $\mathrm{P}(\{n\}) = \frac{1}{7}$ for $n = 0, \dots, 6$. Let $\mathscr{P}_t = (\Omega)$ when $|t| \le 1$ and $\mathscr{P}_t = \mathscr{P}(\Omega)$ when $|t| > 2$. Moreover, set $\mathscr{P}_{(0,2)} = (\{0, 2\}, \{1, 4\}, \{3, 6\}, \{5\})$, $\mathscr{P}_{1,1} = (\{0, 3\}, \{2, 5\}, \{1\}, \{4\}, \{6\})$ and $\mathscr{P}_{(2,0)} = (\{0, 1\}, \{2\}, \{3\}, \{4\}, \{5\}, \{6\})$.

a. Show that the family $(\mathscr{P}_t,\ t \in \mathbb{I}_{(2,2)})$ has Helly's intersection property.

b. For all $t \in \mathbb{I}_{(2,2)}$, let $\mathscr{F}_t$ be the algebra generated by $\mathscr{P}_t$. Show that the intersection graph of the filtration $(\mathscr{F}_t,\ t \in \mathbb{I}_{(2,2)})$ is perfect.

c. Show that CQI does not hold for this filtration.

14. (In the absence of CQI, the optimal stopping problem is not a linear optimization problem) In the context of Section 5.7, assume that $d = 2$ and $u = (2, 2)$. Let $\Omega = \{1, 2, 3\}$, $\mathscr{F} = \mathscr{P}(\Omega)$, and let the probability measure P be such that $\mathrm{P}\{1\} = \mathrm{P}\{2\} = \mathrm{P}\{3\} = \frac{1}{3}$. Set $F_1 = \{2, 3\}$, $F_2 = \{1, 3\}$, and $F_3 = \{1, 2\}$. For $t \in \mathbb{I}_u$, let $\mathscr{F}_t = \{\varnothing, \Omega\}$ if $|t| < 2$, let $\mathscr{F}_{(0,2)}$, $\mathscr{F}_{(1,1)}$, and $\mathscr{F}_{(2,0)}$ be, respectively, the sigma field with generating partition (F_1, F_1^c), (F_2, F_2^c), and (F_3, F_3^c), and let $\mathscr{F}_t = \mathscr{F}$ if $|t| > 2$. Consider the process $(X_t, t \in \mathbb{I}_u)$ defined by $X_t = 0$ if $|t| \neq 2$ and $X_{(0,2)} = 1_{F_1}$, $X_{(1,1)} = 1_{F_2}$ and $X_{(2,0)} = 1_{F_3}$.

a. Check that $(\mathscr{F}_t, t \in \mathbb{I}_u)$ does not satisfy CQI.

b. Write out the matrix of the filtration $(\mathscr{F}_t)$ and the vector **c** associated with (X_t) as in Theorem 1 of Section 5.7.

c. Let $\mathbb{X} = \{\mathbf{x} \in \mathbb{R}_+^n\colon \ \mathbf{M}\mathbf{x} = 1\}$. Show that $\sup_{\mathbf{x} \in \mathbb{X}} \mathbf{c} \cdot \mathbf{x} = 1$ but $\sup_{T \in \mathbf{A}_u} \mathrm{E}(X_T) = \frac{1}{3}$. Notice that in this example, $\mathrm{E}(X_T) \leq \frac{1}{3}$ for any inaccessible stopping point T.

HISTORICAL NOTES

In dimension $d = 2$, Krengel and Sucheston (1981) provided necessary and sufficient conditions on the underlying filtration for all stopping points to be accessible, by isolating condition CQI and proving the theorem of Section 5.3. These authors also gave a related sufficient condition applicable to filtrations indexed by partially ordered sets. Washburn and Willsky (1981) showed for certain (partially) ordered sets and for $\mathbb{N}^2$ that CI is a sufficient condition. This result also appears in Mandelbaum and Vanderbei (1981). For $d > 2$, the two criteria for accessibility provided by the theorems of Section 5.1 are new, as is the theorem of Section 5.2.

The theorem of Section 5.4 is new, and it generalizes results of Mandelbaum (1986), who proved Proposition 2 of that section under a slightly stronger assumption on the filtration, and of Mazziotto and Millet (1987), who proved Proposition 3.

The notion of perfect graph was introduced by C. Berge in thc carly 1960's (see Berge (1973)), and the notion of strongly perfect graph by Berge and Duchet (1984). Graphs that can be equipped with a strong order were introduced by Chvátal (1984), under the name of perfectly orderable graphs. The results on the combinatorial structure of filtrations and the connection between condition CQI and perfect graphs, described in Sections 5.5, 5.6, and 5.7, are part of the results of Chapters 4 and 5 of the doctoral dissertation of Dalang (1987). Numerous additional results in this direction were proved by Dalang, Trotter and de Werra (1988), and by Dalang (1988a). This last reference also contains several equivalent formulations of condition CQI. Lemma 2 of Section 5.7 is a special case of a result due to Fulkerson (1973) and Chvátal (1975); a proof can be found in Golumbic (1980, Theorem 3.19).

Exercises 3, 10, 11, and 13 are taken from Dalang (1987). Exercises 5–9 are due to Dalang, Trotter, and de Werra (1988). The matrix defined in Exercise 12 was the starting point for the research in this last reference. The filtration in Exercise 14 appears in Mazziotto and Millet (1986); it is also considered in Nualart (1992).

CHAPTER 6

Sequential Sampling

Consider a sequence of experiments, each of which consists of observing an individual from one of d populations $\mathcal{P}_1, \dots, \mathcal{P}_d$ with unknown parameters. In each experiment, one of the populations $\mathcal{P}_i$ is chosen, and then an individual is drawn from this population. The result of this draw is identified with the realization of a random variable Y^i. The laws of $Y^1, \dots, Y^d$ are a priori unknown, but it is assumed that the law of $(Y^1, \dots, Y^d)$ belongs to a given set of laws $\{\mu_\theta^1 \times \cdots \times \mu_\theta^d,\ \theta \in \Theta\}$. The ultimate objective of the sequence of experiments is to determine the value of θ, because θ determines the law $\mu_\theta^1 \times \cdots \times \mu_\theta^d$ of $(Y^1, \dots, Y^d)$. Under the assumption that there is a certain cost of experimentation, the theory we shall present leads to an optimal procedure for gathering information concerning the parameter θ. At best, this method will actually determine θ. In order to illustrate the theory, we begin by presenting three examples to which it can be applied.

1. *Best new treatment problem.* Two new treatments are made available to a physician. When a patient arrives for consultation, the physician must decide which treatment to use based on the results observed on previous patients. How should the physician proceed?
2. *Heads with the highest probability.* Consider the following sequential experiment: At each stage, two (possibly biased) coins are tossed; at the nth stage, set $y_n^i = 1$ if coin i falls on heads and $y_n^i = 0$ otherwise $(i = 1, 2)$. The problem consists of identifying the coin that falls on heads with the highest probability. The classical optimal strategy is based on the value of $\sum_{j=1}^n (y_j^1 - y_j^2)$. The experiment we are interested in is significantly different: At the nth stage, only one coin is tossed and the choice of which coin to toss depends on the previous outcomes. The objective remains the same as before but the optimal strategy is different.
3. *Comparison of the parameters of two or more populations.* For example, in the presence of two Gaussian populations $N(\mathsf{m}_1, 1)$ and $N(\mathsf{m}_2, 1)$, where m_1 and m_2 are unknown, a typical problem is to determine whether $\mathsf{m}_1 > \mathsf{m}_2$ or not.

6.1 STATING THE PROBLEM

Implicit in the formulation that we are about to give is a complete probability space $(\Omega, \mathscr{F}, \mathrm{P})$ equipped with a filtration $(\mathscr{F}_t, t \in \mathbb{I})$, which represents the information on the value of the parameter $\omega \in \Omega$ furnished by successive observations from the populations $\mathcal{P}_1, \ldots, \mathcal{P}_d$: When $t = (t^1, \ldots, t^d)$, $\mathscr{F}_t$ represents the information furnished by $|t|$ observations of which t^i are from population $\mathcal{P}_i$, $i = 1, \ldots, d$. The explicit construction of $(\Omega, \mathscr{F}, \mathrm{P})$ and $(\mathscr{F}_t, t \in \mathbb{I})$ will be one of the objectives of Section 6.2.

Bayesian Assumption

In order to obtain an elegant formulation, it is convenient to straightaway restrict the study to the case where Θ is a finite set with cardinality $l > 1$ and to make the assumption (termed *Bayesian assumption*) that θ, rather than denoting the generic element of Θ, denotes a random variable with values in Θ whose law, or *prior distribution*, is known. The elements of Θ will be denoted $\theta_1, \ldots, \theta_l$ and termed *parameters*. A choice (discrimination) will have to be made between the hypotheses $\theta(\omega) = \theta_1, \ldots, \theta(\omega) = \theta_l$. The prior distribution will be denoted by $\pi = (\pi_1, \ldots, \pi_l)$ (where $\pi_j = \mathrm{P}\{\theta = \theta_j\}$). In practice, the experimenter chooses this distribution either on the basis of previous knowledge or just on intuition.

Strategies

This term has already been used above without definition. We are going to give a mathematical meaning to it, which, however, will be slightly different than the one we shall use in Chapter 7. A *strategy* is a triple (Γ, τ, D) formed by a predictable increasing path Γ, a finite stopping time τ relative to the filtration $\mathscr{F}^\Gamma$ and an $\mathscr{F}^\Gamma_\tau$-measurable random variable D with values in $\{1, \ldots, d\}$. The role of D is that of a *decision variable*, in the sense that if $D(\omega) = j$, then the test concludes that $\theta(\omega) = \theta_j$, therefore that the law of $(Y^1, \ldots, Y^d)$ is $\mu^1_{\theta_j} \times \cdots \times \mu^d_{\theta_j}$.

Costs

It is natural to associate with each observation a cost that can depend on the population from which the observation is made and on the value of θ. We will assume that d functions $f_1, \ldots, f_d$ defined on Θ and with values in $\mathbb{R}^*_+$ are given, and that $f_i(\theta_j)$ represents the cost of making one observation from the ith population when $\theta(\omega) = \theta_j$.

On the other hand, since in a finite number of observations it is generally not possible to decide with certitude whether $\theta(\omega) = \theta_j$ or not, it is equally natural to assume that there is a cost $c_{j,k}$ to deciding that $\theta(\omega) = \theta_k$ when in fact $\theta(\omega) = \theta_j$. In other words, we will suppose that a *cost matrix* $(c_{j,k}, 1 \le j, k \le l)$ with entires in $\mathbb{R}_+$ is given, and that this matrix represents the costs of making incorrect decisions.

The *expected cost* associated with a strategy (Γ, τ, D) is the (not necessarily finite) number

$$L(\Gamma, \tau, D) = \mathrm{E}\left(\sum_{i=1}^{d} \Gamma_\tau^i f_i(\theta) \right) + \sum_{j,k=1}^{l} c_{j,k} \, \mathrm{P}\{\theta = \theta_j, D = k\}. \quad (6.1)$$

Of course, in this expression, Γ_τ^i represents the number of times that population $\mathcal{P}_i$ has been observed up to time τ.

Minimizing the Expected Cost

The first problem is to determine whether there exists a strategy (Γ^0, τ^0, D^0) such that $L(\Gamma, \tau, D) \ge L(\Gamma^0, \tau^0, D^0)$ for all other strategies (Γ, τ, D). Such a strategy will be termed *optimal*.

After having reduced this problem to an optimal stopping problem, we will show that the answer to that question is affirmative. As we are about to explain, this reduction begins with an optimal choice of D for each couple (Γ, τ).

Given a strategy (Γ, τ, D), let

$$R_k = \sum_{j=1}^{l} c_{j,k} \, \mathrm{P}\left(\theta = \theta_j | \mathcal{F}_\tau^\Gamma\right)$$

denote the expected cost, given $\mathcal{F}_\tau^\Gamma$, of deciding that $\theta(\omega) = \theta_k$, and let

$$D^o = \inf\left\{k : R_k = \inf_{1 \le j \le l} R_j\right\}. \quad (6.2)$$

Observe that

$$\begin{aligned} \sum_{j,k=1}^{l} c_{j,k} \, \mathrm{P}\{\theta = \theta_j, D = k\} &= \mathrm{E}\left(\sum_{k=1}^{l} 1_{\{D=k\}} \sum_{j=1}^{l} c_{j,k} \, \mathrm{P}\left(\theta = \theta_j | \mathcal{F}_\tau^\Gamma\right) \right) \\ &\ge \mathrm{E}\left(\left(\sum_{k=1}^{l} 1_{\{D=k\}} \right) \left(\inf_{1 \le k \le l} R_k \right) \right) \\ &= \mathrm{E}\left(\inf_{1 \le k \le l} R_k \right). \end{aligned}$$

On the other hand, the inequality becomes an equality if D is replaced by

D^o, which implies that

$$L(\Gamma, \tau, D) \geq L(\Gamma, \tau, D^o).$$

In other words, there is never any advantage to using a decision variable other than D^o.

The Associated Optimal Stopping Problem

Let X denote the process indexed by ▯ and defined by

$$X_t = -\sum_{i=1}^{d} t^i \, \mathrm{E}(f_i(\theta)|\mathscr{F}_t) - g(\pi_1^t, \dots, \pi_l^t), \tag{6.3}$$

where g is the function defined by

$$g(x_1, \dots, x_l) = \inf_{1 \leq k \leq l} \sum_{j=1}^{l} c_{j,k} \, x_j \tag{6.4}$$

and

$$\pi_j^t = \mathrm{P}(\theta = \theta_j|\mathscr{F}_t). \tag{6.5}$$

Clearly, X is an adapted process made up of bounded and nonpositive random variables, and the quantity $-X_t$ represents the conditional expected cost given $\mathscr{F}_t$ of observing population $\mathcal{P}_i$ exactly t^i times, $i = 1, \dots, d$, and then making the optimal decision D^o defined in (6.2). On the other hand, if (Γ, τ, D) is the strategy introduced above and D^o is the random variable defined in (6.2), then taking into account that

$$\inf_{1 \leq k \leq l} R_k = g(\pi_1^{\Gamma_\tau}, \dots, \pi_l^{\Gamma_\tau})$$

and

$$\mathrm{E}\Big(\inf_{1 \leq k \leq l} R_k\Big) = \sum_{j,k=1}^{l} \mathrm{P}\{\theta = \theta_j, D^o = k\},$$

we observe by (6.1) and (6.3) that

$$\begin{aligned} -\mathrm{E}(X_\tau^\Gamma) &= \mathrm{E}\left(\sum_{i=1}^{d} \Gamma_\tau^i \, \mathrm{E}(f_i(\theta)|\mathscr{F}_\tau^\Gamma) + \inf_{1 \leq k \leq l} R_k\right) \\ &= \mathrm{E}\left(\sum_{i=1}^{d} \Gamma_\tau^i \, f_i(\theta) + \inf_{1 \leq k \leq l} R_k\right) \\ &= L(\Gamma, \tau, D^o). \end{aligned}$$

Consequently, minimizing $L(\Gamma, \tau, D)$ is equivalent to determining an optimal accessible stopping point for the reward process X, in other words, a stopping point $T^o \in \mathbf{A}_f$ such that

$$\mathrm{E}(X_{T^o}) = \sup_{T \in \mathbf{A}_f} \mathrm{E}(X_T).$$

Existence of a Solution

Observe that

$$X_t \leq -\sum_{i=1}^{d} t^i \,\mathrm{E}(f_i(\theta)|\mathscr{F}_t) \leq -|t|\,\mathrm{E}\Big(\inf_{1 \leq i \leq d} f_i(\theta)|\mathscr{F}_t\Big) \leq -c|t|,$$

where c denotes the smallest value taken by the functions $f_1, \ldots, f_d$. It follows that

$$\lim_{t \to \bowtie} X_t = -\infty.$$

Since $X \leq 0$, the hypotheses of Theorem 1 of Section 3.4 are satisfied, which implies that an optimal accessible stopping point $T^o \in \mathbf{A}_f$ exists. Notice that this stopping point is integrable in the sense that $\mathrm{E}(|T^o|) < \infty$. In fact, this conclusion holds for all stopping points such that $\mathrm{E}(X_T) > -\infty$, because if c is defined as above, then

$$c\,\mathrm{E}(|T|) \leq \sum_{i=1}^{d} \mathrm{E}(T^i\,\mathrm{E}(f_i(\theta)|\mathscr{F}_T)) \leq -\mathrm{E}(X_T) < \infty.$$

The problem of existence of an optimal strategy is thus solved. Our intention is now to describe such a strategy using a model constructed from data that relate to the examples presented at the beginning of this chapter. But first, here is a remark that will be useful in Section 6.3.

Remark

Because of the particular form of the process X, the value of the game and Snell's envelope Z associated with X can be computed by taking into account only bounded accessible stopping points. More precisely,

$$\sup_{T \in \mathbf{A}_f} \mathrm{E}(X_T) = \sup_{T \in \mathbf{A}_b} \mathrm{E}(X_T)$$

and

$$Z_t = \operatorname*{esssup}_{T \in \mathbf{A}_f^t} \mathrm{E}(X_T|\mathscr{F}_t) = \operatorname*{esssup}_{T \in \mathbf{A}_b^t} \mathrm{E}(X_T|\mathscr{F}_t),$$

where $\mathbf{A}_b$ and $\mathbf{A}'_b$ denote, respectively, the set of bounded accessible stopping points and the set of bounded accessible stopping points $\geq t$. Indeed, if $E(X_T) > -\infty$ (which implies that $E(|T|) < \infty$) and if $(T_n, n \in \mathbb{N})$ is a nondecreasing sequence of stopping points such that $\bigvee_n T_n = T$, then by the dominated convergence theorem,

$$\lim_{n\to\infty} E(X_{T_n}) = E(X_T) \quad \text{and} \quad \lim_{n\to\infty} E(X_{T_n}|\mathscr{F}_t) = E(X_T|\mathscr{F}_t).$$

The assertion follows, because every element of $\mathbf{A}_f$ $[\mathbf{A}'_f]$ is the limit of a nondecreasing sequence of elements of $\mathbf{A}_b$ $[\mathbf{A}'_b]$.

6.2 CONSTRUCTING THE MODEL

A model suitable for describing the examples mentioned in the introduction must take into account certain assumptions that have not yet been stated explicitly. These assumptions are formulated in a, b, and c below.

The starting point is a finite set $\Theta = \{\theta_1, \ldots, \theta_l\}$, a (prior) distribution $\pi = (\pi_1, \ldots, \pi_l)$ on Θ such that $\pi_1 > 0, \ldots, \pi_l > 0$, and a family $(\mu^i_j, 1 \leq i \leq d, 1 \leq j \leq l)$ of probability measures on $\mathbb{R}$. We shall build a complete probability space $(\Omega, \mathscr{F}, P)$, d processes $(Y^1_{t^1}, t^1 \in \mathbb{N}^*), \ldots, (Y^d_{t^d}, t^d \in \mathbb{N}^*)$ and a random variable θ with values in Θ, all defined on this space, in such a way that the following conditions are satisfied:

a. θ is distributed according to the (prior) distribution π.

b. Conditionally with respect to $\{\theta = \theta_j\}$, the random variables $Y^i_1, Y^i_2, \ldots$ are independent and identically distributed with common law μ^i_j ($i = 1, \ldots, d$; $j = 1, \ldots, l$).

c. Conditionally with respect to $\{\theta = \theta_j\}$, the processes $(Y^1_{t^1}, t^1 \in \mathbb{N}^*), \ldots, (Y^d_{t^d}, t^d \in \mathbb{N}^*)$ are independent ($j = 1, \ldots, l$).

Of course, $Y^i_{t^i}$ represents the result of the t^ith draw from population $\mathcal{P}_i$. If the result of the test is that $\theta(\omega) = \theta_j$, then the experimenter will conclude that the law of $(Y^1_1, \ldots, Y^d_1)$ is $\mu^1_j \times \cdots \times \mu^d_j$.

Since we are only given the laws π and μ^i_j, we shall turn to a canonical model. Set

$$\Omega = \Theta \times \tilde{\Omega},$$

where

$$\tilde{\Omega} = \Omega_1 \times \cdots \times \Omega_d \quad \text{and} \quad \Omega_i = \mathbb{R}^{\mathbb{N}^*},$$

and let ω_i, $\tilde{\omega}$ and ω denote the generic element of Ω_i, $\tilde{\Omega}$, and Ω, respectively. By definition, $\omega_i = (\omega_i(t^i), t^i \in \mathbb{N}^*)$, $\tilde{\omega} = (\omega_1, \ldots, \omega_d)$, $\tilde{\omega}(t) = (\omega_1(t^1), \ldots, \omega_d(t^d))$, and $\omega = (\theta_j, \tilde{\omega})$. A useful feature of $\tilde{\Omega}$ and Ω is that we can use the *shift operators* $\tilde{\rho}_u$ and ρ_u ($u \in \mathbb{I}$) defined by

$$(\tilde{\rho}_u(\tilde{\omega}))(t) = \tilde{\omega}(t+u) \quad \text{and} \quad \rho_u(\theta_j, \tilde{\omega}) = (\theta_j, \tilde{\rho}_u(\tilde{\omega})).$$

Let $\mathscr{B}$ denote the Borel sigma field of $\mathbb{R}$, set

$$\tilde{\mathscr{F}} = \mathscr{B}^{\mathbb{N}^*} \times \cdots \times \mathscr{B}^{\mathbb{N}^*} \ (d \text{ factors}),$$

and define probability measures P_j^i on $\mathscr{B}^{\mathbb{N}^*}$, $\tilde{\mathrm{P}}_j$ on $\tilde{\mathscr{F}}$, and P on $\mathscr{P}(\Theta) \times \tilde{\mathscr{F}}$ by

$$\mathrm{P}_j^i = \mu_j^i \times \mu_j^i \times \cdots, \qquad \tilde{\mathrm{P}}_j = \mathrm{P}_j^1 \times \cdots \times \mathrm{P}_j^d,$$
$$\mathrm{P} = \sum_{j=1}^{l} \pi_j\left(\varepsilon_j \times \tilde{\mathrm{P}}_j\right), \tag{6.6}$$

where $\varepsilon_j(\{\theta_k\})$ equals 0 if $k \neq j$ and 1 if $k = j$. Let $\tilde{Y}_{t^i}^i$, $Y_{t^i}^i$ $(t^i \in \mathbb{N}^*)$ and θ denote the random variables defined, respectively, on $\tilde{\Omega}$ and Ω by

$$\tilde{Y}_{t^i}^i(\tilde{\omega}) = \omega_i(t^i), \qquad Y_{t^i}^i(\theta_j, \tilde{\omega}) = \tilde{Y}_{t^i}^i(\tilde{\omega}), \qquad \theta(\theta_j, \tilde{\omega}) = \theta_j.$$

Notice that if $\mathscr{F}$ is the sigma field obtained by completing $\mathscr{P}(\Theta) \times \tilde{\mathscr{F}}$ with respect to P, then $(\Omega, \mathscr{F}, \mathrm{P})$ is a complete probability space, and $(Y_{t^1}^1, t^1 \in \mathbb{N}^*), \ldots, (Y_{t^d}^d, t^d \in \mathbb{N}^*)$ and θ are defined on Ω and satisfy the requirements in a, b, and c. Indeed, a is satisfied, because

$$\mathrm{P}\{\theta = \theta_k\} = \sum_{j=1}^{l} \pi_j\left(\varepsilon_j \times \tilde{\mathrm{P}}\right)\{\theta = \theta_k\} = \sum_{j=1}^{l} \pi_j\, \varepsilon_j(\{\theta_k\}) = \pi_k,$$

and b and c follow directly from the relation

$$\begin{aligned}
&\mathrm{P}\left(\bigcap_{i=1}^{d} \{(Y_1^i, \ldots, Y_{t^i}^i) \in B_1^i \times \cdots \times B_{t^i}^i\} \,\middle|\, \theta = \theta_k\right) \\
&\quad = \left(\varepsilon_k \times \tilde{P}_k\right)\left(\bigcap_{i=1}^{d} \{(Y_1^i, \ldots, Y_{t^i}^i) \in B_1^i \times \cdots \times B_{t^i}^i\}\right) \\
&\quad = \prod_{i=1}^{d} \mu_k^i(B_1^i) \cdots \mu_k^i(B_{t^i}^i),
\end{aligned}$$

which is a direct consequence of (6.6).

The Sets I_t

For all $t \in \mathbb{I}^*$, I_t will denote the set of indices $i \in \{1, \ldots, d\}$ such that $t^i > 0$. Notice that a family indexed by the couples (i, s^i), where $s^i = 1, \ldots, t^i$ and $i \in I_t$, has $|t|$ terms.

The Filtered Probability Space

In the remainder of this chapter, the underlying probability space will be the triple $(\Omega, \mathscr{F}, \mathrm{P})$ that we have just defined. This space will be equipped with the filtration $(\mathscr{F}_t,\ t \in \mathbb{I})$, where $\mathscr{F}_0$ is the sigma field generated by the null sets of $\mathscr{F}$ and, for $t \neq 0$, $\mathscr{F}_t$ denotes the sigma field generated by the family of random variables $(Y^i_{s^i},\ s^i = 1, \ldots, t^i,\ i \in I_t)$. In agreement with one of our conventions, $\mathscr{F}_{\bowtie -}$ will denote the sigma field $\bigvee_{t \in \mathbb{I}} \mathscr{F}_t$. Notice the particular manner in which the sigma fields $\mathscr{F}_t$ increase with t, more precisely, that $\mathscr{F}_{t+e_i}$ is generated by $\mathscr{F}_t$ and $Y^i_{t^i+1}$.

Along with the basic filtration, we will use the noncompleted filtration $(\tilde{\mathscr{F}}_t,\ t \in \mathbb{I})$ associated with the variables $\tilde{Y}^i_{t^i}$. That is, $\tilde{\mathscr{F}}_0$ is the sigma field $\{\varnothing, \tilde{\Omega}\}$, and for all $t \neq 0$, $\tilde{\mathscr{F}}_t$ is the smallest sub-sigma field of $\tilde{\mathscr{F}}$ that contains all sets of the form $\{\tilde{Y}^i_{s^i} \in B\}$, with $s^i = 1, \ldots, t^i$, $i \in I_t$ and $B \in \mathscr{B}$. Obviously, $\mathscr{F}_t$ is the completion of the sigma field $\{\varnothing, \Theta\} \times \tilde{\mathscr{F}}_t$ by the null sets of $\mathscr{F}$, so that any $F \in \mathscr{F}_t$ can be thought of as a set of the form $\Theta \times \tilde{F}$, with $\tilde{F} \in \tilde{\mathscr{F}}_t$.

The Probability Measures P_λ

Let Λ be the set of probability laws λ on Θ. Clearly, $\lambda \in \Lambda$ can be identified with $(\lambda_1, \ldots, \lambda_l)$, where $\lambda_j = \lambda(\{\theta_j\})$, $j = 1, \ldots, l$, and with this identification, Λ becomes a subset of $\mathbb{R}^l$. For all $\lambda \in \Lambda$, we will denote by P_λ the probability measure defined as P in (6.6), but using λ instead of π. This probability measure is first defined on $\mathscr{P}(\Theta) \times \tilde{\mathscr{F}}$, then extended to the completion of this sigma field with respect to P_λ. This completed sigma field contains $\mathscr{F}$ and the inclusion is generally strict, because P_λ is absolutely continuous with respect to P but is not equivalent to P unless $\lambda_1 > 0, \ldots, \lambda_l > 0$. Further on, we will only use P_λ as a probability measure on $\mathscr{F}$, and the fact that $\mathscr{F}$ is not complete for P_λ will not cause any problem. Since any set in $\mathscr{F}$ with probability 0 is also a P_λ-null set, the final remark of the last paragraph concerning sets $F \in \mathscr{F}_t$ also applies to the case where the probability measure under consideration is P_λ. The expression "P_λ-a.s." will mean "except on a P_λ-null set." The operators of expectation, conditional probability, and conditional expectation.0 corresponding to P_λ will be indicated by the usual symbols tagged with the subscript λ. It is clear that for all $F \in \mathscr{F}$ and all bounded and nonnegative random variables ξ, $\lambda \mapsto \mathrm{P}_\lambda(F)$ and $\lambda \mapsto \mathrm{E}_\lambda(\xi)$ are Borel functions.

Defining Y^t and $\tilde{Y}^t$

For each $t \in \mathbb{I}^*$, we will denote the families $(Y^i_{s^i},\ s^i = 1, \ldots, t^i,\ i \in I_t)$ and $(\tilde{Y}^i_{s^i},\ s^i = 1, \ldots, t^i,\ i \in I_t)$ by Y^t and $\tilde{Y}^t$ respectively. It is natural to perceive these families as functions with values in $\mathbb{R}^{|t|}$ whose respective values at ω and $\tilde{\omega}$ are $(Y^i_{s^i}(\omega),\ s^i = 1, \ldots, t^i,\ i \in I_t)$ and $(\tilde{Y}_{s^i}(\tilde{\omega}),\ s^i = 1, \ldots, t^i,\ i \in I_t)$. As such, it is obvious that Y^t and $\tilde{Y}^t$ are, respectively, $\mathscr{F}_t$-measurable and

$\tilde{\mathscr{F}}_t$-measurable. We complete the definition by letting Y^0 and $\tilde{Y}^0$ equal the constant 0. If T is a finite stopping point, Y^T will denote the function $\omega \mapsto Y^{T(\omega)}(\omega)$. Of course, for all $t \neq 0$, Y^t represents the result of the set of draws up to stage t.

Bayes' Formula and the Markov Property

We shall now establish two preliminary results that will play an important role and are also of independent interest.

Define a measure μ^i on $\mathbb{R}$ by setting

$$\mu^i = \mu_1^i + \cdots + \mu_l^i$$

and let δ_j^i denote the density of μ_j^i with respect to μ^i. For all $t \in \mathbb{I}^*$, let φ_j^t denote the Borel function defined on $\mathbb{R}^{|t|}$ by

$$\varphi_j^t(y^t) = \prod_{i \in I_t} \prod_{s^i=1}^{t^i} \delta_j^i(y_{s^i}^i), \tag{6.7}$$

where y^t denotes the element $(y_{s^i}^i, s^i = 1, \ldots, t^i, i \in I_t)$ of $\mathbb{R}^{|t|}$. Moreover, set $\varphi_j^0(0) = 1$, and finally, for all $t \in \mathbb{I}$ and all laws $\lambda \in \Lambda$, let

$$\lambda_j^t = \mathrm{P}_\lambda(\theta = \theta_j | \mathscr{F}_t) \tag{6.8}$$

and

$$\lambda^t = (\lambda_1^t, \ldots, \lambda_l^t). \tag{6.9}$$

It is clear that $\lambda_j^0 = \lambda_j$, and therefore that $\lambda^0 = \lambda$ for all $\lambda \in \Lambda$. Moreover,

$$\lambda^t = \lambda \ \mathrm{P}_\lambda\text{-a.s.} \qquad \text{if } \lambda_j = 0 \text{ for } j \neq k, \text{ and } \lambda_k = 1. \tag{6.10}$$

Indeed, if $F = \Theta \times \tilde{F}$ with $\tilde{F} \in \mathscr{F}_t$, then for $j = 1, \ldots, l$,

$$\int_F \lambda_j^t \, d\mathrm{P}_\lambda = \mathrm{P}_\lambda(\{\theta = \theta_j\} \cap F) = \mathrm{P}_\lambda(\{\theta_j\} \times \tilde{F}) = \varepsilon_k(\{\theta_j\}) \tilde{\mathrm{P}}_k(\tilde{F}) = \mathrm{P}_\lambda(F)\lambda_j,$$

which proves the assertion. Notice that in the case where $\lambda = \pi$, λ_j^t is nothing but the random variable π_j^t defined in (6.5).

Proposition 1 (Bayes' formula). *For all $t \in \mathbb{I}$ and all laws $\lambda \in \Lambda$,*

$$\lambda_j^t = \frac{\lambda_j \, \varphi_j^t(Y^t)}{\sum_{k=1}^{l} \lambda_k \, \varphi_k^t(Y^t)} \quad \mathrm{P}_\lambda\text{-a.s.} \tag{6.11}$$

Proof. The conclusion is obvious when $t = 0$; therefore we assume that $t \neq 0$. Let μ denote the measure on $\mathscr{B}^{|t|}$ defined by

$$\mu = \mathop{\times}_{i \in I_t} \hat{\mu}^i, \qquad \text{where } \hat{\mu}^i = \mu^i \times \cdots \times \mu^i \ (t^i \text{ factors}).$$

We begin by proving that for all sets $B \in \mathscr{B}^{|t|}$,

$$\tilde{\mathrm{P}}_j\{\tilde{Y}^t \in B\} = \int_B \varphi_j^t \, d\mu. \tag{6.12}$$

Clearly, it is sufficient to examine the case where $B = \times_{i \in I_t} \times_{s^i=1}^{t^i} B_{s^i}^i$ and $B_{s^i}^i \in \mathscr{B}$. In this case,

$$\begin{aligned}
\tilde{P}_j\{\tilde{Y}^t \in B\} &= \int_{\tilde{\Omega}} \left(\prod_{i \in I_t} \prod_{s^i=1}^{t^i} 1_{B_{s^i}^i}(\tilde{Y}_{s^i}^i) \right) d\tilde{\mathrm{P}}_j \\
&= \prod_{i \in I_t} \prod_{s^i=1}^{t^i} \mu_j^i(B_{s^i}^i) \\
&= \prod_{i \in I_t} \prod_{s^i=1}^{t^i} \int_{B_{s^i}^i} \delta_j^i \, d\mu^i \\
&= \int_B \varphi_j^t \, d\mu.
\end{aligned}$$

Using (6.12), we now observe that for all sets $B \in \mathscr{B}^{|t|}$,

$$\mathrm{P}_\lambda\{Y^t \in B\} = \sum_{j=1}^{l} \lambda_j \, \tilde{\mathrm{P}}_j\{\tilde{Y}^t \in B\} = \int_B \sum_{j=1}^{l} \lambda_j \, \varphi_j^t \, d\mu,$$

which means that the integrand on the right-hand side is the density (with respect to μ) of Y^t under P_λ. This implies in particular that the denominator on the right-hand side of (6.11) is P_λ-a.s. not zero (as the composition of a random variable with its density function). Now let $F \in \mathscr{F}_t$. By modifying F on a null set, we can assume that $F = \{Y^t \in B\}$ for some $B \in \mathscr{B}^{|t|}$. Using the density of Y^t and (6.12), we can write

$$\int_F \frac{\lambda_j\, \varphi_j^t(Y^t)}{\sum_{k=1}^{l} \lambda_k\, \varphi_k^t(Y^t)}\, \mathrm{dP}_\lambda = \int_B \frac{\lambda_j\, \varphi_j^t}{\sum_{k=1}^{l} \lambda_k\, \varphi_k^t} \left(\sum_{k=1}^{l} \lambda_k\, \varphi_k^t \right) d\mu = \lambda_j \int_B \varphi_j^t\, d\mu,$$

which is equal to

$$\lambda_j \tilde{\mathrm{P}}_j\{\tilde{Y}^t \in B\} = \mathrm{P}_\lambda\big(\{\theta = \theta_j\} \cap \{Y^t \in B\}\big) = \mathrm{P}_\lambda\big(\{\theta = \theta_j\} \cap F\big) = \int_F \lambda_j^t\, \mathrm{dP}_\lambda,$$

which proves (6.11), since F is an arbitrary element of $\mathscr{F}_t$.

Proposition 2 (Markov property). *Let ξ be a bounded or nonnegative function defined on $\Omega \times \Omega$ which is $\mathscr{F}_t \times \mathscr{F}_{\bowtie -}$-measurable. Let ξ_t denote the random variable $\omega \mapsto \xi(\omega, \rho_t(\omega))$, where ρ_t is the shift operator on Ω. Then for all $t \in \mathbb{I}$ and all laws $\lambda \in \Lambda$,*

$$\mathrm{E}_\lambda(\xi_t | \mathscr{F}_t)(\omega) = \int_\Omega \xi(\omega, \omega')\, \mathrm{dP}_{\lambda^t(\omega)}(\omega'), \qquad \text{for } \mathrm{P}_\lambda\text{-almost all } \omega. \quad (6.13)$$

Proof. It is sufficient to prove that if ξ_2 is a bounded or nonnegative $\mathscr{F}_{\bowtie -}$-measurable random variable, then

$$\mathrm{E}_\lambda(\xi_2 \circ \rho_t | \mathscr{F}_t) = \mathrm{E}_{\lambda^t}(\xi_2), \qquad \mathrm{P}_\lambda\text{-a.s.}, \quad (6.14)$$

because this immediately implies the validity of (6.13) in the case where $\xi(\omega, \omega') = \xi_1(\omega)\xi_2(\omega')$, and the general case follows by taking linear combinations and monotone limits of such combinations. Before proving (6.14), observe that if $\tilde{F} \in \tilde{\mathscr{F}}_t$ and $\tilde{G} \in \tilde{\mathscr{F}}$, then

$$\tilde{\mathrm{P}}_j\big(\tilde{F} \cap \tilde{\rho}_t^{-1}(\tilde{G})\big) = \tilde{\mathrm{P}}_j(\tilde{F})\, \tilde{\mathrm{P}}_j(\tilde{G}). \quad (6.15)$$

Indeed, by the definition of $\tilde{\mathrm{P}}_j$, this relation holds for all $\tilde{F}$ and all $\tilde{G}$ of the particular form

$$\tilde{F} = F_1 \times \cdots \times F_d, \qquad \text{where } F_i = \begin{cases} B_1^i \times \cdots \times B_{t^i}^i \times \mathbb{R} \times \cdots, & \text{if } i \in I_t, \\ \Omega_i, & \text{if } i \notin I_t, \end{cases}$$

$$\tilde{G} = G_1 \times \cdots \times G_d, \qquad \text{where } G_i = C_1^i \times \cdots \times C_{u^i}^i \times \mathbb{R} \times \cdots \big(B_{s^i}^i, C_{s^i}^i \in \mathscr{B}\big).$$

Consequently it holds for all $\tilde{F} \in \tilde{\mathscr{F}}_t$ and all $\tilde{G} \in \tilde{\mathscr{F}}$. Now let $F = \Theta \times \tilde{F}$, with $\tilde{F} \in \tilde{\mathscr{F}}_t$, and let ξ_2 be an $\tilde{\mathscr{F}}$-measurable function such that $\xi_2(\theta_j, \tilde{\omega}) = \tilde{\xi}_2(\tilde{\omega})$ for almost all $(\theta_j, \tilde{\omega})$ (therefore for P_λ-almost all $(\theta_j, \tilde{\omega})$, for any λ). Using (6.15) to justify the fourth step below, we see that

$$\begin{aligned}
\int_F E_{\lambda^t}(\xi_2)\, dP_\lambda &= \int_F \left(\sum_{j=1}^{l} \lambda_j^t \int_{\tilde{\Omega}} \tilde{\xi}_2\, d\tilde{P}_j \right) dP_\lambda \\
&= \sum_{j=1}^{l} \int_F P_\lambda(\theta = \theta_j | \mathscr{F}_t)\, dP_\lambda \int_{\tilde{\Omega}} \tilde{\xi}_2\, d\tilde{P}_j \\
&= \sum_{j=1}^{l} \lambda_j\, \tilde{P}_j(\tilde{F}) \int_{\tilde{\Omega}} \tilde{\xi}_2\, d\tilde{P}_j \\
&= \sum_{j=1}^{l} \lambda_j \int_{\tilde{F}} \tilde{\xi}_2 \circ \tilde{\rho}_t\, d\tilde{P}_j \\
&= \int_F \xi_2 \circ \rho_t\, dP_\lambda,
\end{aligned}$$

which establishes (6.14), because F is an arbitrary element of $\mathscr{F}_t$.

6.3 THE REWARD PROCESS AND SNELL'S ENVELOPE

This section proceeds in the same spirit as Section 6.2. The objective is to find a formula that expresses the reward process and Snell's envelope in terms of the observations.

The Reward Process

In addition to the set Θ, the distribution π, and the family $(\mu_j^i, 1 \le i \le d, 1 \le j \le l)$ mentioned at the beginning of Section 6.2, we now need the functions $f_1, \dots, f_d$ and the cost matrix $(c_{j,k}, 1 \le j, k \le l)$ introduced in Section 6.1.

Let Z denote Snell's envelope of the reward process X defined in (6.3). In order to express Z_t as a function of Y^t, it is useful to consider processes X^λ defined analogously to X by replacing P by P_λ. For all $\lambda \in \Lambda$, X^λ will therefore denote the process indexed by $\mathbb{I}$ defined by

$$X_t^\lambda = -\sum_{i=1}^{d} t^i E_\lambda(f_i(\theta) | \mathscr{F}_t) - g(\lambda^t), \tag{6.16}$$

where g is the function defined on Λ by

$$g(\lambda) = g(\lambda_1, \dots, \lambda_l) = \inf_{1 \le k \le l} \sum_{j=1}^{l} c_{j,k} \lambda_j, \tag{6.17}$$

and λ^t is the conditional law defined in (6.8) and (6.9). Note that $X_0^\lambda = -g(\lambda)$. On the other hand, by (6.3)–(6.5), it is clear that $X^\pi = X$.

Our first objective is to determine the relationship between X_t^λ and Y^t. By (6.11), we know that λ^t is a function of Y^t. It is therefore sufficient to relate X_t^λ to λ^t. For this purpose and for further use, we will denote by f^t the function defined on Λ by

$$f^t(\lambda) = f^t(\lambda_1, \dots, \lambda_l) = \sum_{i=1}^{d} t^i \sum_{j=1}^{l} f_i(\theta_j) \lambda_j. \tag{6.18}$$

Proposition (X_t^λ as a function of λ^t). *For all laws $\lambda \in \Lambda$ and all $t \in \mathbb{I}$,*

$$X_t^\lambda = -f^t(\lambda^t) - g(\lambda^t), \qquad \mathrm{P}_\lambda\text{-a.s.} \tag{6.19}$$

Proof. Because

$$\mathrm{E}_\lambda(f_i(\theta)|\mathscr{F}_t) = \sum_{j=1}^{l} f_i(\theta_j)\,\mathrm{P}_\lambda(\theta = \theta_j|\mathscr{F}_t) = \sum_{j=1}^{l} f_i(\theta_j)\,\lambda_j^t, \qquad \mathrm{P}_\lambda\text{-a.s.}, \tag{6.20}$$

it follows from (6.18) that

$$f^t(\lambda^t) = \sum_{i=1}^{d} t^i\,\mathrm{E}_\lambda(f_i(\theta)|\mathscr{F}_t), \qquad \mathrm{P}_\lambda\text{-a.s.} \tag{6.21}$$

It remains to refer to (6.16) to realize that the assertion is proved.

Snell's Envelope

Our objective is now to determine an expression that relates Z_t to the conditional law λ^t, therefore to Y^t thanks to (6.11). As already mentioned, this will be achieved through the use of the processes X^λ. At the same time, the result will apply to Snell's envelope Z^λ of X^λ.

Let h denote the function defined on Λ by

$$h(\lambda) = h(\lambda_1, \dots, \lambda_l) = \sup_{T \in \mathbf{A}_b} \mathrm{E}_\lambda(X_T^\lambda). \tag{6.22}$$

We will see in the next section that h is the limit of a sequence of functions defined by induction starting from g. In this section, we are going to show how h determines Z^{λ}. But first, let us point out some properties of h.

By (6.16) and (6.22), $-g(\lambda) = \mathrm{E}_{\lambda}(X_0^{\lambda}) \leq h(\lambda) \leq 0$; that is, $-g \leq h \leq 0$. On the other hand, if λ is of the particular form indicated in (6.10), then $X_t^{\lambda} \leq -g(\lambda)\,\mathrm{P}_{\lambda}$-a.s., for all $t \in \mathbb{I}$, by (6.16) and (6.10). Consequently $-g(\lambda) \geq h(\lambda)$, and therefore

$$-g(\lambda) = h(\lambda), \qquad \text{if } \lambda_j = 0 \text{ for } j \neq k \text{ and } \lambda_k = 1. \tag{6.23}$$

Let us show that h is a convex function (therefore a Borel function that is continuous in the interior of Λ). By what was established in Section 6.1,

$$h(\lambda) = \sup(-L_{\lambda}(\Gamma, \tau, D)),$$

where the supremum is taken over strategies (Γ, τ, D) such that τ is bounded and $L_{\lambda}(\Gamma, \tau, D)$ is defined as in (6.1) but taking P_{λ} instead of P. By the definition of P_{λ}, $L_{\lambda}(\Gamma, \tau, D)$ is a linear (affine) function of λ. It follows that the supremum is a convex function of λ, which was to be proved.

Moreover, endowed with the metric of uniform convergence, the space of linear (affine) functions on Λ is separable, and therefore every subset of this space is separable. It follows that there exists a countable set $\mathbf{S}$ of strategies (Γ, τ, D), with τ bounded, such that

$$h(\lambda) = \sup_{(\Gamma, \tau, D) \in \mathbf{S}} (-L_{\lambda}(\Gamma, \tau, D)), \qquad \text{for all } \lambda \in \Lambda.$$

Let $\mathbf{D}$ denote the set of stopping points Γ_{τ} such that $(\Gamma, \tau, D) \in \mathbf{S}$. Repeating the computations used to derive the associated optimal stopping problem in Section 6.1, we see that if T denotes Γ_{τ}, then

$$-L_{\lambda}(\Gamma, \tau, D) \leq \mathrm{E}_{\lambda}(X_T^{\lambda}), \qquad \text{for all } \lambda \in \Lambda.$$

Consequently,

$$h(\lambda) \leq \sup_{T \in \mathbf{D}} \mathrm{E}_{\lambda}(X_T^{\lambda}) \leq \sup_{T \in \mathbf{A}_b} \mathrm{E}_{\lambda}(X_T^{\lambda}) = h(\lambda), \qquad \text{for all } \lambda \in \Lambda.$$

We have therefore established that there exists a countable subset $\mathbf{D}$ of $\mathbf{A}_b$ such that

$$h(\lambda) = \sup_{T \in \mathbf{D}} \mathrm{E}_{\lambda}(X_T^{\lambda}), \qquad \text{for all } \lambda \in \Lambda.$$

The following lemma will play a key role in the proof of the theorem that follows.

Lemma. *Let $t \in \mathbb{I}$ and $T \in \mathbf{A}_b^t$. Then there exists a function U defined on $\Omega \times \Omega$ with values in $\mathbb{I}$ that has the following properties:*

a. *U is $\mathscr{F}_t \times \mathscr{F}_{\bowtie -}$-measurable.*
b. *$U(\omega, \cdot)$ is an element of $\mathbf{A}_b$ for all ω.*
c. *$T(\omega) = t + U(\omega, \rho_t(\omega))$ for almost all ω.*

Because the proof of this lemma is rather lengthy, it is deferred to Section 6.7.

Theorem (Z_t^λ as a function of λ^t). *For all laws $\lambda \in \Lambda$, Snell's envelope Z^λ of the process X^λ satisfies*

$$Z_t^\lambda = -f^t(\lambda^t) + h(\lambda^t), \qquad \mathrm{P}_\lambda\text{-a.s.}, \tag{6.24}$$

for all $t \in \mathbb{I}$.

Proof. In order to simplify the notation, we write Φ_j^t instead of $\varphi_j^t(Y^t)$. Let $T \in \mathbf{A}_b^t$ and let U be a function satisfying properties a, b, and c of the lemma. By (6.18) and property c,

$$f^{T(\omega)} = f^t + f^{U(\omega,\, \rho_t(\omega))}. \tag{6.25}$$

On the other hand, by (6.7) and property c,

$$\Phi_j^{T(\omega)}(\omega) = \Phi_j^t(\omega)\Phi_j^{T(\omega)-t}(\rho_t(\omega)) = \Phi_j^t(\omega)\Phi_j^{U(\omega,\, \rho_t(\omega))}(\rho_t(\omega)). \tag{6.26}$$

Let $\bar{f}^t$ and $\bar{g}$ denote, respectively, the extensions of f^t and g to $\mathbb{R}_+^l$ obtained by setting $\bar{f}^t(0) = \bar{g}(0) = 0$, and if $x \neq 0$, $\bar{f}^t(x) = f^t(x/|x|)$ $[\bar{g}(x) = g(x/|x|)]$, where $|x| = \sum_{j=1}^l x_j$. Moreover, let us agree that if r is a function defined on $\{1, \ldots, l\}$ with values in $\mathbb{R}_+^l$, then $f^t(r(j))$ and $g(r(j))$ denote, respectively, $f^t(r(1), \ldots, r(l))$ and $g(r(1), \ldots, r(l))$. By (6.19), (6.11), (6.25), and (6.26), we see that for P_λ-almost all ω,

$$\begin{aligned} X_{T(\omega)}^\lambda(\omega) &= -\bar{f}^{T(\omega)}\big(\lambda_j \Phi_j^{T(\omega)}(\omega)\big) - \bar{g}\big(\lambda_j \Phi_j^{T(\omega)}(\omega)\big) \\ &= -\bar{f}^t\big(\lambda_j \Phi_j^{T(\omega)}(\omega)\big) - \bar{f}^{U(\omega,\, \rho_t(\omega))}\big(\lambda_j \Phi_j^t(\omega)\Phi_j^{U(\omega,\, \rho_t(\omega))}(\rho_t(\omega))\big) \\ &\quad - \bar{g}\big(\lambda_j \Phi_j^t(\omega)\Phi_j^{U(\omega,\, \rho_t(\omega))}(\rho_t(\omega))\big). \end{aligned}$$

By (6.11), (6.18), and (6.20), it follows that for P_λ-almost all ω,

$$X_{T(\omega)}^\lambda(\omega) = -\sum_{i=1}^d t^i\, \mathrm{E}_\lambda(f_i(\theta) | \mathscr{F}_T)(\omega) - \xi(\omega, \rho_t(\omega)),$$

where ξ denotes the nonnegative $\mathscr{F}_t \times \mathscr{F}_{\infty-}$-measurable function defined by

$$\xi(\omega, \omega') = \bar{f}^{U(\omega,\omega')}\big(\lambda_j \Phi_j^t(\omega) \Phi_j^{U(\omega,\omega')}(\omega')\big) + \bar{g}\big(\lambda_j \Phi_j^t(\omega) \Phi_j^{U(\omega,\omega')}(\omega')\big).$$

Now take the conditional expectation relative to $\mathscr{F}_t$ and apply (6.13). The result is that for P_λ-almost all ω,

$$\begin{aligned} E_\lambda(X_T^\lambda | \mathscr{F}_t)(\omega) &= -\sum_{i=1}^{d} t^i \, E_\lambda(f_i(\theta)|\mathscr{F}_t)(\omega) \\ &\quad - \int_\Omega \xi(\omega, \omega') \, dP_{\lambda^t(\omega)}(\omega'). \end{aligned} \tag{6.27}$$

On the other hand, for fixed ω, we see by (6.11) and (6.19) that for $P_{\lambda^t(\omega)}$-almost all ω',

$$-\xi(\omega, \omega') = X_{U(\omega,\omega')}^{\lambda^t(\omega)}(\omega'). \tag{6.28}$$

By (6.21), property b of the lemma, and (6.22), we conclude that for P_λ-almost all ω,

$$\begin{aligned} E_\lambda(X_T^\lambda | \mathscr{F}_t)(\omega) &\le -f^t(\lambda^t(\omega)) + \sup_{S \in \mathbf{A}_f} \int_\Omega X_{S(\omega')}^{\lambda^t(\omega)}(\omega') \, dP_{\lambda^t(\omega)}(\omega') \\ &= -f^t(\lambda^t(\omega)) + h(\lambda^t(\omega)), \end{aligned}$$

which implies, because $T \in \mathbf{A}_b^t$ is arbitrary, that

$$Z_t^\lambda = \operatorname*{ess\,sup}_{T \in \mathbf{A}_b^t} E_\lambda(X_T^\lambda | \mathscr{F}_t) \le -f^t(\lambda^t) + h(\lambda^t), \qquad P_\lambda\text{-a.s.}$$

It now remains to establish the converse inequality. To this end, observe that if S belongs to $\mathbf{A}_b$, and in particular to $\mathbf{D}$, where $\mathbf{D}$ is defined just before the lemma, then $T_S = t + S \circ \rho_t$ belongs to $\mathbf{A}_b^t$. Therefore, by (6.27), (6.28) (applied to the particular case where $U(\omega, \omega') = S(\omega')$), and (6.21), we see that for P_λ-almost all ω,

$$\begin{aligned} Z_t^\lambda(\omega) &= \operatorname*{ess\,sup}_{T \in \mathbf{A}_b^t} E_\lambda(X_T^\lambda | \mathscr{F}_t)(\omega) \\ &\ge \sup_{S \in \mathbf{D}} E_\lambda(X_{T_S}^\lambda | \mathscr{F}_t)(\omega) \\ &= -f^t(\lambda^t(\omega)) + \sup_{S \in \mathbf{D}} \int_\Omega X_{S(\omega')}^{\lambda^t(\omega)}(\omega') \, dP_{\lambda^t(\omega)}(\omega') \\ &= -f^t(\lambda^t(\omega)) + h(\lambda^t(\omega)), \end{aligned}$$

which was to be proved.

6.4 DESCRIBING THE OPTIMAL STRATEGY

The proposition and theorem of Section 6.3 will now be used to describe an optimal strategy. This description will follow from Theorem 1 of Section 3.4, thanks to the representation of the reward process X and Snell's envelope Z obtained from (6.19) and (6.24) by setting $\lambda = \pi$.

Convention

As in the proof of the theorem of the previous section, if q is a function defined on Λ, $\bar{q}$ will denote the extension to $\mathbb{R}_+^l$ obtained by setting $\bar{q}(0) = 0$ and $\bar{q}(x) = q(x/|x|)$ if $x \neq 0$, where $|x| = \sum_{j=1}^{l} x_j$.

Characterizing the Function h

We begin by establishing that the function h is the limit of an increasing sequence of functions defined by induction starting from g. This result is important in two respects: on one hand, it gives a method for computing h, and on the other hand, it shows that the implementation of the optimal strategy that we are about to present can be done without using the probability measures P_λ.

Recall that δ_j^i denotes the density of μ_j^i with respect to μ^i.

Theorem (Recursive construction of h). *Let $(h^{(n)}, n \in \mathbb{N})$ be the sequence of functions defined on Λ by setting $h^{(0)} = -g$ and, by induction on n,*

$$h^{(n+1)} = \sup\Big(-g, \sup_{1 \le i \le d} \delta_i h^{(n)}\Big), \tag{6.29}$$

where δ_i denotes the operator that associates to all measurable and bounded functions q defined on Λ, the function $\delta_i q$ defined on Λ by

$$\delta_i q(\lambda) = \sum_{j=1}^{l} \lambda_j \Big(\int_{\mathbb{R}} \bar{q}(\lambda_1\, \delta_1^i(x), \ldots, \lambda_l\, \delta_l^i(x))\, d\mu_j^i(x) - f_i(\theta_j) \Big). \tag{6.30}$$

Then $h^{(n)} \le h^{(n+1)}$ for all $n \in \mathbb{N}$ and $\lim_{n \to \infty} h^{(n)} = h$. Moreover, h is the smallest solution of the equation

$$q = \sup\Big(-g, \sup_{1 \le i \le d} \delta_i q\Big). \tag{6.31}$$

Proof. We apply the proposition of Section 3.6 to the case where the probability measure is P_λ, the reward process is X^λ, and Snell's envelope is Z^λ. By this proposition and the remark at the end of Section 6.1, the random variables $Z_t^{(n)}$ defined in (3.18) increase P_λ-a.s. to Z_t^λ. Assume for a minute that

$$Z_t^{(n)} = -f^t(\lambda^t) + h^{(n)}(\lambda^t), \qquad P_\lambda\text{-a.s.} \tag{6.32}$$

By taking $t = 0$, we then see that $h^{(n)}(\lambda)$ increases to $h(\lambda)$. In order to establish (6.32), we argue by induction. The relation is true for $n = 0$ because of (6.19). On the other hand, by (3.18),

$$Z_t^{(n+1)} = \sup\left(X_t^\lambda, \sup_{u \in \mathbb{D}_t} E_\lambda\left(Z_u^{(n)} | \mathscr{F}_t \right) \right), \qquad P_\lambda\text{-a.s.} \tag{6.33}$$

By the induction hypothesis,

$$E_\lambda\left(Z_{t+e_i}^{(n)} \middle| \mathscr{F}_t \right) = E_\lambda\left(-f^{t+e_i}(\lambda^{t+e_i}) + h^{(n)}(\lambda^{t+e_i}) \middle| \mathscr{F}_t \right), \qquad P_\lambda\text{-a.s.} \tag{6.34}$$

Moreover, by (6.21),

$$f^{t+e_i}(\lambda^{t+e_i}) = \sum_{k-1}^{d} t^k\, E_\lambda\left(f_k(\theta) \middle| \mathscr{F}_{t+e_i} \right) + E_\lambda\left(f_i(\theta) \middle| \mathscr{F}_{t+e_i} \right), \qquad P_\lambda\text{-a.s.,}$$

which implies, by conditioning with respect to $\mathscr{F}_t$ and using (6.21) and (6.20), that

$$\begin{aligned} E_\lambda\left(f^{t+e_i}(\lambda^{t+e_i}) \middle| \mathscr{F}_t \right) &= \sum_{k=1}^{d} t^k\, E_\lambda\left(f_k(\theta) \middle| \mathscr{F}_t \right) + E_\lambda\left(f_i(\theta) \middle| \mathscr{F}_t \right) \\ &= f^t(\lambda^t) + \sum_{j=1}^{l} f_i(\theta_j)\, \lambda_j^t, \qquad P_\lambda\text{-a.s.} \end{aligned} \tag{6.35}$$

Furthermore,

$$\begin{aligned} &E_\lambda\left(h^{(n)}(\lambda^{t+e_i}) \middle| \mathscr{F}_t \right) \\ &\quad = \sum_{j=1}^{l} \lambda_j^t \int_{\mathbb{R}} \bar{h}^{(n)}\left(\lambda_1^t\, \delta_1^i(x), \ldots, \lambda_l^t\, \delta_l^i(x) \right) d\mu_j^i(x), \quad P_\lambda\text{-a.s.} \end{aligned} \tag{6.36}$$

Indeed, if F is a set of the form $\Theta \times \tilde{F}$, where $\tilde{F} \in \tilde{\mathscr{F}}_t$, then since $\varphi_j^{t+e_i}(Y^{t+e_i}) = \varphi_j^t(Y^t)\,\delta_j^i(Y_{t^i+1}^i)$, we can write, thanks to (6.11),

$$\begin{aligned}
\int_F h^{(n)}(\lambda^{t+e_i})\,\mathrm{dP}_\lambda &= \int_{\Theta\times\tilde{F}} \bar{h}^{(n)}\big(\lambda_1\varphi_1^t(Y^t)\,\delta_1^i(Y_{t^i+1}^i),\ldots,\lambda_l\varphi_l^t(Y^t)\,\delta_l^i(Y_{t^i+1}^i)\big)\,\mathrm{dP}_\lambda \\
&= \sum_{j=1}^{l}\lambda_j\int_{\{\theta_j\}\times\tilde{F}}\bigg(\int_{\mathbb{R}}\bar{h}^{(n)}\big(\lambda_1\varphi_1^t(\tilde{Y}^t)\,\delta_1^i(x),\ldots, \\
&\qquad\qquad \lambda_l\varphi_l^t(\tilde{Y}^t)\,\delta_l^i(x)\big)\,d\mu_j^i(x)\bigg)\,d\big(\varepsilon_j\times\tilde{\mathrm{P}}_j\big) \\
&= \sum_{j=1}^{l}\int_{\{\theta=\theta_j\}\cap F}\bigg(\int_{\mathbb{R}}\bar{h}^{(n)}\big(\lambda_1\varphi_1^t(Y^t)\,\delta_1^i(x),\ldots, \\
&\qquad\qquad \lambda_l\varphi_l^t(Y^t)\,\delta_l^i(x)\big)\,d\mu_j^i(x)\bigg)\,\mathrm{dP}_\lambda \\
&= \sum_{j=1}^{l}\int_F \mathrm{P}_\lambda\big(\theta=\theta_j\big|\mathscr{F}_t\big)\bigg(\int_{\mathbb{R}}\bar{h}^{(n)}\big(\lambda_1\varphi_1^t(Y^t)\,\delta_1^i(x),\ldots, \\
&\qquad\qquad \lambda_l\varphi_l^t(Y^t)\,\delta_l^i(x)\big)\,d\mu_j^i(x)\bigg)\,\mathrm{dP}_\lambda \\
&= \int_F\bigg(\sum_{j=1}^{l}\lambda_j^t\int_{\mathbb{R}}\bar{h}^{(n)}\big(\lambda_1^t\,\delta_1^i(x),\ldots,\lambda_l^t\,\delta_l^i(x)\big)\,d\mu_j^i(x)\bigg)\,\mathrm{dP}_\lambda,
\end{aligned}$$

which implies (6.36). The validity of (6.32) with $n+1$ instead of n follows easily, because by (6.33)–(6.36) and taking (6.19), (6.30), and (6.29) into account,

$$\begin{aligned}
Z_t^{(n+1)} &= \sup\Big(-f^t(\lambda^t)-g(\lambda^t),\ \sup_{1\le i\le d}\big(-f^t(\lambda^t)+\delta_i h^{(n)}(\lambda^t)\big)\Big) \\
&= -f^t(\lambda^t)+\sup\Big(-g(\lambda^t),\ \sup_{1\le i\le d}\delta_i h^{(n)}(\lambda^t)\Big) \\
&= -f^t(\lambda^t)+h^{(n+1)}(\lambda^t), \qquad \mathrm{P}_\lambda\text{-a.s.}
\end{aligned}$$

We now turn to the last part of the theorem. By taking the limit of the right- and left-hand sides of (6.29), we observe that h satisfies (6.31). On the other hand, if q satisfies (6.31), then $q \ge h^{(n)}$ for all $n \in \mathbb{N}$. Indeed, if $q \ge h^{(n)}$, then $\delta_i q \ge \delta_i h^{(n)}$ for $i = 1,\ldots,d$; therefore $q \ge h^{(n+1)}$ by (6.29) and (6.31). Because $q \ge -g = h^{(0)}$, the assertion is thus proved. It remains to pass to the limit to see that $q \ge h$.

The Optimal Strategy

From the conclusions of Section 6.1 and by Theorem 1 of Section 3.4, all strategies (Γ^o, τ^o, D^o) formed by an optimizing increasing path Γ^o, the (necessarily finite) stopping time

$$\tau^o = \inf\{n \in \mathbb{N}: Z_n^{\Gamma^o} = X_n^{\Gamma^o}\}$$

and the decision variable D^o defined in (6.2) are optimal strategies.

In order to construct an optimizing increasing path Γ^o, we use the proposition that will be established below. Moreover, we observe that τ^o is the time of first visit of $\pi^{\Gamma_n^o}$ to

$$B = \{\lambda \in \Lambda: -g(\lambda) = h(\lambda)\}; \tag{6.37}$$

in other words,

$$\tau^o = \inf\{n \in \mathbb{N}: \pi^{\Gamma_n^o} \in B\}.$$

Indeed, by (6.19) and (6.24), $X_t < Z_t$ if and only if $-g(\pi^t) < h(\pi^t)$.

Proposition 1 (Sets regulating the continuation of draws). *For* $i = 1, \ldots, d$, *set*

$$C_i = \{\lambda \in \Lambda: -g(\lambda) < h(\lambda), \delta_i h(\lambda) = \sup(\delta_1 h(\lambda), \ldots, \delta_d h(\lambda))\}, \tag{6.38}$$

where δ_i *is the operator defined in* (6.30). *In order that*

$$X_t < Z_t \quad \text{and} \quad \mathrm{E}(Z_{t+e_i} | \mathcal{F}_t) = \sup_{u \in \mathbb{D}_t} \mathrm{E}(Z_u | \mathcal{F}_t),$$

it is necessary and sufficient that

$$\pi^t = (\pi_1^t, \ldots, \pi_l^t) \in C_i.$$

Proof. We have already noticed that $X_t < Z_t$ if and only if $-g(\pi^t) < h(\pi^t)$. In addition,

$$Z_{t+e_i} = -f^{t+e_i}(\pi^{t+e_i}) + h(\pi^{t+e_i})$$

by (6.24). Taking $\lambda = \pi$ in (6.35) and (6.36) and h instead of $h^{(n)}$ in (6.36), we observe that

$$\mathrm{E}(Z_{t+e_i} | \mathcal{F}_t) = -f^t(\pi^t) + \delta_i h(\pi^t).$$

The proposition follows easily.

Continuation and Stopping Regions

The sets B and $C = \bigcup_{i=1}^{d} C_i$ are called, respectively, the *stopping region* and the *continuation region*. Because g and h do not depend on the choice of the prior distribution π, the same goes for $B, C, C_1, \ldots, C_d$ as well as for the sets $B_1, \ldots, B_l$, which will be introduced below.

Describing the Optimal Strategy

Recall that the experiment consists of a succession of draws, each from one of d populations $\mathcal{P}_1, \ldots, \mathcal{P}_d$, and that each draw is assimilated to the realization of a random variable $Y_{s^i}^i$. The decision to stop as well as the choice of the population to observe at each stage are described in the procedure that follows.

1. Compute π^t using (6.11).
2. If $\pi^t \in B$, stop the test and decide that the value of θ is θ_{D^o}, where D^o is obtained by taking $R_k = \sum_{j=1}^{l} c_{j,k}\pi_j^t$ in (6.2).
3. If $\pi^t \notin B$, choose $i \in \{1, \ldots, d\}$ such that $\pi^t \in C_i$ and draw an individual from population $\mathcal{P}_i$.
4. Return to (1) with $t + e_i$ instead of t.

The procedure we have just described stops with probability 1 after a finite number of steps. If $\bigcup_{i=1}^{d} C_i = \varnothing$, the decision takes place at $t = 0$ and no observation is made.

The Sets B_k

For $k = 1, \ldots, l$, set

$$\begin{aligned} B_k &= B \cap \{\lambda \in \Lambda: g_k(\lambda) = g(\lambda)\} \\ &= \{\lambda \in \Lambda: -g(\lambda) = h(\lambda),\ g_k(\lambda) = g(\lambda)\}, \end{aligned} \tag{6.39}$$

where g_k denotes the function defined by Λ by

$$g_k(\lambda) = \sum_{j=1}^{l} c_{j,k}\,\lambda_j.$$

It is obvious by (6.17) that $\bigcup_{k=1}^{l} B_k = B$. On the other hand, according to step 2 of the above description, the optimal rule is to stop as soon as π^t belongs to one of the sets B_k. The value of D^o is then the smallest k such that B_k contains π^t. The advantage of using the sets $B_1, \ldots, B_l$ comes from the fact that these sets are convex, therefore easier to handle than B.

Proposition 2 (Convexity property). *The sets $B_1, \dots, B_l$ defined by* (6.39) *are convex.*

Proof. Fix k, assume that $\lambda, \lambda' \in B_k$ and that $a \in [0, 1]$. From the definition of B_k, $g_k(\lambda) \le g_j(\lambda)$ and $g_k(\lambda') \le g_j(\lambda')$; therefore $g_k(a\lambda + (1-a)\lambda') \le g_j(a\lambda + (1-a)\lambda')$ for $j = 1, \dots, l$, which implies that

$$g_k(a\lambda + (1-a)\lambda') = \inf_{1 \le j \le l} g_j(a\lambda + (1-a)\lambda') = g(a\lambda + (1-a)\lambda').$$

Moreover, since h is convex,

$$\begin{aligned} h(a\lambda + (1-a)\lambda') &\le ah(\lambda) + (1-a)h(\lambda') \\ &= a(-g(\lambda)) + (1-a)(-g(\lambda')) \\ &= a(-g_k(\lambda) + (1-a)(-g_k(\lambda')) \\ &= -g_k(a\lambda + (1-a)\lambda') \\ &= -g(a\lambda + (1-a)\lambda'). \end{aligned}$$

Because $-g \le h$, it follows that $-g(a\lambda + (1-a)\lambda') = h(a\lambda + (1-a)\lambda')$. Therefore $a\lambda + (1-a)\lambda' \in B_k$, which proves that B_k is convex.

Illustration

Case where $l = 2$. In this case, Λ is represented by the closed segment with extremities $(0, 1)$ and $(1, 0)$, and each of the two sets B_1 and B_2 by a (possibly empty) segment. Moreover, by (6.23), $(1, 0)$ and $(0, 1)$ belong to $B = B_1 \cup B_2$, which means that $\bigcup_{i=1}^{d} C_i$ is also represented by a (possibly empty) segment. Assume now that $c_{1,1} < c_{1,2}$ and $c_{2,2} < c_{2,1}$, which expresses the idea that the cost of making a correct decision is less than the cost of making an incorrect decision. Then $(1, 0)$ belongs to B_1 and $(0, 1)$ belongs to B_2. On the other hand, by comparing $g_1(\lambda)$ to $g_2(\lambda)$, which is the same as solving the inequality $c_{1,1}\lambda_1 + c_{2,1}\lambda_2 \le c_{1,2}\lambda_1 + c_{2,2}\lambda_2$, we see that if λ' denotes the law defined by

$$\lambda'_1 = \frac{c_{2,1} - c_{2,2}}{c_{1,2} - c_{1,1} + c_{2,1} - c_{2,2}}, \qquad \lambda'_2 = 1 - \lambda'_1,$$

then $0 < \lambda'_1 < 1$, $g_1(\lambda') = g_2(\lambda')$, $g_2(\lambda) < g_1(\lambda)$ for all $\lambda \in \Lambda$ such that $\lambda_1 < \lambda'_1$ and $g_1(\lambda) < g_2(\lambda)$ for all $\lambda \in \Lambda$ such that $\lambda_1 > \lambda'_1$. Consequently, the segments representing B_1 and B_2 are closed and have at most the point λ' in common. It follows that there exist two numbers b_1 and b_2 such that

$0 \le b_2 \le b_1 \le 1$, with which step 2 of the description of the optimal strategy becomes:

2a. If $\pi_1^t \ge b_1$, stop the test and decide that the value of θ is θ_1.
2b. If $\pi_1^t \le b_2$ ($< b_2$ in the case where $b_1 = b_2$), stop the test and decide that the value of θ is θ_2.

Case where $l > 2$. In this case, Λ is represented by an l-simplex (a triangle when $l = 3$). This simplex is divided into regions that consist of the convex sets $B_k \neq \varnothing$ and the sets $C_i \neq \varnothing$. By (6.23), the regions of the first kind contain the l points $(1, 0, \ldots, 0), \ldots, (0, \ldots, 0, 1)$. Once these regions have been determined, all decisions are based on the value of π^t.

Approximating the Optimal Strategy

The convergence to h of the functions $h^{(n)}$ defined in the theorem can be used to approximate the sets B_k and C_i. For $k = 1, \ldots, l$, $i = 1, \ldots, d$, $n \in \mathbb{N}$, and $m \in \mathbb{N}^*$, set

$$B_k^n = \{\lambda \in \Lambda: -g(\lambda) = h^{(n)}(\lambda),\ g_k(\lambda) = g(\lambda)\},$$

$$C_i^{n,m} = \Big\{\lambda \in \Lambda: -g(\lambda) < h^{(n)}(\lambda),$$
$$\delta_i h^{(n)}(\lambda) \ge \sup(\delta_1 h^{(n)}(\lambda), \ldots, \delta_d h^{(n)}(\lambda)) - \frac{1}{m}\Big\},$$

where δ_i denotes the operator defined in (6.30). We are going to show that the sets B_k^n and $C_i^{n,m}$ approach, in a sense that has to be made precise, B_k and C_i, respectively. For large m and n, B_k^n and $C_i^{n,m}$ can therefore be used instead of B_k and C_i in order to obtain an approximation to the optimal strategy.

Proposition 3 (Approximating the sets B_k and C_i). *For* $k = 1, \ldots, l$ *and* $i = 1, \ldots, d$,

a. $B_k^n \supset B_k^{n+1}$, for all $n \in \mathbb{N}$, and $\bigcap_{n \in \mathbb{N}} B_k^n = B_k$;
b. $C_i^{n,m} \supset C_i^{n,m+1}$ and $\bigcap_{m \in \mathbb{N}^*} \liminf_{n \to \infty} C_i^{n,m} = C_i$.

Proof. The proof is based on the fact that $-g = h^{(0)}$ and that $h^{(n)}$ increases to h. Since assertion a is obvious, we prove assertion b. Clearly $C_i^{n,m} \supset C_i^{n,m+1}$. Moreover, if $\lambda \in C_i$, then there exists $n_0 \in \mathbb{N}$ such that for all $n \ge n_0$, $-g(\lambda) < h^{(n)}(\lambda)$ and $\delta_i h^{(n)}(\lambda) \ge \delta_i h(\lambda) - 1/(2m)$ for $i = 1, \ldots, d$.

Therefore

$$\begin{aligned}\delta_i h^{(n)}(\lambda) &\geq \delta_i h(\lambda) - \frac{1}{2m}\\ &= \sup(\delta_1 h(\lambda), \dots, \delta_d h(\lambda)) - \frac{1}{2m}\\ &\geq \sup(\delta_1 h^{(n)}(\lambda), \dots, \delta_d h^{(n)}(\lambda)) - \frac{1}{m}.\end{aligned}$$

It follows that $C_i \subset \liminf_{n\to\infty} C_i^{n,m}$ for all $m \in \mathbb{N}^*$. Conversely, if $\lambda \in \liminf_{n\to\infty} C_i^{n,m}$ for all $m \in \mathbb{N}^*$, then $-g(\lambda) < h(\lambda)$ and $\delta_i h(\lambda) \geq \sup(\delta_1 h(\lambda), \dots, \delta_d h(\lambda)) - 1/m$ for all $m \in \mathbb{N}^*$. Therefore $\lambda \in C_i$.

6.5 THE LIKELIHOOD-RATIO TEST

In the particular case where a choice must be made between two hypotheses, it is appropriate to describe and to study the properties of optimal strategies using the notion of likelihood ratio.

Throughout this section, we will study the case where $l = 2$ and suppose that (Γ^o, τ^o, D^o) is an optimal strategy as described in Section 6.4. The stopping point $\Gamma^o_{\tau^o}$ will be denoted by T^o. In order to simplify the presentation, in addition to the hypotheses and conventions made in Section 6.4, we will assume that

1. $c_{1,1} < c_{1,2}$ and $c_{2,2} < c_{2,1}$.
2. For each fixed $i = 1, \dots, d$, the laws μ_1^i and μ_2^i are equivalent.
3. For each $i = 1, \dots, d$, $\mu_1^i \neq \mu_2^i$.

As already pointed out in Section 6.4, condition 1 is justified by the fact that the cost of a correct decision should be less than the cost of an incorrect decision. Condition 2 is equivalent to requiring that the respective densities δ_1^i and δ_2^i of μ_1^i and μ_2^i with respect to $\mu^i = \mu_1^i + \mu_2^i$ are positive μ^i-a.e. By modifying these densities on a null set, we can assume that they are positive everywhere. In this case, it is easy to see that $d\mu_2^i/d\mu_1^i = \delta_2^i/\delta_1^i$. Imposing condition 3 entails no loss of generality, because if $\mu_1^i = \mu_2^i$, then population $\mathcal{P}_i$ plays no role in the test. Indeed, we will see below that in the case where $\delta_2^i/\delta_1^i = 1$, draws from the population $\mathcal{P}_i$ do not modify the likelihood ratio, which means that the optimal strategy will never prescribe a draw from that population.

In the following, we will denote the conditional probability measures $\mathrm{P}(\cdot|\theta = \theta_1)$ and $\mathrm{P}(\cdot|\theta = \theta_2)$ by P_1 and P_2, respectively. Using the definition of P in (6.6), notice that P_1 and P_2 are nothing but the probability measures

$\varepsilon_1 \times \tilde{P}_1$ and $\varepsilon_2 \times \tilde{P}_2$. The expectation operators corresponding to P_1 and P_2 will be denoted E_1 and E_2, respectively.

The Likelihood Ratio

We will term *likelihood ratio*, and denote by r^t, the ratio π_2^t/π_1^t. By (6.8), (6.11), and (6.7),

$$r^t = \frac{P(\theta = \theta_2 | \mathscr{F}_t)}{P(\theta = \theta_1 | \mathscr{F}_t)}$$

$$= \frac{\pi_2\, \varphi_2^t(Y^t)}{\pi_1\, \varphi_1^t(Y^t)} = \begin{cases} \dfrac{\pi_2}{\pi_1} \displaystyle\prod_{i \in I_t} \prod_{s^i=1}^{t^i} \frac{\delta_2^i(Y_{s^i}^i)}{\delta_1^i(Y_{s^i}^i)}, & \text{if } t \neq 0, \\ \dfrac{\pi_2}{\pi_1}, & \text{if } t = 0. \end{cases} \tag{6.40}$$

Observe that according to this definition,

$$r^{t+e_i} = r^t \frac{\delta_2^i(Y_{t^i+1}^i)}{\delta_1^i(Y_{t^i+1}^i)}.$$

The Functions $\hat{g}$ and $\hat{h}$

When using the likelihood ratio, it is preferable to replace the functions g and h by the functions $\hat{g}$ and $\hat{h}$ defined on $\mathbb{R}_+$, respectively, by

$$\hat{g}(y) = \bar{g}(1, y) \quad \text{and} \quad \hat{h}(y) = \bar{h}(1, y).$$

It is then necessary to replace the operator δ_i defined in (6.30) by the operator $\hat{\delta}_i$ which associates to each measurable and bounded function $\hat{q}$ defined on $\mathbb{R}_+$ the function $\hat{\delta}_i \hat{q}$ defined by

$$\hat{\delta}_i \hat{q}(y) = \int_{\mathbb{R}} \hat{q}\left(y \frac{\delta_2^i(x)}{\delta_1^i(x)} \right) \left(1 + y \frac{\delta_2^i(x)}{\delta_1^i(x)} \right) d\mu_1^i(x) - f_i(\theta_1) - y f_i(\theta_2). \tag{6.41}$$

By the theorem of Section 6.4, if $(\hat{h}^{(n)}, n \in \mathbb{N})$ denotes the sequence of functions defined by setting $\hat{h}^{(0)} = -\hat{g}$ and, by induction on n,

$$\hat{h}^{(n+1)} = \sup\left(-\hat{g}, \sup_{1 \leq i \leq d} \hat{\delta}_i \hat{h}^{(n)} \right), \tag{6.42}$$

then $\hat{h}^{(n)} \leq \hat{h}^{(n+1)}$ and $\lim_{n \to \infty} \hat{h}^{(n)} = \hat{h}$. Moreover, by the same theorem, $\hat{h}$ is the smallest solution of the equation

$$\hat{q} = \sup\left(-g, \sup_{1 \le i \le d} \hat{\delta}_i \hat{q}\right).$$

Describing the Optimal Strategy Using the Likelihood Ratio

According to the description of the optimal strategy given in Section 6.4 (steps 2a and 2b included), the stopping and continuation regions adapted to the use of the likelihood ratio are determined by two numbers a_1 and a_2 such that $0 \le a_1 \le a_2 \le \infty$. The continuation region $\bigcup_{i=1}^{d} C_i = \{y \in \mathbb{R}_+ : -\hat{g}(y) < \hat{h}(y)\}$ is empty if $a_1 = a_2$ and nonempty if $a_1 < a_2$. In the second case, this region is the open interval with extremities a_1 and a_2.

The optimal strategy is achieved by using the following procedure.

1. Compute r^t using (6.40).
2a. If $r^t \le a_1$, stop the test and decide that the value of θ is θ_1.
2b. If $r^t \ge a_2$, stop the test and decide that the value of θ is θ_2.
3. If $a_1 < r^t < a_2$, choose $i \in \{1, \dots, d\}$ such that $\hat{\delta}_i \hat{h}(r^t) = \sup(\hat{\delta}_1 \hat{h}(r^t), \dots, \hat{\delta}_d \hat{h}(r^t))$ and draw from the population $\mathcal{P}_i$ ($\hat{\delta}_i$ denotes the operator defined in (6.41)).
4. Return to 1 with $t + e_i$ instead of t.

Excluding the Case Where $a_1 = a_2$

From now on, we will exclude the uninteresting case where $a_1 = a_2$ by assuming, in addition to the hypotheses 1–3 at the beginning of the section, that

4. $a_1 < a_2$.

Because the values of δ_1^i and δ_2^i are positive, we deduce from (6.40) that $0 < r^t < \infty$. Given that the test stops with probability 1, it is therefore not possible that both $a_1 = 0$ and $a_2 = \infty$. In fact, neither $a_1 = 0$ nor $a_2 = \infty$ are possible, but for technical reasons, we will only establish this further on.

The Log-Likelihood Ratio and a Random Walk

Given the role that the family $(r^t, t \in \mathbb{D})$ plays in the determination of an optimal strategy, any information concerning its behavior can be very useful. By (6.40),

$$\log r^t = \sum_{i=1}^{d} W_{t^i}^i, \tag{6.43}$$

where

$$W_{t^i}^i = \begin{cases} \dfrac{1}{d}\log\dfrac{\pi_2}{\pi_1} + \displaystyle\sum_{s^i=1}^{t^i} \log\dfrac{\delta_2^i(Y_{s^i}^i)}{\delta_1^i(Y_{s^i}^i)}, & \text{if } t^i \neq 0, \\ \dfrac{1}{d}\log\dfrac{\pi_2}{\pi_1}, & \text{if } t^i = 0. \end{cases} \tag{6.44}$$

On the other hand, conditionally with respect to $\{\theta = \theta_j\}$, the random variables

$$\log\frac{\delta_2^i(Y_{s^i}^i)}{\delta_1^i(Y_{s^i}^i)}, \qquad s^i \in \mathbb{N}, \quad i = 1, \dots, d,$$

are independent and their law only depends on i. Consequently, under P_j, $(W_{t^1}^1, t^1 \in \mathbb{N}), \dots, (W_{t^d}^d, t^d \in \mathbb{N})$ are random walks, or better, $((W_{t^1}^1, \dots, W_{t^d}^d), (t^1, \dots, t^d) \in \mathbb{D})$ is a *random d-walk*. By the description of the optimal strategy, the optimal rule is to stop the test as soon as this random d-walk exits the region of continuation

$$\{(x_1, \dots, x_d) \in \mathbb{R}^d: \log a_1 < x_1 + \cdots + x_d < \log a_2\}.$$

Illustration

In the case where $d = 2$, the region of continuation for the random 2-walk introduced above is the (open) strip determined by the two straight lines with equations $x_1 + x_2 = \log a_1$ and $x_1 + x_2 = \log a_2$, respectively. If the exit point is located in the lower left [upper right] half plane of the complement of this strip, the optimal decision is that the value of θ is θ_1 [θ_2].

Kullback–Leibler Information

For $i = 1, \dots, d$ and $j = 1, 2$, set

$$I_j^i = \mathrm{E}_j\left(\log\frac{\delta_2^i(Y_1^i)}{\delta_1^i(Y_1^i)}\right) = \int_{\mathbb{R}} \log\left(\frac{\delta_2^i}{\delta_1^i}\right) \delta_j^i \, d\mu^i. \tag{6.45}$$

Thanks to the elementary inequality

$$\log\frac{\delta_2^i}{\delta_1^i} \leq \frac{\delta_2^i}{\delta_1^i} - 1$$

and taking into account that the assumption $\mu'_1 \neq \mu'_2$ translates into

$$\mu^i\left\{\log\frac{\delta_2^i}{\delta_1^i} < \frac{\delta_2^i}{\delta_1^i} - 1\right\} = \mu^i\{\delta_1^i \neq \delta_2^i\} > 0,$$

we see that the integral is well defined and that

$$I_1^i < \int_{\mathbb{R}}\left(\frac{\delta_2^i}{\delta_1^i} - 1\right)\delta_1^i\,d\mu^i = 0 \quad \text{and} \quad -I_2^i < \int_{\mathbb{R}}\left(\frac{\delta_1^i}{\delta_2^i} - 1\right)\delta_2^i\,d\mu^i = 0.$$

Thus

$$-\infty \leq I_1^i < 0 \quad \text{and} \quad 0 < I_2^i \leq \infty. \tag{6.46}$$

Observe that I_j^i represents the expected amplitude of one step of the random walk $(W_{t^i}^i, t^i \in \mathbb{N})$ under P_j. In relation to the illustration above, we can therefore deduce from (6.46) that under P_1 [P_2], it is more likely that the exit point from the strip of continuation will be located in the lower left [upper right] half plane.

In the statistical literature, the number I_j^i is called the *Kullback–Leibler information*. It represents the information in favor of the decision $D(\omega) = j$ contained in one draw from population $\mathcal{P}_i$ under P_j.

Proof that $a_1 > 0$ and $a_2 < \infty$

We will only prove the second inequality, because the first is obtained in a similar manner by using $1/r^t$ instead of r^t. By the strong law of large numbers and the fact that $I_2^i > 0$,

$$\lim_{t^i \to \infty} \sum_{s^i=1}^{t^i} \log\frac{\delta_2^i(Y_{s^i}^i)}{\delta_1^i(Y_{s^i}^i)} = \infty, \qquad P_2\text{-a.s.}$$

Consequently, there exists $\alpha \in]0,1[$, which we can choose independently from i, such that

$$P_2\left\{\inf_{t^i \in \mathbb{N}^*} \sum_{s^i=1}^{t^i} \log\frac{\delta_2^i(Y_{s^i}^i)}{\delta_1^i(Y_{s^i}^i)} > \log\alpha\right\} > 0.$$

Taking (6.43) and (6.44) into account, we see that

$$\begin{aligned}
\mathrm{P}_2\left\{\inf_{t\in\mathbb{I}} r^t > \alpha\left(\frac{\pi_2}{\pi_1}\right)^{-d}\right\} &= \mathrm{P}_2\left\{\inf_{t\in\mathbb{I}} \log r^t > \log\left(\alpha\left(\frac{\pi_2}{\pi_1}\right)^{-d}\right)\right\} \\
&= \mathrm{P}_2\left\{\inf_{t\in\mathbb{I}} \sum_{i=1}^{d} W_{t^i}^i > \log\left(\alpha\left(\frac{\pi_2}{\pi_1}\right)^{-d}\right)\right\} \\
&\geq \mathrm{P}_2\left(\bigcap_{i=1}^{d}\left\{\inf_{t^i\in\mathbb{N}^*} \sum_{s^i=1}^{t^i} \log\frac{\delta_2^i(Y_{s^i}^i)}{\delta_1^i(Y_{s^i}^i)} > \log\alpha\right\}\right) \\
&= \prod_{i=1}^{d} \mathrm{P}_2\left\{\inf_{t^i\in\mathbb{N}^*} \sum_{s^i=1}^{t^i} \log\frac{\delta_2^i(Y_{s^i}^i)}{\delta_1^i(Y_{s^i}^i)} > \log\alpha\right\} > 0.
\end{aligned}$$

Now assume that $a_2 = \infty$ and show that this leads to a contradiction. By choosing the law π in such a way that $\alpha\,(\pi_2/\pi_1)^{-d} > a_1$, we see that

$$\mathrm{P}\left\{\inf_{t\in\mathbb{I}} r^t > a_1\right\} \geq \pi_2\, \mathrm{P}_2\left\{\inf_{t\in\mathbb{I}} r^t > a_1\right\} > 0.$$

But $\tau^o = \infty$ on $\{\inf_{t\in\mathbb{I}} r^t > a_1\}$, which is impossible because τ^o is finite. Therefore $a_2 < \infty$.

Quality of the Optimal Strategy

In order to evaluate the quality of the optimal strategy (Γ^o, τ^o, D^o), one typically uses the error probabilities

$$\begin{aligned}
\alpha_{1,2} &= \mathrm{P}_1\{D^o = 2\} = \mathrm{P}_1\{r^{T^o} \geq a_2\}, \\
\alpha_{2,1} &= \mathrm{P}_2\{D^o = 1\} = \mathrm{P}_2\{r^{T^o} \leq a_1\}.
\end{aligned} \tag{6.47}$$

Clearly, if $\pi_2/\pi_1 \leq a_1$ $[\geq a_2]$, then $\alpha_{1,2} = 0$ and $\alpha_{2,1} = 1$ $[\alpha_{1,2} = 1$ and $\alpha_{2,1} = 0]$. Conversely, if $a_1 < \pi_2/\pi_1 < a_2$, then

$$0 < \alpha_{1,2} < 1 \quad \text{and} \quad 0 < \alpha_{2,1} < 1. \tag{6.48}$$

We will only establish the inequality $\alpha_{1,2} > 0$, since the other inequalities are obtained in a similar manner. Because $\mu_1^i \neq \mu_2^i$, it follows that $\mu_1^i\{\delta_2^i > \delta_1^i\} > 0$, therefore that $\mu_1^i\{\log(\delta_2^i/\delta_1^i) > 0\} > 0$. Consequently, there exists $\alpha > 0$, which we can choose independently from i, such that

$$\mathrm{P}_1\left\{\log\frac{\delta_2^i(Y_1^i)}{\delta_1^i(Y_1^i)} > \alpha\right\} > \alpha.$$

Let n denote the integer part of $(\log a_2 - \log(\pi_2/\pi_1))/\alpha$. By (6.43) and (6.44), we see that

$$\{r^{T^o} \geq a_2\} = \{\log r^{T^o} \geq \log a_2\} \supset \bigcap_{i=1}^{d} \bigcap_{s^i=1}^{n+1} \left\{\log \frac{\delta_2^i(Y_{s^i}^i)}{\delta_1^i(Y_{s^i}^i)} > \alpha\right\}.$$

Therefore,

$$\alpha_{1,2} = \mathrm{P}_1\{r^{T^o} \geq a_2\} \geq \prod_{i=1}^{d} \prod_{s^i=1}^{n+1} \mathrm{P}_1\left\{\log \frac{\delta_2^i(Y_{s^i}^i)}{\delta_1^i(Y_{s^i}^i)} > \alpha\right\} > \alpha^{d(n+1)} > 0,$$

which was to be proved.

The next proposition gives an approximation of $\alpha_{1,2}$ and $\alpha_{2,1}$ in terms of a_1 and a_2, under the assumption that the values of r^{T^o} are close to a_1 or a_2. Of course, a_1 and a_2 may only be evaluated approximately, using the theorem of Section 6.4.

Proposition 1 (Estimating $\alpha_{1,2}$ and $\alpha_{2,1}$). *The error probabilities $\alpha_{1,2}$ and $\alpha_{2,1}$ satisfy the two inequalities*

$$\pi_1\, a_2\, \alpha_{1,2} \leq \pi_2\,(1 - \alpha_{2,1}) \quad \text{and} \quad \pi_2\, \alpha_{2,1} \leq \pi_1\, a_1\,(1 - \alpha_{1,2}). \tag{6.49}$$

These inequalities are equalities if $\mathrm{P}\{r^{T^o} \in \{a_1, a_2\}\} = 1$ *and approximate equalities (denoted $\simeq$) if*

$$r^{T^o} \cong a_1 \quad \text{on } \{r^{T^o} \leq a_1\} \quad \text{and} \quad r^{T^o} \cong a_2 \quad \text{on } \{r^{T^o} \geq a_2\}. \tag{6.50}$$

In this last case,

$$\alpha_{1,2} \cong \frac{\pi_2 - \pi_1 a_1}{\pi_1(a_2 - a_1)} \quad \text{and} \quad \alpha_{2,1} \cong \frac{\pi_1 a_1 a_2 - \pi_2 a_1}{\pi_2(a_2 - a_1)}. \tag{6.51}$$

Proof. We will only prove the first inequality in (6.49), because the proof of the second is similar. By the definition of P in (6.6) and the description of the optimal strategy,

$$\alpha_{1,2} = (\varepsilon_1 \times \tilde{\mathrm{P}}_1)\{r^{T^o} \geq a_2\} = \sum_{n \in \mathbb{N}} \sum_{t:\,|t|=n} (\varepsilon_1 \times \tilde{\mathrm{P}}_1)\{r^T \geq a_2,\, \Gamma_n^o = t,\, \tau^o = n\}.$$

If $t \neq 0$ and $|t| = n$, then letting B denote a Borel subset of $\mathbb{R}^{|t|}$ such that

$\{\Gamma_n^o = t, \tau^o = n\} = \Theta \times \{\tilde{Y}^t \in B\}$, we see by (6.12) and (6.40) that

$$
\begin{aligned}
\left(\varepsilon_1 \times \tilde{\mathrm{P}}_1\right)&\left\{r^t \geq a_2, \Gamma_n^o = t, \tau^o = n\right\} \\
&= \tilde{\mathrm{P}}_1\left(\left\{\frac{\pi_2\, \varphi_2^t(\tilde{Y}^t)}{\pi_1\, \varphi_1^t(\tilde{Y}_t)} \geq a_2\right\} \cap \{\tilde{Y}^t \in B\}\right) \\
&= \int_{\{\pi_2 \varphi_2^t/(\pi_1 \varphi_1^t) \geq a_2\} \cap B} \varphi_1^t \, d\mu \\
&\leq \int_{\{\pi_2 \varphi_2^t/(\pi_1 \varphi_1^t) \geq a_2\} \cap B} \frac{\pi_2}{\pi_1\, a_2} \varphi_2^t \, d\mu \\
&= \frac{\pi_2}{\pi_1\, a_2}\left(\varepsilon_2 \times \tilde{\mathrm{P}}_2\right)\left\{r^t \geq a_2, \Gamma^o = t, \tau^o = n\right\}.
\end{aligned}
$$

Since the inequality between the first and last terms is also valid in the case where $t = 0$, it follows that

$$
\begin{aligned}
\alpha_{1,2} &\leq \frac{\pi_2}{\pi_1\, a_2} \sum_{n \in \mathbb{N}} \sum_{t:\, |t| = n} \left(\varepsilon_2 \times \tilde{\mathrm{P}}_2\right)\left\{r^t \geq a_2, \Gamma_n^o = t, \tau^o = n\right\} \\
&= \frac{\pi_2}{\pi_1\, a_2}\left(\varepsilon_2 \times \tilde{\mathrm{P}}_2\right)\left\{r^{T^o} \geq a_2\right\} \\
&= \frac{\pi_2}{\pi_1\, a_2}(1 - \alpha_{2,1}),
\end{aligned}
$$

which establishes the first inequality in (6.49). At the same time, we observe that this inequality is an equality if $r^{T^o} = a_2$ on $\{r^{T^o} \geq a_2\}$ and is approximately an equality if $r^{T^o} \cong a_2$ on $\{r^{T^o} \geq a_2\}$. In order to establish (6.51), it remains to replace $\leq$ by $\cong$ in (6.49) and to solve with respect to $\alpha_{1,2}$ and $\alpha_{2,1}$.

Remarks

1. If $a_1 < \pi_2/\pi_1 < a_2$, which amounts to saying that $\tau^o > 0$, then $r^{T_-^o} < a_2 \leq r^{T^o}$ on $\{r^{T^o} \geq a_2\}$ and $r^{T^o} \leq a_1 < r^{T_-^o}$ on $\{r^{T^o} \leq a_1\}$, where T_-^o denotes $\Gamma_{\tau^o - 1}^o$. Consequently, the distance between $\log r^{T^o}$ and the open interval with extremities $\log a_1$ and $\log a_2$ is less than the length of the τ^oth jump of the random d-walk $((W_{t^1}^1, \ldots, W_{t^d}^d), (t^1, \ldots, t^d) \in \mathbb{I})$. It is therefore to be expected that (6.50) will often occur in practice.
2. It is interesting to observe that the approximate values of the error probabilities do not depend on the choice of the optimizing increasing path.

3. Assume that the observation costs depend neither on the choice of the population nor on θ, in other words, that $f_i(\theta_1) = f_i(\theta_2) = c$ for $i = 1, \ldots, d$. By introducing into the expression

$$L(\Gamma^o, \tau^o, D^o) - c\,\mathrm{E}(\tau^o)$$
$$= \pi_1 c_{1,1} + \pi_2 c_{2,2} + \pi_1(c_{1,2} - c_{1,1})\alpha_{1,2} + \pi_2(c_{2,1} - c_{2,2})\alpha_{2,1}$$

(which follows from (6.1)) the approximate values of $\alpha_{1,2}$ and $\alpha_{2,1}$ indicated in (6.51), we conclude that under the hypothesis (6.50),

$$L(\Gamma^o, \tau^o, D^o) - c\,\mathrm{E}(\tau^o) \cong \pi_1 c_{1,1} + \pi_2 c_{2,2} + (c_{1,2} - c_{1,1})\frac{\pi_2 - \pi_1 a_1}{a_2 - a_1}$$
$$+ (c_{2,1} - c_{2,2})\frac{\pi_1 a_1 a_2 - \pi_2 a_1}{a_2 - a_1}.$$

Bounds on the Expected Number of Observations

In order to simplify the notation, we will assume in this paragraph and the next that $d = 2$.

Proposition 2 (Lower bound on the expected number of observations). *Suppose that* $a_1 < \pi_2/\pi_1 < a_2$ *(which implies that* (6.48) *is satisfied). Set*

$$M_1 = \alpha_{1,2} \log\frac{\alpha_{1,2}}{1 - \alpha_{2,1}} + (1 - \alpha_{1,2}) \log\frac{1 - \alpha_{1,2}}{\alpha_{2,1}},$$

$$M_2 = \alpha_{2,1} \log\frac{\alpha_{2,1}}{1 - \alpha_{1,2}} + (1 - \alpha_{2,1}) \log\frac{1 - \alpha_{2,1}}{\alpha_{1,2}}.$$

Then

a. $-\inf(I_1^1, I_1^2)\,\mathrm{E}_1(\tau^o) \geq M_1$;

b. $\sup(I_2^1, I_2^2)\,\mathrm{E}_2(\tau^o) \geq M_2$;

c. $\mathrm{E}(\tau^o) \geq -\dfrac{\pi_1 M_1}{\inf(I_1^1, I_1^2)} + \dfrac{\pi_2 M_2}{\inf(I_2^1, I_2^2)}$.

Proof. It is clear that c follows from a and b. Since the proof of b is similar to that of a, we will only establish a. If $\inf(I_1^1, I_1^2) = -\infty$, then the inequality is trivially satisfied. We can therefore assume that I_1^1 and I_1^2 are finite. By Wald's equality (see Exercise 2), which applies because $\mathrm{E}(|T^o|) < \infty$,

$$\mathrm{E}_1\left(W^1_{T^{o1}} + W^2_{T^{o2}}\right) - \log\frac{\pi_2}{\pi_1} = I_1^1\,\mathrm{E}_1(T^{o1}) + I_1^2\,\mathrm{E}_1(T^{o2}).$$

On the other hand, by (6.43),

$$E_1(W^1_{T^{o1}} + W^2_{T^{o2}}) = \mathrm{E}_1(\log r^{T^o}) = \int_{\{r^{T^o} \le a_1\}} \log r^{T^o}\, \mathrm{dP}_1 + \int_{\{r^{T^o} \ge a_2\}} \log r^{T^o}\, \mathrm{dP}_1.$$

Let P_1' and P_1'' denote the conditional probability measures $\mathrm{P}_1(\cdot\,|r^{T^o} \le a_1)$ and $\mathrm{P}_1(\cdot\,|r^{T^o} \ge a_2)$. Taking (6.47) into account, the right-hand side above can be written as

$$(1 - \alpha_{1,2}) \int_\Omega \log r^{T^o}\, \mathrm{dP}_1' + \alpha_{1,2} \int_\Omega \log r^{T^o}\, \mathrm{dP}_1''.$$

Since the function log is concave, we apply Jensen's inequality to both integrals to see that this is less than

$$\begin{aligned}
&(1 - \alpha_{1,2}) \log\left(\int_\Omega r^{T^o}\, \mathrm{dP}_1'\right) + \alpha_{1,2} \log\left(\int_\Omega r^{T^o}\, \mathrm{dP}_1''\right) \\
&\quad = (1 - \alpha_{1,2}) \log\left(\frac{1}{1 - \alpha_{1,2}} \int_{\{r^{T^o} \le a_1\}} r^{T^o}\, \mathrm{dP}_1\right) \\
&\qquad + \alpha_{1,2} \log\left(\frac{1}{\alpha_{1,2}} \int_{\{r^{T^o} \ge a_2\}} r^{T^o}\, \mathrm{dP}_1\right).
\end{aligned}$$

Assume for a moment that the integrals on the right-hand side are, respectively, equal to $\pi_2\alpha_{2,1}/\pi_1$ and $\pi_2(1 - \alpha_{2,1})/\pi_1$. We can then conclude that

$$\begin{aligned}
\inf(I_1^1, I_1^2)\, \mathrm{E}_1(\tau^o) &\le I_1^1\, \mathrm{E}_1(T^{o1}) + I_1^2\, \mathrm{E}_1(T^{o2}) \\
&= \mathrm{E}_1(W^1_{T^{o1}} + W^2_{T^{o2}}) - \log\frac{\pi_2}{\pi_1} \\
&\le (1 - \alpha_{1,2})\log\frac{\alpha_{2,1}}{1 - \alpha_{1,2}} + \alpha_{1,2} \log\frac{1 - \alpha_{2,1}}{\alpha_{1,2}},
\end{aligned}$$

which establishes a. We now return to the two integrals. Taking (6.47) into account, we observe that their values are indeed those indicated, because for all $a \in \mathbb{R}_+$,

$$\int_{\{r^{T^o} \le a\}} r^{T^o}\, \mathrm{dP}_1 = \frac{\pi_2}{\pi_1} \mathrm{P}_2\{r^{T^o} \le a\}.$$

The proof of this equality uses (6.12) and (6.40), by decomposing $\{r^{T^o} \le a\}$ as in the proof of (6.49). Details are left to the reader (cf. Exercise 3).

An Approximately Optimal Strategy

In this paragraph, we assume again that $d = 2$. Moreover, we assume that I_1^1, I_2^1, I_1^2, and I_2^2 are finite, that $a_1 < \pi_2/\pi_1 < a_2$ and that $f_i(\theta_1) = f_i(\theta_2) = 1$, $i = 1, 2$.

In order to minimize the total number of observations, it is natural to choose at each stage the direction that seems most likely to allow the random 2-walk $((W_{t^1}^1, W_{t^2}^2),\ (t^1, t^2) \in \mathbb{I})$ to leave the continuation strip as quickly as possible. Because the slope of the lines that bound this strip is -1, the only advantage of one direction over the other is that on average, the steps of the random walk in that direction may be larger than in the other.

Let Γ be the deterministic increasing path defined by $\Gamma_n = (n, 0)$, and set

$$\tau_1 = \inf\{n \in \mathbb{N}: (W_n^1, W_0^2) \in B\},$$

where $B = \mathbb{R}^2 - \{(x_1, x_2): \log a_1 < x_1 + x_2 < \log a_2\}$. By (6.46), the elementary renewal theorem implies that $E_j(\tau_1) < \infty$. We can therefore apply Wald's equality and deduce that

$$\mathrm{E}_j(W_{\tau_1}^1) - \frac{1}{2}\log\frac{\pi_2}{\pi_1} = I_j^1\ \mathrm{E}_j(\tau_1).$$

By assuming that the law (π_1, π_2) is close to $(\frac{1}{2}, \frac{1}{2})$ and that $W_{\tau_1}^1 \cong \log a_1$ on $\{W_{\tau_1}^1 \le \log a_1\}$ and $W_{\tau_1}^1 \cong \log a_2$ on $\{W_{\tau_1}^1 \ge \log a_2\}$, we deduce that

$$\mathrm{E}_j(\tau_1) \cong \frac{1}{I_j^1}\Big(\mathrm{P}_j\{W_{\tau_1}^1 \le \log a_1\}\log a_1 + \mathrm{P}_j\{W_{\tau_1}^1 \ge \log a_2\}\log a_2\Big).$$

Moreover, by assuming that $\mathrm{P}_1\{W_{\tau_1}^1 \le \log a_1\} \cong 1$ and $\mathrm{P}_2\{W_{\tau_1}^1 \ge \log a_2\} \cong 1$, we conclude that

$$\mathrm{E}(\tau_1) = \pi_1\,\mathrm{E}_1(\tau_1) + \pi_2\,\mathrm{E}_2(\tau_1) \cong \frac{\pi_1}{I_1^1}\log a_1 + \frac{\pi_2}{I_2^1}\log a_2.$$

The same argument applied to the case where $\Gamma_n = (0, n)$ and $\tau_2 = \inf\{n \in \mathbb{N}: (W_0^1, W_n^2) \in B\}$ shows that

$$\mathrm{E}(\tau_2) = \pi_1\,\mathrm{E}_1(\tau_2) + \pi_2\,\mathrm{E}_2(\tau_2) \cong \frac{\pi_1}{I_1^2}\log a_1 + \frac{\pi_2}{I_2^2}\log a_2.$$

Thus, when the assumptions that we have made are realized, we obtain an *approximately optimal strategy* by replacing step 3 of the description of the optimal strategy by

3′. If $a_1 < r^t < a_2$, draw from population $\mathcal{P}_i$ if $h_i(r^t) = \inf(h_1(r^t), h_2(r^t))$, where, for $i = 1, 2$, h_i denotes the function defined by

$$h_i(y) = \frac{1}{I_1^i}\log a_1 + \frac{y}{I_2^i}\log a_2 \qquad \left(h_i(y) = -\frac{1}{I_1^i} + \frac{y}{I_2^i} \text{ if } a_1 \cong \frac{1}{a_2}\right).$$

6.6 APPLICATIONS

We are now going to present two concrete examples to which the theory developed in the previous sections applies.

Heads with the Highest Probability

The problem is to determine which of two (possibly biased) coins falls on heads with the highest probability.

We identify heads with the value 1 and tails with the value 0 and assume that for $i = 1, 2$ and $j = 1, 2$,

$$\mu_j^i = \left(1 - p_j^i\right)\varepsilon_0 + p_j^i \varepsilon_1,$$

where ε_0 and ε_1 denote the laws concentrated on 0 and on 1, respectively, and p_j^i is a number such that $0 < p_j^i < 1$. This is equivalent to saying that conditionally with respect to $\{\theta = \theta_j\}$, coin i falls on heads with probability p_j^i and on tails with probability $1 - p_j^i$.

Since

$$\frac{\delta_2^i(0)}{\delta_1^i(0)} = \frac{1 - p_2^i}{1 - p_1^i} \quad \text{and} \quad \frac{\delta_2^i(1)}{\delta_1^i(1)} = \frac{p_2^i}{p_1^i},$$

formula (6.41) can be written as

$$\begin{aligned} \hat{\delta}_i \hat{q}(y) = \hat{q}\left(y \frac{1 - p_2^i}{1 - p_1^i}\right)\left(1 + y \frac{1 - p_2^i}{1 - p_1^i}\right) \\ + \hat{q}\left(y \frac{p_2^i}{p_1^i}\right)\left(1 + y \frac{p_2^i}{p_1^i}\right) - f_i(\theta_1) - y f_i(\theta_2). \end{aligned}$$

This can be used to compute $\hat{h}$ and the extremities a_1 and a_2 of the interval of continuation by approximation, using the functions $\hat{h}^{(n)}$ defined in (6.42).

The calculation of $r^{t+e_i}(\omega)$ given $r^t(\omega)$ is done as follows:

$$r^{t+e_i}(\omega) = \begin{cases} r^t(\omega) \dfrac{p_2^i}{p_1^i}, & \text{if coin } i \text{ falls on heads,} \\ r^t(\omega) \dfrac{1 - p_2^i}{1 - p_1^i}, & \text{if coin } i \text{ falls on tails.} \end{cases}$$

In order to get a feeling for the optimal strategy, we assume that we are in a situation where the approximately optimal strategy of the previous section applies. We carry out the calculations in the case where $p_1^1 = 0.9$, $p_1^2 = 0.4$, $p_2^1 = 0.4$ and $p_2^2 = 0.8$. Using the formula

$$I_j^i = \log\left(\frac{\delta_2^i(0)}{\delta_1^i(0)}\right)(1 - p_j^i) + \log\left(\frac{\delta_2^i(1)}{\delta_1^i(1)}\right)p_j^i$$

$$= \log\left(\frac{1 - p_2^i}{1 - p_1^i}\right)(1 - p_j^i) + \log\left(\frac{p_2^i}{p_1^i}\right)p_j^i,$$

we obtain

$$I_1^1 = -0.5506, \quad I_1^2 = -0.3822, \quad I_2^1 = 0.7506, \quad \text{and} \quad I_2^2 = 0.3347.$$

Because

$$\frac{1}{0.5506} + \frac{y}{0.7506} \leq \frac{1}{0.3822} + \frac{y}{0.3347}$$

if and only if

$$y \geq 0.4834,$$

the rule given by 3′ is to toss coin 1 if $r^t(\omega) \geq 0.4834$ and coin 2 if $r^t(\omega) \leq 0.4834$. The calculation of $r^{t+e_i}(\omega)$ is done as follows:

$$r^{t+e_1}(\omega) = \begin{cases} \frac{4}{9} r^t(\omega), & \text{if coin 1 falls on heads,} \\ 6r^t(\omega), & \text{if coin 1 falls on tails,} \end{cases}$$

$$r^{t+e_2}(\omega) = \begin{cases} 2r^t(\omega), & \text{if coin 2 falls on heads,} \\ \frac{1}{3} r^t(\omega), & \text{if coin 2 falls on tails.} \end{cases}$$

If $r^{T^o}(\omega) \leq a_1$ [$r^{T^o}(\omega) \geq a_2$], the decision is that coins 1 and 2 fall on heads with probabilities 0.9 and 0.4 [0.4 and 0.8], respectively.

Comparing Two Gaussian Populations

Consider two Gaussian populations $N(\mathsf{m}_1, 1)$ and $N(\mathsf{m}_2, 1)$. We would like to compare their means m_1 and m_2. We are going to indicate how the likelihood-ratio test can be used to decide which population has the largest mean. Notice that the best new treatment problem mentioned at the beginning of the chapter is generally of this type. Indeed, it suffices to consider that the degree of efficiency of treatment i is the mean m_i of a Gaussian population.

In order that the problem be of the type our theory describes, we choose two numbers α_1 and α_2 such that $\alpha_1 < \alpha_2$ and assume that

$$d\mu_1^1(x) = \frac{1}{\sqrt{2\pi}} \exp\left(-\frac{1}{2}(x-\alpha_1)^2\right) dx,$$

$$d\mu_1^2(x) = \frac{1}{\sqrt{2\pi}} \exp\left(-\frac{1}{2}(x-\alpha_2)^2\right) dx,$$

$$d\mu_2^1(x) = \frac{1}{\sqrt{2\pi}} \exp\left(-\frac{1}{2}(x-\alpha_2)^2\right) dx,$$

$$d\mu_2^2(x) = \frac{1}{\sqrt{2\pi}} \exp\left(-\frac{1}{2}(x-\alpha_1)^2\right) dx.$$

In other words, if $\{\theta = \theta_j\}$ is realized, then it is population $\mathcal{P}_j$ that has the smallest mean.

It is clear that

$$\frac{\delta_2^1(x)}{\delta_1^1(x)} = \frac{\exp\left(-\frac{1}{2}(x-\alpha_2)^2\right)}{\exp\left(-\frac{1}{2}(x-\alpha_1)^2\right)}, \quad \frac{\delta_2^2(x)}{\delta_1^2(x)} = \frac{\exp\left(-\frac{1}{2}(x-\alpha_1)^2\right)}{\exp\left(-\frac{1}{2}(x-\alpha_2)^2\right)},$$

so that the likelihood ratio is expressed by

$$r^t = \frac{\pi_2}{\pi_1} \exp\left((\alpha_2 - \alpha_1)\left(\sum_{s^1=1}^{t^1} Y_{s^1}^1 - \sum_{s^2=1}^{t^2} Y_{s^2}^2\right) + \frac{1}{2}(t^2 - t^1)(\alpha_2^2 - \alpha_1^2)\right), \tag{6.52}$$

with the convention that the sum over s^i is replaced by 0 if $t^i = 0$. The computation of r^t is therefore particularly easy.

Formula (6.41) becomes

$$\hat{\delta}_1 \hat{q}(y) = \frac{1}{\sqrt{2\pi}} \int_{\mathbb{R}} \hat{q}\left(y \frac{\exp\left(-\frac{1}{2}(x-\alpha_2)^2\right)}{\exp\left(-\frac{1}{2}(x-\alpha_1)^2\right)}\right)\left(1 + y \frac{\exp\left(-\frac{1}{2}(x-\alpha_2)^2\right)}{\exp\left(-\frac{1}{2}(x-\alpha_1)^2\right)}\right)$$
$$\times \exp\left(-\frac{1}{2}(x-\alpha_1)^2\right) dx - f_1(\theta_1) - y f_1(\theta_2),$$

$$\hat{\delta}_2 \hat{q}(y) = \frac{1}{\sqrt{2\pi}} \int_{\mathbb{R}} \hat{q}\left(y \frac{\exp\left(-\frac{1}{2}(x-\alpha_1)^2\right)}{\exp\left(-\frac{1}{2}(x-\alpha_2)^2\right)}\right)\left(1 + y \frac{\exp\left(-\frac{1}{2}(x-\alpha_1)^2\right)}{\exp\left(-\frac{1}{2}(x-\alpha_2)^2\right)}\right)$$
$$\times \exp\left(-\frac{1}{2}(x-\alpha_2)^2\right) dx - f_2(\theta_1) - y f_2(\theta_2).$$

Using the change of variables $x = \alpha_1 + \alpha_2 - x'$, it is easy to see that both integrals have the same value, so that $\hat{\delta}_1\hat{q}(y) \leq \hat{\delta}_2\hat{q}(y)$ if and only if

$$f_2(\theta_1) - f_1(\theta_1) \leq y(f_1(\theta_2) - f_2(\theta_2)). \tag{6.53}$$

In the special case where the costs of experimentation do not depend on the population, that is $f_1(\theta_j) = f_2(\theta_j)$ for $j = 1, 2$, then in the interior of the interval of continuation, there is no advantage to choosing to draw from one population rather than from the other. In other words, all predictable increasing paths are optimizing increasing paths. The rule for stopping (which is the same for all paths) is the one indicated in points 2a and 2b of the description of the optimal strategy.

Assume now that $f_1(\theta_1) < f_2(\theta_1)$ and $f_2(\theta_2) < f_1(\theta_2)$. In the setting of the best new treatment problem, this hypothesis means that giving the least effective medication to the patient implies larger overall costs. Notice that under this hypothesis, the optimal strategy will favor a path that minimizes the expected number of observations from the population with smallest mean. Taking (6.53) into account, in the interior of the interval of continuation the rule is to draw from population $\mathcal{P}_1$ or $\mathcal{P}_2$ according as

$$r^t \geq \frac{f_2(\theta_1) - f_1(\theta_1)}{f_1(\theta_2) - f_2(\theta_2)} \quad \text{or} \quad r^t \leq \frac{f_2(\theta_1) - f_1(\theta_1)}{f_1(\theta_2) - f_2(\theta_2)}.$$

In the particular case where $f_2(\theta_1) - f_1(\theta_1) = f_1(\theta_2) - f_2(\theta_2)$, this alternative is expressed by

$$r^t \geq 1 \quad \text{or} \quad r^t \leq 1.$$

When $\pi_1 = \pi_2 = \frac{1}{2}$ and $\alpha_1 = -\alpha_2$, we see using (6.52) (including the convention on sums) that this is equivalent to

$$\sum_{s^1=1}^{t^1} Y^1_{s^1} \geq \sum_{s^2=1}^{t^2} Y^2_{s^2} \quad \text{or} \quad \sum_{s^1=1}^{t^1} Y^1_{s^1} \leq \sum_{s^2=1}^{t^2} Y^2_{s^2}.$$

The rule for stopping is the one indicated in 2a and 2b of the description of the optimal strategy.

6.7 COMPLEMENT

This section is entirely devoted to the proof of the lemma of Section 6.3. This lemma is of independent interest, because in the setting of the model constructed in Section 6.2, it furnishes a characterization of accessible stopping points that are greater than a given point t. For the convenience of the reader, we repeat the statement of the lemma here.

Lemma. *Let* $t \in \mathbb{I}$ *and* $t \in \mathbf{A}_{\mathrm{b}}^{t}$. *Then there exists a function* U *defined on* $\Omega \times \Omega$ *with values in* $\mathbb{I}$ *that has the following properties*:

a. U *is* $\mathscr{F}_t \times \mathscr{F}_{\bowtie -}$*-measurable.*
b. $U(\omega, \cdot)$ *is an element of* $\mathbf{A}_{\mathrm{b}}$ *for all* ω.
c. $T(\omega) = t + U(\omega, \rho_t(\omega))$ *for almost all* ω.

In addition to the basic sigma field and filtration, we will use in the proof of this lemma the uncompleted sigma field $\dot{\mathscr{F}} = \{\varnothing, \Theta\} \times \tilde{\mathscr{F}}$ and the uncompleted filtration $(\dot{\mathscr{F}}_t,\ t \in \mathbb{I})$, where $\dot{\mathscr{F}}_t = \{\varnothing, \Theta\} \times \tilde{\mathscr{F}}_t$. We will also need ad hoc versions of the notions of stopping point and of predictable increasing path. We will term *special stopping point* all functions $\dot{T}$ defined on Ω with values in $\mathbb{I}$ such that $\{\dot{T} = t\} \in \tilde{\mathscr{F}}_t$ for all $t \in \mathbb{I}$. We will term *special predictable increasing path*, and will denote by $\dot{\Gamma}$, all sequences $(\dot{\Gamma}_n,\ n \in \mathbb{N})$ of special stopping points satisfying the following properties: $\dot{\Gamma}_0(\omega) = 0$ for all ω, $\dot{\Gamma}_{n+1}(\omega) \in \mathbb{D}_{\dot{\Gamma}_n}(\omega)$ for all ω and all $n \in \mathbb{N}$, and $\dot{\Gamma}_{n+1}$ is $\dot{\mathscr{F}}_{\dot{\Gamma}_n}$-measurable for all $n \in \mathbb{N}$. Of course, $\tilde{\mathscr{F}}_{\dot{\Gamma}_n}$ is the sigma field consisting of subsets F of Ω such that $F \cap \{\dot{\Gamma}_n = t\} \in \tilde{\mathscr{F}}_t$ for all $t \in \mathbb{I}$ such that $|t| = n$ (thus for all $t \in \mathbb{I}$).

Throughout this section, $t \in \mathbb{I}$ and $T \in \mathbf{A}_{\mathrm{b}}^{t}$ will be fixed. Moreover, $N \in \mathbb{N}$ will denote an integer larger than $|T|$ and Γ will denote a predictable increasing path that passes through t and T.

For all $n \in \mathbb{N}$, G_n will denote the set of finite sequences $(\gamma_0, \ldots, \gamma_n)$ of elements of $\mathbb{I}$ such that $\gamma_0 = 0$ and $\gamma_{m+1} \in \mathbb{D}_{\gamma_m}$ for all $m \in \mathbb{N}_{n-1}$ (if $n > 0$). Temporarily, the letter γ will be used to denote the generic element $(\gamma_0, \ldots, \gamma_n)$ of G_n.

Lemma 1. *There exists a special predictable increasing path* $\dot{\Gamma}$ *such that* $\dot{\Gamma}_{|t|}(\omega) = t$ *for all* ω *and* $\dot{\Gamma}_n(\omega) = \Gamma_n(\omega)$ *for almost all* ω *and all* $n \in \mathbb{N}$. *Moreover, there exists a special stopping point* $\dot{T}$ *such that* $\dot{T}(\omega) = T(\omega)$ *for almost all* ω, $\dot{\Gamma}_{|\dot{T}(\omega)|}(\omega) = \dot{T}(\omega)$ *for all* ω *and* $|t| \le |\dot{T}(\omega)| \le N$ *for all* ω.

Proof. We establish the first part of the lemma by constructing $\dot{\Gamma}$ using induction on n. Set $\dot{\Gamma}_0(\omega) = 0$ for all ω and assume that $\dot{\Gamma}_1, \ldots, \dot{\Gamma}_n$ are special stopping points satisfying the following conditions: $\dot{\Gamma}_{m+1}(\omega) \in \mathbb{D}_{\dot{\Gamma}_m(\omega)}$ for all ω and all $m \in \mathbb{N}_{n-1}$, $\dot{\Gamma}_{m+1}$ is $\dot{\mathscr{F}}_{\dot{\Gamma}_m}$-measurable for all $m \in \mathbb{N}_{n-1}$, and $\dot{\Gamma}_m(\omega) = \Gamma_m(\omega)$ for almost all ω and all $m \in \mathbb{N}_n$. For all $\gamma \in \mathrm{G}_n$, set $\dot{F}(\gamma) = \{\dot{\Gamma}_0 = \gamma_0, \ldots, \dot{\Gamma}_n = \gamma_n\}$. It is clear that $(\dot{F}(\gamma),\ \gamma \in \mathrm{G}_n)$ is a partition of Ω. For all $\gamma \in \mathrm{G}_n$ and $i = 1, \ldots, d$, set $F_i(\gamma) = \dot{F}(\gamma) \cap \{\Gamma_{n+1} = \dot{\Gamma}_n + e_i\}$. Because $F_i(\gamma) \in \mathscr{F}_{\gamma_n}$, there exists $\dot{F}_i(\gamma) \in \dot{\mathscr{F}}_{\gamma_n}$ such that $\mathrm{P}(\dot{F}_i(\gamma) \,\Delta\, F_i(\gamma)) = 0$, where Δ denotes symmetric difference. Since it is easy to make $\dot{F}_i(\gamma), \ldots, \dot{F}_d(\gamma)$ pairwise disjoint, we assume that this is the case and for all

$\gamma \in G_n$, we set

$$\dot{\Gamma}_{n+1}(\omega) = \begin{cases} \gamma_n + e_i & \text{if } \omega \in \dot{F}(\gamma) \cap \dot{F}_i(\gamma), \\ \gamma_n + e_1, & \text{if } \omega \in \dot{F}(\gamma) - \bigcup_{i=1}^{d} \dot{F}_i(\gamma). \end{cases}$$

We thus define a stopping point $\dot{\Gamma}_{n+1}$ such that $\dot{\Gamma}_1, \ldots, \dot{\Gamma}_{n+1}$ satisfy the conditions given at the beginning of the proof, but with $n+1$ instead of n. We are now going to refine the construction of $\dot{\Gamma}$ so that $\dot{\Gamma}_{|t|}(\omega) = t$ for all ω. Set $\dot{\sigma}(\omega) = \sup\{n \in \mathbb{N}: \dot{\Gamma}_n(\omega) \leq t\}$. If $s < t$ and $\{\dot{\Gamma}_{\dot{\sigma}} = s\} \neq \emptyset$, we modify $\dot{\Gamma}$ so that $\dot{\Gamma}_{\dot{\sigma}(\omega)+1}(\omega) \leq t$ for all ω such that $\dot{\Gamma}_{\dot{\sigma}(\omega)}(\omega) = s$ (for example, by replacing $\dot{\Gamma}_n(\omega)$ by $\dot{\Gamma}_{\dot{\sigma}(\omega)}(\omega) + (n - \dot{\sigma}(\omega))\, e_i$ if $n > \dot{\sigma}(\omega)$, where e_i is chosen such that $s + e_i \leq t$). By repeating this type of modification at most a finite number of times, we obtain a special predictable increasing path that satisfies the requirements of the first part of the lemma. Let us now turn to the second part. For all $n \in \mathbb{N}$ such that $|t| \leq n \leq N$ and for all $\gamma \in G_n$, set

$$F_n(\gamma) = \left\{\dot{\Gamma}_0 = \gamma_0, \ldots, \dot{\Gamma}_n = \gamma_n, |T| = n\right\}.$$

Clearly $F_n(\gamma) \in \mathscr{F}_{\gamma_n}$, so there exists $F_n'(\gamma) \in \dot{\mathscr{F}}_{\gamma_n}$ such that $P(F_n'(\gamma)\,\Delta F_n(\gamma)) = 0$. For the same n, set

$$\dot{F}_n = \bigcup_{\gamma \in G_n} \left(\left\{\dot{\Gamma}_0 = \gamma_0, \ldots, \dot{\Gamma}_n = \gamma_n\right\} \cap F_n'(\gamma)\right)$$

and let $\dot{\tau}$ be defined by

$$\dot{\tau}(\omega) = \begin{cases} |t|, & \text{if } \omega \in \dot{F}_{|t|}, \\ n, & \text{if } |t| < n \leq N \quad \text{and} \quad \omega \in \dot{F}_n - \bigcup_{m=|t|}^{n-1} \dot{F}_m, \\ N, & \text{if } \omega \in \Omega - \bigcup_{m=|t|}^{N} \dot{F}_m. \end{cases}$$

Because $\dot{F}_n \in \dot{\mathscr{F}}_{\dot{\Gamma}_n}$, it is clear that $\dot{\tau}$ is a stopping time relative to the filtration $(\dot{\mathscr{F}}_{\dot{\Gamma}_n}, n \in \mathbb{N})$. By setting $\dot{T}(\omega) = \dot{\Gamma}_{\dot{\tau}(\omega)}(\omega)$, we thus define a special stopping point $\dot{T}$ such that $\dot{\Gamma}_{|\dot{T}(\omega)|}(\omega) = \dot{T}(\omega)$ for all ω. It is clear that $\dot{T}(\omega) = T(\omega)$ for almost all ω and that $|t| \leq |\dot{T}(\omega)| \leq N$ for all ω.

Lemma 2. *Let $t \in \mathbb{I}$ and let ψ_t be the map defined on $\Omega \times \Omega$ with values in Ω by*

$$\psi_t\big((\theta_j, \omega_1, \ldots, \omega_d), (\theta_k, \omega_1', \ldots, \omega_d')\big) = \left(\theta_j, \omega_1'', \ldots, \omega_d''\right),$$

where ω_i'' is the element $(\omega_i''(s^i),\ s^i \in \mathbb{N}^*)$ of Ω_i defined by

$$\omega_i''(s^i) = \begin{cases} \omega_i(s^i), & \text{if } 1 \le s^i \le t^i, \\ \omega_i'(s^i - t^i), & \text{if } s^i > t^i. \end{cases}$$

Then for all $s \in \mathbb{I}$ such that $s \ge t$ and all $F \in \dot{\mathscr{F}}_s$, $\psi_t^{-1}(F) \in \dot{\mathscr{F}}_t \times \dot{\mathscr{F}}_{s-t}$. In particular, for all $F \in \dot{\mathscr{F}}$, $\psi_t^{-1}(F) \in \dot{\mathscr{F}}_t \times \dot{\mathscr{F}}$.

Proof. It is sufficient to consider the case where $F = \Theta \times \tilde{F}$ with

$$\tilde{F} = F_1 \times \cdots \times F_d \quad \text{and} \quad F_i = \begin{cases} B_1^i \times \cdots \times B_{s^i}^i \times \mathbb{R} \times \cdots, & \text{if } s^i > 0, \\ \Omega_i, & \text{if } s^i = 0. \end{cases}$$

The conclusion is then clear.

Lemma 3. *Let $\dot{\Gamma}$ be a special predictable increasing path such that $\dot{\Gamma}_{|t|}(\omega) = t$ for all ω. For all $n \in \mathbb{N}$, let us define Δ_n by setting, for all (ω, ω'),*

$$\Delta_n(\omega, \omega') = \dot{\Gamma}_{|t|+n}(\psi_t(\omega, \omega')) - t,$$

where ψ_t denotes the map defined in Lemma 2. Then the following properties are satisfied:

a. *Δ_n is $\dot{\mathscr{F}}_t \times \dot{\mathscr{F}}$-measurable for all $n \in \mathbb{N}$.*
b. *$(\Delta_n(\omega, \cdot),\ n \in \mathbb{N})$ is a special predictable increasing path for all ω.*
c. *$\dot{\Gamma}(\omega) = t + \Delta_{n-|t|}(\omega, \rho_t(\omega))$ for all ω and all $n \ge |t|$.*

Proof. Property c is immediate, since $\psi_t(\omega, \rho_t(\omega)) = \omega$ for all ω. Property a follows from Lemma 2, because if $s \in \mathbb{I}$, then

$$\{(\omega, \omega'): \Delta_n(\omega, \omega') = s\} = \psi_t^{-1}\Big(\big\{\omega'': \dot{\Gamma}_{|t|+n}(\omega'') = s + t\big\}\Big) \in \dot{\mathscr{F}}_t \times \dot{\mathscr{F}}_s.$$

It therefore remains to establish b. To begin with, it is clear that $(\Delta_n(\omega, \omega'),\ n \in \mathbb{N})$ is a deterministic increasing path. Moreover, again by Lemma 2, for all $n \in \mathbb{N}$, all $s \in \mathbb{I}$, and $i = 1, \ldots, d$,

$$\begin{aligned} &\{(\omega, \omega'): \Delta_n(\omega, \omega') = s,\ \Delta_{n+1}(\omega, \omega') = s + e_i\} \\ &\quad = \psi_t^{-1}\Big(\big\{\omega'': \dot{\Gamma}_{|t|+n}(\omega'') = s + t,\ \dot{\Gamma}_{|t|+n}(\omega'') = s + t + e_i\big\}\Big) \in \dot{\mathscr{F}}_t \times \dot{\mathscr{F}}_s, \end{aligned}$$

because $\{\omega'': \dot{\Gamma}_{|t|+n}(\omega'') = s + t,\ \dot{\Gamma}_{|t|+n}(\omega'') = s + t + e_i\} \in \dot{\mathscr{F}}_{s+t}$. By taking the ω-section, we see therefore that

$$\{\omega' : \Delta_n(\omega, \omega') = s,\ \Delta_{n+1}(\omega, \omega') = s + e_i\} \in \dot{\mathscr{F}}_s,$$

which completes the proof of b.

Proof of the Lemma of Section 6.3

Let $\dot{\Gamma}$ and $\dot{T}$ be, respectively, a special predictable increasing path and a special stopping point that satisfy the properties of Lemma 1. Let $(\Delta_n, n \in \mathbb{N})$ be the sequence defined in Lemma 3. For all (ω, ω'), set

$$U(\omega, \omega') = \Delta_{|\dot{T}(\psi_t(\omega, \omega'))| - |t|}(\omega, \omega').$$

The function U thus defined takes its values in $\mathbb{I}$. Moreover, $|U(\omega, \omega')| = |\dot{T}(\psi_t(\omega, \omega'))| - |t| \leq N$ for all (ω, ω'). We are going to show that U satisfies properties a, b, and c of the lemma of Section 6.3. Property a follows from Lemma 2 and from a of Lemma 3, since for all $s \in \mathbb{I}$,

$$\begin{aligned}&\{(\omega, \omega') : U(\omega, \omega') = s\}\\ &\quad = \bigcup_n \left\{(\omega, \omega') : \Delta_n(\omega, \omega') = s, \left|\dot{T}(\psi_t(\omega, \omega'))\right| = |t| + n\right\} \in \dot{\mathscr{F}}_t \times \dot{\mathscr{F}}.\end{aligned}$$

To prove b, it suffices to observe that for fixed ω, $|\dot{T}(\psi_t(\omega, \cdot))| - |t|$ is a stopping time relative to the filtration $(\dot{\mathscr{F}}_{\Delta_n(\omega, \cdot)}, n \in \mathbb{N})$. Indeed, for all $n \in \mathbb{N}$ and all $s \in \mathbb{I}$, the set

$$\left\{\omega' : \left|\dot{T}\ (\psi_t(\omega, \omega'))\right| - |t| = n,\ \Delta_n(\omega, \omega') = s\right\}$$

belongs to $\dot{\mathscr{F}}_s$, because it is the ω-section of the set

$$\begin{aligned}&\left\{(\omega, \omega') : \left|\dot{T}\ (\psi_t(\omega, \omega'))\right| - |t| = n,\ \Delta_n(\omega, \omega') = s\right\}\\ &\quad = \psi_t^{-1}\left(\left\{\omega'' : \left|\dot{T}\ (\omega'')\right| = |t| + n,\ \dot{\Gamma}_{|t|+n}(\omega'') = s + t\right\}\right),\end{aligned}$$

which belongs to $\dot{\mathscr{F}}_t \times \dot{\mathscr{F}}_s$ by Lemma 2 and the fact that $\{\omega'' : |\dot{T}(\omega'')| = |t| + n,\ \dot{\Gamma}_{|t|+n} = s + t\}$ belongs to $\dot{\mathscr{F}}_{s+t}$. It remains to establish c. Since $\psi_t(\omega, \rho_t(\omega)) = \omega$ for all ω, by Lemma 1, c of Lemma 3, and the definition of U, we conclude that for almost all ω,

$$\begin{aligned}T(\omega) &= \dot{T}\ (\omega) = \dot{\Gamma}_{|\dot{T}(\omega)|}(\omega) = t + \Delta_{|\dot{T}(\omega)| - |t|}(\omega, \rho_t(\omega))\\ &= t + \Delta_{|\dot{T}(\psi_t(\omega, \rho_t(\omega)))| - |t|}(\omega, \rho_t(\omega)) = t + U(\omega, \rho_t(\omega)),\end{aligned}$$

which was to be proved.

EXERCISES

1. In the context of Section 6.5 (in particular, under assumption 4 of that section), let Γ be a predictable increasing path and τ the number of draws required by the optimal strategy described there.

a. Show that there exist two positive constants b and c such that $c < 1$ and $\mathrm{P}\{\tau > n\} \leq bc^n$ for all $n \in \mathbb{N}$. (*Hint*. Set $a = \log a_2 - \log a_1$ and suppose that for $i = 1, \ldots, d$,

$$\mathrm{P}_j\left\{\left|\log \frac{\delta_2^i(Y_1^i)}{\delta_1^i(Y_1^i)}\right| \leq a\right\} = p_j < 1, \qquad j = 1, 2. \tag{6.54}$$

Show that on $\{\tau > n,\ \Gamma_n^i = t^i\}$,

$$\left|\log \frac{\delta_2^i(Y_{s^i}^i)}{\delta_1^i(Y_{s^i}^i)}\right| \leq a, \qquad \text{for } s^i = 1, \ldots, t^i \text{ and } i = 1, \ldots, d,$$

and therefore that $\mathrm{P}_j\{\tau > n\} \leq p_j^n$. If (6.54) is not satisfied, conclude using the fact that for $i = 1, \ldots, d$, there exists $t^i \in \mathbb{N}^*$ such that

$$\mathrm{P}_j\left\{\left|\sum_{s^i=1}^{t^i} \log \frac{\delta_2^i(Y_{s^i}^i)}{\delta_1^i(Y_{s^i}^i)}\right| \leq a\right\} < 1, \quad j = 1, 2.)$$

b. Deduce from a that there is $x \in \mathbb{R}_+^*$ such that $\mathrm{E}(e^{x\tau}) < \infty$, which implies that $\mathrm{E}(\tau^k) < \infty$ for all $k \in \mathbb{N}$.

2. (Wald's equality for stopping points)

a. For $i = 1, \ldots, d$, let $(X_{t^i}^i,\ t^i \in \mathbb{N})$ be a family of independent and identically distributed random variables with finite mean. Assume these d families are independent. Let T be a stopping point relative to the natural filtration of the random d-walk $((W_{t^1}^1, \ldots, W_{t^d}^d), (t^1, \ldots, t^d) \in \mathbb{I})$, where

$$W_{t^i}^i = \begin{cases} \sum_{s^i=1}^{t^i} X_{s^i}^i, & \text{if } t^i > 0, \\ 0, & \text{if } t^i = 0. \end{cases}$$

Show that if $\mathrm{E}(|T|) < \infty$, then $\mathrm{E}(\sum_{i=1}^d W_{T^i}^i) = \sum_{i=1}^d \mathrm{E}(T^i)\mathrm{E}(X_0^i)$.

b. Give an example where each $X_{t^i}^i$ only takes two values and T is such that $\mathrm{E}(|T|) = \infty$ but $\mathrm{E}(\sum_{i=1}^d W_{T^i}^i) < \infty$.

3. Prove the last statement in the proof of Proposition 2 of Section 6.5.

HISTORICAL NOTES

The sequential probability ratio test for a single population was introduced by Wald (1947), and extensively studied by Arrow, Blackwell, and Girshick (1949) as well as by Wald and Wolfowitz (1948, 1950), Robbins (1952), Chernoff (1959), and Haggstrom (1966). It has become one of the fundamental sequential statistical tests, and there is now a huge body of literature on the subject. Classical books that treat this topic include Blackwell and Girshick (1954), Lehman (1959), Chow, Robbins, and Siegmund (1971), Shiryayev (1978), and more recently, Siegmund (1985), which contains numerous statistical applications. Part of the interest in this test is due to the fact that it is optimal in the following sense: among all tests for discriminating between two simple hypotheses with prescribed error probabilities, it requires the minimal expected number of observations (see Lehman (1959, Theorem 8 of Section 3.10).

The Best new treatment problem described in the introduction is widely applicable (see Armitage (1985)). The formulation of the sequential probability ratio test in Sections 6.1 and 6.2 is new, though all the main mathematical ideas are contained in the books of Lehman and Shiryayev mentioned above. As opposed to most presentations, we have not assumed that the measures μ_j^i are equivalent. The expression for Snell's envelope given in the theorem of Section 6.3 is well known at least in the case of a single population (see Neveu (1975)), but the proof given here is new. In the case of a finite probability space, the optimal strategy is described in Blackwell and Girshick (1954), where several explicit examples are also solved. The likelihood ratio test in Section 6.5 is the case most authors consider.

The notion of Kullback–Leibler information is studied in detail in Kullback (1968). The bounds in Propositions 1 and 2 of Section 6.5 are also given in Shiryayev (1978) in the case of a single population. Using the Kullback–Leibler information numbers to define strategies that are close to optimal is suggested in Bradt and Karlin (1956). The elementary renewal theorem and its proof can be found in Chow and Teicher (1988, Chapter 5). The approximately optimal strategy of Section 6.5 was given in Keener (1980). As shown by Keener, part of the interest of this heuristic strategy is that it is asymptotically optimal (see also Hayre (1982)). The application to the comparison of two Gaussian populations was first presented in Robbins and Siegmund (1974) and extensions are considered in Louis (1975, 1977). Other interesting applications are considered for instance in Lalley and Lorden (1986). The content of Section 6.7 is new and extends to dimensions $d > 1$ Galmarino's test for stopping times (see Dellacherie and Meyer (1978)).

Exercise 1 is taken from Lehman (1959). Exercise 2 establishes a result that is classical in the case $d = 1$ and can be found in Krengel and Sucheston (1981).

CHAPTER 7

Optimal Sequential Control

In Chapter 3, we described a general theory of optimal stopping in which the emphasis was on accessible stopping points, even though the predictable increasing paths on which they are located play almost as important a role. This approach is justified by the fact that in general, there are several paths that lead to the point where the observation stops and there is no advantage in that problem to choosing one path over another. The situation is very different when a running reward is added to the terminal reward, because choosing the best path becomes as important as choosing the best stopping point. This type of problem is part of a vast subject known as *optimal sequential control*.

In this chapter, we will solve a general control problem that encompasses many applications. Our approach consists of reducing this problem to an optimal stopping problem with values in an appropriate set. This approach will shed new light on the optimization problems dealt with in this book.

We again let $(\Omega, \mathscr{F}, \mathrm{P})$ and $(\mathscr{F}_t)$ denote the general probability space and filtration of Section 1.1.

7.1 AN EXAMPLE

In Chapter 3 we introduced a process, which we denote here by $(\tilde{X}_t, t \in \bar{\mathbb{I}})$, representing the rewards $\tilde{X}_t$ that a player would win when quitting the game at each stage t. Imagine now that the game is modified to include running rewards, more precisely, quantities $Y_{s,t}$ representing the player's reward for going from stage s to a stage t that is a direct successor of s. Assume moreover that future rewards are discounted in such a way that a unit reward is worth α^n units after n stages, where α is a discounting rate belonging to $]0, 1]$. If Γ is a predictable increasing path and τ is a stopping time relative to the filtration $\mathscr{F}^\Gamma$, then the reward that the player will receive by following the path Γ and by terminating the game at stage Γ_τ will be

$$X_{(\Gamma,\tau)} = \begin{cases} \tilde{X}_0 & \text{on } \{\tau = 0\}, \\ \sum_{n=1}^{\tau} \alpha^n \, Y_{\Gamma_{n-1},\Gamma_n} + \alpha^\tau \tilde{X}_{\Gamma_\tau} & \text{on } \{\tau > 0\}, \end{cases} \tag{7.1}$$

where, by convention, $\alpha^\infty = 0$ when $\alpha < 1$ and $\alpha^\infty = 1$ when $\alpha = 1$. A couple (Γ, τ) formed by a predictable increasing path Γ and a stopping time τ relative to the filtration $\mathscr{F}^\Gamma$ is termed a *strategy*. It is natural to ask whether there exists a strategy $(\Gamma, \tau)^o$ such that

$$\mathrm{E}(X_{(\Gamma,\tau)^o}) = \sup_{(\Gamma,\tau)} \mathrm{E}(X_{(\Gamma,\tau)}). \tag{7.2}$$

Such a strategy is termed *optimal*. If the processes $(\tilde{X}_t, t \in \bar{\mathbb{I}})$ and $(Y_{s,t}, s \in \mathbb{I}, t \in \mathbb{D}_s)$ satisfy certain hypotheses that will be indicated further on, we will show that an optimal strategy does exist and can be determined in a constructive way.

In the example just presented, the reward associated with the passage from stage s to stage t does not depend on how the player reached stage s, but only on the couple (s, t). Similarly, the terminal reward does not depend on the choice of the path but only on the point where it stops. In Sections 7.4 and 7.5, we will formulate and study a more general problem where the total reward will be a function of the whole path.

7.2 PRELIMINARIES

The couple path/stopping point mentioned above can be identified with a stopped increasing path. In the theory we shall present, paths of this type will play a key role.

The Sets $\mathbb{G}$, $\mathbb{G}_{\mathrm{f}}$, and $\mathbb{G}^\infty$

Let $\mathbb{G}$ denote the set of sequences $(\gamma_n, n \in \mathbb{N})$ of elements of $\mathbb{I}$ such that

a. $\gamma_0 = 0$;
b. $\gamma_{n+1} \in \mathbb{D}_{\gamma_n}$, if $\gamma_{n+1} \neq \gamma_n$;
c. $\gamma_{n+k} = \gamma_n$ for all $k > 1$, if $\gamma_{n+1} = \gamma_n$.

We will write γ instead of $(\gamma_n, n \in \mathbb{N})$ and will also use the letter δ in the same role as γ. The element $(0, 0, \dots)$ of $\mathbb{G}$ will be denoted 0. For all $\gamma \in \mathbb{G}$ and all $m \in \overline{\mathbb{N}}$, $\gamma^{[m]}$ will denote the element $(\gamma_n^{[m]}, n \in \mathbb{N})$ of $\mathbb{G}$ defined by

$\gamma_n^{[m]} = \gamma_{n \wedge m}$. Clearly, $\gamma^{[\infty]} = \gamma$ and $(\gamma^{[m]})^{[p]} = \gamma^{[m \wedge p]}$. For $\gamma \in \mathbb{G}$, we set

$$l(\gamma) = \inf\{n \in \mathbb{N}: \gamma_{n+1} = \gamma_n\}. \tag{7.3}$$

An element γ of $\mathbb{G}$ is *finite* if $l(\gamma) < \infty$. The set of finite elements of $\mathbb{G}$ will be denoted $\mathbb{G}_f$, and the set of elements $\gamma \in \mathbb{G}$ such that $l(\gamma) = \infty$ will be denoted $\mathbb{G}^\infty$. Clearly $\mathbb{G}_f$ is countable and $\mathbb{G}^\infty$ is uncountable. Observe that $\gamma^{[m]} = \gamma$ for all $m \geq l(\gamma)$, that $l(\gamma^{[m]}) = l(\gamma) \wedge m$, and that if γ and δ are elements of $\mathbb{G}$ such that $\gamma^{[m]} = \delta$ for some $m > l(\delta)$, then $\gamma = \delta$.

Equipping $\mathbb{G}$ with an Order

The set $\mathbb{G}$ will be equipped with the order $\prec$ defined by

$$\gamma \prec \delta \quad \text{if} \quad \gamma = \delta^{[l(\gamma)]}. \tag{7.4}$$

Clearly, $\gamma \prec \delta$ if and only if $\gamma^{[m]} \prec \delta^{[m]}$ for all $m \in \mathbb{N}$, and $\gamma \prec \delta$ implies $l(\gamma) \leq l(\delta)$; moreover, if $\delta \in \mathbb{G}_f - \{0\}$, then $\gamma = \delta$ if and only if $\gamma \prec \delta$ and $\gamma \not\prec \delta^{[l(\delta)-1]}$. Notice that if $\gamma \in \mathbb{G}^\infty$ and $\gamma^{[m]} \prec \delta$ for all $m \in \mathbb{N}$, then $\gamma = \delta$.

For all $\gamma \in \mathbb{G}$, we let $\mathbb{G}^\gamma$ denote the set of elements $\delta \in \mathbb{G}$ such that $\gamma \prec \delta$. Clearly $\mathbb{G}^\gamma = \{\gamma\}$ if $\gamma \in \mathbb{G}^\infty$.

We will term *direct successor* of $\gamma \in \mathbb{G}_f$ each element $\delta \in \mathbb{G}^\gamma$ such that $l(\delta) = l(\gamma) + 1$. The set of direct successors of $\gamma \in \mathbb{G}_f$ will be denoted $\mathbb{D}_\gamma$. Observe that card $\mathbb{D}_\gamma = d$.

Upper Semicontinuity

A function $\gamma \mapsto x_\gamma$ defined on $\mathbb{G}$ and with values in $\overline{\mathbb{R}}$ is *upper semicontinuous* provided

$$\lim_{n \to \infty} \sup_{\delta \in \mathbb{G}^{\gamma[n]}} x_\delta = x_\gamma, \qquad \text{for all } \gamma \in \mathbb{G}^\infty. \tag{7.5}$$

This function is *continuous* if the two functions $\gamma \mapsto x_\gamma$ and $\gamma \mapsto -x_\gamma$ are upper semicontinuous.

According to this definition, a function is upper semicontinuous [continuous] if and only if it is upper semicontinuous [continuous] relative to the topology on $\mathbb{G}$ having for (countable) base the singletons $\{\gamma\}$, with $\gamma \in \mathbb{G}_f$, and the sets $\mathbb{G}^\gamma$, also with $\gamma \in \mathbb{G}_f$.

A Sigma Field on $\mathbb{G}$

The set $\mathbb{G}$ will be equipped with the separable sigma field $\mathscr{B}$ generated by the sets in the topological base described above. Thus equipped, $\mathbb{G}$ becomes a measurable space.

Filtrations Indexed by $\mathbb{G}_f$

A *filtration indexed by* $\mathbb{G}_f$ is a family $(\mathscr{G}_\gamma,\ \gamma \in \mathbb{G}_f)$ of sub-sigma fields of $\mathscr{F}$ such that

a. $\mathscr{G}_0$ contains all null sets of $\mathscr{F}$, and

b. $\mathscr{G}_\gamma \subset \mathscr{G}_\delta$ if $\gamma \prec \delta$.

From now on, we will assume that such a filtration $(\mathscr{G}_\gamma,\ \gamma \in \mathbb{G}_f)$ is given and fixed and will set $\mathscr{G}_\gamma = \mathscr{G}$ for all $\gamma \in \mathbb{G}^\infty$, where $\mathscr{G}$ is an arbitrary sigma field such that $\bigvee_{\gamma \in \mathbb{G}_f} \mathscr{G}_\gamma \subset \mathscr{G} \subset \mathscr{F}$, also considered as given and fixed. As was already the case for the sigma field $\mathscr{F}_{\bowtie}$, the choice of $\mathscr{G}$ is not important as long as these inclusions are satisfied, and it will become clear that there is no loss of generality in assuming that for $\gamma \in \mathbb{G}^\infty$, $\mathscr{G}_\gamma$ does not depend on γ. In order to simplify the notation, we will write $(\mathscr{G}_\gamma)$ instead of $(\mathscr{G}_\gamma,\ \gamma \in \mathbb{G}_f)$.

Assuming that $\mathscr{G}_\gamma$ (with $\gamma \in \mathbb{G}_f$) represents the information available at stage γ and that this information is acquired step by step, that is, by going from one stage to another chosen among its direct successors, we observe that there is only one way to acquire the information contained in $\mathscr{G}_\gamma$, contrary to the setup involving the filtration $(\mathscr{F}_t)$ when the dimension d is > 1. From this point of view, the ordered set $\mathbb{G}$ may appear simpler than $\bar{\mathbb{I}}$. However, $\mathbb{G}$ contains an infinite number of nonfinite elements, whereas $\bar{\mathbb{I}}$ contains only one such element.

7.3 CONTROLS

A *control* is a function Γ defined on Ω with values in $\mathbb{G}$ such that

$$\{\Gamma^{[n+1]} = \gamma\} \in \mathscr{G}_{\gamma^{[n]}}, \qquad \text{for all } n \in \mathbb{N} \text{ and all } \gamma \in \mathbb{G}_f. \tag{7.6}$$

Of course, $\Gamma^{[n]}$ denotes the function $\omega \mapsto (\Gamma(\omega))^{[n]}$. Observe that $\{\Gamma^{[n+1]} = \gamma\} = \varnothing$ if $l(\gamma) > n + 1$, which means that (7.6) only needs to be checked for those γ for which $l(\gamma) \leq n + 1$. Notice also that we can replace $=$ by $\prec$ in (7.6) without changing the definition. This follows from the fact that for all functions Δ defined on Ω with values in $\mathbb{G}$ and for all $\gamma \in \mathbb{G}_f$,

$$\{\Delta = \gamma\} = \begin{cases} \{\Delta \prec \gamma\}, & \text{if } \gamma = 0, \\ \{\Delta \prec \gamma\} - \{\Delta \prec \gamma^{[l(\gamma)-1]}\}, & \text{if } \gamma \neq 0, \end{cases}$$

and

$$\{\Delta \prec \gamma\} = \bigcup_{m=0}^{l(\gamma)} \{\Delta = \gamma^{[m]}\}.$$

The set of controls will be denoted by **G** and its elements by the letters Γ or Δ. We write $\Gamma \prec \Delta$ to express that $\Gamma(\omega) \prec \Delta(\omega)$ for almost all ω. It is clear that $\prec$ is an order relation on **G**. We will say that a control Γ is *finite* if $\Gamma(\omega) \in \mathbb{G}_f$ for almost all ω and that it is *bounded* if there exists $n \in \mathbb{N}$ such that $\Gamma^{[n]} = \Gamma$.

Any constant function Γ with values in $\mathbb{G}$ is a control; therefore we can consider $\mathbb{G}$ to be a subset of **G**.

Any control Γ is a random variable with values in $\mathbb{G}$, more precisely, a $(\bigvee_{\gamma \in \mathbb{G}_f} \mathscr{G}_\gamma)$-measurable random variable. Indeed, for all $\gamma \in \mathbb{G}_f$,

$$\{\Gamma = \gamma\} = \{\Gamma^{[l(\gamma)+1]} = \gamma\} \in \mathscr{G}_\gamma \quad \text{and} \quad \{\gamma \prec \Gamma\} = \{\Gamma^{[l(\gamma)]} = \gamma\} \in \mathscr{G}_\gamma.$$

Sigma Fields Associated with Controls

To each control Γ we associate the sigma field $\mathscr{G}_\Gamma$ consisting of the sets $F \in \mathscr{G}$ such that

$$F \cap \{\Gamma = \gamma\} \in \mathscr{G}_\gamma, \qquad \text{for all } \gamma \in \mathbb{G}_f. \tag{7.7}$$

As with property (7.6), the meaning of (7.7) is unchanged if the symbol $=$ is replaced by $\prec$. It is then clear that $\mathscr{G}_\Gamma \subset \mathscr{G}_\Delta$ if $\Gamma \prec \Delta$. Also, observe that $\mathscr{G}_\Gamma = \mathscr{G}_\gamma$ if Γ is equal to the constant $\gamma \in \mathbb{G}$, and that Γ is $\mathscr{G}_\Gamma$-measurable. Moreover, for all $\gamma \in \mathbb{G}$ and all integrable random variables X, $\mathrm{E}(X|\mathscr{G}_\Gamma) = \mathrm{E}(X|\mathscr{G}_\gamma)$ on $\{\Gamma = \gamma\}$.

Properties

Here are several useful properties of controls.

1. If Γ is a control, so is $\Gamma^{[n]}$, and $\Gamma^{[n+1]}$ is $\mathscr{G}_{\Gamma^{[n]}}$-measurable for all $n \in \mathbb{N}$. The first assertion follows directly from (7.6). In order to check the second assertion, it suffices to observe that for all couples (γ, δ) of elements of $\mathbb{G}_f$,

$$\{\Gamma^{[n+1]} = \gamma\} \cap \{\Gamma^{[n]} = \delta\} = \begin{cases} \{\Gamma^{[n+1]} = \gamma\}, & \text{if } \delta = \gamma^{[n]}, \\ \varnothing, & \text{if } \delta \neq \gamma^{[n]}, \end{cases}$$

which shows that the intersection on the left-hand side belongs to $\mathscr{G}_\delta$.

2. If Γ is a control, then $l(\Gamma)$ is a stopping time relative to the filtration $(\mathscr{G}_{\Gamma^{[n]}}, n \in \mathbb{N})$. In particular, $l(\Gamma)$ is $\mathscr{G}_\Gamma$-measurable. Indeed, since

$$\{l(\Gamma) = n\} = \begin{cases} \{\Gamma^{[1]} = 0\}, & \text{if } n = 0, \\ \{\Gamma^{[n-1]} \neq \Gamma^{[n]} = \Gamma^{[n+1]}\}, & \text{if } n \in \mathbb{N}^*, \end{cases}$$

the conclusion follows from 1.

3. If Γ is a control and τ is a stopping time relative to the filtration $(\mathscr{G}_{\Gamma^{[n]}}, n \in \mathbb{N})$, then $\Gamma^{[\tau]}$ (that is, the function $\omega \mapsto \Gamma^{[\tau(\omega)]}(\omega)$) is a control. Indeed, for all $\gamma \in \mathbb{G}_f$,

$$\left\{\left(\Gamma^{[\tau]}\right)^{[n+1]} = \gamma\right\} = \left\{\Gamma^{[\tau \wedge (n+1)]} = \gamma\right\}$$
$$= \bigcup_{m=0}^{n} \left(\{\Gamma^{[m]} = \gamma\} \cap \{\tau = m\}\right)$$
$$\cup \left(\{\Gamma^{[n+1]} = \gamma\} \cap \{\tau > n\}\right).$$

But for all $m \leq n$,

$$\{\Gamma^{[m]} = \gamma\} \cap \{\tau = m\} = \{\Gamma^{[m]} = \gamma\} \cap \{\Gamma^{[m]} = \gamma^{[m]}\} \cap \{\tau = m\},$$

and the set on the right-hand side belongs to $\mathscr{G}_{\gamma^{[m]}}$, therefore to $\mathscr{G}_{\gamma^{[n]}}$, by (7.6) and the fact that $\{\tau = m\}$ belongs to $\mathscr{G}_{\gamma^{[m]}}$. On the other hand,

$$\{\Gamma^{[n+1]} = \gamma\} \cap \{\tau > n\} = \{\Gamma^{[n+1]} = \gamma\} \cap \{\Gamma^{[n]} = \gamma^{[n]}\} \cap \{\tau > n\},$$

and the set on the right-hand side also belongs to $\mathscr{G}_{\gamma^{[n]}}$, by (7.6) and the fact that $\{\tau > n\}$ belongs to $\mathscr{G}_{\gamma^{[n]}}$. Thus $\{(\Gamma^{[\tau]})^{[n+1]} = \gamma\}$ belongs to $\mathscr{G}_{\gamma^{[n]}}$, which was to be proved.

4. If $(\Gamma(n), n \in \mathbb{N})$ is a sequence of finite controls such that $\Gamma(0) = 0$ and $\Gamma(n+1) \in \mathbb{D}_{\Gamma(n)}$ for all $n \in \mathbb{N}$, and if τ is a finite stopping time relative to the filtration $(\mathscr{G}_{\Gamma(n)}, n \in \mathbb{N})$, then $\Gamma(\tau)$ (that is, the function $\omega \mapsto \Gamma(\tau(\omega))(\omega)$) is a control. Indeed, the unique function Γ defined on Ω with values in $\mathbb{G}^\infty$ such that $\Gamma^{[n]} = \Gamma(n)$ for all $n \in \mathbb{N}$ is a control, and $\Gamma^{[\tau]} = \Gamma(\tau)$, which shows that the conclusion follows from 3.

5. If Γ, Δ', and Δ'' are controls such that $\Gamma \prec \Delta'$ on F and $\Gamma \prec \Delta''$ on F^c, where F belongs to $\mathscr{G}_\Gamma$, then

$$\Delta = \begin{cases} \Delta' & \text{on } F, \\ \Delta'' & \text{on } F^c, \end{cases}$$

is a control. In order to establish this, consider an element $\gamma \in \mathbb{G}_f$ and write

$$\{\Delta^{[n+1]} = \gamma\} = \left(\{(\Delta')^{[n+1]} = \gamma\} \cap F\right) \cup \left(\{(\Delta'')^{[n+1]} = \gamma\} \cap F^c\right)$$
$$= \bigcup_{m=0}^{n} \left(\left(\{(\Delta')^{[n+1]} = \gamma\} \cap \{\Gamma = \gamma^{[m]}\} \cap F\right)\right.$$
$$\left.\cup \left(\{(\Delta'')^{[n+1]} = \gamma\} \cap \{\Gamma = \gamma^{[m]}\} \cap F^c\right)\right)$$
$$\cup \left(\{(\Delta')^{[n+1]} = \gamma\} \cap \{l(\Gamma) > n\} \cap F\right)$$
$$\cup \left(\{(\Delta'')^{[n+1]} = \gamma\} \cap \{l(\Gamma) > n\} \cap F^c\right).$$

The set on the first line of the right-hand side belongs to $\mathscr{G}_{\gamma^{[n]}}$, since Δ' and Δ'' are controls and $F \in \mathscr{G}_\Gamma$. As for the set on the second line, it is equal to

$$\begin{aligned}\{\Gamma^{[n+1]} = \gamma\} &\cap \{l(\Gamma) > n\} \\ &= \{\Gamma^{[n+1]} = \gamma\} \cap \{\Gamma^{[n]} = \gamma^{[n]}\} \cap \{l(\Gamma) \le n\}^c,\end{aligned}$$

because $(\Delta')^{[n+1]} = \Gamma^{[n+1]}$ on $\{l(\Gamma) > n\} \cap F$ and $(\Delta'')^{[n+1]} = \Gamma^{[n+1]}$ on $\{l(\Gamma) > n\} \cap F^c$. Since $\{l(\Gamma) \le n\} \in \mathscr{G}_{\Gamma^{[n]}}$, this set also belongs to $\mathscr{G}_{\gamma^{[n]}}$.

The Relationship Between Controls and Accessible Stopping Points

The notion of predictable increasing path is not needed in the present setting. However, proceeding by analogy with what was done in Section 1.4, it can be shown that controls are exactly accessible stopping points taking values in $\mathbb{G}$. By *stopping point with values in* $\mathbb{G}$, we mean a random variable Γ with values in $\mathbb{G}$ such that $\{\Gamma = \gamma\} \in \mathscr{G}_\gamma$ for all $\gamma \in \mathbb{G}_{\mathrm{f}}$ (cf. Exercise 1).

The Relationship Between Controls and Strategies

Starting with a filtration $(\mathscr{F}_t)$, it is possible to define a filtration indexed by $\mathbb{G}_{\mathrm{f}}$ by setting $\mathscr{G}_\gamma = \mathscr{F}_{\gamma_{l(\gamma)}}$ for all $\gamma \in \mathbb{G}_{\mathrm{f}}$. There is then a one-to-one correspondence between controls (relative to the filtration $(\mathscr{G}_\gamma)$) and strategies as they are defined in Section 7.1. Let us denote the controls by Γ and the strategies by $(\tilde{\Gamma}, \tilde{\tau})$. The one-to-one map associates to the control Γ the strategy $(\tilde{\Gamma}, \tilde{\tau})$ defined by

$$\tilde{\Gamma}_n = \begin{cases} \Gamma_n & \text{on } \{l(\Gamma) > n\}, \\ \Gamma_{l(\Gamma)} + (n - l(\Gamma))e_1 & \text{on } \{l(\Gamma) \le n\}, \end{cases} \qquad \text{and} \quad \tilde{\tau} = l(\Gamma). \quad (7.8)$$

Of course, $\Gamma_n(\omega)$ denotes the nth component of the element $\Gamma(\omega)$ of $\mathbb{G}$. We will only show that $(\tilde{\Gamma}_n, n \in \mathbb{N})$ is a predictable increasing path and that $\tilde{\tau}$ is a stopping time relative to the filtration $(\mathscr{F}_{\tilde{\Gamma}_n}, n \in \mathbb{N})$, because it is straightforward to check that the map is one to one (cf. Exercise 2). To begin with, it is clear that $\tilde{\Gamma}_0 = 0$ and that $\tilde{\Gamma}_{n+1} \in \mathbb{D}_{\tilde{\Gamma}_n}$. On the other hand, if $s \in \mathbb{I}$ and $t \in \mathbb{D}_s$, then

$$\begin{aligned}\{\tilde{\Gamma}_n = s\} \cap \{\tilde{\Gamma}_{n+1} = t\} \cap \{l(\Gamma) > n\} &= \{\Gamma_n = s\} \cap \{\Gamma_{n+1} = t\} \cap \{l(\Gamma) > n\} \\ &= \bigcup_{\gamma \in \mathbb{G}'} (\{\Gamma^{[n+1]} = \gamma\} \cap \{l(\Gamma) > n\})\end{aligned}$$

where $\mathbb{G}' = \{\gamma \in \mathbb{G}\colon l(\gamma) = n + 1, \gamma_n = s, \gamma_{n+1} = t\}$. It follows that the set on the left-hand side belongs to $\mathscr{F}_s$. Indeed, $\{l(\Gamma) > n\} \in \mathscr{G}_{\Gamma^{[n]}}$, because Γ is a control, and therefore for all $\gamma \in \mathbb{G}'$,

$$\{\Gamma^{[n+1]} = \gamma\} \cap \{l(\Gamma) > n\} = \{\Gamma^{[n+1]} = \gamma\} \cap \{\Gamma^{[n]} = \gamma^{[n]}\} \cap \{l(\Gamma) > n\}$$
$$\in \mathscr{G}_{\gamma^{[n]}} = \mathscr{F}_{\gamma_n} = \mathscr{F}_s,$$

since $l(\gamma^{[n]}) = n$. On the other hand, observing that

$$\{\tilde{\Gamma}_n = s\} \cap \{\tilde{\Gamma}_{n+1} = t\} \cap \{l(\Gamma) \leq n\}$$
$$= \begin{cases} \bigcup_{m=0}^{n} \{\Gamma_m + (n-m)e_1 = s\} \cap \{l(\Gamma) = m\}, & \text{if } t = s + e_1, \\ \varnothing, & \text{if } t \neq s + e_1, \end{cases}$$

we conclude in the same way that the left-hand side of this equality also belongs to $\mathscr{F}_s$. Thus $\{\tilde{\Gamma}_n = s\} \cap \{\tilde{\Gamma}_{n+1} = t\}$ belongs to $\mathscr{F}_s$, which proves that $(\tilde{\Gamma}_n,\ n \in \mathbb{N})$ is a predictable increasing path. It remains to establish that $\{\tilde{\tau} = n\} \in \mathscr{F}_{\tilde{\Gamma}_n}$. If $t \in \mathbb{I}$, then

$$\{\tilde{\tau} = n\} \cap \{\tilde{\Gamma}_n = t\} = \{l(\Gamma) = n\} \cap \{\Gamma_n = t\}$$
$$= \bigcup_{\gamma \in \mathbb{G}''} \left(\{l(\Gamma) = n\} \cap \{\Gamma^{[n]} = \gamma\}\right),$$

where $\mathbb{G}'' = \{\gamma \in \mathbb{G}: l(\gamma) = n,\ \gamma_n = t\}$. But for all $\gamma \in \mathbb{G}''$,

$$\{l(\Gamma) = n\} \cap \{\Gamma^{[n]} = \gamma\} \in \mathscr{G}_\gamma = \mathscr{F}_{\gamma_{l(\gamma)}} = \mathscr{F}_{\gamma_n} = \mathscr{F}_t;$$

therefore $\{\tilde{\tau} = n\} \cap \{\tilde{\Gamma}_n = t\}$ belongs to $\mathscr{F}_t$, which was to be proved.

Stochastic Processes Indexed by $\mathbb{G}$

A *stochastic process indexed by* $\mathbb{G}$, also termed more simply a *process indexed by* $\mathbb{G}$, is any measurable map X defined on $\mathbb{G} \times \Omega$ (equipped with the sigma field $\mathscr{B} \times \mathscr{G}$) with values in $\overline{\mathbb{R}}$. The value of X at (γ, ω) will be denoted $X_\gamma(\omega)$. If X is a process indexed by $\mathbb{G}$, then for all $\gamma \in \mathbb{G}$, the partial map $\omega \mapsto X_\gamma(\omega)$, which we denote X_γ, is a $\mathscr{G}$-measurable random variable.

A process X indexed by $\mathbb{G}$ will be termed

a. *adapted* if X_γ is $\mathscr{G}_\gamma$-measurable for all $\gamma \in \mathbb{G}_f$;

b. *integrable* if X_γ is integrable for all $\gamma \in \mathbb{G}$;

c. *upper semicontinuous* [*continuous*] if the map $\gamma \mapsto X_\gamma(\omega)$ is upper semicontinuous [continuous] (as defined in Section 7.2) for almost all ω.

If X and Y are two processes indexed by $\mathbb{G}$, we will say that they are equal, and will write $X = Y$, if for almost all ω, $X_\gamma(\omega) = Y_\gamma(\omega)$ for all $\gamma \in \mathbb{G}$. The set of processes indexed by $\mathbb{G}$ will be equipped with the order relation $X \leq Y$, which means that for almost all ω, $X_\gamma(\omega) \leq Y_\gamma(\omega)$ for all $\gamma \in \mathbb{G}$. The usual terminology for order relations applies. In particular, the positive part of X, denoted X^+, is the process $\sup(X, 0)$, where 0 denotes the zero process.

If X is a process indexed by $\mathbb{G}$ and Γ is a control, then X_Γ denotes the $\mathscr{G}$-measurable random variable $\omega \mapsto X_{\Gamma(\omega)}(\omega)$. Notice that if X is adapted, then X_Γ is $\mathscr{G}_\Gamma$-measurable, because

$$\{X_\Gamma \in B\} \cap \{\Gamma = \gamma\} = \{X_\gamma \in B\} \cap \{\Gamma = \gamma\} \in \mathscr{G}_\gamma$$

for all Borel sets B and all $\gamma \in \mathbb{G}_{\mathrm{f}}$.

A process X indexed by $\mathbb{G}$ is of *class* $\mathrm{D}_{\mathbf{G}}$ if the family $(X_\Gamma,\ \Gamma \in \mathbf{G})$ is uniformly integrable. This condition is clearly satisfied when there exists an integrable random variable Y such that $\sup_{\gamma \in \mathbb{G}} |X_\gamma(\omega)| \leq Y(\omega)$ for almost all ω.

Supermartingales and Martingales Indexed by $\mathbb{G}$

A process X indexed by $\mathbb{G}$ is a *supermartingale* [*martingale*] if it is adapted, integrable, and

$$\mathrm{E}(X_\delta | \mathscr{G}_\gamma) \leq X_\gamma \qquad [\mathrm{E}(X_\delta | \mathscr{G}_\gamma) = X_\gamma]$$

for all couples (γ, δ) of elements of $\mathbb{G}$ such that $\gamma \prec \delta$.

Clearly, in order that a process X indexed by $\mathbb{G}$ be a supermartingale [martingale], it is necessary and sufficient that $(X_{\gamma[n]},\ n \in \overline{\mathbb{N}})$ be a supermartingale [martingale] relative to the filtration $(\mathscr{G}_{\gamma[n]},\ n \in \overline{\mathbb{N}})$, for all $\gamma \in \mathbb{G}$.

If X is a supermartingale [martingale] and Γ is a control, then $(X_{\Gamma[n]},\ n \in \mathbb{N})$ is a supermartingale [martingale] relative to the filtration $(\mathscr{G}_{\Gamma[n]},\ n \in \mathbb{N})$. The proof of this assertion is similar to the proof of Theorem 1 of Section 1.6 (cf. Exercise 3).

7.4 OPTIMIZATION

A *reward process* now refers to an adapted integrable process X indexed by $\mathbb{G}$ whose positive part X^+ is of class $\mathrm{D}_{\mathbf{G}}$.

Given a reward process X, a control Γ^o is *optimal* if

$$\mathrm{E}(X_{\Gamma^o}) = \sup_{\Gamma \in \mathbf{G}} \mathrm{E}(X_\Gamma). \tag{7.9}$$

As in Chapter 3, the controls Γ for which $E(X_{\Gamma})$ equals $-\infty$ play no role in the search for optimal controls, because the right-hand side of (7.9) is finite due to the inequalities

$$-\infty < E(X_0) \leq \sup_{\Gamma \in \mathbf{G}} E(X_\Gamma) \leq \sup_{\Gamma \in \mathbf{G}} E(X_\Gamma^+) < \infty.$$

In the remainder of this section, we will assume that a reward process X is given and fixed. Our objective is to prove that if X is upper semicontinuous, then an optimal control exists and can be determined in a constructive manner.

A Special Case

In the special case where $X_{(\Gamma, \tau)}$ is defined by (7.1), the existence and construction of an optimal strategy are obtained using the one-to-one map between controls (relative to the filtration $(\mathscr{G}_\gamma)$ defined from $(\mathscr{F}_t)$) and strategies. Notice that in this case, the reward process X is defined by

$$X_\gamma = \begin{cases} \tilde{X}_0, & \text{if } l(\gamma) = 0, \\ \sum_{n=1}^{l(\gamma)} \alpha^n Y_{\gamma_{n-1}, \gamma_n} + \alpha^{l(\gamma)} \tilde{X}_{l(\gamma)}, & \text{if } l(\gamma) > 0. \end{cases}$$

Definition (7.1) is meaningful (in the case where τ is not finite) and the measurability and integrability properties required of a reward process are satisfied in particular when the four conditions below are satisfied:

a. $(\tilde{X}_t,\ t \in \bar{\mathbb{I}})$ is adapted, integrable, and its positive part is of class $D_{\mathbf{A}}$;
b. $(Y_{s,t},\ s \in \mathbb{I},\ t \in \mathbb{D}_s)$ is adapted, that is, $Y_{s,t}$ is $\mathscr{F}_t$-measurable for all $s \in \mathbb{I}$ and all $t \in \mathbb{D}_s$;
c. $E(\sum_{n \in \mathbb{N}^*} \alpha^n Y^-_{\gamma_{n-1}, \gamma_n}) < \infty$, for all $\gamma \in \mathbb{G}^\infty$;
d. $E(\sup_{s \in \mathbb{I}} \sup_{t \in \mathbb{D}_s} Y^+_{s,t}) < \infty$ if $\alpha < 1$, and
$E(\sum_{n \in \mathbb{N}} \sup_{s \in \mathbb{I}:\ |s| = n} \sup_{t \in \mathbb{D}_s} Y^+_{s,t}) < \infty$ if $\alpha = 1$,

where α is the discounting rate in (7.1).

Under condition d, if $(\tilde{X}_t,\ t \in \bar{\mathbb{I}})$ satisfies condition (3.3), then X (as a process indexed by $\mathbb{G}$) is upper semicontinuous (cf. Exercise 5).

The Sets $\mathbf{G}^\Gamma$, $\mathbf{G}_X$, and $\mathbf{G}_X^\Gamma$

For all $\Gamma \in \mathbf{G}$, $\mathbf{G}^\Gamma$ will denote the set of $\Delta \in \mathbf{G}$ such that $\Gamma \prec \Delta$, and $\mathbf{G}_X^\Gamma$ the (possibly empty) set of $\Delta \in \mathbf{G}^\Gamma$ such that $E(X_\Delta^-) < \infty$. We will write $\mathbf{G}_X$ instead of $\mathbf{G}_X^0$. Clearly, $\mathbf{G}_X$ contains all controls that only take a finite number of values, and in particular all bounded controls. Observe that if $\Delta \in \mathbf{G}^{\Gamma}$, then $\Delta = \Gamma$ on $\{l(\Gamma) = \infty\}$; in particular, $\mathbf{G}^{\gamma} = \{\gamma\}$ for all $\gamma \in \mathbb{G}^{\infty}$.

Snell's Envelope

We will term *Snell's envelope of X*, or simply *Snell's envelope*, the process Z indexed by $\mathbb{G}$ defined by

$$Z_{\gamma} = \begin{cases} \operatorname{esssup}_{\Delta \in \mathbf{G}^{\gamma}} \mathrm{E}\left(X_{\Delta} | \mathscr{G}_{\gamma}\right), & \text{if } \gamma \in \mathbb{G}_{\mathrm{f}}, \\ X_{\gamma}, & \text{if } \gamma \in \mathbb{G}^{\infty}. \end{cases} \tag{7.10}$$

It should be emphasized that the second line of this definition means that $Z_{\gamma}(\omega) = X_{\gamma}(\omega)$ for all ω. This guarantees, as it must, that the map $(\gamma, \omega) \mapsto Z_{\gamma}(\omega)$ is measurable and allows us to extend (7.10) to controls: For all $\Gamma \in \mathbf{G}$,

$$Z_{\Gamma} = \operatorname*{esssup}_{\Delta \in \mathbf{G}^{\Gamma}} \mathrm{E}\left(X_{\Delta} | \mathscr{G}_{\Gamma}\right). \tag{7.11}$$

In order to prove this assertion, let us begin by noticing that the equality in (7.11) holds on $\{l(\Gamma) = \infty\}$, because on this set $Z_{\Gamma} = X_{\Gamma}$ and $\mathrm{E}(X_{\Delta} | \mathscr{G}_{\Gamma}) = \mathrm{E}(X_{\Gamma} | \mathscr{G}_{\Gamma}) = X_{\Gamma}$ for all $\Delta \in \mathbf{G}^{\Gamma}$. It therefore remains to establish that the same equality also holds on $\{\Gamma = \gamma\}$ for all $\gamma \in \mathbb{G}_{\mathrm{f}}$. For all $\Delta \in \mathbf{G}^{\Gamma}$ and all $\gamma \in \mathbb{G}_{\mathrm{f}}$,

$$\mathrm{E}\left(X_{\Delta} | \mathscr{G}_{\Gamma}\right) = \mathrm{E}\left(X_{\Delta} | \mathscr{G}_{\gamma}\right) = \mathrm{E}\left(X_{\Delta} 1_{\{\Gamma = \gamma\}} | \mathscr{G}_{\gamma}\right) \qquad \text{on } \{\Gamma = \gamma\}.$$

On the other hand,

$$\operatorname*{esssup}_{\Delta \in \mathbf{G}^{\gamma}} \mathrm{E}\left(X_{\Delta} 1_{\{\Gamma = \gamma\}} | \mathscr{G}_{\gamma}\right) = \operatorname*{esssup}_{\Delta \in \mathbf{G}^{\Gamma}} \mathrm{E}\left(X_{\Delta} 1_{\{\Gamma = \gamma\}} | \mathscr{G}_{\Gamma}\right).$$

Indeed, by Property 5 of controls, if $\Delta \in \mathbf{G}^{\gamma}\,[\mathbf{G}^{\Gamma}]$, there exists $\tilde{\Delta} \in \mathbf{G}^{\Gamma}\,[\mathbf{G}^{\gamma}]$ such that $\tilde{\Delta} = \Delta$ on $\{\Gamma = \gamma\}$, because if Δ belongs to $\mathbf{G}^{\gamma}\,[\mathbf{G}^{\Gamma}]$, then

$$\tilde{\Delta} = \begin{cases} \Delta & \text{on } \{\Gamma = \gamma\}, \\ \Gamma & \text{on } \{\Gamma \neq \gamma\}, \end{cases} \qquad \left[\tilde{\Delta} = \begin{cases} \Delta & \text{on } \{\Gamma = \gamma\}, \\ \gamma & \text{on } \{\Gamma \neq \gamma\}, \end{cases} \right]$$

is equal to Δ on $\{\Gamma = \gamma\}$ and belongs to $\mathbf{G}^{\Gamma}\,[\mathbf{G}^{\gamma}]$. To check this last statement, set

$$\tau = \sup\left\{n \in \mathbb{N} \colon \Gamma^{[n]} \prec \gamma\right\}.$$

Then τ is a stopping time relative to the filtration $(\mathscr{G}_{\Gamma^{[n]}}, n \in \mathbb{N})$. Indeed, for all $n \in \mathbb{N}$ and $\delta \in \mathbb{G}_f$,

$$[\tau = n\} \cap \{\Gamma^{[n]} = \delta\} = \begin{cases} \bigcup_{\delta' \in \mathbb{D}'_\delta} \{\Gamma^{[n]} = \delta\} \cap \{\Gamma^{[n+1]} = \delta'\}, & \text{if } \delta \prec \gamma, \\ \varnothing, & \text{if } \delta \not\prec \gamma, \end{cases}$$

where $\mathbb{D}'_\delta = \mathbb{D}_\delta - \{\delta' \in \mathbb{G}: \delta' \prec \gamma\}$, and the right-hand side belongs to $\mathscr{G}_\delta$ by (7.6). Set $\tilde{\Gamma} = \Gamma^{[\tau]}$. Then $\tilde{\Gamma}$ is a control such that $\tilde{\Gamma} \prec \Delta$, $\tilde{\Gamma} \prec \Gamma$, and $\tilde{\Gamma} \prec \gamma$. Notice now that $\{\Gamma = \gamma\}$ belongs to $\mathscr{G}_{\tilde{\Gamma}}$; therefore the conclusion follows from Property 5 of Section 7.3.

The Upwards-Directed Property of Conditional Rewards

For all $\Gamma \in \mathbf{G}$, the family $(E(X_\Delta | \mathscr{G}_\Gamma), \Delta \in \mathbf{G}^\Gamma)$ is upwards directed. Indeed, if Δ' and Δ'' are two elements of $\mathbf{G}^\Gamma$ and if F is the set $\{E(X_{\Delta'} | \mathscr{G}_\Gamma) \geq E(X_{\Delta''} | \mathscr{G}_\Gamma)\}$, then

$$\Delta = \begin{cases} \Delta' & \text{on } F, \\ \Delta'' & \text{on } F^c, \end{cases}$$

belongs to $\mathbf{G}^\Gamma$, by Property 5 mentioned above, and

$$E(X_\Delta | \mathscr{G}_\Gamma) = E(X_{\Delta'} | \mathscr{G}_\Gamma) 1_F + E(X_{\Delta''} | \mathscr{G}_\Gamma) 1_{F^c} = \sup(E(X_{\Delta'} | \mathscr{G}_\Gamma), E(X_{\Delta''} | \mathscr{G}_\Gamma)),$$

which establishes the assertion.

By the theorem of Section 1.2, it follows that for all $\Gamma \in \mathbf{G}$, there exists a sequence $(\Delta(n), n \in \mathbb{N})$ of elements of $\mathbf{G}^\Gamma$ such that $E(X_{\Delta(n)} | \mathscr{G}_\Gamma)$ increases a.s. to Z_Γ; moreover, if $\mathbf{G}_X^\Gamma$ is nonempty, then the $\Delta(n)$ can be chosen in $\mathbf{G}_X^\Gamma$, and in this case Z_Γ is integrable, $\mathbf{G}_X^\Gamma$ can be substituted to $\mathbf{G}^\Gamma$ in (7.11) and

$$E(Z_\Gamma) = \sup_n E(X_{\Delta(n)}) = \sup_{\Delta \in \mathbf{G}^\Gamma} E(X_\Delta). \tag{7.12}$$

In particular,

$$E(Z_0) = \sup_{\Delta \in \mathbf{G}} E(X_\Delta). \tag{7.13}$$

Properties of Snell's Envelope

The properties of Snell's envelope that we shall need are similar to those established in Sections 3.2 and 3.4. We will gather them together in a single proposition.

A process Y indexed by $\mathbb{G}$ will be termed a *regular supermartingale* (relative to $\mathbf{G}_X$) if it is adapted, if Y_Δ is integrable for all $\Delta \in \mathbf{G}_X$ and if

$$E(Y_\Delta | \mathscr{G}_\gamma) \le Y_\gamma, \qquad \text{for all } \gamma \in \mathbb{G}_{\mathrm{f}} \text{ and all } \Delta \in \mathbf{G}_X^\gamma.$$

A regular supermartingale is clearly a supermartingale.

Proposition 1 (Properties of Snell's envelope). *Snell's envelope Z has the following properties*:

1. *For all couples (Γ, Δ) of controls such that $\Gamma \prec \Delta$ and $\mathbf{G}_X^\Delta$ is nonempty, Z_Γ and Z_Δ are integrable and $E(Z_\Delta | \mathscr{G}_\Gamma) \le Z_\Gamma$; in particular, Z is a regular supermartingale.*
2. *Z is the smallest regular supermartingale that dominates X.*
3. *For all finite controls Γ,*

$$Z_\Gamma = \sup\Big(X_\Gamma, \sup_{\delta \in \mathbb{D}_\Gamma} E(Z_\delta | \mathscr{G}_\Gamma)\Big). \tag{7.14}$$

4. *Z^+ is of class $D_{\mathbf{G}}$.*
5. *If Γ is a control such that $\mathbf{G}_X^\Gamma$ is nonempty, then $(Z_{\Gamma[n]}, n \in \overline{\mathbb{N}})$ is a uniformly integrable supermartingale (relative to the filtration $(\mathscr{G}_{\Gamma[n]}, n \in \overline{\mathbb{N}})$); if in addition, X is upper semicontinuous, then the limit $Z_{\Gamma-}$ of $Z_{\Gamma[n]}$ satisfies*

$$Z_{\Gamma-} = E\Big(Z_\Gamma | \bigvee_{n \in \mathbb{N}} \mathscr{G}_{\Gamma[n]}\Big). \tag{7.15}$$

Proof. Since $\mathbf{G}_X^\Delta$ is nonempty, Z_Γ and Z_Δ are integrable, as was pointed out just before (7.12). On the other hand, if $(\Delta(n), n \in \mathbb{N})$ is a sequence of elements of $\mathbf{G}_X^\Delta$ such that $E(X_{\Delta(n)} | \mathscr{G}_\Delta)$ increases a.s. to Z_Δ, then

$$E(Z_\Delta | \mathscr{G}_\Gamma) = \lim_{n \to \infty} E(X_{\Delta(n)} | \mathscr{G}_\Gamma) \le \operatorname*{esssup}_{\Delta' \in \mathbf{G}^\Delta} E(X_{\Delta'} | \mathscr{G}_\Gamma) \le Z_\Gamma,$$

which completes the proof of 1. As for 2, it is clear that Z dominates X; moreover, if Z' is another regular supermartingale that dominates X, then $E(X_\Delta | \mathscr{G}_\gamma) \le E(Z'_\Delta | \mathscr{G}_\gamma) \le Z'_\gamma$ for all $\gamma \in \mathbb{G}_{\mathrm{f}}$ and all $\Delta \in \mathbf{G}_X^\gamma$. Therefore $Z_\gamma = \operatorname{esssup}_{\Delta \in \mathbf{G}_X^\gamma} E(X_\Delta | \mathscr{G}_\gamma) \le Z'_\gamma$ for all $\gamma \in \mathbb{G}_{\mathrm{f}}$, which proves that Z' dominates Z, since $Z_\gamma = X_\gamma$ (everywhere) if $\gamma \in \mathbb{G}^\infty$. Having thus established 2, let us prove 3. It is sufficient to consider the special case where Γ is the constant $\gamma \in \mathbb{G}_{\mathrm{f}}$, since the general case follows immediately. Because Z is a supermartingale that dominates X, $Z_\gamma \ge \sup(X_\gamma, \sup_{\delta \in \mathbb{D}_\gamma} E(Z_\delta | \mathscr{G}_\gamma))$. It remains therefore to prove the converse inequality. Take $\Delta \in \mathbf{G}_X^\gamma$, and for all

$\delta \in \mathbb{D}_\gamma$, set $F_\delta = \{\Delta^{[l(\gamma)+1]} = \delta\} \cap \{\Delta \neq \gamma\}$. Clearly $F_\delta \in \mathscr{G}_\gamma$ and $\bigcup_{\delta \in \mathbb{D}_\gamma} F_\delta = \{\Delta \neq \gamma\}$. For all $\delta \in \mathbb{D}_\gamma$, set

$$\Delta(\delta) = \begin{cases} \Delta & \text{on } F_\delta, \\ \delta & \text{on } F_\delta^c \end{cases}.$$

By Property 5 of controls, $\Delta(\delta)$ is a control. Moreover, $\delta \prec \Delta(\delta)$ and $X_{\Delta(\delta)}$ is integrable; therefore $\Delta(\delta) \in \mathbf{G}_X^\delta$. On the other hand,

$$X_\Delta = X_\gamma 1_{\{\Delta=\gamma\}} + \sum_{\delta \in \mathbb{D}_\gamma} X_{\Delta(\delta)} 1_{F_\delta};$$

therefore

$$\begin{aligned} \mathrm{E}(X_\Delta | \mathscr{G}_\gamma) &= X_\gamma 1_{\{\Delta=\gamma\}} + \sum_{\delta \in \mathbb{D}_\gamma} \mathrm{E}(X_{\Delta(\delta)} | \mathscr{G}_\gamma) 1_{F_\delta} \\ &\leq X_\gamma 1_{\{\Delta=\gamma\}} + \sum_{\delta \in \mathbb{D}_\gamma} \mathrm{E}(\mathrm{E}(Z_{\Delta(\delta)} | \mathscr{G}_\delta) | \mathscr{G}_\gamma) 1_{F_\delta} \\ &\leq X_\gamma 1_{\{\Delta=\gamma\}} + \sup_{\delta \in \mathbb{D}_\gamma} \mathrm{E}(Z_\delta | \mathscr{G}_\gamma) 1_{\{\Delta \neq \gamma\}} \\ &\leq \sup\Big(X_\gamma, \sup_{\delta \in \mathbb{D}_\gamma} \mathrm{E}(Z_\delta | \mathscr{G}_\gamma)\Big). \end{aligned}$$

In order to obtain the desired inequality, it only remains to take the essential supremum over $\Delta \in \mathbf{G}_X^\gamma$ on the left-hand side. The proof of 4 is identical to the proof of Proposition 2 of Section 3.2. The first assertion of 5 is a direct consequence of 1 and 4. As for the second assertion, notice that passing to the limit in the inequality $Z_{\Gamma^{[n]}} \geq \mathrm{E}(Z_\Gamma | \mathscr{G}_{\Gamma^{[n]}})$ leads to the inequality $Z_{\Gamma_-} \geq \mathrm{E}(Z_\Gamma | \bigvee_{n \in \mathbb{N}} \mathscr{G}_{\Gamma^{[n]}})$. Because $Z_{\Gamma_-} = Z_\Gamma$ on $\{l(\Gamma) < \infty\}$, in order to obtain (7.15) it is sufficient to prove that $\mathrm{E}(Z_{\Gamma_-}; l(\Gamma) = \infty) \leq \mathrm{E}(Z_\Gamma; l(\Gamma) = \infty)$. Let $\varepsilon > 0$. For all $n \in \mathbb{N}$, there exists $\Delta(n) \in \mathbf{G}_X^{\Gamma^{[n]}}$ such that $\mathrm{E}(Z_{\Gamma^{[n]}}; l(\Gamma^{[n]}) = n) - \varepsilon \leq \mathrm{E}(X_{\Delta(n)}; l(\Gamma^{[n]}) = n)$. If X is upper semicontinuous, then using Fatou's lemma and the fact that $\{l(\Gamma^{[n]}) = n\} \supset \{l(\Gamma^{[n+1]}) = n+1\}$ and $\bigcap_{n \in \mathbb{N}} \{l(\Gamma^{[n]}) = n\} = \{l(\Gamma) = \infty\}$, we see that

$$\begin{aligned} \mathrm{E}(Z_{\Gamma_-}; l(\Gamma) = \infty) - \varepsilon &= \lim_{n \to \infty} \mathrm{E}(Z_{\Gamma^{[n]}}; l(\Gamma^{[n]}) = n) - \varepsilon \\ &\leq \limsup_{n \to \infty} \mathrm{E}(X_{\Delta(n)}; l(\Gamma^{[n]}) = n) \\ &\leq \mathrm{E}\Big(\limsup_{n \to \infty} X_{\Delta(n)}; l(\Gamma) = \infty\Big) \\ &\leq \mathrm{E}(X_\Gamma; l(\Gamma) = \infty) \\ &= \mathrm{E}(Z_\Gamma; l(\Gamma) = \infty). \end{aligned}$$

Because ε is arbitrary, the conclusion follows.

Optimality Criterion

The following proposition gives the appropriate tool for recognizing optimal controls.

Proposition 2 (Optimality criterion). *A control Γ is optimal if and only if*

a. $Z_\Gamma = X_\Gamma$, *and*
b. $\mathrm{E}(Z_\Gamma) = \mathrm{E}(Z_0)$.

The proof is identical to that of Proposition 1 of Section 3.3.

Optimization

For each $\gamma \in \mathbb{G}_{\mathrm{f}}$, we equip $\mathbb{D}_\gamma$ with an arbitrary total order and denote by $(\Gamma(n),\ n \in \mathbb{N})$ the sequence of random variables with values in $\mathbb{G}_{\mathrm{f}}$ defined inductively by setting $\Gamma(0) = 0$ and, for all $\gamma \in \mathbb{G}_{\mathrm{f}}$ such that $\{\Gamma(n) = \gamma\} \neq \varnothing$,

$$\Gamma(n+1) = \inf\left\{\delta \in \mathbb{D}_\gamma : \mathrm{E}\big(Z_\delta | \mathscr{G}_\gamma\big) = \sup_{\delta' \in \mathbb{D}_\gamma} \mathrm{E}\big(Z_{\delta'} | \mathscr{G}_\gamma\big)\right\} \qquad \text{on } \{\Gamma(n) = \gamma\}. \tag{7.16}$$

Clearly, $\Gamma(n+1) \in \mathbb{D}_{\Gamma(n)}$ for all $n \in \mathbb{N}$, and in particular, $l(\Gamma(n)) = n$ for all $n \in \mathbb{N}$. Moreover, by assuming that $\{\Gamma(n) = \gamma\} \in \mathscr{G}_\gamma$ for all γ such that $l(\gamma) = n$, we deduce from (7.16) that $\{\Gamma(n+1\} = \delta\} \in \mathscr{G}_\gamma$ for all γ such that $l(\gamma) = n$ and all $\delta \in \mathbb{D}_\gamma$. Thus

$$\{\Gamma(n+1) = \gamma\} \in \mathscr{G}_{\gamma^{[n]}}, \qquad \text{for all } n \in \mathbb{N} \text{ and all } \gamma \in \mathbb{G}_{\mathrm{f}}. \tag{7.17}$$

Let Γ be the function defined on Ω with values in $\mathbb{G}^\infty$ such that $\Gamma^{[n]} = \Gamma(n)$ for all $n \in \mathbb{N}$. By (7.17), Γ is a control. Moreover,

$$\mathrm{E}\big(Z_{\Gamma^{[n+1]}} | \mathscr{G}_{\Gamma^{[n]}}\big) = \sup_{\gamma \in \mathbb{D}_{\Gamma^{[n]}}} \mathrm{E}\big(Z_\gamma | \mathscr{G}_{\Gamma^{[n]}}\big), \qquad \text{for all } n \in \mathbb{N}. \tag{7.18}$$

We will term *optimizing control* any control Γ with values in $\mathbb{G}^\infty$ for which (7.18) is satisfied.

Theorem (Existence of optimal controls). *Let X be an adapted integrable process indexed by $\mathbb{G}$ whose positive part X^+ is of class $\mathrm{D}_{\mathbf{G}}$. Let Z be Snell's envelope of X and let Γ be an optimizing control. Set*

$$\sigma = \inf\{n \in \mathbb{N} : Z_{\Gamma^{[n]}} = X_{\Gamma^{[n]}}\} \tag{7.19}$$

and

$$\tau = \inf\{n \in \mathbb{N}: Z_{\Gamma[n]} - \mathrm{E}(Z_{\Gamma[n+1]} | \mathscr{G}_{\Gamma[n]}) > 0\}. \tag{7.20}$$

If X is upper semicontinuous or if σ is finite, then $\Gamma^{[\sigma]}$ is an optimal control, more precisely the smallest optimal control $\prec \Gamma$. If X is upper semicontinuous or if τ is finite, then $\Gamma^{[\tau]}$ is also an optimal control, more precisely the largest optimal control $\prec \Gamma$.

Proof. Since σ and τ are stopping times relative to the filtration $(\mathscr{G}_{\Gamma[n]}, n \in \mathbb{N})$, $\Gamma^{[\sigma]}$ and $\Gamma^{[\tau]}$ are controls. Let us prove that $\Gamma^{[\sigma]}$ is optimal. By the definition of σ and because $\Gamma^{[\infty]} = \Gamma$ and $Z_\Gamma = X_\Gamma$, we see that $Z_{\Gamma[\sigma]} = X_{\Gamma[\sigma]}$. By Proposition 2, it therefore suffices to prove that $\mathrm{E}(Z_{\Gamma[\sigma]}) = \mathrm{E}(Z_0)$. By (7.13), $\mathrm{E}(Z_{\Gamma[\sigma]}) = \mathrm{E}(X_{\Gamma[\sigma]}) \le \mathrm{E}(Z_0)$, so it is only necessary to prove the converse inequality.

Again by the definition of σ, $Z_{\Gamma[n]} > X_{\Gamma[n]}$ on $\{\sigma > n\}$; therefore $Z_{\Gamma[n]} = \sup_{\gamma \in \mathbb{D}_{\Gamma[n]}} \mathrm{E}(Z_\gamma | \mathscr{G}_{\Gamma[n]})$ on $\{\sigma > n\}$ by (7.14). By (7.18), it follows that $\mathrm{E}(Z_{\Gamma[n+1]} | \mathscr{G}_{\Gamma[n]}) = Z_{\Gamma[n]}$ on $\{\sigma > n\}$; therefore $\mathrm{E}(Z_{\Gamma[n \wedge \sigma]}) = \mathrm{E}(Z_0)$ for all $n \in \mathbb{N}$.

If σ is finite, then $Z_{\Gamma[\sigma]} = \limsup_{n \to \infty} Z_{\Gamma[n \wedge \sigma]}$, and since Z^+ is of class $\mathrm{D}_{\mathbf{G}}$, Fatou's lemma applies and lets us conclude that

$$\mathrm{E}(Z_{\Gamma[\sigma]}) \ge \limsup_{n \to \infty} \mathrm{E}(Z_{\Gamma[n \wedge \sigma]}) = \mathrm{E}(Z_0),$$

which was to be proved.

If the hypothesis that σ is finite is not satisfied, we assume that X is upper semicontinuous. Let $\varepsilon > 0$. For each $n \in \mathbb{N}$, $\Gamma^{[n \wedge \sigma]}$ only takes a finite number of values; therefore $\mathbf{G}_X^{\Gamma[n \wedge \sigma]}$ contains $\Gamma^{[n \wedge \sigma]}$ and thus is not empty. By (7.12), it follows that there exists $\Delta(n) \in \mathbf{G}_X^{\Gamma[n \wedge \sigma]}$ such that $\mathrm{E}(Z_{\Gamma[n \wedge \sigma]}) - \varepsilon \le \mathrm{E}(X_{\Delta(n)})$. Because $\{\sigma \le n\}$ belongs to $\mathscr{G}_{\Gamma[n \wedge \sigma]}$, Property 5 of Section 7.3 implies that

$$\Delta'(n) = \begin{cases} \Gamma^{[\sigma]} & \text{on } \{\sigma \le n\}, \\ \Delta(n) & \text{on } \{\sigma > n\}, \end{cases}$$

is a control. Notice that on $\{\sigma \le n\}$,

$$\mathrm{E}(X_{\Delta'(n)} | \mathscr{G}_{\Gamma[n \wedge \sigma]}) = \mathrm{E}(X_{\Gamma[\sigma]} | \mathscr{G}_{\Gamma[n \wedge \sigma]}) = X_{\Gamma[\sigma]} = Z_{\Gamma[\sigma]} \ge \mathrm{E}(X_{\Delta(n)} | \mathscr{G}_{\Gamma[n \wedge \sigma]})$$

by (7.11). Because $\Delta'(n) = \Delta(n)$ on $\{\sigma > n\}$, we conclude that $\mathrm{E}(Z_{\Delta'(n)}) \ge \mathrm{E}(Z_{\Delta(n)})$. Since X is upper semicontinuous,

$$\limsup_{n \to \infty} X_{\Delta'(n)} \le X_{\Gamma[\sigma]} 1_{\{\sigma < \infty\}} + X_\Gamma 1_{\{\sigma = \infty\}} = X_{\Gamma[\sigma]},$$

and therefore $\limsup_{n\to\infty} X_{\Delta'(n)}$ is semi-integrable. Since X^+ is of class $D_{\mathbf{G}}$, Fatou's lemma applies and lets us conclude that

$$E(Z_0) - \varepsilon \le \limsup_{n\to\infty} E(X_{\Delta'(n)}) \le E\Big(\limsup_{n\to\infty} X_{\Delta'(n)}\Big) \le E(X_{\Gamma^{[\sigma]}}).$$

Since ε is arbitrary, $E(X_{\Gamma^{[\sigma]}}) \ge E(Z_0)$, which was to be proved. In view of Proposition 2, it is clear that $\Gamma^{[\sigma]}$ is the smallest optimal control $\prec \Gamma$.

We now turn to the assertion concerning $\Gamma^{[\tau]}$. By the definition of τ, $Z_{\Gamma^{[n]}} > E(Z_{\Gamma^{[n+1]}} | \mathscr{G}_{\Gamma^{[n]}})$ on $\{\tau = n\}$ for all $n \in \mathbb{N}$. By (7.18) and (7.14), it follows that $Z_{\Gamma^{[n]}} = X_{\Gamma^{[n]}}$ on $\{\tau = n\}$ for all $n \in \mathbb{N}$. Because $\Gamma^{[\tau]} = \Gamma$ on $\{\tau = \infty\}$ and $Z_\Gamma = X_\Gamma$, it follows that $Z_{\Gamma^{[\tau]}} = X_{\Gamma^{[\tau]}}$. By Proposition 2, it remains to prove that $E(Z_{\Gamma^{[\tau]}}) = E(Z_0)$. By (7.13), $E(Z_{\Gamma^{[\tau]}}) = E(X_{\Gamma^{[\tau]}}) \le E(Z_0)$, so it is only necessary to establish the converse inequality. By (7.18), (7.14), and the definition of τ, $E(Z_{\Gamma^{[n+1]}} | \mathscr{G}_{\Gamma^{[n]}}) = Z_{\Gamma^{[n]}}$ on $\{\tau > n\}$; therefore $E(Z_{\Gamma^{[n \wedge \sigma]}}) = E(Z_0)$ for all $n \in \mathbb{N}$. The remainder of the proof is obtained by replacing σ by τ in the corresponding portion of the proof of the inequality $E(Z_{\Gamma^{[\sigma]}}) \ge E(Z_0)$ above.

Now let Δ be any optimal control such that $\Delta \prec \Gamma$. Then $E(Z_\Delta) = E(Z_0)$ by b of Proposition 2; therefore $(Z_{\Delta^{[n]}}, n \in \overline{\mathbb{N}})$ is a martingale by 1 of Proposition 1 or, equivalently, $(Z_{\Gamma^{[n \wedge l(\Delta)]}}, n \in \overline{\mathbb{N}})$ is a martingale. In other words, $E(Z_{\Gamma^{[n+1]}} | \mathscr{G}_{\Gamma^{[n]}}) = Z_{\Gamma^{[n]}}$ on $\{l(\Delta) > n\}$. By (7.20), this implies that $\tau \ge l(\Delta)$, and therefore $\Delta \prec \Gamma^{[\tau]}$. Consequently, $\Gamma^{[\tau]}$ is the largest optimal control $\prec \Gamma$.

7.5 OPTIMIZATION OVER FINITE CONTROLS

The optimization problem we have just studied can be reformulated using only elements of $\mathbb{G}_f$ and finite controls (the analogous problem for stopping points was considered in Section 3.4). We will develop this idea here.

The set of finite controls will be denoted $\mathbf{G}_f$. A consequence of replacing $\mathbb{G}$ by $\mathbb{G}_f$ is that definitions related to the index set must be modified. A process indexed by $\mathbb{G}_f$ is simply a family $(X_\gamma, \gamma \in \mathbb{G}_f)$ of $\mathscr{G}$-measurable random variables. Such a process is adapted [integrable] if X_γ is $\mathscr{G}_\gamma$-measurable [integrable] for all $\gamma \in \mathbb{G}_f$. The statements concerning order and composition with a control remain valid for processes indexed by $\mathbb{G}_f$. A process indexed by $\mathbb{G}_f$ is of *class* $D_{\mathbf{G}_f}$ if the family $(X_\Gamma, \Gamma \in \mathbf{G}_f)$ is uniformly integrable. The notions of supermartingale and martingale are obtained by replacing $\mathbb{G}$ by $\mathbb{G}_f$ in the paragraph of Section 7.3 concerned with these subjects. The validity of that paragraph is unaffected by this substitution.

A *reward process* is now an adapted integrable process X indexed by $\mathbb{G}_f$ whose positive part X^+ is of class $D_{\mathbf{G}_f}$. An *optimal* control is a finite control Γ^o that satisfies (7.9) with $\mathbf{G}_f$ instead of $\mathbf{G}$. Snell's envelope of X is the process Z indexed by $\mathbb{G}_f$ and defined by

$$Z_\gamma = \operatorname*{esssup}_{\Delta \in \mathbf{G}_f^\gamma} \mathrm{E}\big(X_\Delta \mid \mathscr{G}_\gamma\big), \tag{7.21}$$

where $\mathbf{G}_f^\gamma = \mathbf{G}^\gamma \cap \mathbf{G}_f$. By replacing $\mathbf{G}$, $\mathbf{G}^\Gamma$, and $\mathbf{G}_X^\Gamma$, respectively, by $\mathbf{G}_f$, $\mathbf{G}^\Gamma \cap \mathbf{G}_f$, and $\mathbf{G}_X^\Gamma \cap \mathbf{G}_f$, it is easy to verify that the content of Section 7.4 applies to the new situation. Since the definition of the control Γ just after (7.17) does not quite fit into this framework, it is appropriate to replace the notion of optimizing control by the notion of *optimizing sequence*, that is, an increasing sequence $(\Gamma(n),\ n \in \mathbb{N})$ of finite controls such that $\Gamma(0) = 0$, $\Gamma(n+1) \in \mathbb{D}_{\Gamma(n)}$ and

$$\mathrm{E}\big(Z_{\Gamma(n+1)} \mid \mathscr{G}_{\Gamma(n)}\big) = \sup_{\gamma \in \mathbb{D}_{\Gamma(n)}} \mathrm{E}\big(Z_\gamma \mid \mathscr{G}_{\Gamma(n)}\big), \qquad \text{for all } n \in \mathbb{N}.$$

The induction formula in (7.16) indicates how such a sequence can be constructed. By replacing the optimizing control Γ by an optimizing sequence $(\Gamma(n),\ n \in \mathbb{N})$ (and therefore $\Gamma^{[n]}$ by $\Gamma(n)$) in the theorem of the previous section, one can easily see that the conclusions in the case where σ and τ are finite remain valid.

Here are the results implied by the hypotheses made in this section.

Theorem (Existence of optimal finite controls). *Let X be an adapted integrable process indexed by $\mathbb{G}_f$ whose positive part X^+ is of class $\mathrm{D}_{\mathbf{G}_f}$. Let Z be Snell's envelope of X and $(\Gamma(n),\ n \in \mathbb{N})$ an optimizing sequence. Let σ and τ be defined by* (7.19) *and* (7.20), *but using $\Gamma(n)$ instead of $\Gamma^{[n]}$. If σ is finite [if τ is finite], then $\Gamma(\sigma)$ [$\Gamma(\tau)$] is an optimal control, more precisely the smallest [largest] optimal control of the form $\Gamma(\rho)$, where ρ is a stopping time relative to the filtration $(\mathscr{G}_{\Gamma(n)},\ n \in \mathbb{N})$. In particular, σ is finite if* $\mathrm{E}(\sup_{\gamma \in \mathbb{G}_f} X_\gamma^+) < \infty$ *and* $\limsup_{l(\gamma) \to \infty} X_\gamma = -\infty$.

Proof. The proof of the first part of the theorem is identical to the proof of the corresponding assertions in the theorem of the previous section, with $\Gamma^{[n]}$ replaced by $\Gamma(n)$. The last part is proved in a manner similar to the corresponding assertion of Theorem 1 of Section 3.4, using an appropriate version of Proposition 3 of the same section (cf. part a of Exercise 6).

Case Where the Reward Process Is of Class $\mathrm{D}_{\mathbf{G}_f}$

If not only X^+ but also X is of class $\mathrm{D}_{\mathbf{G}_f}$, then any supermartingale that dominates X is a regular supermartingale (cf. Exercise 3). In this case, Snell's envelope Z of X is the smallest supermartingale that dominates X.

Moreover, Z is the limit of the increasing sequence $(Z^{(n)}, n \in \mathbb{N})$ of processes (indexed by $\mathbb{G}_f$) defined recursively by

$$Z_\gamma^{(0)} = X_\gamma \quad \text{and} \quad Z_\gamma^{(n+1)} = \sup\left(X_\gamma, \sup_{\delta \in \mathbb{D}_\gamma} \mathrm{E}\left(Z_\delta^{(n)} \middle| \mathscr{G}_\gamma\right)\right). \tag{7.22}$$

The proof of this last assertion is similar to that of Proposition 4 of Section 3.2.

7.6 CASE WHERE THE INDEX SET IS FINITE

A situation analogous to the one studied in Section 3.6 is obtained by considering the index set $\mathbb{G}_n$ formed by the elements γ of $\mathbb{G}_f$ such that $l(\gamma) \leq n$, where $n \in \mathbb{N}^*$ is given and fixed. In this setting, the filtration $(\mathscr{G}_\gamma)$ is indexed by $\mathbb{G}_n$ and the controls take their values in $\mathbb{G}_n$. We denote this set of controls by $\mathbf{G}_n$, and for $\gamma \in \mathbb{G}_n$, we denote the set $\{\delta \in \mathbb{G}_n: \gamma \prec \delta\}$ $[\{\Gamma \in \mathbf{G}_n: \gamma \prec \Gamma\}]$ by $\mathbb{G}_n^\gamma$ $[\mathbf{G}_n^\gamma]$.

Let X be an adapted integrable process indexed by $\mathbb{G}_n$. Define a process Z indexed by $\mathbb{G}_n$ using (backwards) induction as follows:

a. $Z_\gamma = X_\gamma$ for all γ such that $l(\gamma) = n$;
b. if $l(\gamma) < n$ and Z_δ has been defined for all $\delta \in \mathbb{D}_\gamma$, then

$$Z_\gamma = \sup\left(X_\gamma, \sup_{\delta \in \mathbb{D}_\gamma} \mathrm{E}\left(Z_\delta | \mathscr{G}_\gamma\right)\right). \tag{7.23}$$

It clearly follows from this definition that Z is a supermartingale indexed by $\mathbb{G}_n$ that dominates X. Moreover, it is easy to show by (backwards) induction that Z is the smallest supermartingale indexed by $\mathbb{G}_n$ that dominates X.

We equip each $\mathbb{D}_\gamma$ with an arbitrary total order, and for all γ such that $l(\gamma) < n$, we set

$$D(\gamma) = \inf\left\{\delta \in \mathbb{D}_\gamma: \mathrm{E}\left(Z_\delta | \mathscr{G}_\gamma\right) = \sup_{\delta' \in \mathbb{D}_\gamma} \mathrm{E}\left(Z_{\delta'} | \mathscr{G}_\gamma\right)\right\}. \tag{7.24}$$

Let $(\Gamma(\gamma), \gamma \in \mathbb{G}_n)$ be the family of random variables with values in $\mathbb{G}$ defined by (backwards) induction as follows:

a. $\Gamma(\gamma) = \gamma$ for all γ such that $l(\gamma) = n$;
b. if $l(\gamma) < n$ and $\Gamma(\delta)$ has been defined for all $\delta \in \mathbb{D}_\gamma$, then

$$\Gamma(\gamma) = \begin{cases} \gamma & \text{on } \{Z_\gamma = X_\gamma\}, \\ \Gamma(D(\gamma)) & \text{on } \{Z_\gamma > X_\gamma\}. \end{cases} \tag{7.25}$$

Theorem (Optimal control in the case where the index set is $\mathbb{G}_n$). *For all $\gamma \in \mathbb{G}_n$, $\Gamma(\gamma)$ belongs to $\mathbf{G}_n^\gamma$ and*

$$Z_\gamma = \mathrm{E}\big(X_{\Gamma(\gamma)} | \mathscr{G}_\gamma\big) \geq \mathrm{E}\big(X_\Delta | \mathscr{G}_\gamma\big), \tag{7.26}$$

for all $\Delta \in \mathbf{G}_n^\gamma$. In particular, Z is Snell's envelope of X (that is, the process defined in (7.21) with $\mathbf{G}_n^\gamma$ instead of $\mathbf{G}_f^\gamma$), and

$$\mathrm{E}(Z_0) = \mathrm{E}\big(X_{\Gamma(0)}\big) \geq \mathrm{E}(X_\Delta) \tag{7.27}$$

for all $\Delta \in \mathbf{G}_n$. In particular, $\Gamma(0)$ is an optimal control (in $\mathbf{G}_n$).

Proof. It is sufficient to establish the assertions in the first part of the theorem, which are clearly true in the case where $l(\gamma) = n$, because $\mathbf{G}_n^\gamma = \{\gamma\}$ in this case. Fix γ such that $l(\gamma) < n$ and assume that these assertions are true for all elements of $\mathbb{D}_\gamma$. We are going to prove that they are also true for γ. By (7.25) and (7.24), $D(\gamma) \prec \Gamma(D(\gamma))$, so we deduce that $\gamma \prec \Gamma(\gamma)$. Let us show that $\Gamma(\gamma)$ is a control, or equivalently, that $\{\Gamma(\gamma)^{[m+1]} = \delta\}$ belongs to $\mathscr{G}_{\delta^{[m]}}$ for all $m \geq l(\gamma)$ and all $\delta \in \mathbb{G}_n^\gamma$. Taking (7.25) and (7.24) into account, we see that

$$\big\{\Gamma(\gamma)^{[m+1]} = \delta\big\} \cap \{Z_\gamma = X_\gamma\} = \big\{\gamma^{[m+1]} = \delta\big\} \cap \{Z_\gamma = X_\gamma\}$$

and that

$$\begin{aligned}\big\{\Gamma(\gamma)^{[m+1]} = \delta\big\} &\cap \{Z_\gamma > X_\gamma\} \\ &= \bigcup_{\delta' \in \mathbb{D}_\gamma} \Big(\{D(\gamma) = \delta'\} \cap \big\{\Gamma(\delta')^{[m+1]} = \delta\big\} \cap \{Z_\gamma > X_\gamma\}\Big).\end{aligned}$$

The right-hand side of the first equality clearly belongs to $\mathscr{G}_\gamma$, therefore to $\mathscr{G}_{\delta^{[m]}}$, because $\gamma \prec \delta^{[m]}$, and the right-hand side of the second equality belongs to $\mathscr{G}_{\delta^{[m]}}$, because $D(\gamma)$ is $\mathscr{G}_\gamma$-measurable by (7.24) and $\Gamma(\delta')$ is a control by hypothesis. We conclude that $\{\Gamma(\gamma)^{[m+1]} = \delta\}$ belongs to $\mathscr{G}_{\delta^{[m]}}$, which was to be proved.

We now turn to the proof of the equality in (7.26). Let F be an arbitrary element of $\mathscr{G}_\gamma$. By (7.25),

$$\int_F Z_\gamma \, \mathrm{dP} = \int_{F \cap \{Z_\gamma = X_\gamma\}} X_{\Gamma(\gamma)} \, \mathrm{dP} + \int_{F \cap \{Z_\gamma > X_\gamma\}} Z_\gamma \, \mathrm{dP},$$

and by hypothesis, $Z_{\delta'} = \mathrm{E}(X_{\Gamma(\delta')} | \mathscr{G}_{\delta'})$ for all $\delta' \in \mathbb{D}_\gamma$. On the other hand, by (7.23) and (7.24), $Z_\gamma = \mathrm{E}(Z_{\delta'} | \mathscr{G}_\gamma)$ on $\{Z_\gamma > X_\gamma\} \cap \{D(\gamma) = \delta'\}$. It follows that

$Z_\gamma = \mathrm{E}(X_{\Gamma(\delta')} | \mathscr{G}_\gamma)$ on $\{Z_\gamma > X_\gamma\} \cap \{D(\gamma) = \delta'\}$, and therefore that $Z_\gamma = \mathrm{E}(X_{\Gamma(D(\gamma))} | \mathscr{G}_\gamma)$ on $\{Z_\gamma > X_\gamma\}$. By (7.25), we conclude that

$$\int_F Z_\gamma \, \mathrm{dP} = \int_F X_{\Gamma(\gamma)} \, \mathrm{dP},$$

which establishes the equality in (7.26).

It remains to establish the inequality in (7.26). Fix $\Delta \in \mathbf{G}_n^\gamma$. For all $\delta' \in \mathbb{D}_\gamma$, set

$$\Delta(\delta') = \begin{cases} \Delta & \text{on } \left\{\Delta^{[l(\gamma)+1]} = \delta'\right\}, \\ \delta' & \text{on } \left\{\Delta^{[l(\gamma)+1]} \neq \delta'\right\}. \end{cases}$$

Since Δ is a control, $\{\Delta^{[l(\gamma)+1]} = \delta'\}$ belongs to $\mathscr{G}_\gamma$ because $\delta'^{[l(\gamma)]} = \gamma$. By Property 5 of Section 7.3, $\Delta(\delta')$ is a control and therefore $\Delta(\delta') \in \mathbf{G}_n^{\delta'}$. Let F be an arbitrary element of $\mathscr{G}_\gamma$. Since Z is a supermartingale that dominates X and since the inequality in (7.26) holds with γ replaced by δ' and Δ by $\Delta(\delta')$, we can write

$$\begin{aligned} \int_F Z_\gamma \, \mathrm{dP} &\geq \int_{F \cap \{\Delta = \gamma\}} Z_\Delta \, \mathrm{dP} + \sum_{\delta' \in \mathbb{D}_\gamma} \int_{F \cap \{\Delta \neq \gamma\} \cap \{\Delta^{[l(\gamma)+1]} = \delta'\}} Z_{\delta'} \, \mathrm{dP} \\ &\geq \int_{F \cap \{\Delta = \gamma\}} X_\Delta \, \mathrm{dP} + \sum_{\delta' \in \mathbb{D}_\gamma} \int_{F \cap \{\Delta \neq \gamma\} \cap \{\Delta^{[l(\gamma)+1]} = \delta'\}} X_{\Delta(\delta')} \, \mathrm{dP} \\ &= \int_{F \cap \{\Delta = \gamma\}} X_\Delta \, \mathrm{dP} + \int_{F \cap \{\Delta \neq \gamma\}} X_\Delta \, \mathrm{dP} \\ &= \int_F X_\Delta \, \mathrm{dP}, \end{aligned}$$

which establishes the inequality in (7.26).

ε-Optimal Controls for Processes Indexed by $\mathbb{G}_\mathrm{f}$

Let X be an adapted process indexed by $\mathbb{G}_\mathrm{f}$ and of class $\mathrm{D}_{\mathbf{G}_\mathrm{f}}$. Let Z be Snell's envelope of X. From what was observed at the end of Section 7.5, Z is the smallest supermartingale indexed by $\mathbb{G}_\mathrm{f}$ that dominates X. For all $n \in \mathbb{N}^*$, let X^n denote the restriction of X to $\mathbb{G}_n$, and let Z^n be Snell's envelope of X^n. It is clear that $Z_\gamma^m \leq Z_\gamma^n \leq Z_\gamma$ if $l(\gamma) \leq m \leq n$. Consequently, $Z^\infty \leq Z$, where Z^∞ is the process $(Z_\gamma^\infty, \gamma \in \mathbb{G}_\mathrm{f})$ defined by $Z_\gamma^\infty = \lim_{n \to \infty} Z_\gamma^n$. Moreover, $Z_\gamma^n \geq X_\gamma^n$ for all $n \geq l(\gamma)$; therefore $Z^\infty \geq X$. On the other hand, Z^∞ is clearly a supermartingale; therefore $Z^\infty \geq Z$, because Z is the smallest supermartingale that dominates X. In conclusion, $Z^\infty = Z$.

What we have just established provides a method for constructing an ε-optimal control Γ^ε. Indeed, if $\varepsilon > 0$ is given, it suffices to choose n large enough so that $E(Z_0^n) \geq E(Z_0) - \varepsilon$ and to set $\Gamma^\varepsilon = \Gamma(0)$, where $\Gamma(0)$ is the control shown to be optimal in the theorem. This control Γ^ε is indeed ε-optimal, because by (7.27) and (7.13),

$$E(X_{\Gamma^\varepsilon}) = E(Z_0^n) \geq E(Z_0) - \varepsilon = \sup_{\Delta \in \mathbf{G}_f} E(X_\Delta) - \varepsilon.$$

7.7 EXTENSION TO GENERAL INDEX SETS

In this section, we indicate how the content of Sections 7.1 to 7.5 can be extended to a setting in which the role of $\mathbb{I}$ is played by a more general index set. We also indicate how a similar modification can be made in Sections 3.1 to 3.4 and will conclude with a brief description of the problem that motivates the theory of optimal sequential control.

General Index Sets

The structure of the index set $\mathbb{I}$ has never been fully utilized. It is therefore useful to isolate the properties of $\mathbb{I}$ that are essential to the elaboration of the theory we have presented. This amounts to showing that in this theory, $\mathbb{I}$ can be replaced by a countable (partially) ordered set $\mathbb{P}$ that satisfies the following conditions:

a. $\mathbb{P}$ has a smallest element.

b. Each element of $\mathbb{P}$ only has a finite number of predecessors.

c. The set of direct successors of each element of $\mathbb{P}$ is finite and nonempty.

Before indicating the changes that are necessary in order to replace $\mathbb{I}$ by $\mathbb{P}$, some preliminary comments are necessary. The elements of $\mathbb{P}$ are denoted in the same way as elements of $\mathbb{I}$, that is, by the letters s, t, or u; the smallest element is denoted 0 and is termed *zero element*. The order relation on $\mathbb{P}$ is denoted $\leq$; we write $s < t$ if $s \leq t$ and $s \neq t$; s is a *predecessor* of t, or t is a *successor* of s, if $s < t$; t is a *direct successor* of s if it is a successor of s and if $\{u \in \mathbb{P}: s < u < t\} = \varnothing$. The set of direct successors of s is denoted $\mathbb{D}_s$. For all couples (s, t) of elements of $\mathbb{P}$ such that $s < t$, there exists a sequence $(t_0, \ldots, t_n)$ of elements of $\mathbb{P}$ such that $t_0 = s$, $t_n = t$, and $t_{m+1} \in \mathbb{D}_{t_m}$ for $m = 0, \ldots, n-1$. The set obtained by adding to $\mathbb{P}$ an infinite element $\bowtie$ is denoted by $\overline{\mathbb{P}}$. By convention, $t < \bowtie$ for all elements $t \in \mathbb{P}$. If $t \mapsto x_t$ is a function defined on $\mathbb{P}$ with values in $\overline{\mathbb{R}}$, then the symbol $\limsup_{t \to \bowtie} x_t$ denotes the number $\inf_{t \in \mathbb{P}} \sup_{u \in \mathbb{P} - \mathbb{P}_t} x_u$, where $\mathbb{P}_t = \{s \in \mathbb{P}: s \leq t\}$. The symbol $\liminf_{t \to \bowtie} x_t$ is defined analogously, and we write $\lim_{t \to \bowtie} x_t$ when lim sup and lim inf are equal. Unless otherwise indicated, all notions that involve the index set $\mathbb{P}$ are defined by literal transposition of the corresponding notions involving $\mathbb{I}$.

From now on, we will term *general index set* any countable ordered set $\mathbb{P}$ that satisfies conditions a, b, and c above.

Optimal Control

The two theorems in Sections 7.4 and 7.5, as well as the subject matter elaborated in order to establish them, are not at all affected by replacing $\mathbb{I}$ by a general index set $\mathbb{P}$ in the definition of $\mathbb{G}$. It should, however, be observed that in the special case where property c of $\mathbb{P}$ is replaced by "each element of $\mathbb{P}$ has exactly d direct successors," there is a one-to-one correspondence between $\mathbb{G}$ defined using $\mathbb{P}$ and $\mathbb{G}$ defined using $\mathbb{I}$ (cf. Exercise 7), and any optimal control problem defined using $\mathbb{P}$ is easily identified with an equivalent problem using $\mathbb{I}$. It follows that the only added generality achieved by allowing a general index set is due to the fact that the cardinality of the set of direct successors of $\gamma \in \mathbb{G}$ may depend on γ.

Optimal Stopping

The theory of optimal stopping can be considered to be a special case of the theory of optimal control (cf. Exercise 8). On the other hand, it is interesting to observe that Sections 3.1 to 3.4 apply almost literally to the case where the index set $\mathbb{I}$ is replaced by a general index set $\mathbb{P}$. It is only necessary to make the following modifications. The symbol $\mathbf{A}^T$ is defined differently: for all $T \in \mathbf{A}[\mathbf{A}_f]$, $\mathbf{A}^T [\mathbf{A}_f^T]$ denotes the set of $U \in \mathbf{A}[\mathbf{A}_f]$ that are accessible through T. By definition, U is *accessible through* T if $T \leq U$ and if there exists a predictable increasing path passing through T and U. By definition, a predictable increasing path Γ *passes through* a stopping point T if there exists a stopping time τ relative to the filtration $\mathscr{F}^\Gamma$ such that $\Gamma_\tau = T$. Such a stopping time is necessarily unique. It is easy to check that in the case where the index set is $\mathbb{I}$, this definition of $\mathbf{A}^T$ is equivalent to the other. Another modification that has to be made concerns the symbol $|t|$ or $|T|$. This symbol appears only in connection with a predictable increasing path Γ passing through t or T. It must therefore be replaced by the stopping time τ such that $\Gamma_\tau = t$ or T.

Digression

The rigorous setting in which is placed the theory of optimal sequential control that we have just developed is the result of a reduction to formal structures of the following typical situation, which we described informally in Section 1.1. An observer (or controller) is in the presence of a random process whose evolution he can influence. At each stage of this evolution, the observer chooses among a finite number of control actions that will affect the evolution of the process. If a reward is associated with each action, or conversely, if each action entails a cost, then the observer will act in such a way as to maximize the reward or minimize the cost. As mentioned in Section 1.1, it is reasonable to index information and rewards or costs by finite or infinite sequences $(a_1, a_2, \dots)$ of actions. This set is naturally equipped with the order $\prec$ introduced in Section 7.2. A concrete example will be studied in the following chapter.

EXERCISES

1. Prove that any control (as defined in Section 7.3) is an accessible stopping point with values in $\mathbb{G}$, and conversely.

2. a. Prove that the correspondence defined in (7.8) is one to one.

b. In the setting of this correspondence, prove that $\mathscr{G}_{\Gamma[n]} = \mathscr{F}_{\bar{\Gamma}_n}$ for all controls $\Gamma \in \mathbf{G}$ and all $n \in \mathbb{N}$.

3. Let X be a supermartingale (indexed by $\mathbb{G}$) and (Γ, Δ) a couple of controls such that $\Gamma \prec \Delta$. Prove that

$$E(X_\Delta | \mathscr{G}_\Gamma) \leq X_\Gamma$$

in each of the three following cases:

a. Δ is bounded.

b. Δ is finite and $(X_{\overline{\Delta[n]}}, n \in \mathbb{N})$ is uniformly integrable.

c. X is continuous and $(X_{\overline{\Delta[n]}}, n \in \mathbb{N})$ is uniformly integrable.

4. Prove that a function $\gamma \mapsto x_\gamma$ defined on $\mathbb{G}$ with values in $\overline{\mathbb{R}}$ is upper semicontinuous if and only if for all sequences $(\gamma(k), k \in \mathbb{N})$ of elements of $\mathbb{G}$ that converge to an element $\gamma \in \mathbb{G}^\infty$,

$$\limsup_{k \to \infty} x_{\gamma(k)} \leq x_\gamma.$$

5. Let $(y_{s,t}, s \in \mathbb{I}, t \in \mathbb{D}_s)$ be a family of elements of $\overline{\mathbb{R}}$ and let $\alpha \in]0, 1]$. Assuming that

a. $\sup_{s \in \mathbb{I}} \sup_{t \in \mathbb{D}_s} y_{s,t}^+ < \infty$ if $\alpha < 1$,

b. $\Sigma_{n \in \mathbb{N}} \sup_{s \in \mathbb{I}:\ |s| = n} \sup_{t \in \mathbb{D}_s} y_{s,t}^+ < \infty$ if $\alpha = 1$,

set

$$x_\gamma = \begin{cases} 0 & \text{if } \gamma = 0, \\ \sum_{n=1}^{l(\gamma)} \alpha^n \, y_{\gamma_{n-1}, \gamma_n}, & \text{if } \gamma \in \mathbb{G} - \{0\}, \end{cases}$$

and prove, using the previous exercise, that the function $\gamma \mapsto x_\gamma$ is upper semicontinuous.

6. Let X be an adapted integrable process indexed by $\mathbb{G}_f$ such that $E(\sup_{\gamma \in \mathbb{G}_f} X_\gamma^+) < \infty$. Let Z be Snell's envelope of X and let $(\Gamma(n), n \in \mathbb{N})$ be an optimizing sequence.

a. Prove that if

$$\lim_{n \to \infty} \sup_{\gamma \in \mathbb{G}_f:\, \Gamma(n) \prec \gamma} X_\gamma = \limsup_{n \to \infty} X_{\Gamma(n)} \tag{7.28}$$

(in particular, if $\lim_{l(\gamma) \to \infty} X_\gamma$ exists (in $\overline{\mathbb{R}}$)), then

$$\limsup_{n \to \infty} Z_{\Gamma(n)} = \limsup_{n \to \infty} X_{\Gamma(n)}. \tag{7.29}$$

(*Hint.* Follow the proof of Proposition 3 of Section 3.4.)

b. Given $\varepsilon > 0$, set

$$\tau^\varepsilon = \begin{cases} \inf\{n \in \mathbb{N}: Z_{\Gamma(n)} \le X_{\Gamma(n)} + \varepsilon\}, & \text{if } \{\ \} \ne \emptyset, \\ \infty, & \text{if } \{\ \} = \emptyset, \end{cases}$$

and prove using (7.29) that if (7.28) is satisfied, then τ^ε is finite and $\Gamma(\tau^\varepsilon)$ is an ε-optimal control (that is, $E(X_{\Gamma(\tau^\varepsilon)}) \ge \sup_{\Gamma \in \mathbf{G}_f} E(X_\Gamma) - \varepsilon$). (*Hint.* Follow the proof of Theorem 2 of Section 3.4.)

7. Given a general index set $\mathbb{P}$ (as defined in Section 7.7), let $\mathbb{G}(\mathbb{P})$ [$\mathbf{G}(\mathbb{P})$] be the set defined in the same way as $\mathbb{G}$ [$\mathbf{G}$] in Section 7.2 [7.3], but with $\mathbb{I}$ replaced by $\mathbb{P}$. Assume that each element of $\mathbb{P}$ has exactly d direct successors.

a. Prove that there is a one-to-one correspondence between $\mathbb{G}(\mathbb{P})$ and $\mathbb{G}(\mathbb{I})$.

b. Let $(\mathcal{G}_\gamma, \gamma \in \mathbb{G}(\mathbb{P}))$ be a filtration. Define a filtration indexed by $\mathbb{G}(\mathbb{I})$ so that the correspondence defined in part a extends to a one-to-one correspondence between controls of $\mathbf{G}(\mathbb{P})$ and controls of $\mathbf{G}(\mathbb{I})$.

c. If X is a reward process indexed by $\mathbb{G}(\mathbb{P})$, define a reward process $\tilde{X}$ indexed by $\mathbb{G}(\mathbb{I})$ in such a way that there is a one-to-one correspondence between optimal controls in $\mathbf{G}(\mathbb{P})$ (for X) and optimal controls in $\mathbf{G}(\mathbb{I})$ (for $\tilde{X}$).

8. Let $(X_t, t \in \bar{\mathbb{I}})$ be a reward process as considered in Sections 3.1, 3.2, and 3.3. For all $\gamma \in \mathbb{G}$, set $\mathcal{G}_\gamma = \mathcal{F}_{\gamma_{l(\gamma)}}$ and $\tilde{X}_\gamma = X_{\gamma_{l(\gamma)}}$.

a. Show that if $(X_t, t \in \bar{\mathbb{I}})$ satisfies condition (3.3), then $(\tilde{X}_\gamma, \gamma \in \mathbb{G})$ is upper semicontinuous (according to the definition of Section 7.3).

b. Prove that finding an optimal control Γ^o (for $(\tilde{X}_\gamma, \gamma \in \mathbb{G})$) is equivalent to finding an optimal accessible stopping point T^o (for $(X_t, t \in \bar{\mathbb{I}})$).

c. Suppose that $(X_t, t \in \bar{\mathbb{I}})$ satisfies the following condition (which is stronger than condition (3.6)): For almost all ω and all deterministic increasing paths γ, $\limsup_{n \to \infty} X_n^\gamma(\omega) \leq X_{\bowtie}(\omega)$. Show that $(\tilde{X}_\gamma, \gamma \in \mathbb{G})$ is upper semicontinuous (in the sense recalled above) and conclude that there exists an accessible stopping point T^o (for $(X_t, t \in \bar{\mathbb{I}})$).

HISTORICAL NOTES

The example in Section 7.1 was studied in Lawler and Vanderbei (1983). The set $\mathbb{G}$ is a generalization of an index set introduced by Haggstrom (1966), though in the spirit of Exercise 7, problems formulated using $\mathbb{G}$ can also be formulated in Haggstrom's framework. The notion of *control* is analogous to the notion of *control variable* introduced by this author. This setup allows very general types of reward processes to be considered. The formalism is quite easy to manipulate and makes possible the extension to general index sets considered in Section 7.7. Two objectives that have been achieved here are the unification of the problems of optimal stopping and of optimal sequential control, and the identification of the minimal assumptions under which optimal controls exist. Other possible formalizations of control problems include those proposed by Brown (1977) or the early formulations used, for instance, in Blackwell and Girshick (1954).

The theorems of Sections 7.4, 7.5, and 7.6 generalize results of Haggstrom (1966). In particular, the boundedness assumption on the reward process used by this author is replaced by the "almost necessary" assumption of the theorem of Section 7.4.

CHAPTER 8

Multiarmed Bandits

In this chapter, we shall study a specific class of optimal control problems. One example in this class is the following. Imagine a player who plays one of d slot machines (with unknown characteristics) at a time, deciding after each play on which machine to play the next game. It is assumed that the outcome of the game at time n does not affect the state of the remaining idle machines, which is equivalent to saying that the d machines are assumed to be independent of each other. Moreover, the player is allowed to play any one of the machines an infinite number of times, a finite number of times, or even not at all. Let $Y_{t^i}^i$ be the (random) reward that machine i yields when it is played for the t^ith time. Fix a discounting rate $\alpha \in]0, 1[$ and assume that the effective reward from machine i when it is played for the t^ith time is $\alpha^n Y_{t^i}^i$ if this play occurs at time n. With the introduction of such a rate, it is clear that the order in which the machines are played is important, because the reward from each play is discounted differently for each order. The problem that confronts the player is to determine a strategy that yields the maximal expected reward. This problem is the basic example of a *d-armed bandit problem*, since each machine is likened to an arm that activates the playing mechanism.

Throughout this chapter, we will use the notation and terminology of Chapter 7 and will assume that $d > 1$.

8.1 FORMULATING THE PROBLEM

For $i = 1, \ldots, d$, let $(\mathscr{F}_{t^i}^i,\ t^i \in \mathbb{N})$ be a filtration and $(Y_{t^i}^i,\ t^i \in \mathbb{N})$ be a real-valued process adapted to this filtration and such that $Y_0^i = 0$. In the player's terminology, $Y_{t^i}^i(\omega)$ represents the undiscounted reward that machine i yields the t^ith time it is played. The player is aware of the (joint) distribution of all the variables $Y_{t^i}^i$. The sigma field $\mathscr{F}_{t^i}^i$ represents the information on the value of the parameter ω that the player will acquire during the first t^i plays on this machine.

Throughout this chapter, we fix $\alpha \in]0, 1[$, and for all $\gamma \in \mathbb{G}$, we set

$$X_\gamma = \sum_{n=1}^{\infty} \alpha^n R(\gamma, n), \qquad \text{where } R(\gamma, n) = \sum_{i=1}^{d} Y^i_{\gamma^i_n} \left(\gamma^i_n - \gamma^i_{n-1}\right). \quad (8.1)$$

We will introduce below a hypothesis that, among other things, will guarantee the convergence of the series in (8.1). Notice that if $\gamma^i_n - \gamma^i_{n-1} = 1$, then $\gamma^j_n - \gamma^j_{n-1} = 0$ for all $j \neq i$, and therefore $R(\gamma, n) = Y^i_{\gamma^i_n}$, which expresses the fact that if machine i is played at time n, then the player receives the discounted reward $\alpha^n Y^i_{\gamma^i_n}$. Notice also that $R(\gamma, n) = 0$ for all $n > l(\gamma)$, which is equivalent to saying that the player quits the game at time $l(\gamma)$. Therefore, X_γ represents the effective reward obtained by applying the (deterministic) control γ.

After having played machines $1, \dots, d$, respectively, $t^1, \dots, t^d$ times, the information available to the player about the state of all d machines is represented by the sigma field $\mathscr{F}^1_{t^1} \vee \cdots \vee \mathscr{F}^d_{t^d}$. Therefore, it is natural to set

$$\mathscr{G}_\gamma = \mathscr{F}^1_{\gamma^1_{l(\gamma)}} \vee \cdots \vee \mathscr{F}^d_{\gamma^d_{l(\gamma)}}, \qquad \text{for all } \gamma \in \mathbb{G}_f,$$

and to let $(\mathscr{G}_\gamma, \gamma \in \mathbb{G}_f)$ be the filtration that plays the role of the general filtration introduced in Section 7.2. For all $\gamma \in \mathbb{G}^\infty$, we set

$$\mathscr{G}_\gamma = \mathscr{G} = \mathscr{F}^1_\infty \vee \cdots \vee \mathscr{F}^d_\infty, \qquad \text{where } \mathscr{F}^i_\infty = \bigvee_{t^i \in \mathbb{N}} \mathscr{F}^i_{t^i}, \qquad i = 1, \dots, d.$$

In the sequel, the filtration $(\mathscr{F}^i_{t^i}, t^i \in \mathbb{N})$ will be denoted more simply by $\mathscr{F}^i$. For the time being, we will not assume that $\mathscr{F}^1, \dots, \mathscr{F}^d$ are independent.

The Reward Process

In order to ensure that (8.1) is meaningful and that the theorem of Section 7.4 applies, we will assume from now on that

$$\sum_{n=1}^{\infty} \alpha^n \operatorname{E}\left(|Y^i_n|\right) < \infty, \qquad \text{for } i = 1, \dots, d. \quad (8.2)$$

The map $X: (\gamma, \omega) \mapsto X_\gamma(\omega)$ from $\mathbb{G} \times \Omega$ into $\mathbb{R}$ is then a reward process as defined at the beginning of Section 7.4. Indeed, X is $\mathscr{B} \times \mathscr{G}$-measurable, adapted, and for all $\Gamma \in \mathbf{G}$, satisfies

$$
\begin{aligned}
|X_\Gamma| &\le \sum_{n=1}^{\infty} \alpha^n |R(\Gamma, n)| \\
&\le \sum_{i=1}^{d} \sum_{n=1}^{\infty} \alpha^n \left|Y^i_{\Gamma^i_n}\right| \left(\Gamma^i_n - \Gamma^i_{n-1}\right) \\
&\le \sum_{i=1}^{d} \sum_{n=1}^{\infty} \alpha^{\Gamma^i_n} \left|Y^i_{\Gamma^i_n}\right| \left(\Gamma^i_n - \Gamma^i_{n-1}\right) \\
&\le \sum_{i=1}^{d} \sum_{n=1}^{\infty} \alpha^n \left|Y^i_n\right|,
\end{aligned}
\tag{8.3}
$$

which implies by (8.2) that

$$\mathrm{E}\Big(\operatorname*{esssup}_{\Gamma \in \mathbf{G}} |X_\Gamma|\Big) < \infty,$$

and therefore that X is of class $\mathrm{D}_{\mathbf{G}}$.

Convergence

By (8.2), the inequalities in (8.3) show in particular that the series in (8.1), as well as the series $\sum_{n=1}^{\infty} \alpha^n Y_n^1, \ldots, \sum_{n=1}^{\infty} \alpha^n Y_n^d$, converge absolutely a.s. and in L^1. We will often use this conclusion without recalling its justification.

Another Expression for X

In order to show that X is continuous (according to the definition given in Section 7.3), we are going to express X in a form that will also be useful further on. For all $\gamma \in \mathbb{G}$, all $k \in \mathbb{N}$ and $i = 1, \ldots, d$, set

$$\lambda^i_k(\gamma) = \inf\{n \in \mathbb{N}: \gamma^i_n = k\}. \tag{8.4}$$

The nondecreasing function $k \mapsto \lambda^i_k(\gamma)$ thus defined is precisely the generalized inverse of the function $n \mapsto \gamma^i_n$. It is clear that $\lambda^i_0(\gamma) = 0$ and $\lambda^i_k(\gamma) \ge k$. Moreover, $\lambda^i_k(\gamma) = \infty$ if $k > \sup_n \gamma^i_n$, therefore in particular if $k > l(\gamma)$. Notice by (8.4) that $\gamma^i_{\lambda^i_k(\gamma)} = k$ if $\lambda^i_k(\gamma) < \infty$. The interpretation of this equality is that $\lambda^i_k(\gamma)$ is the time at which machine i is played for the kth time. With this interpretation, it is easy to see that X_γ can be written

$$X_\gamma = \sum_{i=1}^{d} \sum_{k=1}^{\infty} \alpha^{\lambda^i_k(\gamma)} Y^i_k, \tag{8.5}$$

where $\alpha^\infty = 0$ by convention.

Continuity

We are going to prove that the process X is continuous, which is equivalent to showing that for almost all ω,

$$\lim_{n\to\infty} \sup_{\delta\in\mathbb{G}^{\gamma[n]}} \left|X_\gamma(\omega) - X_\delta(\omega)\right| = 0, \qquad \text{for all } \gamma \in \mathbb{G}^\infty.$$

By (8.2), for almost all ω,

$$\sum_{k=1}^{\infty} \alpha^k \left|Y_k^i(\omega)\right| < \infty, \qquad \text{for } i = 1,\dots,d. \tag{8.6}$$

Assume that ω satisfies (8.6) and let $\gamma \in \mathbb{G}^\infty$. If $\delta \in \mathbb{G}^{\gamma[n]}$, then $\gamma_k^i = \delta_k^i$ for all $k \le n$ and $i = 1,\dots,d$; therefore $\lambda_k^i(\gamma) = \lambda_k^i(\delta)$ for all $k \le \gamma_n^i$ and $i = 1,\dots,d$. By (8.5), it is therefore sufficient to prove that for $i = 1,\dots,d$,

$$\lim_{n\to\infty} \sup_{\delta\in\mathbb{G}^{\gamma[n]}} \left| \sum_{k=\gamma_n^i+1}^{\infty} \left(\alpha^{\lambda_k^i(\gamma)} - \alpha^{\lambda_k^i(\delta)}\right) Y_k^i(\omega) \right| = 0. \tag{8.7}$$

In the case where $\lim_{n\to\infty}\gamma_n^i = \infty$, this follows immediately from (8.6) and the fact that $\lambda_k^i(\gamma) \ge k$ and $\lambda_k^i(\delta) \ge k$. In the case where $\sup_n \gamma_n^i < \infty$, set $n_0 = \inf\{n\colon \gamma_n^i = \sup_m \gamma_m^i\}$ and observe that if $n > n_0$ and $\delta \in \mathbb{G}^{\gamma[n]}$, then $\lambda_k^i(\gamma) = \infty$ and $\lambda_k^i(\delta) \ge n + k - \gamma_{n_0}^i$ for all $k > \gamma_{n_0}^i$. It follows that for $n > n_0$, the supremum on the left-hand side of (8.7) is bounded by

$$\sum_{k=\gamma_n^i+1}^{\infty} \alpha^{n+k-\gamma_{n_0}^i} \left|Y_k^i(\omega)\right| \le \alpha^{n-\gamma_{n_0}^i} \sum_{k=1}^{\infty} \alpha^k \left|Y_k^i(\omega)\right|,$$

which by (8.6) shows that this supremum converges to 0 when $n \to \infty$.

Optimization

From the above, we conclude that the hypotheses of the theorem of Section 7.4 are satisfied, and so the conclusions of this theorem apply. Therefore there exists an optimal control, that is a control Γ^o which satisfies $\mathrm{E}(X_{\Gamma^o}) = \sup_{\Gamma\in\mathbf{G}} \mathrm{E}(X_\Gamma)$, where X is defined in (8.1).

Optimization Over Infinite Controls

Let $\mathbf{G}^\infty$ denote the set of controls with values in $\mathbb{G}^\infty$. Let X again be defined by (8.1), but only taking into account the elements $\gamma \in \mathbb{G}^\infty$. Under assumption (8.2), one can ask whether there exists a control $\Gamma^o \in \mathbf{G}^\infty$ such that

$$\mathrm{E}(X_{\Gamma^o}) = \sup_{\Gamma\in\mathbf{G}^\infty} \mathrm{E}(X_\Gamma). \tag{8.8}$$

We are going to show that the answer to this question is affirmative (it should be pointed out that the supremum over $\Gamma \in \mathbf{G}^\infty$ may be strictly smaller than the supremum over $\Gamma \in \mathbf{G}$). Set

$$\tilde{X}_\gamma = \begin{cases} \mathrm{E}(X_{\tilde{\gamma}} | \mathscr{G}_\gamma) - 1, & \text{if } \gamma \in \mathbb{G}_f, \\ X_\gamma, & \text{if } \gamma \in \mathbb{G}^\infty, \end{cases}$$

where $\tilde{\gamma}$ is the element of $\mathbb{G}^\infty$ defined by

$$\tilde{\gamma}_n = \begin{cases} \gamma_n, & \text{if } n \le l(\gamma), \\ \gamma_{l(\gamma)} + (n - l(\gamma))e_1, & \text{if } n > l(\gamma). \end{cases}$$

Using the properties of X established above, it is easy to check that the map $(\gamma, \omega) \mapsto \tilde{X}_\gamma(\omega)$, which we denote $\tilde{X}$, is a reward process according to the definition of Section 7.4. By applying the theorem of Section 7.4 to $\tilde{X}$, we can therefore conclude that there exists a control Γ^o such that

$$\mathrm{E}(\tilde{X}_{\Gamma^o}) = \sup_{\Gamma \in \mathbf{G}} \mathrm{E}(\tilde{X}_\Gamma). \tag{8.9}$$

On the other hand, for any $\Gamma \in \mathbf{G}$,

$$\begin{aligned} \mathrm{E}(\tilde{X}_\Gamma) &= \mathrm{E}(\tilde{X}_\Gamma; l(\Gamma) < \infty) + \mathrm{E}(\tilde{X}_\Gamma; l(\Gamma) = \infty) \\ &= \mathrm{E}(\mathrm{E}(X_{\tilde{\Gamma}} | \mathscr{G}_\Gamma) - 1; l(\Gamma) < \infty) + \mathrm{E}(X_{\tilde{\Gamma}}; l(\Gamma) = \infty) \\ &= \mathrm{E}(X_{\tilde{\Gamma}}) - \mathrm{P}\{l(\Gamma) < \infty\}, \end{aligned}$$

where $\tilde{\Gamma}$ is the element of $\mathbf{G}^\infty$ defined by

$$\tilde{\Gamma}_n = \begin{cases} \Gamma_n & \text{on } \{n < l(\Gamma)\}, \\ \Gamma_{l(\Gamma)} + (n - l(\Gamma))e_1 & \text{on } \{n > l(\Gamma)\}, \end{cases}$$

and which is therefore equal to Γ on $\{l(\Gamma) = \infty\}$. It follows first of all that

$$\sup_{\Gamma \in \mathbf{G}} \mathrm{E}(\tilde{X}_\Gamma) = \sup_{\Gamma \in \mathbf{G}^\infty} \mathrm{E}(\tilde{X}_\Gamma) = \sup_{\Gamma \in \mathbf{G}^\infty} \mathrm{E}(X_\Gamma),$$

and second that the control Γ^o in (8.9) must belong to $\mathbf{G}^\infty$, which implies that this control satisfies (8.8). Another conclusion that can be drawn from the theorem of Section 7.4 is that an optimizing control Γ^o for $\tilde{X}$ necessarily satisfies (8.9), therefore (8.8), hence it is optimal for X in $\mathbf{G}^\infty$.

8.2 INDEX CONTROLS

When the filtrations $\mathscr{F}^1, \ldots, \mathscr{F}^d$ are independent, the optimal controls for the optimization problem over infinite controls considered at the end of the previous section can be described using the notion of index control. In this section, we will define this notion and will draw the first consequences. The hypothesis that stipulates independence of the filtrations $\mathscr{F}^1, \ldots, \mathscr{F}^d$ will only be made as of the next section.

Controls

Our goal is to describe the controls Γ^o in $\mathbf{G}^\infty$ that satisfy (8.8). Therefore in the remainder of this chapter, the term *control* will always refer to an element of $\mathbf{G}^\infty$. Notice that such an element can be identified with a predictable increasing path (Γ_n) relative to the filtration $(\mathscr{F}_t,\ t \in \mathbb{I})$ defined by

$$\mathscr{F}_t = \mathscr{F}^1_{t^1} \vee \cdots \vee \mathscr{F}^d_{t^d}, \qquad t = \left(t^1, \ldots, t^d\right). \tag{8.10}$$

Indeed, by what was established in Section 7.3, there is a one-to-one map between the set of predictable increasing paths and $\mathbf{G}^\infty$. Given this identification, it becomes possible to use both the terminology related to controls and that related to increasing paths. We will exploit this possibility whenever it improves the clarity of exposition. For example, it will be useful to be able to say that a control Γ passes through a stopping point T instead of saying that the corresponding increasing path (Γ_n) passes through T. Of course, in the remainder of this chapter, the term *stopping point* will always refer to a stopping point relative to the filtration $(\mathscr{F}_t,\ t \in \mathbb{I})$ defined in (8.10).

Index Controls

For $i = 1, \ldots, d$ and all $t^i \in \mathbb{N}$, let $I_i(t^i)$ be a real-valued $\mathscr{F}^i_{t^i}$-measurable random variable. For the moment, this measurability property is the only link with the data of the problem, but starting with Section 8.3, $I_1(t^1), \ldots, I_d(t^d)$ will be defined in terms of the processes $(Y^1_{t^1},\ t^1 \in \mathbb{N}), \ldots, (Y^d_{t^d},\ t^d \in \mathbb{N})$. For all $t = (t^1, \ldots, t^d) \in \mathbb{I}$, set

$$I(t) = \sup_i I_i(t^i). \tag{8.11}$$

The following notion will be the key to our description of the controls that satisfy (8.8). An *index control* (relative to the variables $I_i(t^i)$) is any control Γ

such that

$$I(\Gamma_n) = I_i(\Gamma_n^i) \quad \text{on} \quad \{\Gamma_{n+1} = \Gamma_n + e_i\},$$
$$\text{for all } n \in \mathbb{N} \text{ and } i = 1, \dots, d. \qquad (8.12)$$

This property can be used to construct an index control: Set $\Gamma_0 = 0$ and construct Γ_{n+1} from Γ_n by choosing i such that $I(\Gamma_n) = I_i(\Gamma_n^i)$ and setting $\Gamma_{n+1} = \Gamma_n + e_i$; in fact, once i has been chosen, one can take $\Gamma_{n+m} = \Gamma_n + me_i$ for all $m \in \mathbb{N}^*$ such that $m \le \tau$, where τ is the smallest $k \in \mathbb{N}^*$ satisfying $I_i(\Gamma_n^i + k) < I_i(\Gamma_n^i)$ if such a k exists, and $\tau = \infty$ otherwise. Indeed, $I_i(\Gamma_n^i + k) \ge I_i(\Gamma_n^i) \ge \sup_{j \ne i} I_j(\Gamma_n^j)$, for all $k \in \mathbb{N}^*$ such that $k < \tau$, which implies that $I(\Gamma_n + ke_i) = I_i(\Gamma_n^i + k)$ for these k.

Stages that Index Controls Must Visit

In order to describe properties of $I_i(t^i)$, $I(t)$ and the associated index controls, we will use the auxiliary random variables $\tilde{I}_i(t^i)$ and $\tilde{I}(t)$ defined for all $t^i \in \mathbb{N}$ and all $t = (t^1, \dots, t^d) \in \mathbb{I}$ by

$$\tilde{I}_i(t^i) = \inf_{n \le t^i} I_i(n) \quad \text{and} \quad \tilde{I}(t) = \sup_i \tilde{I}_i(t^i). \qquad (8.13)$$

Clearly, $\tilde{I}_i(t^i)$ and $\tilde{I}(t)$ are nonincreasing functions of t^i and t. Moreover, $\tilde{I}_i(t^i) \le I_i(t^i)$ and $\tilde{I}(t) \le I(t)$.

For all $x \in \mathbb{R}$ and $i = 1, \dots, d$, set

$$J_i(x) = \inf\{n \in \mathbb{N} : \tilde{I}_i(n) < x\}. \qquad (8.14)$$

Note that $\tilde{I}_i(n)$ can be replaced by $I_i(n)$ without modifying $J_i(x)$, and that $x \mapsto J_i(x)$ is nonincreasing. Moreover,

$$\tilde{I}_i(J_i(x)) = I_i(J_i(x)) \qquad \text{on } \{J_i(x) < \infty\}. \qquad (8.15)$$

Indeed, $J_i(x)$ is the point where the value of the function $n \mapsto \tilde{I}_i(n)$ is less than x for the first time, and at this point, the equality in (8.15) holds.

Let $(T_n, n \in \mathbb{N})$ and $(T_n^i, n \in \mathbb{N})$, $i = 1, \dots, d$, be the sequences defined inductively by setting

$$T_0 = (T_0^1, \dots, T_0^d) = 0,$$
$$T_{n+1}^i = \begin{cases} J_i(\tilde{I}(T_n)) & \text{on } \{T_n < \bowtie\}, \\ \infty & \text{on } \{T_n = \bowtie\}, \end{cases}$$
$$T_{n+1} = \begin{cases} (T_{n+1}^1, \dots, T_{n+1}^d) & \text{on } \bigcap_{i=1}^d \{T_{n+1}^i < \infty\}, \\ \bowtie & \text{on } \bigcup_{i=1}^d \{T_{n+1}^i = \infty\}. \end{cases} \qquad (8.16)$$

It is clear that the sequence $(T_n,\ n \in \mathbb{N})$ is also determined by the induction formula

$$T_{n+1} = \begin{cases} J(\tilde{I}(T_n)) & \text{on } \{T_n < \bowtie\}, \\ \bowtie & \text{on } \{T_n = \bowtie\}, \end{cases} \tag{8.17}$$

where for all $x \in \mathbb{R}$,

$$J(x) = \begin{cases} (J_1(x), \dots, J_d(x)) & \text{on } \bigcap_{i=1}^{d} \{J_i(x) < \infty\}, \\ \bowtie & \text{on } \bigcup_{i=1}^{d} \{J_i(x) = \infty\}. \end{cases} \tag{8.18}$$

Note that $T_n^i(\omega)$ is indeed the ith coordinate of $T_n(\omega)$ if $T_n(\omega)$ is finite, but that $T_n^i(\omega)$ can be finite even though $T_n(\omega)$ is not, if n is equal to the smallest integer m such that $T_m(\omega) = \bowtie$.

The sequence $(T_n,\ n \in \mathbb{N})$ plays a central role in the description of index controls. We will show in particular that index controls pass through all the T_n, but first we shall show that the functions $t \mapsto I(t)$ and $t \mapsto \tilde{I}(t)$ coincide at each T_n, and describe how these functions behave for $T_n \le t < T_{n+1}$.

Proposition 1 (Properties). *$(T_n,\ n \in \mathbb{N})$ is a nondecreasing sequence of stopping points. Moreover, for any $n \in \mathbb{N}$ and $t = (t^1, \dots, t^d) \in \mathbb{I}$,*

(a) $\{T_n = T_{n+1}\} \subset \{T_n = \bowtie\}$;
(b) $\tilde{I}(T_n) = \tilde{I}(t) = \tilde{I}_i(t^i)$ *on* $\{T_n \le t < T_{n+1}\} \cap \{t^i < T_{n+1}^i\}$;
(c) $\tilde{I}(T_n) = \tilde{I}(t) > \tilde{I}_i(t^i)$ *on* $\{T_n \le t < T_{n+1}\} \cap \{t^i = T_{n+1}^i\}$;
(d) $\tilde{I}_i(T_n^i) = I_i(T_n^i)$ *on* $\{T_n^i < \infty\}$;
(e) $\tilde{I}(T_n) = I(T_n)$ *on* $\{T_n < \bowtie\}$.

Proof. Assume that $T_{n-1} \le T_n$ for some $n \in \mathbb{N}^*$. Then on $\{T_n < \bowtie\}$, $\tilde{I}(T_{n-1}) \ge \tilde{I}(T_n)$, and therefore $J(\tilde{I}(T_{n-1})) \le J(\tilde{I}(T_n))$. By (8.17), it follows that $T_n \le T_{n+1}$. Since $T_0 \le T_1$, the sequence $(T_n,\ n \in \mathbb{N})$ is a nondecreasing. Let us show that each term of this sequence is a stopping point. For $n = 0$, this is obvious because $T_0 = 0$. Suppose that the assertion is true for some $n \in \mathbb{N}$ and observe that for all $t \in \mathbb{I}$,

$$\begin{aligned} \{T_{n+1} \le t\} &= \{T_n \le t\} \cap \left\{J_1(\tilde{I}(T_n)) \le t^1\right\} \cap \cdots \cap \left\{J_d(\tilde{I}(T_n)) \le t^d\right\} \\ &= \{T_n \le t\} \cap \left\{\tilde{I}_1(t^1) < \tilde{I}(T_n)\right\} \cap \cdots \cap \left\{\tilde{I}_d(t^d) < \tilde{I}(T_n)\right\} \\ &= \bigcup_{s \in \mathbb{I}_t} \left(\{T_n = s\} \cap \left\{\tilde{I}_1(t^1) < \tilde{I}(s)\right\} \cap \cdots \cap \left\{\tilde{I}_d(t^d) < \tilde{I}(s)\right\}\right). \end{aligned}$$

This shows that T_{n+1} is a stopping point, because $\{T_n = s\} \in \mathscr{F}_s \subset \mathscr{F}_t$ and

$$\left\{\tilde{I}_i(t^i) < \tilde{I}(s)\right\} = \bigcup_{j=1}^{d} \left\{\tilde{I}_i(t^i) < \tilde{I}_j(s^j)\right\} \in \mathscr{F}^i_{t^i} \vee \mathscr{F}^j_{s^j} \subset \mathscr{F}_t.$$

We now prove (a). On $\{T_n < \bowtie\} \cap \{\tilde{I}(T_n) = \tilde{I}_i(T_n^i)\}$,

$$T_n^i < \inf\left\{m \in \mathbb{N}: \tilde{I}_i(m) < \tilde{I}_i(T_n^i)\right\} = J_i\left(\tilde{I}_i(T_n^i)\right) = J_i\left(\tilde{I}(T_n)\right) = T_{n+1}^i;$$

therefore $T_n < T_{n+1}$, which implies (a) because

$$\{T_n < \bowtie\} = \bigcup_{i=1}^{d} \left(\{T_n < \bowtie\} \cap \left\{\tilde{I}(T_n) = \tilde{I}_i(T_n^i)\right\}\right).$$

We now turn to (b) and (c). Since on $\{T_n < \bowtie\}$,

$$T_{n+1}^i = J_i\left(\tilde{I}(T_n)\right) = \inf\left\{m \in \mathbb{N}: \tilde{I}_i(m) < \tilde{I}(T_n)\right\},$$

we deduce that on $\{T_n \le t < T_{n+1}\}$,

$$\tilde{I}_i(t^i) \ge \tilde{I}(T_n) \text{ or } \tilde{I}_i(t^i) < \tilde{I}(T_n) \quad \text{according as} \quad t^i < T_{n+1}^i \text{ or } t^i = T_{n+1}^i.$$

But on $\{T_n \le t\}$,

$$\tilde{I}_i(t^i) \le \tilde{I}(t) \le \tilde{I}(T_n),$$

therefore (b) follows. In order to complete the proof of (c), it remains to establish the equality $\tilde{I}(T_n) = \tilde{I}(t)$. But his follows from (b), because

$$\{T_n \le t < T_{n+1}\} = \bigcup_{j=1}^{d} \left(\{T_n \le t < T_{n+1}\} \cap \left\{t^j < T_{n+1}^j\right\}\right).$$

We now prove (d). The case where $n = 0$ is immediate, since $\tilde{I}_i(0) = I_i(0)$. The case where $n \in \mathbb{N}^*$ follows from (8.16) and (8.15), because on $\{T_n^i < \infty\}$,

$$\tilde{I}_i(T_n^i) = \tilde{I}_i\left(J_i\left(\tilde{I}(T_{n-1})\right)\right) = I_i\left(J_i\left(\tilde{I}(T_{n-1})\right)\right) = I_i(T_n^i).$$

It is clear that (e) is a direct consequence of (d).

Proposition 2 (Controls passing through all the T_n). *Every index control passes through T_n, for all $n \in \mathbb{N}$. In order that a control Γ pass through T_n for all $n \in \mathbb{N}$, it is necessary and sufficient that*

$$\tilde{I}(\Gamma_n) = \tilde{I}_i(\Gamma_n^i) \quad on \quad \{\Gamma_{n+1} = \Gamma_n + e_i\}, \qquad for\ all\ n \in \mathbb{N}\ and\ i = 1, \ldots, d. \tag{8.19}$$

Proof. Let Γ be a control that satisfies (8.19). Suppose that Γ passes through T_n for some $n \in \mathbb{N}$ and set

$$\tau = \begin{cases} \sup\{m \in \mathbb{N}: m > |T_n|, \Gamma_m \leq T_{n+1}\} & \text{on } \{T_n < \bowtie\}, \\ \infty & \text{on } \{T_n = \bowtie\}. \end{cases}$$

We are going to prove that $\Gamma_\tau = T_{n+1}$. Assume that the contrary is true, that is, that $\mathrm{P}\{\Gamma_\tau \neq T_{n+1}\} > 0$. We can then choose $t \in \mathbb{I}$ and i in such a way that the set $F = \{\Gamma_\tau = t, \Gamma_{\tau+1} = t + e_i \nleq T_{n+1}\}$ has positive probability. But on F, on the one hand $\tilde{I}(t) = \tilde{I}_i(t^i)$ by (8.19), and on the other hand $\tilde{I}(t) > \tilde{I}_i(t^i)$ by (c) of Proposition 1. This is a contradiction, which proves that $\Gamma_\tau = T_{n+1}$, therefore that Γ passes through T_{n+1}.

This conclusion remains true under the hypothesis that Γ satisfies (8.12) instead of (8.19). Indeed, the same arguments, together with (d) of Proposition 1, imply that on F,

$$I(t) = I_i(t^i) = I_i(T_{n+1}^i) = \tilde{I}_i(T_{n+1}^i) = \tilde{I}_i(t^i) < \tilde{I}(t) \leq I(t),$$

which is impossible. Thus all index controls pass through all the T_n.

Conversely, suppose that Γ is a control that passes through all the T_n. Then for all $n \in \mathbb{N}$ and $i = 1, \ldots, d$,

$$\{\Gamma_{n+1} = \Gamma_n + e_i\} = \bigcup_{t \in \mathbb{I}} \bigcup_{m \in \mathbb{N}} (\{\Gamma_n = t, \Gamma_{n+1} = t + e_i\} \cap \{T_m \leq t < T_{m+1}\}).$$

But on each set of this double union, $t^i < T_{m+1}^i$ because Γ passes through T_{m+1}; therefore $\tilde{I}(\Gamma_n) = \tilde{I}_i(\Gamma_n^i)$ by (b) of Proposition 1. Condition (8.19) is thus satisfied.

Stages Where Index Controls Change Direction

Assume that each term of the sequence $(T_n, n \in \mathbb{N})$ is finite and differs from the next by only one coordinate. By Proposition 2, there is in this case only one control that passes through every T_n, and this control is an index control. However, if there exists an n such that T_n differs from T_{n+1} by several coordinates, then there are generally several controls that pass through all the T_n and they are not all index controls. We are now going to describe the points where a control changes direction and show that in the case of an index control, the values of the functions $t \mapsto \tilde{I}(t)$ and $t \mapsto I(t)$ coincide at these points.

Returning to the case where no restrictions on the sequence $(T_n, n \in \mathbb{N})$ are assumed, consider a control Γ passing through all the T_n. Set $\tau_n = |T_n|$ for all $n \in \mathbb{N}$ and let $(\sigma_m, m \in \mathbb{N})$ denote the sequence defined inductively by

$$\sigma_0 = 0,$$

$$\sigma_{m+1} = \begin{cases} \sup\{k \in \mathbb{N}: k > \sigma_m, \Gamma_k - \Gamma_{\sigma_m} /\!/ \Gamma_{\sigma_m+1} - \Gamma_{\sigma_m}\} \wedge \tau_{n+1}, & \\ \qquad\qquad \text{on } \{\tau_n \leq \sigma_m < \tau_{n+1}\}, & \\ \infty \qquad\qquad \text{on } \{\sigma_m = \infty\}, & \end{cases} \tag{8.20}$$

where for s and t in $\mathbb{I}$, we write $s /\!/ t$ to indicate that s is a multiple of t. Clearly, σ_m is a stopping time relative to the filtration $\mathscr{F}^\Gamma$ and $\sigma_m \leq \sigma_{m+1}$ for all $m \in \mathbb{N}$. Set

$$S_m = \Gamma_{\sigma_m}, \qquad \text{for all } m \in \mathbb{N}. \tag{8.21}$$

This defines an increasing sequence $(S_m, m \in \mathbb{N})$ of stopping points, which enumerates both the points T_n and the points where Γ changes direction.

Proposition 3 (Property). *If Γ is an index control and $(S_m, m \in \mathbb{N})$ is the associated sequence defined in* (8.21), *then*

$$\tilde{I}(S_m) = I(S_m) \qquad \text{on } \{S_m < \bowtie\}, \qquad \text{for all } m \in \mathbb{N}. \tag{8.22}$$

Proof. It suffices to show that for each $n \in \mathbb{N}$, the equality in (8.22) occurs on $\{T_n = S_k \leq S_m < T_{n+1}\}$ for all couples $(k, m) \in \mathbb{N}^2$ such that $n \leq k \leq m$. We fix n and k and carry out an induction argument on m. For $m = k$, the conclusion is justified by (e) of Proposition 1. Assume that this conclusion is true for $m - 1 \geq k$ and show that it is also true for m. Set

$$F_{i,j} = \{T_n = S_k \leq S_m < T_{n+1}\} \cap \{\Gamma_{\sigma_m} - \Gamma_{\sigma_m - 1} = e_i, \Gamma_{\sigma_m+1} - \Gamma_{\sigma_m} = e_j\}.$$

Because

$$\{T_n = S_k \leq S_m < T_{n+1}\} = \bigcup_{(i,j):\, i \neq j} F_{i,j},$$

it suffices to consider a single set $F_{i,j}$ with $i \neq j$. By (8.20) and (8.21),

$$S^j_{m-1} = S^j_m < T^j_{n+1} \qquad \text{on } F_{i,j}.$$

On the other hand, by hypothesis, since $F_{i,j} \subset \{T_n = S_k \leq S_{m-1} < T_{n+1}\}$,

$$\tilde{I}(S_{m-1}) = I(S_{m-1}) \qquad \text{on } F_{i,j}.$$

By (b) of Proposition 1 and (8.12), we deduce that

$$\begin{aligned}\tilde{I}(T_n) &= \tilde{I}(S_{m-1}) = I(S_{m-1}) \geq I_j(S^j_{m-1})\\ &= I_j(S^j_m) = I(S_m) \geq \tilde{I}(S_m) = \tilde{I}(T_n) \qquad \text{on } F_{i,j},\end{aligned}$$

which proves that $\tilde{I}(S_m) = I(S_m)$ on $F_{i,j}$.

8.3 GITTINS INDICES

Throughout the remainder of the chapter, we will assume that the filtrations $\mathscr{F}^1, \dots, \mathscr{F}^d$ are independent.

For $i = 1, \dots, d$ and all $t^i \in \mathbb{N}$, set

$$\tilde{\mathscr{F}}^i_{t^i} = \mathscr{F}^i_{t^i} \vee \left(\bigvee_{j \neq i} \mathscr{F}^j_\infty \right). \tag{8.23}$$

An immediate consequence of the assumption that the filtrations are independent is that if Y is an $\mathscr{F}^i_\infty$-measurable and integrable random variable, then

$$\mathrm{E}(Y | \tilde{\mathscr{F}}^i_{t^i}) = \mathrm{E}(Y | \mathscr{F}^i_{t^i}), \qquad \text{for all } t^i \in \mathbb{N}. \tag{8.24}$$

In particular,

$$\mathrm{E}(Y | \mathscr{F}_t) = \mathrm{E}(Y | \mathscr{F}^i_{t^i}), \qquad \text{for all } t = (t^1, \dots, t^d) \in \mathbb{I}. \tag{8.25}$$

A second consequence is that all optional increasing paths are predictable, by Proposition 2 of Section 5.4. Indeed, the filtration $(\mathscr{F}_t,\ t \in \mathbb{I})$ defined in (8.10) satisfies CQI because for all $t \in \mathbb{I}$ and all choices of distinct elements i and j of $\{1, \dots, d\}$, $\mathscr{F}_{t+e_i}$ and $\mathscr{F}_{t+e_j}$ are conditionally independent, therefore conditionally qualitatively independent, given $\mathscr{F}_t$. Via (8.24), a third and important consequence of the independence assumption is the first equality in (8.32) below.

For $i = 1, \dots, d$, the filtration $(\tilde{\mathscr{F}}^i_{t^i},\ t^i \in \mathbb{N})$ defined by (8.23) will from now on be denoted $\tilde{\mathscr{F}}^i$. Moreover, for all $n \in \mathbb{N}$, $\mathbf{T}^{n,i}$ and $\tilde{\mathbf{T}}^{n,i}$ will, respectively, denote the set of stopping times τ relative to the filtration $\mathscr{F}^i$ such that $\tau \geq n$ and the set of stopping times τ relative to the filtration $\tilde{\mathscr{F}}^i$ that satisfy the same inequality.

The Simplest Case

In the case where $d = 2$, imagine that Y_1^1 is a constant y_1, Y_1^2 a constant y_2, and that $Y_{t^1}^1 = 0$ if $t^1 > 1$ and $Y_{t^2}^2 = 0$ if $t^2 > 1$. Assume that $y_1 > y_2$. In the player's terminology, the problem reduces to deciding whether to use machine 1 first and then machine 2, or vice versa. In the first case, the reward is

$$r_1 = \alpha y_1 + \alpha^2 y_2 = \alpha(y_1 + \alpha y_2),$$

and in the second case, it is

$$r_2 = \alpha y_2 + \alpha^2 y_1 = \alpha(y_2 + \alpha y_1).$$

Now

$$r_1 - r_2 = \alpha(y_1 - y_2 + \alpha(y_2 - y_1)) = \alpha(1 - \alpha)(y_1 - y_2) > 0;$$

therefore the correct decision is to use machine 1 first, that is, the machine with the higher reward. This conclusion is indeed natural, because it is preferable to apply the higher discounting factor α^2 to the smaller of the rewards.

Equivalent Deterministic Machines

Assume that at time n, the control the player is using prescribes the use of machine i, and assume that the d machines have already been used, respectively, $t^1, \dots, t^d$ times. Machine i will then be used several times in a row, say $\tau - t^i$ times, where τ is a stopping time relative to the filtration $\mathscr{F}^i$ such that $\tau \geq t^i + 1$. At time n, the conditional expected reward from these $\tau - t^i$ plays is

$$\mathrm{E}\left(\sum_{k=t^i+1}^{\tau} \alpha^{n+k-t^i} Y_k^i \middle| \mathscr{F}_t \right) = \mathrm{E}\left(\sum_{k=t^i+1}^{\tau} \alpha^{n+k-t^i} Y_k^i \middle| \mathscr{F}_{t^i}^i \right), \tag{8.26}$$

where $t = (t^1, \dots, t^d)$. It is natural to ask whether there exists (conditionally with respect to $\mathscr{F}_t$) a deterministic machine that yields a fixed reward y_i at each play, such that there is no advantage to using this deterministic machine or machine i during the interval $[t^i + 1, \tau]$. If such a machine exists for each τ, then it will be advantageous for the player to choose τ so that y_i is maximal. By substituting y_i for Y_k^i in (8.26), we see that the conditional expected reward from the deterministic machine is

$$y_i \,\mathrm{E}\left(\sum_{k=t^i+1}^{\tau} \alpha^{n+k-t^i} \middle| \mathscr{F}^i_{t^i} \right).$$

The player will therefore prefer the deterministic machine if and only if $y_i \geq Q_i(t^i, \tau)$, where

$$Q_i(t^i, \tau) = \frac{\mathrm{E}\left(\sum_{k=t^i+1}^{\tau} \alpha^k Y^i_k | \mathscr{F}^i_{t^i} \right)}{\mathrm{E}\left(\sum_{k=t^i+1}^{\tau} \alpha^k | \mathscr{F}^i_{t^i} \right)}. \tag{8.27}$$

The ratio $Q_i(t^i, \tau)$ should therefore be maximized.

Gittins Indices

For $i = 1, \ldots, d$ and all $t^i \in \mathbb{N}$, set

$$I_i(t^i) = \operatorname*{esssup}_{\tau \in \mathbf{T}^{t^i+1, i}} Q_i(t^i, \tau). \tag{8.28}$$

It is clear that $I_i(t^i)$ is an $\mathscr{F}^i_{t^i}$-measurable random variable which is integrable, by (8.2). From now on, the symbols $I_1(t^1), \ldots, I_d(t^d)$ will denote exclusively the random variables defined in (8.28). These random variables are termed *Gittins indices*.

The simplest case discussed above suggests that at each stage, an optimal control will prescribe one of the directions e_i where the Gittins index is maximal. This conjecture will be confirmed by the theorem of Section 8.4. Our goal for the time being is to prove a preliminary result that establishes in particular the existence of a stopping time that achieves the essential supremum in (8.28).

Lemma. *Fix i, and for all $m \in \mathbb{N}$, let $(Y(m, n),\ n \in \overline{\mathbb{N}})$ denote the process defined by*

$$Y(m, n) = \begin{cases} 0, & \text{if } n \leq m, \\ \displaystyle\sum_{k=m+1}^{n} a^k \left(Y^i_k - I_i(m) \right), & \text{if } n > m. \end{cases} \tag{8.29}$$

If τ belongs to $\mathbf{T}^{m+1,i}$ then

(a) $E(Y(m,\tau)|\mathscr{F}^i_m) \leq 0$;
(b) $E(Y(m,\tau)|\mathscr{F}^i_m) = 0$ *if and only if* $I_i(m) = Q_i(m,\tau)$;
(c) $\operatorname{esssup}_{\tau \in \mathbf{T}^{m+1,i}} \mathrm{E}(Y(m,\tau)|\mathscr{F}^i_m) = 0$.

Proof. By (8.29) and (8.27),

$$\begin{aligned} \mathrm{E}\big(Y(m,\tau)|\mathscr{F}^i_m\big) &= \mathrm{E}\left(\sum_{k=m+1}^{\tau} \alpha^k Y^i_k \middle| \mathscr{F}^i_m \right) - I_i(m)\,\mathrm{E}\left(\sum_{k=m+1}^{\tau} \alpha^k \middle| \mathscr{F}^i_m \right) \\ &= \big(Q_i(m,\tau) - I_i(m)\big)\,\mathrm{E}\left(\sum_{k=m+1}^{\tau} \alpha^k \middle| \mathscr{F}^i_m \right). \end{aligned} \tag{8.30}$$

This implies (b), as well as (a), because $Q_i(m,\tau) \leq I_i(m)$. Moreover, it also follows that

$$\mathrm{E}\big(Y(m,\tau)|\mathscr{F}^i_m\big) \geq \big(Q_i(m,\tau) - I_i(m)\big) \sum_{k=m+1}^{\infty} \alpha^k$$

for all $\tau \in \mathbf{T}^{m+1,i}$. Because the essential supremum over $\tau \in \mathbf{T}^{m+1,i}$ of the last right-hand side is 0, it follows that $\operatorname{esssup}_{\tau \in \mathbf{T}^{m+1,i}} \mathrm{E}(Y(m,\tau)|\mathscr{F}^i_m) \geq 0$. As the converse inequality is a direct consequence of (a), (c) follows.

Theorem (Definition and properties of $\tau_i(t^i)$). *For $i = 1,\dots,d$ and all $t^i \in \mathbb{N}$, let*

$$\tau_i(t^i) = \inf\{n \in \mathbb{N}: n > t^i,\ I_i(n) < I_i(t^i)\}. \tag{8.31}$$

Then $\tau_i(t^i)$ belongs to $\mathbf{T}^{t^i+1,i}$ and

$$\mathrm{E}\big(Y(t^i,\tau_i(t^i))|\mathscr{F}^i_{t^i}\big) = \operatorname*{esssup}_{\tau \in \tilde{\mathbf{T}}^{t^i+1,i}} \mathrm{E}\big(Y(t^i,\tau)|\tilde{\mathscr{F}}^i_{t^i}\big) = 0. \tag{8.32}$$

In particular,

$$I_i(t^i) = Q_i\big(t^i,\tau_i(t^i)\big). \tag{8.33}$$

Proof. Fix i, and for all $t^i \in \mathbb{N}$, consider the process $(Y(t^i,n),\ n \in \overline{\mathbb{N}})$ defined in (8.29). This process is adapted to the filtration $\mathscr{F}^i$ and satisfies

$$E\left(\sup_{n \in \overline{\mathbb{N}}} \big|Y(t^i,n)\big| \right) \leq \sum_{k=1}^{\infty} \alpha^k\, \mathrm{E}\big(|Y^i_k|\big) + \mathrm{E}\big(|I_i(t^i)|\big) \sum_{k=1}^{\infty} \alpha^k < \infty.$$

Moreover, $\lim_{n \to \infty} Y(t^i, n) = Y(t^i, \infty)$; therefore the conditions under which the results of Section 3.3 apply are satisfied. Let Z be Snell's envelope of $(Y(t^i, n),\ n \in \overline{\mathbb{N}})$ and $\tilde{Z}$ Snell's envelope of the same process, but taken relative to the filtration $\tilde{\mathscr{F}}^i$. We are going to show that these two envelopes are identical. In order to do this, we refer to Proposition 4 of Section 3.2. By this proposition, it suffices to check that $Z^{(m)} = \tilde{Z}^{(m)}$ for all $m \in \mathbb{N}$. This equality is trivially true for $m = 0$. On the other hand, by assuming that it is true for some $m \in \mathbb{N}$, we see, thanks to this proposition and to (8.24), that for all $n \in \mathbb{N}$,

$$\begin{aligned}
\tilde{Z}_n^{(m+1)} &= \sup\left(Y(t^i, n), \mathrm{E}\left(Y(t^i, \infty) | \tilde{\mathscr{F}}_n^i\right), \mathrm{E}\left(\tilde{Z}_{n+1}^m \middle| \tilde{\mathscr{F}}_n^i\right)\right) \\
&= \sup\left(Y(t^i, n), \mathrm{E}\left(Y(t^i, \infty) | \mathscr{F}_\infty^i\right), \mathrm{E}\left(Z_{n+1}^m \middle| \mathscr{F}_n^i\right)\right) \\
&= Z_n^{m+1}.
\end{aligned}$$

We have thus proved that $Z = \tilde{Z}$. We now apply the corollary of Section 3.3, together with relation (3.12) (according to which $Z_n = \sup(Y(t^i, n), \mathrm{E}(Z_{n+1} | \mathscr{F}_n^i))$ for all $n \in \mathbb{N}$), to deduce that

$$\rho(t^i) = \inf\left\{n \in \mathbb{N}: n > t^i, Y(t^i, n) - \mathrm{E}\left(Z_{n+1} \middle| \mathscr{F}_n^i\right) > 0\right\} \tag{8.34}$$

is a stopping time relative to the filtration $\mathscr{F}^i$ that is optimal in $\tilde{\mathbf{T}}^{t^i+1,\, i}$ (for the process $(Y(t^i, n), n \in \overline{\mathbb{N}})$). Using (8.24) and Proposition 1 of Section 3.3, we conclude that

$$\begin{aligned}
\mathrm{E}\left(Y(t^i, \rho(t^i)) \middle| \mathscr{F}_{t^i}^i\right) &= \mathrm{E}\left(Y(t^i, \rho(t^i)) \middle| \tilde{\mathscr{F}}_{t^i}^i\right) \\
&= \mathrm{E}\left(\tilde{Z}_{\rho(t^i)} \middle| \tilde{\mathscr{F}}_{t^i}^i\right) \\
&= \tilde{Z}_{t^i} \\
&= Z_{t^i} \\
&= \operatorname*{esssup}_{\tau \in \mathbf{T}^{t^i+1, i}} \mathrm{E}\left(Y(t^i, \tau) \middle| \mathscr{F}_{t^i}^i\right).
\end{aligned} \tag{8.35}$$

But the right-hand side of the last equality is 0, by (c) of the lemma; therefore the proof will be complete once we have shown that $\rho(t^i) = \tau_i(t^i)$. By the definition of Snell's envelope and the theorem of Section 1.2, for all $n \in \mathbb{N}$ such that $n > t^i$,

$$\begin{aligned}
\mathrm{E}\left(Z_{n+1} \middle| \mathscr{F}_n^i\right) &= \operatorname*{esssup}_{\tau \in \mathbf{T}^{n+1, i}} \mathrm{E}\left(Y(t^i, \tau) \middle| \mathscr{F}_n^i\right) \\
&= Y(t^i, n) + \operatorname*{esssup}_{\tau \in \mathbf{T}^{n+1, i}} \mathrm{E}\left(Y(t^i, \tau) - Y(t^i, n) \middle| \mathscr{F}_n^i\right);
\end{aligned}$$

therefore, by (8.29),

$$Y(t^i, n) - \mathrm{E}\left(Z_{n+1} \middle| \mathscr{F}_n^i\right) \tag{8.36}$$

$$= - \operatorname*{esssup}_{\tau \in \mathbf{T}^{n+1,i}} \mathrm{E}\left(Y(n,\tau) \middle| \mathscr{F}_n^i\right) + \left(I_i(n) - I_i(t^i)\right) \mathrm{E}\left(\sum_{k=n+1}^{\tau} \alpha^k \middle| \mathscr{F}_n^i \right).$$

Let $n \in \mathbb{N}$ be such that $n > t^i$ and $\mathrm{P}\{\rho(t^i) = n\} > 0$. By (8.34) and (8.36),

$$\mathrm{E}\left(Y(n,\tau) \middle| \mathscr{F}_n^i\right) + \left(I_i(n) - I_i(t^i)\right) \mathrm{E}\left(\sum_{k=n+1}^{\tau} \alpha^k \middle| \mathscr{F}_n^i \right) < 0 \qquad \text{on } \{\rho(t^i) = n\}$$

for all $\tau \in \mathbf{T}^{n+1,i}$, therefore in particular for $\tau = \rho(n)$. But $\mathrm{E}(Y(n, \rho(n))|\mathscr{F}_n^i) = 0$ by (8.35); therefore $I_i(n) < I_i(t^i)$ on $\{\rho(t^i) = n\}$ and consequently $\tau_i(t^i) \le n$ on this set. On the other hand, let $n \in \overline{\mathbb{N}}$ and $m \in \mathbb{N}$ be such that $t^i < m < n$. Again by (8.34) and (8.36),

$$\operatorname*{esssup}_{\tau \in \mathbf{T}^{m+1,i}} \left(\mathrm{E}\left(Y(m,\tau) \middle| \mathscr{F}_m^i\right)\right) + \left(I_i(m) - I_i(t^i)\right) \mathrm{E}\left(\sum_{k=m+1}^{\tau} \alpha^k \middle| \mathscr{F}_m^i \right) \ge 0$$

$$\text{on } \{\rho(t^i) = n\};$$

therefore $I_i(m) \ge I_i(t^i)$ on this set, because $\mathrm{E}(Y(m,\tau)|\mathscr{F}_m^i) \le 0$ for all $\tau \in \mathbf{T}^{m+1,i}$ by (a) of the lemma. It follows that $\tau_i(t^i) \ge n$ on $\{\rho(t^i) = n\}$ and finally that $\rho(t^i) = \tau_i(t^i)$, which was to be proved.

8.4 CHARACTERIZING OPTIMAL CONTROLS

We shall now combine the developments of the two previous sections to establish the main result of this chapter, namely, that all optimal controls are index controls and conversely. This result is doubly useful, because on the one hand, it sheds new light on the structure of optimal controls, and on the other hand, it lets us take advantage of the information we have acquired concerning index controls to refine the procedures available for constructing optimal controls.

Recall that $I_1(t^1), \ldots, I_d(t^d)$ denote the Gittins indices and that index controls are defined using these indices. In order to avoid repetitions, we agree once and for all that T_n and T_n^i are the symbols defined in (8.16).

We begin by establishing four lemmas.

Lemma 1. *Let Γ be an index control and let $(S_m, m \in \mathbb{N})$ be the associated sequence defined in* (8.21). *Then for all $n \in \mathbb{N}$, all $t = (t^1, \ldots, t^d) \in \mathbb{I}$, and $i = 1, \ldots, d$,*

$$T^i_{n+1} = \tau_i(t^i) \qquad \text{on } \{T_n \le S_m < T_{n+1}\} \cap \{S_m = t,\, t^i < S^i_{m+1}\}. \quad (8.37)$$

In particular,

$$T^i_{n+1} = \tau_i(t^i) \qquad \text{on } \{T_n = t,\, t^i < T^i_{n+1}\}. \quad (8.38)$$

Proof. On the set indicated in (8.37), $\tilde{I}(T_n) = \tilde{I}(t) = I(t) = I_i(t^i)$ by (b) of Proposition 1 of Section 8.2, (8.22), and (8.12). By (8.16), we see therefore that on the same set, $T^i_{n+1} = J_i(I_i(t^i))$. Taking into account the fact that replacing $\tilde{I}_i(n)$ by $I_i(n)$ in (8.14) does not change the value of $J_i(x)$, (8.37) follows by comparing (8.14) and (8.31). As for (8.38), it can be deduced from (8.37) or established directly with the same argument, using (b) and (d) of Proposition 1 of Section 8.2 to justify the equality $\tilde{I}(T_n) = I_i(t^i)$.

Lemma 2. *Let Γ and $(S_m, m \in \mathbb{N})$ be as in Lemma* 1. *Then for all $m \in \mathbb{N}$ and $i = 1, \ldots, d$,*

$$\mathrm{E}\left(\sum_{k=S^i_m+1}^{S^i_{m+1}} \alpha^k Y^i_k \,\middle|\, \tilde{\mathscr{F}}^i_{S^i_m} \right) = I_i(S^i_m)\, \mathrm{E}\left(\sum_{k=S^i_m+1}^{S^i_{m+1}} \alpha^k \,\middle|\, \tilde{\mathscr{F}}^i_{S^i_m} \right) \quad (8.39)$$
$$\text{on } \{S_m < \bowtie,\, S^i_m < S^i_{m+1}\},$$

where $S^i_{m+1} = \infty$ if $S_{m+1} = \bowtie$ by convention.

Proof. It suffices to prove that for all $t = (t^1, \ldots, t^d) \in \mathbb{I}$, equality (8.39) is true on $\{S_m = t,\, t^i < S^i_{m+1}\}$. Let F denote this set and observe that it can be written $\{S_m = t,\, \Gamma_{\sigma_m+1} = t + e_i\}$. This shows in particular that it belongs to $\mathscr{F}_t$, therefore to $\tilde{\mathscr{F}}^i_{t^i}$. By (8.29), it is therefore sufficient to prove that $\mathrm{E}(Y(t^i, S^i_{m+1}) | \tilde{\mathscr{F}}^i_{t^i}) = 0$ on F. Let Z be Snell's envelope of $(Y(t^i, n), n \in \overline{\mathbb{N}})$. By (8.37), $S^i_{m+1} \le \tau_i(t^i)$ on F. On the other hand, $\tau_i(t^i)$ is optimal in $\tilde{\mathbf{T}}^{i+1,i}$ by (8.32). Consequently,

$$\mathrm{E}\left(Z_{S^i_{m+1}} \middle| \tilde{\mathscr{F}}^i_{t^i}\right) = \mathrm{E}\left(Z_{t^i+1} \middle| \tilde{\mathscr{F}}^i_{t^i}\right) \qquad \text{on } F.$$

But again by (8.32), $\mathrm{E}(Z_{t^i+1} | \tilde{\mathscr{F}}^i_{t^i}) = 0$; therefore $\mathrm{E}(Z_{S^i_{m+1}} | \tilde{\mathscr{F}}^i_{t^i}) = 0$ on F. Because $Z_n \ge Y(t^i, n)$ for all $n \in \overline{\mathbb{N}}$, it remains to prove that $Z_{S^i_{m+1}} \le Y(t^i, S^i_{m+1})$ on F or, equivalently, that

$$\mathrm{E}\left(Y(t^i, \tau) - Y(t^i, S^i_{m+1}) \middle| \tilde{\mathscr{F}}^i_{S^i_{m+1}}\right) \le 0 \qquad \text{on } F$$

for all $\tau \in \tilde{\mathbf{T}}^{S^i_{m+1}, i}$. Take such a τ and observe that this inequality is trivially satisfied on $\{\tau = S^i_{m+1}\}$. On the other hand, on $\{\tau > S^i_{m+1}\}$, its left-hand side is equal to

$$\mathrm{E}\left(Y(S^i_{m+1}, \tau)\middle|\tilde{\mathscr{F}}^i_{S^i_{m+1}}\right) + \left(I_i(S^i_{m+1}) - I_i(t^i)\right)\mathrm{E}\left(\sum_{k=S^i_{m+1}+1}^{\tau} \alpha^k \middle| \tilde{\mathscr{F}}^i_{S^i_{m+1}}\right).$$

By (8.32), $\mathrm{E}(Y(n, \tau)|\tilde{\mathscr{F}}^i_n) \le 0$ on $\{\tau > n\} \cap \{S^i_{m+1} = n\}$ for all $n \in \overline{\mathbb{N}}$; therefore $\mathrm{E}(Y(S^i_{m+1}, \tau)|\tilde{\mathscr{F}}^i_{S^i_{m+1}}) \le 0$ on $\{\tau > S^i_{m+1}\}$. In order to complete the proof, it suffices to prove that $I_i(S^i_{m+1}) \le I_i(t^i)$ on $F \cap \{S_{m+1} < \bowtie\}$. By (8.11), (8.22), and (8.12), we see that on $F \cap \{S_{m+1} < \bowtie\}$,

$$I_i(S^i_{m+1}) \le I(S_{m+1}) = \tilde{I}(S_{m+1}) \le \tilde{I}(S_m) = I(S_m) = I_i(S^i_m) = I_i(t^i),$$

which was to be proved.

Lemma 3. *If Γ and Δ are index controls, then* $\mathrm{E}(X_\Gamma) = \mathrm{E}(X_\Delta)$; *in other words, all index controls yield the same expected reward.*

Proof. Let Γ and Δ be two index controls. By Proposition 2 of Section 8.2, Γ and Δ pass through all the T_n; therefore $\{T_n \le \Gamma_k < T_{n+1}\} = \{T_n \le \Delta_k < T_{n+1}\}$ for any $k \in \mathbb{N}$ and $n \in \mathbb{N}$. By (b) and (c) of Proposition 1 of Section 8.2, it follows that $\tilde{I}(\Gamma_k) = \tilde{I}(\Delta_k)$ for all $k \in \mathbb{N}$. It suffices therefore to show that for all index controls Γ,

$$\mathrm{E}(X_\Gamma) = \mathrm{E}\left(\sum_{k=1}^{\infty} \alpha^k \tilde{I}(\Gamma_{k-1})\right). \tag{8.40}$$

Let Γ be such a control and let $(S_m, m \in \mathbb{N})$ be the associated sequence defined in (8.21). By (8.1) and the definition of S_m,

$$\begin{aligned}
\mathrm{E}(X_\Gamma) &= \sum_{n=1}^{\infty} \alpha^n R(\Gamma, n) \\
&= \sum_{m=0}^{\infty} \mathrm{E}\left(\sum_{k=\sigma_m+1}^{\sigma_{m+1}} \alpha^k R(\Gamma, k); S_m < \bowtie\right) \\
&= \sum_{i=1}^{d} \sum_{m=0}^{\infty} \mathrm{E}\left(\sum_{k=\sigma_m+1}^{\sigma_{m+1}} \alpha^k Y^i_{\Gamma^i_k}; S_m < \bowtie, S^i_m < S^i_{m+1}\right) \\
&= \sum_{i=1}^{d} \sum_{m=0}^{\infty} \mathrm{E}\left(\alpha^{\sigma_m - S^i_m} \sum_{k=S^i_m+1}^{S^i_{m+1}} \alpha^k Y^i_k; S_m < \bowtie, S^i_m < S^i_{m+1}\right),
\end{aligned}$$

where we have made the same convention concerning S^i_{m+1} as in Lemma 2. By (8.39), the last right-hand side is equal to

$$\sum_{i=1}^{d}\sum_{m=0}^{\infty} \mathrm{E}\left(\alpha^{\sigma_m - S^i_m} I_i(S^i_m) \sum_{k=S^i_m+1}^{S^i_{m+1}} \alpha^k;\, S_m < \bowtie,\, S^i_m < S^i_{m+1}\right).$$

On the other hand, by (8.12) and (8.22), $I_i(S^i_m) = I(S_m) = \tilde{I}(S_m)$ on $\{S_m < \bowtie,\, S^i_m < S^i_{m+1}\}$. Moreover, by (b) and (c) of Proposition 1 of Section 8.2, $\tilde{I}(S_m) = \tilde{I}(\Gamma_k)$ on $\{S_m < \bowtie\}$ for all k such that $\sigma_m \le k < \sigma_{m+1}$. We can now conclude that

$$\begin{aligned}
\mathrm{E}(X_\Gamma) &= \sum_{m=0}^{\infty} \mathrm{E}\left(\tilde{I}(S_m) \sum_{k=\sigma_m+1}^{\sigma_{m+1}} \alpha^k;\, S_m < \bowtie\right) \\
&= \mathrm{E}\left(\sum_{m=0}^{\infty} \sum_{k=\sigma_m+1}^{\sigma_{m+1}} \alpha^k \tilde{I}(\Gamma_{k-1});\, S_m < \bowtie\right) \\
&= \mathrm{E}\left(\sum_{k=1}^{\infty} \alpha^k \tilde{I}(\Gamma_{k-1})\right),
\end{aligned}$$

which establishes (8.40).

Lemma 4. *For fixed i and $t^i \in \mathbb{N}$, let $(V_k,\, k \in \mathbb{N}_{t^i+1})$ be a process such that V_k is $\tilde{\mathscr{F}}^i_{k-1}$-measurable and $0 \le V_{k+1} \le V_k \le 1$ for all k. Then*

$$\mathrm{E}\left(\sum_{k=t^i+1}^{\infty} \alpha^k V_k\big(Y^i_k - I_i(t^i)\big) \,\Big|\, \tilde{\mathscr{F}}^i_{t^i}\right) \le 0. \tag{8.41}$$

Proof. Define $Y(t^i, n)$ by (8.29), take an $m \in \mathbb{N}$ such that $m > t^i$ and assume for the time being that $V_{m+1} = 0$. Then

$$\begin{aligned}
\sum_{n=t^i+1}^{m} (V_n - V_{n+1}) Y(t^i, n) &= \sum_{n=t^i+1}^{m} (V_n - V_{n+1}) \sum_{k=t^i+1}^{n} \alpha^k\big(Y^i_k - I_i(t^i)\big) \\
&= \sum_{k=t^i+1}^{m} \alpha^k\big(Y^i_k - I_i(t^i)\big) \sum_{n=k}^{m} (V_n - V_{n+1}) \\
&= \sum_{k=t^i+1}^{m} \alpha^k V_k\big(Y^i_k - I_i(t^i)\big).
\end{aligned}$$

It is therefore sufficient to prove that $\mathrm{E}(\sum_{n=t^i+1}^{m} (V_n - V_{n+1}) Y(t^i, n) | \tilde{\mathscr{F}}^i_{t^i}) \le 0$.

For all $x \in [0, 1]$, set

$$\tau(x) = \inf\{n \in \mathbb{N}: t^i < n \le m, V_{t^i+1} - V_{n+1} \ge x\}.$$

Then $\tau(x)$ belongs to $\tilde{\mathbf{T}}^{t^i+1,i}$ and

$$\sum_{n=t^i+1}^{m} (V_n - V_{n+1}) Y(t^i, n) = \int_0^{V_{t^i+1}} Y(t^i, \tau(x))\, dx.$$

Consequently, since V_{t^i+1} is $\tilde{\mathscr{F}}^i_{t^i}$-measurable,

$$\mathrm{E}\left(\sum_{n=t^i+1}^{m} (V_n - V_{n+1}) Y(t^i, n) \middle| \tilde{\mathscr{F}}^i_{t^i} \right) = \int_0^{V_{t^i+1}} \mathrm{E}\left(Y(t^i, \tau(s)) \middle| \tilde{\mathscr{F}}^i_{t^i} \right) dx.$$

But $\mathrm{E}(Y(t^i, \tau(x)) | \tilde{\mathscr{F}}^i_{t^i}) \le 0$ by (8.32); therefore $\mathrm{E}(\sum_{k=t^i+1}^{m} \alpha^k V_k (Y^i_k - I_i(t^i)) | \tilde{\mathscr{F}}^i_{t^i}) \le 0$. It now remains to let m tend to infinity and to use the dominated convergence theorem.

Theorem (Optimality of index controls). *If the filtrations $\mathscr{F}^1, \ldots, \mathscr{F}^d$ are independent, then the set of optimal controls and the set of index controls are equal.*

Proof. We are going to show that all optimal controls are index controls. This will establish the theorem, because the set of optimal controls is not empty by what was proved in the last paragraph of Section 8.1, and because all index controls yield the same expected reward by Lemma 3. Let Γ be a control that is not an index control. Then there exists $t \in \mathbb{I}$ and two distinct indices i and j such that the set

$$F = \{\Gamma_{|t|} = t, \Gamma_{|t|+1} = t + e_j, I(t) = I_i(t^i) > I_j(t^j)\}$$

has positive probability. Without loss of generality, we will assume that $i = 1$ and $j = 2$. Our goal is to construct a control Δ such that $\mathrm{E}(X_\Delta) > \mathrm{E}(X_\Gamma)$, which will establish the theorem. Informally, the control Δ can be described as follows. Let $\tau_1(t^1)$ be the stopping time defined in (8.31). On F^c, Δ coincides with Γ, and on F, Δ coincides with Γ up to time $|t|$, then goes in direction e_1 exactly $\tau_1(t^1) - t^1$ times, and finally joins up with Γ at time $\tau = \inf\{n: \Gamma^1_n \ge \tau_1(t^1)\}$ after duplicating the increments of Γ between times $|t|$ and τ, which are parallel to any one of the directions $e_2, \ldots, e_d$. To be more specific, set

$$\tau = \inf\{n: \Gamma^1_n \ge \tau_1(t^1)\},$$

and for all $m \in \mathbb{N}$, let

$$\nu(m) = \inf\left\{n \in \mathbb{N}: n \geq |t|, \sum_{i=2}^{d} \Gamma_n^i \geq m + \sum_{i=2}^{d} t^i\right\}$$
$$= \inf\{n \in \mathbb{N}: n \geq |t|, n - \Gamma_n^1 \geq m + |t| - t^1\}.$$

Notice that $\nu(m) - |t|$ is precisely the number of steps, starting from time $|t|$, that are necessary for Γ to make m steps in the directions $e_2, \ldots, e_d$. Also notice that τ and $\nu(m)$ are stopping times relative to the filtration $\mathscr{F}^\Gamma$, because $\tau_1(t^1)$ is a stopping time relative to the filtration $\mathscr{F}^1$, and if $n \in \mathbb{N}^*$, $s \in \mathbb{I}$, and $k = s^1$ or $s^1 - 1$, then $\{\Gamma_{n-1}^1 = k\} \cap \{\Gamma_n = s\} \in \mathscr{F}_s$. For all $n \in \mathbb{N}$, let $\tilde{\Gamma}_n$ denote the random variable with values in $\mathbb{I}$ defined by $\tilde{\Gamma}_n = \Gamma_n$ if $n < |t|$ and by the following formula if $n \geq |t|$:

$$\tilde{\Gamma}_n = \begin{cases} t + (n - |t|)e_1 & \text{on } \{n - |t| + t^1 < \tau_1(t^1)\}, \\ \left(\tilde{\Gamma}_n^1, \ldots, \tilde{\Gamma}_n^d\right) & \text{on } \{n - |t| + t^1 \geq \tau_1(t^1), n < \tau\}, \\ \quad \text{where } \tilde{\Gamma}_n^1 = \tau_1(t^1) \text{ and } \tilde{\Gamma}_n^i = \Gamma_{\nu(n-|t|+t^1-\tau_1(t^1))}^i & \text{if } 2 \leq i \leq d, \\ \Gamma_n & \text{on } \{n \geq \tau\}. \end{cases}$$

Note that on $\{\tau_1(t^1) = \infty\}$, the definition of $\tilde{\Gamma}_n$ for $n \geq |t|$ reduces to the first of the three cases. For all $n \in \mathbb{N}$, set

$$\Delta_n = \begin{cases} \tilde{\Gamma}_n & \text{on } F, \\ \Gamma_n & \text{on } F^c. \end{cases}$$

We are going to prove that $(\Delta_n, n \in \mathbb{N})$ is a control. Let us begin by proving that $(\Delta_n(\omega), n \in \mathbb{N})$ is a deterministic increasing path for all fixed ω (possibly taken outside of a null set). This is obvious if $\omega \in F^c$, so assume that $\omega \in F$. It is clear that $\Delta_0(\omega) = 0$ and that $\Delta_{n+1}(\omega) \in \mathbb{D}_{\Delta_n(\omega)}$ if $0 \leq n < |t|$ or if $\tau(\omega) < \infty$ and $n \geq \tau(\omega)$. Moreover, this conclusion is also true if $n \geq |t|$ and $n - |t| + t^1 < \tau_1(t^1)(\omega)$, because in this case $\Delta_n(\omega) = t + (n - |t|)e_1$. It remains therefore to examine the case where $n \geq |t|$, $n - |t| + t^1 \geq \tau_1(t^1)(\omega)$ and $n < \tau(\omega)$. To this end, observe that the function $m \mapsto \nu(m)$ is nondecreasing, which implies that the function $n \mapsto \Delta_n(\omega) = \tilde{\Gamma}_n(\omega)$ is also nondecreasing. On the other hand,

$$\begin{aligned} |\Delta_n(\omega)| &= \left|\tilde{\Gamma}_n(\omega)\right| \\ &= \tau_1(t^1)(\omega) + \sum_{i=2}^{d} \Gamma_{\nu(n-|t|+t^1-\tau_1(t^1))}^i(\omega) \\ &= \tau_1(t^1)(\omega) + n - \tau_1(t^1)(\omega) \\ &= n. \end{aligned}$$

Taking into account the fact that $\Delta_{\tau(\omega)-1} \le \Delta_{\tau(\omega)}$ if $\tau(\omega) < \infty$, we conclude that $\Delta_{n+1}(\omega) \in \mathbb{D}_{\Delta_n(\omega)}$ for all n that satisfy the indicated inequalities. Thus $(\Delta_n(\omega), n \in \mathbb{N})$ is a deterministic increasing path. We now show that Δ_n is a stopping point for all $n \in \mathbb{N}$. If $n \le |t|$, this follows from the fact that $\Delta_n = \Gamma_n$. We therefore assume that $n > |t|$. Because $F \in \mathscr{F}_t$, $\Gamma_n \ge t$ and $\tilde{\Gamma}_n \ge t$, it is sufficient by Property 5 of Section 1.3 to check that $\tilde{\Gamma}_n$ is a stopping point. Take $s \ge t$ and notice that $\{\tilde{\Gamma}_n = s\} = F_1 \cup F_2 \cup F_3$, where

$$F_1 = \{\Gamma_{|t|} + (n - |t|)e_1 = s\} \cap \{n - |t| + t^1 < \tau_1(t^1)\}$$
$$= \{t + (n - |t|)e_1 = s\} \cap \{s^1 < \tau_1(t^1)\},$$

$$F_2 = \{\tilde{\Gamma}_n = s\} \cap \{n - |t| + t^1 \ge \tau_1(t^1)\} \cap \{n < \tau\},$$
$$F_3 = \{\Gamma_n = s\} \cap \{n \ge \tau\}.$$

It is clear that $F_1 \in \mathscr{F}_s$, since $\{s^1 < \tau_1(t^1)\} \in \mathscr{F}^1_{s^1} \subset \mathscr{F}_s$. Moreover, $F_3 \in \mathscr{F}_s$ since τ is a stopping time relative to the filtration $\mathscr{F}^\Gamma$. As for F_2, the only elements s that need to be considered satisfy the equality $n = |s|$, because $n = |\tilde{\Gamma}_n| = |s|$ on F_2, and the inequality $n - |t| + t^1 \ge s^1$, because $\tilde{\Gamma}_n^1 = \tau_1(t^1) = s^1$ on F_2. For these s, we can write

$$F_2 = \bigcup_{r=t^1}^{s^1-1} \Bigg(\{\tau_1(t^1) = s^1\} \cap \{\Gamma_{r-s^1+|s|} = (r, s^2, \dots, s^d)\}$$
$$\cap \{\nu(n - |t| + t^1 - s^1) = r - s^1 + |s|\} \cap \bigcup_{u=r}^{s^1-1} \{\Gamma^1_{u-s^1+|s|+1} = \Gamma^1_{u-s^1+|s|}\}\Bigg).$$

Indeed, for each $r \in [t^1, s^1[$, $\Gamma^1_{r-s^1+n} = r$ on the set in parentheses, and Γ moves at least once in a direction other than e_1 between times $r - s^1 + n$ and $n - 1$, therefore at least $s^1 - r + 1$ steps are needed after time $r - s^1 + n$ before Γ^1_m can be $\ge \tau_1(t^1) = s^1$. Equivalently, $\Gamma^1_m < s^1$ if $m < r - s^1 + n + (s^1 - r + 1) = n + 1$, which means that $n < \tau$ on the set in parentheses. Now $\nu(n - |t| + t^1 - s^1)$ is a stopping time relative to the filtration $\mathscr{F}^\Gamma$; therefore $\{\nu(n - |t| + t^1 - s^1) = r - s^1 + |s|\} \in \mathscr{F}^\Gamma_{r-s^1+|s|}$. Taking into account the fact that $\{\tau_1(t^1) = s^1\} \in \mathscr{F}^1_{s^1} \subset \mathscr{F}_s$, it follows that the set on the right-hand side above belongs to $\mathscr{F}_s$. By Proposition 2 of Section 5.4, we have thus established that Δ is a control.

It remains to prove that $\mathrm{E}(X_\Delta) > \mathrm{E}(X_\Gamma)$. Set $\Gamma^i_\tau = \infty$ on $\{\tau = \infty\}$. By (8.5),

$$X_\Gamma - X_\Delta = 1_F \sum_{i=1}^{d} \sum_{k=t^i+1}^{\Gamma^i_\tau} (\alpha^{\lambda^i_k(\Gamma)} - \alpha^{\lambda^i_k(\Delta)}) Y^i_k \, 1_{\{\Gamma^i_\tau > t^i\}}.$$

If γ is a deterministic increasing path passing through t, it is easy to check that for all $n \in \mathbb{N}$ such that $n > |t|$,

$$\sum_{i:\, \gamma_n^i > t^i} \sum_{k=t^i+1}^{\gamma_n^i} \alpha^{\lambda_k^i(\gamma)} = \sum_{i:\, \gamma_n^i > t^i} \sum_{k=|t|+1}^{n} \alpha^k \left(\gamma_k^i - \gamma_{k-1}^i\right) = \sum_{k=|t|+1}^{n} \alpha^k,$$

which shows that the left-hand side does not depend on γ. This conclusion also holds for $n = \infty$ if we set $\gamma_\infty^i = \infty$ (because $\lambda_k^i(\gamma) = \infty$ if $k > \sup_n \gamma_n^i$). Consequently,

$$X_\Gamma - X_\Delta = 1_F \sum_{i=1}^{d} \sum_{k=t^i+1}^{\Gamma_\tau^i} \left(\alpha^{\lambda_k^i(\Gamma)} - \alpha^{\lambda_k^i(\Delta)}\right)\left(Y_k^i - I_i(t^i)\right) 1_{\{\Gamma_\tau^i > t^i\}}.$$

Observe now that on F,

$$\Gamma_\tau^1 > t^1, \qquad \Gamma_\tau^2 > t^2,$$

$$\lambda_k^i(\Delta) \geq \lambda_k^i(\Gamma), \qquad \text{for all } i > 1 \text{ and all } k \in \mathbb{N} \text{ such that } k > t^i,$$

$$I_1(t^1) > I_2(t^2) \text{ and } I_1(t^1) \geq I_i(t^i), \qquad \text{for all } i > 2.$$

Observe moreover that the bound Γ_τ^1 in the second summation can be replaced by $\tau_1(t^1)$, because $\lambda_k^1(\Gamma) = \lambda_k^1(\Delta) = \infty$ on $\{k > \tau_1(t^1),\ \tau = \infty\} \cap F$, and also that $\lambda_k^1(\Delta) = k + |t| - t^1$ on $\{k \leq \tau_1(t^1)\} \cap F$ if $k > t^1$. Therefore on F,

$$\begin{aligned} X_\Gamma - X_\Delta < & \sum_{k=t^1+1}^{\tau_1(t^1)} \left(\alpha^{\lambda_k^1(\Gamma)} - \alpha^{k+|t|-t^1}\right)\left(Y_k^1 - I_1(t^1)\right) \\ & + \sum_{k=t^2+1}^{\Gamma_\tau^2} \left(\alpha^{\lambda_k^2(\Gamma)} - \alpha^{\lambda_k^2(\Delta)}\right)\left(Y_k^2 - I_2(t^2)\right) \\ & + \sum_{i=3}^{d} \sum_{k=t^i+1}^{\infty} \left(\alpha^{\lambda_k^i(\Gamma)} - \alpha^{\lambda_k^i(\Delta)}\right)\left(Y_k^i - I_i(t^i)\right) 1_{\{k \leq \Gamma_\tau^i\}}, \end{aligned}$$

where the last term is absent in the case where $d = 2$. Define the processes $(V_k^1,\ k \in \mathbb{N}_{t^1+1}), \ldots, (V_k^d,\ k \in \mathbb{N}_{t^d+1})$ by

$$V_k^i = \begin{cases} \alpha^{\lambda_k^1(\Gamma)-k}\, 1_{\{k \leq \tau_1(t^1)\}}, & \text{if } i = 1 \text{ and } k > t^1, \\ \alpha^{-k}\left(\alpha^{\lambda_k^i(\Gamma)} - \alpha^{\lambda_k^i(\Delta)}\right) 1_{\{k \leq \Gamma_\tau^i\}}, & \text{if } i = 2, \ldots, d \text{ and } k > t^i. \end{cases}$$

The right-hand side of the last inequality above can be written

$$-\alpha^{|t|-t^1} \sum_{k=t^1+1}^{\tau_1(t^1)} \alpha^k \big(Y_k^1 - I_1(t^1)\big) + \sum_{i=1}^{d} \sum_{k=t^i+1}^{\infty} \alpha^k V_k^i \big(Y_k^i - I_i(t^i)\big).$$

In order to establish that $\mathrm{E}(X_\Gamma) < \mathrm{E}(X_\Delta)$, it suffices to show that

$$-\alpha^{|t|-t^1} \mathrm{E}\left(\sum_{k=t^1+1}^{\tau_1(t^1)} \alpha^k \big(Y_k^1 - I_1(t^1)\big); F \right) + \mathrm{E}\left(\sum_{i=1}^{d} \sum_{k=t^i+1}^{\infty} \alpha^k V_k^i \big(Y_k^i - I_i(t^i)\big); F \right) \le 0.$$

Since $F \in \mathscr{F}_t$, the first expectation is zero by (8.29), (8.32), and (8.25). As for the second expectation, it is ≤ 0 by Lemma 4, since $F \in \tilde{\mathscr{F}}_{t^i}^i$ for all i and the hypotheses of this lemma are satisfied, as we are about to see. It is clear that $\{k \le \Gamma_\tau^i\} = \{\Gamma_\tau^i < k\}^c \in \tilde{\mathscr{F}}_{k-1}^i$, $i = 2, \dots, d$. Similarly, $\{k \le \tau_1(t^1)\} \in \tilde{\mathscr{F}}_{k-1}^1$. On the other hand, for any control Γ, $\lambda_k^i(\Gamma)$ is $\tilde{\mathscr{F}}_{k-1}^i$-measurable. Indeed, by (8.4),

$$\begin{aligned} \{\lambda_k^i(\Gamma) = n\} &= \{\Gamma_{n-1}^i = k-1,\ \Gamma_n^i = k\} \\ &= \bigcup_{s \in \mathbb{I}:\ s^i = k-1} \{\Gamma_{n-1} = s,\ \Gamma_n = s + e_i\} \\ &\in \tilde{\mathscr{F}}_{k-1}^i. \end{aligned}$$

It follows that V_k^i is $\tilde{\mathscr{F}}_{k-1}^i$-measurable. We now show that if $k > t^i$, then $0 \le V_{k+1}^i \le V_k^i \le 1$. This is clear for $i = 1$, because $\lambda_k^i(\Gamma) - k \le \lambda_{k+1}^i(\Gamma) - (k+1)$. For $i = 2, \dots, d$, it is sufficient to write

$$\alpha^{-k}\left(\alpha^{\lambda_k^i(\Gamma)} - \alpha^{\lambda_k^i(\Delta)}\right) = \alpha^{\lambda_k^i(\Gamma)-k}\left(1 - \alpha^{(\lambda_k^i(\Delta) - \lambda_k^i(\Gamma))}\right),$$

and to observe that $\lambda_{k+1}^i(\Delta) - \lambda_{k+1}^i(\Gamma) \le \lambda_k^i(\Delta) - \lambda_k^i(\Gamma)$.

8.5 EXAMPLES

We shall now use the theory developed in the previous sections to solve explicitly two concrete examples.

Researcher's Problem

A researcher can work on one of d projects at a time, and at each time n can choose on which project to work next. The amounts of time required to complete each of these projects are random and independent of each other. If project i is completed at time n, then the researcher receives a reward $\alpha^n v_i$, where $\alpha \in]0, 1[$ is a discounting rate and $v_i \in \mathbb{R}_+$. How should the researcher proceed in order to maximize the expected reward? It is assumed that switching from one project to another entails neither cost nor loss of time, and moreover, that the only information acquired over time by the researcher is whether or not a project has been completed (no new information that would, for instance, help estimate the remaining amount of time required to complete the various projects becomes available).

Formulating the Problem

We will assume that the amount of time required to complete project i is a random variable σ_i with values in $\mathbb{N}$ and that the d random variables $\sigma_1, \ldots, \sigma_d$ are independent. In order to formulate the problem as a d-armed bandit problem, we define filtrations $\mathcal{F}^i$ and processes $(Y^i_{t^i},\ t^i \in \mathbb{N})$ as follows. For $i = 1, \ldots, d$ and $t^i \in \mathbb{N}$, let $Y^i_{t^i} = v_i 1_{\{\sigma_i = t^i\}}$ and let $\mathcal{F}^i_{t^i}$ denote the sigma field generated by the family $(Y^i_{s^i},\ s^i \in \mathbb{N}_{t^i})$. With these definitions, solving the researcher's problem is equivalent to finding a control $\Gamma^o \in \mathbf{G}^\infty$ that satisfies (8.8), with $(X_\gamma,\ \gamma \in \mathbb{G}^\infty)$ defined by (8.1).

Solving the Problem

By the theorem of Section 8.4, it is sufficient to determine the index controls. This will be achieved by expressing the Gittins index $I_i(t^i)$ defined in (8.28) in terms of the law F_i of σ_i.

The lemma below will be the key to solving the problem, because it will make possible the evaluation of the essential supremum that appears in the definition of $I_i(t^i)$.

Lemma. *For each $i = 1, \ldots, d$ and all $n \in \mathbb{N}$, set*

$$\tau_n = \begin{cases} \sigma_i & \text{on } \{\sigma_i \leq n\}, \\ n & \text{on } \{\sigma_i > n\}, \end{cases} \tag{8.42}$$

and $\tau_\infty = \sigma_i$. Let τ be a stopping time relative to the filtration $\mathcal{F}^i$. Then there exists $n \in \overline{\mathbb{N}}$ such that $\sigma_i \wedge \tau = \tau_n$.

Proof. If $\sigma_i \leq \tau$, then $\sigma_i \wedge \tau = \sigma_i$ and it suffices to take $n = \infty$. Assume that $P\{\sigma_i > \tau\} > 0$. For all $n \in \mathbb{N}$, set $F_n = \{\sigma_i > n\} \cap \{\tau = n\}$. Then $F_m \cap F_n = \emptyset$ if $m \neq n$. Observe that $\mathcal{F}^i_{t^i}$ is generated by the finite partition of Ω

$$(\{\sigma_i = 0\}, \ldots, \{\sigma_i = t^i\},\ \{\sigma_i > t^i\}). \tag{8.43}$$

Because $F_n \in \mathscr{F}_n^i$, it is therefore easy to see that if $\mathrm{P}(F_n) > 0$, then $\{\sigma_i > n\} \subset \{\tau = n\}$. Assume that there exists a couple (m, n) such that $m < n$, $\mathrm{P}(F_m) > 0$ and $\mathrm{P}(F_n) > 0$. Then $\{\sigma_i > n\} \subset \{\sigma_i > m\} \subset \{\tau = m\}$, and so $F_n = \{\sigma_i > n\} \cap \{\tau = n\} \subset \{\tau = m\} \cap \{\tau = n\} = \emptyset$, which contradicts the fact that $\mathrm{P}(F_n) > 0$. Since $\{\sigma_i > \tau\} = \bigcup_{n \in \mathbb{N}} (\{\sigma_i > n\} \cap \{\tau = n\})$, there exists therefore a unique $n \in \mathbb{N}$ such that $\mathrm{P}(F_n) > 0$. For that n, $\{\sigma_i > \tau\} = \{\sigma_i > n\} \cap \{\tau = n\} = \{\sigma_i > n\}$, which implies that $\sigma_i \wedge \tau = \tau_n$.

Proposition 1 (An explicit formula for the Gittins indices). *For $i = 1, \ldots, d$ and all $t^i \in \mathbb{N}$, the Gittins index $I_i(t^i)$ is expressed by*

$$I_i(t^i) = v_i \sup_{n > t^i} \frac{\sum_{k=t^i+1}^{n} \alpha^k \mathrm{P}\{\sigma_i = k\}}{\sum_{k=t^i+1}^{n} \alpha^k (1 - F_i(k-1))} 1_{\{\sigma_i > t^i\}}. \tag{8.44}$$

Proof. By definition, $I_i(t^i) = \operatorname{esssup}_{\tau \in \mathbf{T}^{t^i+1, i}} Q_i(t^i, \tau)$, where

$$Q_i(t^i, \tau) = v_i \frac{\mathrm{E}\left(\sum_{k=t^i+1}^{\tau} \alpha^k 1_{\{\sigma_i = k\}} \middle| \mathscr{F}_{t^i}^i \right)}{\mathrm{E}\left(\sum_{k=t^i+1}^{\tau} \alpha^k \middle| \mathscr{F}_{t^i}^i \right)}. \tag{8.45}$$

Since $\{\sigma_i \le t^i\}$ belongs to $\mathscr{F}_{t^i}^i$, it is clear that $Q_i(t^i, \tau) = 0$ on $\{\sigma_i \le t^i\}$, and we only have to evaluate $Q_i(t^i, \tau)$ on $\{\sigma_i > t^i\}$. It is easy to see that

$$Q_i(t^i, \tau) \le Q_i\big(t^i, (\sigma_i \wedge \tau) \vee (t^i + 1)\big),$$

because replacing τ by $(\sigma_i \wedge \tau) \vee (t^i + 1)$ on the right-hand side of (8.45) leaves the numerator unchanged and makes the denominator decrease. By the lemma, there exists $n \in \overline{\mathbb{N}}$ such that $\sigma_i \wedge \tau = \tau_n$, where τ_n is defined in (8.42). For this n, $(\sigma_i \wedge \tau) \vee (t^i + 1) = \tau_n \vee (t^i + 1)$. But

$$\tau_n \vee (t^i + 1) = \begin{cases} t^i + 1, & \text{if } n \le t^i + 1, \\ \rho_n, & \text{if } n > t^i + 1, \end{cases}$$

where

$$\rho_n = \begin{cases} \sigma_i \vee (t^i + 1) & \text{on } \{\sigma_i \le n\}, \\ n & \text{on } \{\sigma_i > n\}. \end{cases}$$

Thus

$$Q_i(t^i, \tau)\, 1_{\{\sigma_i > t^i\}} \le \sup_{n > t^i} Q_i(t^i, \rho_n)\, 1_{\{\sigma_i > t^i\}},$$

and therefore

$$I_i(t^i) = \sup_{n > t^i} Q_i(t^i, \rho_n)\, 1_{\{\sigma_i > t^i\}}.$$

Using (8.45), notice that the supremum can be taken indifferently over n in $\mathbb{N}$ or n in $\overline{\mathbb{N}}$. Also, given the form of the generating partition of $(\mathscr{F}^i_{t^i})$ indicated in (8.43), observe that on $\{\sigma^i > t^i\}$, the numerator that appears in the quotient that defines $Q_i(t^i, \rho_n)$ is equal to

$$v_i \sum_{k=t^i+1}^{n} \alpha^k\, \mathrm{P}(\sigma_i = k \mid \sigma_i > t^i) = v_i \sum_{k=t^i+1}^{n} \alpha^k \frac{\mathrm{P}\{\sigma_i = k\}}{1 - F_i(t^i)},$$

and the denominator to

$$\mathrm{E}\left(\sum_{j=t^i+1}^{n} 1_{\{\sigma_i = j\}} \sum_{k=t^i+1}^{j} \alpha^k + 1_{\{\sigma_i > n\}} \sum_{k=t^i+1}^{n} \alpha^k \,\middle|\, \sigma_i > t^i \right)$$

$$= \sum_{k=t^i+1}^{n} \alpha^k \sum_{j=k}^{n} \mathrm{P}(\sigma_i = j \mid \sigma_i > t^i) + \mathrm{P}(\sigma_i > n \mid \sigma_i > t^i) \sum_{k=t^i+1}^{n} \alpha^k.$$

The conditional probabilities on the last right-hand side are, respectively, equal to

$$\frac{F_i(n) - F_i(k-1)}{1 - F_i(t^i)} \quad \text{and} \quad \frac{1 - F_i(n)}{1 - F_i(t^i)},$$

and the denominator is therefore equal to

$$\frac{1}{1 - F_i(t^i)} \sum_{k=t^i+1}^{n} \alpha^k (1 - F_i(k-1)).$$

Formula (8.44) is therefore established.

Remark

For numerous specific laws F_i, the supremum in (8.44) can be evaluated explicitly (cf. Exercise 3), and in such cases the researcher's problem is completely solved.

The Problem of Choosing a Job

A person is faced with d employment opportunities that can be investigated at a rate of one per day. The investigation of a job entails a cost (related, for instance, to the purchase of tools or of a uniform), and informs the person of the daily wage rate. The person may or may not find job i to be congenial, and associates an additional positive value v_i to the daily wage when the job is congenial, and a negative value $-v_i$ when it is not. The person does not know in advance how long it will take to realize if job i is congenial, and must decide each evening at which job to work the next day or which new job to investigate. In what order and when should the d jobs be investigated if the objective is to maximize the expected long-term reward? It is assumed as in the researcher's problem that the present value of future rewards is discounted at rate $\alpha \in]0, 1[$.

Formulating and Solving the Problem

We shall assume that the daily wage when working at job i is a random variable X_i with values in $\mathbb{R}_+$. The value of X_i is revealed on the first day of work at job i. The investigation cost for this day is c_i, where $c_i \in \mathbb{R}_+$. The number of days of work at job i that are required before the person realizes that the job is congenial or not is identified with the realization of a random variable N_i with values in $\mathbb{N}^*$, and the fact that job i turns out to be congenial or not is identified with the realization of a random variable ε_i with values in $\{-1, 1\}$, which takes value 1 if the job is congenial and -1 otherwise. We set $p_i = \mathrm{P}\{\varepsilon_i = 1\}$ and shall assume that the random variables $X_1, \ldots, X_d, N_1, \ldots, N_d, \varepsilon_1, \ldots, \varepsilon_d$ are independent.

Let

$$Y_n^i = \begin{cases} 0, & \text{if } n = 0, \\ X_i + \varepsilon_i v_i 1_{\{N_i < n\}} - c_i 1_{\{1\}}(n), & \text{if } n \in \mathbb{N}^*. \end{cases}$$

Moreover, let $\mathscr{F}_0^i$ be the sigma field generated by the null sets, and for $n \in \mathbb{N}^*$, let $\mathscr{F}_n^i$ be the sigma field generated by X_i and by the finite partition of Ω

$$(\{N_i = 1, \varepsilon_i = -1\}, \{N_i = 1, \varepsilon_i = 1\}, \ldots, \\ \{N_i = n, \varepsilon_i = -1\}, \{N_i = n, \varepsilon_i = 1\}, \{N_i > n\}). \tag{8.46}$$

Set $\mathscr{F}_\infty^i = \bigwedge_{n \in \mathbb{N}} \mathscr{F}_n^i$. Clearly, $\mathscr{F}_\infty^i$ is the sigma field generated by the three random variables X_i, N_i and ε_i, and for $n \in \mathbb{N}^*$, Y_n^i represents the value of working at job i on the nth day of work at that job, and $\mathscr{F}_n^i$ the information acquired during the first n days of work at job i. The filtration $(\mathscr{F}_n^i, n \in \mathbb{N})$ will be denoted $\mathscr{F}^i$.

Maximizing the expected reward is now equivalent to finding a control $\Gamma^{\circ} \in \mathbf{G}^{\infty}$ that satisfies (8.8), with $(X_{\gamma},\ \gamma \in \mathbb{G}^{\infty})$ defined by (8.1). By the theorem of Section 8.4, this problem will be solved once we compute the Gittins indices $I_i(t^i)$. The computation depends on whether $t^i = 0$ or $t^i > 0$, because in the first case, the most important information to be learned is the value of X_i, whereas in the latter case, X_i is known and only the congeniality of job i remains to be determined. We begin with the latter case.

Case Where $t^i > 0$

The expression for the Gittins indices involves the conditional probabilities

$$p_i^k = \mathrm{P}\left(N_i = k \,\middle|\, N_i > t^i\right).$$

Proposition 2 (Gittins indices for $t^i > 0$). *If $t^i > 0$, then the Gittins index $I_i(t^i)$ is given by*

$$I_i(t^i) = \begin{cases} X_i - v_i & \text{on } \{N_i \le t^i,\ \varepsilon_i = -1\}, \\ X_i + v_i & \text{on } \{N_i \le t^i,\ \varepsilon_i = 1\}, \\ X_i + v_i p_i \dfrac{\sum\limits_{k=t^i+1}^{\infty} p_i^k \alpha^{k+1}}{\sum\limits_{k=t^i+1}^{\infty} p_i^k\left(\alpha^{t^i+1} - (1-p_i)\alpha^{k+1}\right)} & \text{on } \{N_i > t^i\}. \end{cases} \tag{8.47}$$

Moreover, it satisfies the relation

$$I_i(t^i) = Q_i(t^i, \tilde{\tau}), \tag{8.48}$$

where $\tilde{\tau}$ is the stopping time relative to $\mathscr{F}^i$ defined by

$$\tilde{\tau} = \begin{cases} N_i \vee (t^i + 1) & \text{on } \{\varepsilon_i = -1\}, \\ \infty & \text{on } \{\varepsilon_i = 1\}. \end{cases} \tag{8.49}$$

Further, on $\{N_i > t^i\}$,

$$\mathrm{E}\left(\sum_{k=t^i+1}^{\tilde{\tau}} \alpha^k \,\middle|\, \mathscr{F}^i_{t^i}\right) = \frac{1}{1-\alpha} \sum_{k=t^i+1}^{\infty} p_i^k\left(\alpha^{t^i+1} - (1-p_i)\alpha^{k+1}\right) \tag{8.50}$$

and

$$\mathrm{E}\left(\sum_{k=t^i+1}^{\tilde{\tau}} \alpha^k Y_i^k \middle| \mathscr{F}_{t^i}^i \right) \tag{8.51}$$

$$= \frac{X_i}{1-\alpha} \sum_{k=t^i+1}^{\infty} p_i^k \left(\alpha^{t^i+1} - (1-p_i)\alpha^{k+1} \right) + v_i p_i \sum_{k=t^i+1}^{\infty} p_i^k \, \alpha^{k+1}.$$

Proof. By (8.28) and (8.27),

$$I_i(t^i) = \operatorname*{esssup}_{\tau \in \mathbf{T}^{t^i+1,i}} \frac{\mathrm{E}\left(\sum_{k=t^i+1}^{\tau} \alpha^k \left(X_i + \varepsilon_i v_i 1_{\{N_i < k\}} - c_i 1_{\{1\}}(k) \right) \middle| \mathscr{F}_{t^i}^i \right)}{\mathrm{E}\left(\sum_{k=t^i+1}^{\tau} \alpha^k \middle| \mathscr{F}_{t^i}^i \right)}.$$

Because $t^i \geq 1$, X_i is $\mathscr{F}_{t^i}^i$-measurable and $1_{\{1\}}(k) = 0$ for $k \geq t^i + 1$. Consequently,

$$I_i(t^i) = X_i + v_i \operatorname*{esssup}_{\tau \in \mathbf{T}^{t^i+1,i}} \frac{\mathrm{E}\left(\sum_{k=t^i+1}^{\tau} \alpha^k \varepsilon_i 1_{\{N_i < k\}} \middle| \mathscr{F}_{t^i}^i \right)}{\mathrm{E}\left(\sum_{k=t^i+1}^{\tau} \alpha^k \middle| \mathscr{F}_{t^i}^i \right)}. \tag{8.52}$$

Fix $n \leq t^i$. On the set $\{N_i = n, \varepsilon_i = -1\}$, which belongs to $\mathscr{F}_{t^i}^i$, we now apply Remark 1 of Section 1.2 to obtain

$$I_i(t^i) = X_i - v_i \operatorname*{esssup}_{\tau \in \mathbf{T}^{t^i+1,i}} \frac{\mathrm{E}\left(\sum_{k=t^i+1}^{\tau} \alpha^k \middle| \mathscr{F}_{t^i}^i \right)}{\mathrm{E}\left(\sum_{k=t^i+1}^{\tau} \alpha^k \middle| \mathscr{F}_{t^i}^i \right)} = X_i - v_i.$$

In the same way, we see that $I_i(t^i) = X_i + v_i$ on the set $\{N_i = n, \varepsilon_i = 1\}$. It now remains to evaluate $I_i(t^i)$ on $\{N_i > t^i\}$. Using formula (8.52), we see that on this set,

$$I_i(t^i) = X_i + v_i \operatorname*{esssup}_{\tau \in \mathbf{T}^{t^i+1,i}} \frac{\mathrm{E}\left(1_{\{\tau > N_i\}} \sum_{n=N_i+1}^{\tau} \alpha^n \varepsilon_i \middle| \mathscr{F}_{t^i}^i \right)}{\mathrm{E}\left(\sum_{n=t^i+1}^{\tau} \alpha^n \middle| \mathscr{F}_{t^i}^i \right)}.$$

Because τ is a stopping time relative to $\mathscr{F}^i$, τ is an $\mathscr{F}_\infty^-$ measurable random variable, and because this sigma field is generated by X_i, N_i, and ε_i, there exists a Borel function φ defined on $\mathbb{R}_+ \times \mathbb{N}^* \times \{-1, 1\}$ such that $\tau = \varphi(X_i, N_i, \varepsilon_i)$. The quotient above can now be written

$$\frac{p_i \sum_{k=t^i+1}^{\infty} p_i^k 1_{\{\varphi(X_i,k,1)>k\}} \sum_{n=k+1}^{\varphi(X_i,k,1)} \alpha^n - (1-p_i) \sum_{k=t^i+1}^{\infty} p_i^k 1_{\{\varphi(X_i,k,-1)>k\}} \sum_{n=k+1}^{\varphi(X_i,k,-1)} \alpha_n}{p_i \sum_{k=t^i+1}^{\infty} p_i^k \sum_{n=t^i+1}^{\varphi(X_i,k,1)} \alpha^n + (1-p_i) \sum_{k=t^i+1}^{\infty} p_i^k \sum_{n=t^i+1}^{\varphi(X_i,k,-1)} \alpha^n}.$$

The numerator increases and the denominator decreases when $\varphi(X_i, k, -1)$ is replaced by $k \vee (t^i + 1)$. After this substitution, the new quotient again increases if $\varphi(X_i, k, 1)$ is replaced by ∞. By (8.49), we see that on $\{N_i > t^i\}$,

$$I_i(t^i) = Q(t^i, \tilde{\tau}) = X_i + v_i p_i \frac{\sum_{k=t^i+1}^{\infty} p_i^k \sum_{n=k+1}^{\infty} \alpha^n}{\sum_{k=t^i+1}^{\infty} p_i^k \left(\sum_{n=t^i+1}^{\infty} p_i \alpha^n + \sum_{n=t^i+1}^{k} (1-p_i)\alpha^n \right)}. \tag{8.53}$$

After evaluating the geometric series, the validity of (8.47) and (8.48) on $\{N_i > t^i\}$ follows. The validity of (8.47) on $\{N_i \leq t^i\}$ follows from (8.49), and that of (8.48) on the same set is a consequence of the arguments immediately after (8.52). Formula (8.50) is obtained by proceeding as we did above to transform the denominator in the quotient at the beginning of the proof, the function $\varphi(x, k, j)$ associated with $\tilde{\tau}$ being now equal to the constant ∞ when $j = 1$ and to $k \vee (t^i + 1)$ when $j = -1$. Equality (8.51) follows from the fact that

$$\mathrm{E}\left(\sum_{k=t^i+1}^{\tilde{\tau}} \alpha^k Y_k^i \middle| \mathscr{F}_{t^i}^i \right) = X_i \, \mathrm{E}\left(\sum_{k=t^i+1}^{\tilde{\tau}} \alpha^k \middle| \mathscr{F}_{t^i}^i \right) + \frac{v_i p_i}{1-\alpha} \sum_{k=t^i+1}^{\infty} p_i^k \alpha^{k+1}.$$

The proposition is now proved.

Case Where $t^i = 0$: Computing $I_i(0)$ in a Special Case

Consider the case where $p_i = \frac{1}{2}$ and $N_i - 1$ is a geometric random variable; that is, there exists $\lambda_i \in]0, 1[$ such that $\mathrm{P}\{N_i = n\} = \lambda_i (1 - \lambda_i)^{n-2}$ for all $n \geq 2$. This hypothesis implies in particular that it takes at least two days of work at job i to realize whether this job is congenial or not, and so the only information to be learned on the first day of work at this job is the value of

X_i. For $t^i > 0$, the Gittins index $I_i(t^i)$ has a particularly simple expression, which will make it possible to compute $I_i(0)$.

An elementary calculation shows that $p_i^k = \lambda_i(1 - \lambda_i)^{k-1-t^i}$ for $k \geq t^i + 1$. After evaluating the geometric series that appear in the expression for $Q_i(t^i, \tilde{\tau})$ given in (8.53), we see that on $\{N_i > t^i\}$,

$$I_i(t^i) = X_i + \frac{v_i \lambda_i \alpha}{2 - 2\alpha + \lambda_i \alpha}. \tag{8.54}$$

Moreover, on $\{N_i > t^i\}$, formulas (8.50) and (8.51) become, respectively,

$$\mathrm{E}\left(\sum_{k=t^i+1}^{\tilde{\tau}} \alpha^k \middle| \mathscr{F}^i_{t^i} \right) = \frac{\alpha^{t^i+1}}{1 - (1 - \lambda_i)\alpha} \left(1 + \frac{\lambda_i \alpha}{2(1 - \alpha)} \right) \tag{8.55}$$

and

$$\mathrm{E}\left(\sum_{k=t^i+1}^{\tilde{\tau}} \alpha^k Y^i_k \middle| \mathscr{F}^i_{t^i} \right) = \frac{\alpha^{t^i+1}}{1 - (1 - \lambda_i)\alpha} \left(X_i + (X_i + v_i) \frac{\lambda_i \alpha}{2(1 - \alpha)} \right). \tag{8.56}$$

We shall now give a formula for $I_i(0)$ that does not contain an essential supremum. Set

$$b_i = \frac{v_i \lambda_i \alpha}{2 - 2\alpha + \lambda_i \alpha}$$

and observe that $0 < b_i < v_i$. For all $a \in \overline{\mathbb{R}}_+$ such that $a \geq v_i$, set

$$\tau(0) = \begin{cases} 1 & \text{on } \{X_i + b_i < I_i(0)\}, \\ N_i & \text{on } \{X_i - v_i < I_i(0) \leq X_i + b_i,\ \varepsilon_i = -1\}, \\ \infty & \text{on } \{X_i - v_i < I_i(0) \leq X_i + b_i,\ \varepsilon_i = 1\} \cup \{I_i(0) \leq X_i - b_i\}, \end{cases}$$

It is clear that τ^a is a stopping time relative to the filtration $\mathscr{F}^i$. This stopping time can be described as follows: After learning on the first day of work at job i the value of the daily wage X_i, quit this job if $X_i < a - b_i$; otherwise work exactly N_i days in a row at this job then quit the job if it turns out not to be congenial and $X_i < a + v_i$; otherwise remain permanently at work at this job.

The proposition that follows shows that computing $I_i(0)$ reduces to maximizing a function of a real variable.

Proposition 3 (An explicit formula for $I_i(0)$). *Let μ_i be the law of X_i. If $p_i = \frac{1}{2}$ and $N_i - 1$ is a geometric random variable with parameter $\lambda_i \in]0, 1[$, then*

$$I_i(0) = \sup_{a \geq v_i} Q_i(0, \tau^a) \tag{8.57}$$

and $Q_i(0, \tau^a)$ is equal to

$$\frac{-c_i + \mathrm{E}(X_i) + \dfrac{\alpha}{1-(1-\lambda_i)\alpha}\displaystyle\int_{a-b_i}^{a+v_i}\left(x + \frac{\lambda_i \alpha (x + v_i)}{2(1-\alpha)}\right) d\mu_i(x) + \dfrac{\alpha}{1-\alpha}\displaystyle\int_{a+v_i}^{\infty} x\, d\mu_i(x)}{1 + \dfrac{\alpha}{1-(1-\lambda_i)\alpha}\left(1 + \dfrac{\lambda_i \alpha}{2(1-\alpha)}\right)\mu_i([a - b_i, a + v_i[) + \dfrac{\alpha}{1-\alpha}\mu_i([a + v_i, \infty[)}. \tag{8.58}$$

Proof. Observe that for all stopping times τ relative to the filtration $\mathscr{F}^i$, $Q_i(0, \tau)$ is an $\mathscr{F}_0^i$-measurable random variable, therefore a constant, which implies that

$$I_i(0) = \operatorname*{ess\,sup}_{\tau \in \mathbf{T}^{0,i}} Q_i(0, \tau) = \sup_{\tau \in \mathbf{T}^{0,i}} Q_i(0, \tau) \geq \sup_{a \geq v_i} Q_i(0, \tau^a).$$

In order to establish (8.57), it therefore suffices to exhibit a value of a such that $I_i(0) = Q_i(0, \tau^a)$. By the theorem of Section 8.3, the stopping time $\tau_i(0) = \inf\{n \in \mathbb{N}^*: I_i(n) < I_i(0)\}$ satisfies the equality $I_i(0) = Q_i(0, \tau_i(0))$. It is therefore sufficient to show that $\tau(0) = \tau^{I_i(0)}$. By (8.49) and (8.54), for all $n \in \mathbb{N}^*$,

$$I_i(n) = \begin{cases} X_i + b_i & \text{on } \{N_i > n\}, \\ X_i - v_i & \text{on } \{N_i \leq n, \varepsilon_i = -1\}, \\ X_i + v_i & \text{on } \{N_i \leq n, \varepsilon_i = 1\}, \end{cases}$$

and consequently,

$$\tau(0) = \begin{cases} 1 & \text{on } \{X_i + b_i < I_i(0)\}, \\ N_i & \text{on } \{X_i - v_i < I_i(0) \leq X_i + b_i, \varepsilon_i = -1\}, \\ \infty & \text{on } \{X_i - v_i < I_i(0) \leq X_i + b_i, \varepsilon_i = 1\} \cup \{I_i(0) \leq X_i - b_i\}, \end{cases}$$

which is also the definition of $\tau^{I_i(0)}$. In order to check (8.58), observe using (8.27) that

$$Q_i(0,\tau^a) = \frac{E\left(\sum_{k=1}^{\tau^a} \alpha^k Y_k^i\right)}{E\left(\sum_{k=1}^{\tau^a} \alpha^k\right)} = \frac{\alpha(-c_i + E(X_i)) + E\left(1_{\{\tau^a > 1\}} \sum_{k=2}^{\tau^a} \alpha^k Y_k^i\right)}{\alpha + E\left(1_{\{\tau^a > 1\}} \sum_{k=2}^{\tau^a} \alpha^k\right)}.$$

Let $\tilde{\tau}$ be the stopping time defined in (8.49) with $t^i = 1$. Observe that

$$\{\tau^a > 1\} = \{X_i \geq a - b_i\}$$

and

$$\tau^a = \begin{cases} \tilde{\tau} & \text{on } \{a - b_i \leq X_i < a + v_i\}, \\ \infty & \text{on } \{a + v_i \leq X_i\}. \end{cases}$$

Since X_i and ε_i are independent, we see that $Q_i(0,\tau^a)$ is equal to

$$\frac{(-c_i + E(X_i))\alpha + \int_{a-b_i}^{a+v_i} E\left(\sum_{k=2}^{\tilde{\tau}} \alpha^k Y_k^i \middle| X_i = x\right) d\mu_i(x) + \int_{a+v_i}^{\infty} E\left(\sum_{k=2}^{\infty} \alpha^k Y_k^i \middle| X_i = x\right) d\mu_i(x)}{\alpha + \int_{a-b_i}^{a+v_i} E\left(\sum_{k=2}^{\tilde{\tau}} \alpha^k \middle| X_i = x\right) d\mu_i(x) + \int_{a+v_i}^{\infty} E\left(\sum_{k=2}^{\infty} \alpha^k \middle| X_i = x\right) d\mu_i(x)}$$

The formula in the theorem now follows straightforwardly from (8.55) and (8.56).

EXERCISES

1. With the notation of Section 8.1, fix $i \in \{1,\ldots,d\}$ and for all $m \in \mathbb{N}$, $n \in \overline{\mathbb{N}}$, and $x \in \mathbb{R}$, set

$$Y(m,n,x) = \begin{cases} 0, & \text{if } n \leq m, \\ \sum_{k=m+1}^{n} \alpha^k (Y_k^i - x), & \text{if } n > m, \end{cases}$$

and let

$$Z(m,x) = \operatorname*{esssup}_{\tau \in \mathbf{T}^{m,i}} E\left(Y(m,\tau,x) \middle| \mathcal{F}_m^i\right).$$

a. Show that for each fixed m, the function $x \mapsto Z(m, x)$ is convex, nonincreasing and such that $\lim_{x \to \infty} Z(m, x) = 0$.

b. For $m \in \mathbb{N}$ and $x \in \mathbb{R}$, let $\rho(m, x) = \inf\{n \geq m\colon Z(n, x) = 0\}$. Show that for each fixed m, the function $x \mapsto \rho(m, x)$ is nonincreasing, right continuous and such that

$$Z(m, x) = \mathrm{E}\big(Y(m, \rho(m, x), x) \,|\, \mathscr{F}_m^i\big).$$

(*Hint.* Show that Snell's envelope (relative to the filtration $(\mathscr{F}_{m+k}^i, k \in \mathbb{N})$) of the process $(Y(m, m+k, x),\ k \in \overline{\mathbb{N}})$ is the process $(\tilde{Z}_k, k \in \overline{\mathbb{N}})$ defined by

$$\tilde{Z}_k = Y(m, m+k, x) + Z(m+k, x).$$

Then use the theorem of Section 3.3 and the comments which precede the corollary of Section 3.3.)

c. Let $\partial^+/\partial x$ denote the operation that consists of taking the derivative on the right with respect to the variable x. Show that for each m and x,

$$\frac{\partial^+}{\partial x} Z(m, x) = -\frac{\alpha^{m+1}}{1-\alpha}\big(1 - \mathrm{E}\big(\alpha^{\rho(m, x) - m} \,|\, \mathscr{F}_m^i\big)\big). \tag{8.59}$$

(*Hint.* To prove that (8.59) holds with $=$ replaced by $\leq$, check that for any $\delta > 0$,

$$Z(m, x) \leq Z(m, x+\delta) + \delta\, \mathrm{E}\left(\sum_{k=m+1}^{\rho(m, x)} \alpha^k 1_{\{\rho(m, x) > m\}} \middle| \mathscr{F}_m^i \right).$$

To prove the converse inequality, observe that $\rho(m, x+\delta) \leq \rho(m, x)$; then use (b) of Proposition 1 of Section 3.3 to get

$$\begin{aligned} Z(m, x) &= \mathrm{E}\big(Y(m, \rho(m, x+\delta), x) + Z(\rho(m, x+\delta), x) \,\big|\, \mathscr{F}_m^i\big) \\ &= Z(m, x+\delta) + \mathrm{E}\big(Z(\rho(m, x+\delta), x) \,\big|\, \mathscr{F}_m^i\big) \\ &\quad + \delta\, \mathrm{E}\left(\sum_{k=m+1}^{\rho(m, x+\delta)} \alpha^k 1_{\{\rho(m, x+\delta) > m\}} \middle| \mathscr{F}_m^i \right). \end{aligned}$$

2. a. In the context of Section 8.2, show that if Γ is an index control, then for any $n \geq 1$,

$$\big\{\tilde{I}(\Gamma_{n-1}) \geq x\big\} = \left\{ \sum_{i=1}^{d} J_i(x) \geq n \right\}.$$

(*Hint*. Notice that $\Sigma_{i=1}^{d} \Gamma_{n-1}^{i} = n - 1$, so the event on the right-hand side, on which the total time until all indices are $< x$ is at least n, can be written $\bigcup_{i=1}^{d} \{J_i(x) > \Gamma_{n-1}^{i}\}$.)

b. Show that

$$\sup_{\Gamma \in \mathbf{G}^{\infty}} \mathrm{E}(X_{\Gamma}) = \frac{\alpha}{1-\alpha} \int_0^{\infty} \left(1 - \prod_{i=1}^{d} \mathrm{E}\left(\alpha^{J_i(x)}\right)\right) dx.$$

(*Hint*. Use the fact that the left-hand side is equal to (8.40), and that the right-hand side of (8.40) can be written

$$\mathrm{E}\left(\sum_{n=1}^{\infty} \alpha^n \int_0^{\infty} 1_{\{x \le \bar{I}(\Gamma_{n-1})\}}\right) dx;$$

then use a.)

3 a. In the context of the researcher's problem considered in Section 8.5, show that if the function

$$k \mapsto \frac{\mathrm{P}\{\sigma_i = k\}}{1 - F_i(k-1)}$$

is nondecreasing, then formula (8.44) becomes

$$I_i(t^i) = v_i \frac{\mathrm{P}\{\sigma_i = t^i + 1\}}{1 - F_i(t^i)} 1_{\{\sigma_i > t^i\}}.$$

b. In the same context, give an example of a law F_i for which the supremum in (8.44) is not attained at $n = t^i + 1$.

HISTORICAL NOTES

The multiarmed bandit problem with geometric discounting considered in this chapter was first solved in an article by Gittins and Jones (1974) (see the account in Gittins (1979)). At the present time, there is an extensive literature on the subject, which culminates in the books of Whittle (1982), Berry and Fristedt (1985), Gittins (1989), and Presman and Sonin (1990). Our resolution follows the approach developed by Varaiya, Walrand, and Buyukkoc (1985), translated into the framework of multiparameter processes by Mandelbaum (1986). However, our assumptions on the bandit process (e.g., (8.2)) are weaker than those generally used in the literature. The two examples of Section 8.5 are borrowed from Gittins's book. Thc solutions given in this reference have a more intuitive flavor and can be advantageously consulted by the reader.

Recent developments in the study of the multiarmed bandit problem include papers of Keener (1986), Lai (1987), Ishikida and Varaiya (1993), and El Karoui and Karatzas (1993). This last reference is particularly interesting in that it contains a solution to the problem (for bounded reward processes) that uses Snell's envelope. This solution is therefore an interesting complement to the one presented here.

There is also an important literature concerning multiarmed bandit problems in continuous time, including Karatzas (1984), Mandelbaum (1987), Mazziotto and Millet (1987), Dalang (1990), and El Karoui and Karatzas (1994).

The statements in Exercise 1 are due to El Karoui and Karatzas (1993). The integral formula in Exercise 2 is known as Whittle's formula; the derivation given in this exercise is due to A. Mandelbaum. Exercise 3 is borrowed from Gittins (1989).

CHAPTER 9

The Markovian Case

Markovian systems model a type of evolution in which the future, once a state of the system is reached, depends only on this state and not on previously occupied states. It turns out that the general results on optimal stopping and optimal control gain in clarity and in simplicity once they are placed in a Markovian setting. Moreover, such a setting offers significant advantages for carrying out explicit computations.

This chapter deals primarily with two problems. In the first problem, a function f is defined on the state space E of a Markovian system and the objective is to maximize the expected reward, assuming that there is a payoff $f(x)$ when the evolution stops at state x. In the second problem, a subset D of E and a function f defined on D are given; the goal is again to maximize the expected payoff, assuming now that there is a reward $f(x)$ in the case where the evolution of the system reaches the set D for the first time at state x and stops at x.

9.1 MARKOV CHAINS AND SUPERHARMONIC FUNCTIONS

The subject matter we will develop in this chapter builds on the notion of Markov chain. In fact, we will deal with Markov chains indexed either by $\mathbb{I}$ or by $\mathbb{G}_f$, and in order to unify these two cases, we are naturally led to use an index set $\mathbb{P}$ that retains the essential features of both of these sets. This set will be a general index set according to the definition given in Section 7.6 but will satisfy a reinforced version of property c. More precisely, $\mathbb{P}$ will be a countable ordered set satisfying the following three properties:

a. $\mathbb{P}$ has a smallest element.

b. Each element of $\mathbb{P}$ admits only a finite number of predecessors.

c. Each element of $\mathbb{P}$ admits exactly d direct successors.

It is obvious that $\mathbb{I}$ and $\mathbb{G}_f$ (equipped with the order $\prec$) are special cases of the set $\mathbb{P}$. The same is true for $\mathbb{G}_f$ defined starting with $\mathbb{P}$ (instead of $\mathbb{I}$).

We will conform to all that was said in Section 7.6 regarding $\mathbb{P}$, and in particular, we will use the terminology and notation mentioned there. Moreover, for each $t \in \mathbb{P}$, we order the direct successors of t and denote them $\delta_1(t), \dots, \delta_d(t)$.

When the index set is $\mathbb{P}$, it will be assumed from now on that the basic filtration of Section 1.1 is replaced by a filtration $(\mathscr{F}_t, t \in \mathbb{P})$.

Markov Chains

In this chapter, E will denote a countable set. This set will be equipped with the sigma field $\mathscr{P}(\mathsf{E})$.

We will say that E is the *state space* of a process X indexed by $\mathbb{P}$ if the random variables that make up this process take their values in E. The set of $x \in \mathsf{E}$ such that there exists $t \in \mathbb{P}$ with $\mathrm{P}\{X_t = x\} > 0$ is then termed the *minimal state space* of X.

We will term *transition function* (on E) any family $P = (P(x, y), (x, y) \in \mathsf{E} \times \mathsf{E})$ of nonnegative numbers such that $\Sigma_{y \in \mathsf{E}} P(x, y) = 1$ for all $x \in \mathsf{E}$.

We will say that a process X indexed by $\mathbb{P}$ and with state space E is a *Markov chain* (with stationary transitions) if it is adapted and if there exist d transition functions $P_1, \dots, P_d$ such that

$$\mathrm{P}\left(X_{\delta_i(t)} = y | \mathscr{F}_t\right) = P_i(X_t, y) \tag{9.1}$$

for all $t \in \mathbb{P}$, $y \in \mathsf{E}$, and $i = 1, \dots, d$.

Notice that (9.1) implies that

$$\mathrm{P}\left(X_{\delta_i(t)} = y | X_t = x\right) = P_i(x, y) \tag{9.2}$$

for all $t \in \mathbb{P}$, all $x \in \mathsf{E}$ such that $\mathrm{P}\{X_t = x\} > 0$, all $y \in \mathsf{E}$, and $i = 1, \dots, d$. It is easy to check that (9.1) also implies the following more general property (cf. Exercise 1): For all $n \in \mathbb{N}^*$ and all sequences $(t_0, \dots, t_n)$ of elements of $\mathbb{P}$ such that $t_0 = t$ and $t_j = \delta_{i_j}(t_{j-1})$, $j = 1, \dots, n$,

$$\mathrm{P}\left(X_{t_1} = y_1, \dots, X_{t_n} = y_n | \mathscr{F}_t\right) = P_{i_1}(X_t, y_1) P_{i_2}(y_1, y_2) \cdots P_{i_n}(y_{n-1}, y_n),$$

for all elements $y_1, \dots, y_n$ of E. This property will, however, not be used.

Notice that if X is a Markov chain with state space E and transition functions $P_1, \dots, P_d$, and if E' is the minimal state of space of X, then the restrictions $P_1', \dots, P_d'$ of $P_1, \dots, P_d$ to $\mathsf{E}' \times \mathsf{E}'$ are transition functions (on E') and X is a Markov chain with state space E' and transition functions $P_1', \dots, P_d'$.

Throughout this section, X will denote a Markov chain with state space E and transition functions $P_1, \dots, P_d$.

An Example

Let $(Y^1_{t^1}, t^1 \in \mathbb{N})$ and $(Y^2_{t^2}, t^2 \in \mathbb{N})$ be two independent sequences of random variables with values in $\mathbb{Z}$. If the variables in each sequence are independent and identically distributed, then the process X defined by

$$X_t = \left(\sum_{s^1 \le t^1} Y^1_{s^1}, \sum_{s^2 \le t^2} Y^2_{s^2} \right), \qquad t = (t^1, t^2) \in \mathbb{N}^2,$$

is a Markov chain with state space $\mathbb{Z}^2$. Here, the filtration $(\mathcal{F}_t)$ is the natural filtration of the process X. The transition functions P_1 and P_2 are defined by

$$P_1(x, y) = \begin{cases} \mathrm{P}\{Y^1_0 = y^1 - x^1\}, & \text{if } x^2 = y^2, \\ 0, & \text{if } x^2 \neq y^2, \end{cases}$$

$$P_2(x, y) = \begin{cases} \mathrm{P}\{Y^2_0 = y^2 - x^2\}, & \text{if } x^1 = y^1, \\ 0, & \text{if } x^1 \neq y^1, \end{cases}$$

where $x = (x^1, x^2)$ and $y = (y^1, y^2)$.

The Operators P_i

In this section, the term *function* will be reserved for functions defined on E. If f is a nonnegative function, then $P_i f$ will denote the function defined by

$$P_i f(x) = \sum_{y \in \mathsf{E}} P_i(x, y) f(y).$$

If f is a function such that $P_i f^-(x)$ is finite, then we set

$$P_i f = P_i f^+ - P_i f^-.$$

Under the assumption that $P_i f^-$ is finite, it follows from (9.1) that

$$\begin{aligned} \mathrm{E}\big(f(X_{\delta_i}(t)) \mid \mathcal{F}_t\big) &= \sum_{y \in \mathsf{E}} \mathrm{P}\big(X_{\delta_i(t)} = y \mid \mathcal{F}_t\big) f(y) \\ &= \sum_{y \in \mathsf{E}} P_i(X_t, y) f(y) \qquad (9.3) \\ &= P_i f(X_t) \end{aligned}$$

for all $t \in \mathbb{P}$ and $i = 1, \ldots, d$.

From now on, for all functions f, $f(X)$ will denote the process $(f(X_t), t \in \mathbb{P})$. Notice that in the case where E is the minimal state space of X, the integrability of $f(X)$, or even only of $f^-(X)$, implies that $P_i f^-$ is finite for $i = 1, \ldots, d$. Indeed, by (9.3),

$$\mathrm{E}\big(f^-(X_{\delta_i(t)});\, X_t = x\big) = P_i f^-(x)\, \mathrm{P}\{X_t = x\}$$

for all $x \in \mathsf{E}$, from which we conclude that $P_i f^-(x) < \infty$, because there exists $t \in \mathbb{P}$ such that $\mathrm{P}\{X_t = x\} > 0$.

Superharmonic Functions

We will say that a function f is *superharmonic* if $P_i f^-$ is finite and $P_i f \leq f$ for $i = 1, \ldots, d$. Clearly, if f and g are superharmonic, then $\inf(f, g)$ is also superharmonic. In addition, if $(f_n,\ n \in \mathbb{N})$ is an increasing sequence of superharmonic functions, then the limit f of this sequence is superharmonic.

Proposition 1 (Criterion for superharmonicity). *Let f be a function. If f is superharmonic and $f(X)$ is integrable, then $f(X)$ is a supermartingale. Conversely, if $f(X)$ is a supermartingale and if E is the minimal state space of X, then f is superharmonic.*

Proof. Taking into account the comment that follows (1.11), the first part of the proposition is a direct consequence of (9.3). As for the second part, observe that if $f(X)$ is a supermartingale, then $P_i f$ is defined and $P_i f(X_t) \leq f(X_t)$ for $i = 1, \ldots, d$, by (9.3). Given $x \in \mathsf{E}$, it remains therefore to choose $t \in \mathbb{P}$ such that $\mathrm{P}\{X_t = x\} > 0$ in order to conclude that $P_i f(x) \leq f(x)$.

Superharmonic Envelopes

We will say that g is a *superharmonic majorant* of a function f if g is a superharmonic function such that $g \geq f$. When the set of superharmonic majorants of f has a smallest element, we term this element the *superharmonic envelope* of f.

Proposition 2 (Existence of the superharmonic envelope). *If f is a function such that $P_i f^-$ is finite for $i = 1, \ldots, d$, then the superharmonic envelope g of f exists and is equal to the limit of the nondecreasing sequence $(g_n,\ n \in \mathbb{N})$ defined inductively by*

$$g_0 = f \quad \text{and} \quad g_{n+1} = \sup\Big(f,\, \sup_i P_i g_n\Big). \tag{9.4}$$

Moreover,

$$g = \sup\Big(f,\, \sup_i P_i g\Big). \tag{9.5}$$

Proof. By (9.4),

$$g_0 = f \le \sup\Big(f, \sup_i P_i g_0\Big) = g_1.$$

Furthermore, also by (9.4), if $g_{n-1} \le g_n$ for some $n \in \mathbb{N}^*$, then

$$g_n = \sup\Big(f, \sup_i P_i g_{n-1}\Big) \le \sup\Big(f, \sup_i P_i g_n\Big) = g_{n+1},$$

and therefore the sequence $(g_n, n \in \mathbb{N})$ is nondecreasing. Set $g = \lim_{n\to\infty} g_n$. By passing to the limit in (9.4), we see that (9.5) is satisfied. It follows that $f \le g$ and $P_i g \le g$ for $i = 1, \dots, d$; in other words g is a superharmonic majorant of f. Assume that $\tilde{g}$ is also a superharmonic majorant of f. Then $\sup(f, \sup_i P_i \tilde{g}) \le \tilde{g}$. If $g_n \le \tilde{g}$ for some $n \in \mathbb{N}^*$, then

$$g_{n+1} = \sup\Big(f, \sup_i P_i g_n\Big) \le \sup\Big(f, \sup_i P_i \tilde{g}\Big) \le \tilde{g}.$$

Because $g_0 = f \le \tilde{g}$, it follows that $g_n \le \tilde{g}$ for all $n \in \mathbb{N}$, and therefore that $g \le \tilde{g}$.

Computing the Superharmonic Envelope

In the case where E is finite, formula (9.4) provides a method for computing an approximation to g. Indeed, the sequence $(g_n, n \in \mathbb{N})$ converges generally very rapidly to g and the iterative evaluation of the terms of this sequence is easy to carry out.

Again when E is finite, Proposition 2 shows that the calculation of g can also be achieved using methods from linear programming (the simplex algorithm, Karmarkar's algorithm). Indeed, given $f \in \mathbb{R}^{\mathsf{E}}$, the problem is to find $g \in \mathsf{S}$ such that

$$\sum_{x \in \mathsf{E}} g(x) = \inf_{h \in \mathsf{S}} \sum_{x \in \mathsf{E}} h(x), \tag{9.6}$$

where S denotes the subset of $\mathbb{R}^{\mathsf{E}}$ formed by all elements h such that $h \ge f$ and $\sup_i P_i h \le h$.

9.2 OPTIMAL CONTROL OF A MARKOV CHAIN

The results of Sections 3.4 and 7.5 take a particularly simple form in the Markovian setting. With appropriate identifications, they can even be presented in a unified way in the form of a general theorem on optimal stopping for Markov chains indexed by a set $\mathbb{P}$ of the type introduced in Section 9.1.

We prefer, however, to separate optimal stopping from optimal control. In fact, we will only present the conclusions concerning optimal control, leaving it to the reader to translate these conclusions in terms of optimal stopping.

We place ourselves in the setting of Section 7.5. Because the index set is $\mathbb{G}_f$, the concept of Markov chain will now be applied to the case where $\mathbb{P} = \mathbb{G}_f$ and where $(\mathcal{G}_\gamma)$ replaces $(\mathcal{F}_t)$. We will make the reader aware of this situation by saying that the *Markov chain is indexed by* $\mathbb{G}_f$. The direct successors of $\gamma \in \mathbb{G}_f$ are $\delta_i(\gamma) = (\delta_i(\gamma)_n, n \in \mathbb{N})$, $i = 1, \ldots, d$, defined by

$$\delta_i(\gamma)_n = \begin{cases} \gamma_n, & \text{if } n \leq l(\gamma), \\ \gamma_n + e_i, & \text{if } n > l(\gamma). \end{cases}$$

Let X be a Markov chain indexed by $\mathbb{G}_f$ with state space E. If f is a function defined on E, then in order that $f(X)$ be a reward process according to the definition of Section 7.5, it is necessary and sufficient that $f(X)$ be integrable and that $f^+(X)$ be of class $D_{\mathbf{G}_f}$. Under these conditions, the theorem of Section 7.5 applies and solves the problem of controlling X in order to maximize $E(f(X_\Gamma))$. The case where not only $f^+(X)$ but also $f(X)$ is of class $D_{\mathbf{G}_f}$ is particularly interesting, because in this case Snell's envelope of the process $f(X)$ is the process $g(X)$, where g is the superharmonic envelope of f. We will detail the main results concerning this case in the theorem below.

But first we need a definition that will also be useful in the following sections. Let $(Y_n, n \in \mathbb{N})$ be a process with state space E, and let D be a subset of E. We will term *time of first visit* of $(Y_n, n \in \mathbb{N})$ to D the random variable τ defined by

$$\tau = \inf\{n \in \mathbb{N}: Y_n \in \mathsf{D}\}. \tag{9.7}$$

Notice that if $(Y_n, n \in \mathbb{N})$ is adapted to a given filtration, then τ is a stopping time relative to this filtration.

Theorem (Optimal control of a Markov chain). *Let X be a Markov chain indexed by $\mathbb{G}_f$ with minimal state space E and transition functions $P_1, \ldots, P_d$. Let f be a function defined on E such that $f(X)$ is of class $D_{\mathbf{G}_f}$ (this condition is obviously satisfied if f is bounded). Let g denote the superharmonic envelope of f and set*

$$\begin{aligned} \mathsf{E}_0 &= \{x \in \mathsf{E}: f(x) = g(x)\}, \\ \mathsf{E}_i &= \left\{x \in \mathsf{E}: P_i g(x) = \sup_j P_j g(x)\right\}, \qquad i = 1, \ldots, d. \end{aligned} \tag{9.8}$$

Let $(\Gamma(n), n \in \mathbb{N})$ *be the sequence of elements of* $\mathbf{G}_f$ *defined inductively by*

$$\Gamma(0) = 0 \quad \text{and} \quad \Gamma(n+1) = \begin{cases} \delta_1(\Gamma(n)) & \text{on } \{X_{\Gamma(n)} \in \mathsf{E}_1\}, \\ \delta_i(\Gamma(n)) & \text{on } \left\{X_{\Gamma(n)} \in \mathsf{E}_i - \bigcup_{j:\, 1 \leq j < i} \mathsf{E}_j\right\}, \\ & i = 2, \ldots, d \end{cases} \tag{9.9}$$

and let τ *denote the time of first visit of* $(X_{\Gamma(n)}, n \in \mathbb{N})$ *to* E_0. *If* τ *is finite, then* $\Gamma(\tau)$ *is an optimal control (that is,* $\mathrm{E}(f(X_{\Gamma(\tau)})) = \sup_{\Gamma \in \mathbf{G}_f} \mathrm{E}(f(X_\Gamma)))$.

Proof. Because E is minimal and $f(X)$ is integrable, $P_i f^-$ is finite for $i = 1, \ldots, d$, by what was established after (9.3). The conclusions of Proposition 2 are therefore valid. By the last paragraph of Section 7.5, Snell's envelope of $f(X)$ is the smallest supermartingale that dominates $f(X)$. Let Z be any supermartingale that dominates $f(X)$, and let $(g_n, n \in \mathbb{N})$ be the sequence of functions defined in (9.4). Assume that $g_n(X) \leq Z$ for some $n \in \mathbb{N}$. By (9.3), we see that

$$P_i g_n(X_\gamma) = \mathrm{E}\left(g_n(X_{\delta_i(\gamma)}) | \mathscr{G}_\gamma\right) \leq \mathrm{E}\left(Z_{\delta_i(\gamma)}) | \mathscr{G}_\gamma\right) \leq Z_\gamma$$

for all $\gamma \in \mathbb{G}_f$ and $i = 1, \ldots, d$. By (9.4), it follows that

$$g_{n+1}(X) = \sup\left(f(X), \sup_i P_i g_n(X)\right) \leq Z.$$

Because $g_0(X) = f(X) \leq Z$, this shows that $g_n(X) \leq Z$ for all $n \in \mathbb{N}$ and therefore that $g(X) \leq Z$. But because g dominates f and $f(X)$ is integrable, $g(X)$ is an integrable process and therefore a supermartingale by Propositions 2 and 1. It follows that $g(X)$ is the smallest supermartingale that dominates $f(X)$ and therefore is Snell's envelope of $f(X)$. Taking (9.3) into account, the remainder of the theorem is a direct consequence of the theorem of Section 7.5.

9.3 THE SPECIAL CASE OF A RANDOM WALK

When $d = 1$, the problem studied in the preceding section reduces to a problem of optimal stopping. We are going to illustrate this with a simple example involving a random walk. This illustration will prepare for a more substantial example that we will study in Chapter 10.

In this section and in Chapter 10, it will be useful to have a symbol which denotes the *intervals* of $\mathbb{N}$, that is the subsets of $\mathbb{N}$ of the form $\{x\colon a \leq x \leq b\}$, where a and b are nonnegative integers such that $a \leq b$. We will use the symbol $[a, b]$ and term a and b the *extremities* of $[a, b]$. The intervals $[a+1, b]$, $[a, b-1]$, and $[a+1, b-1]$ may also be denoted, respectively, by $]a, b]$, $[a, b[$ and $]a, b[$. The *length* of the interval $[a, b]$ is the number $b - a$. Notice that an interval of length n contains $n + 1$ elements.

The fact that intervals of $\mathbb{N}$ are denoted using symbols that up to now referred to intervals of $\mathbb{R}$ will not cause any confusion, because intervals of $\mathbb{R}$ will only appear occasionally in the sequel and their presence will be indicated explicitly.

Concave Functions

Let f be a real-valued function defined on an interval $[a, b]$ of $\mathbb{N}$ with length > 1. We will say that f is *concave* if

$$\tfrac{1}{2}f(x+1) + \tfrac{1}{2}f(x-1) \leq f(x), \qquad \text{for all } x \in]a, b[. \tag{9.10}$$

We will say that f is *linear* (*affine*) if the inequality is replaced by an equality. We are going to prove that if f is concave [linear], then

$$\frac{b-x}{b-a}f(a) + \frac{x-a}{b-a}f(b) \leq f(x)\ [=f(x)], \qquad \text{for all } x \in [a, b]. \tag{9.11}$$

Notice that this relation is trivially satisfied by $x = a$ and $x = b$. Our assertion is obviously true in the case where $[a, b]$ is of length 1 or 2. Suppose that it is true for all intervals $[a, b]$ of length $\leq n$ $(n > 2)$ and consider an interval $[a, b]$ of length $n + 1$. We have already observed that for $x = a$ or $x = b$ there is nothing to prove. Let us therefore take an $x \in [a+1, b-1]$ and apply (9.11) to both intervals $[a+1, b]$ and $[a, x]$. It follows that

$$\frac{b-x}{b-a-1}f(a+1) + \frac{x-a-1}{b-a-1}f(b) \leq f(x)\ [=f(x)],$$

$$\frac{x-a-1}{x-a}f(a) + \frac{1}{x-a}f(x) \leq f(a+1)\ [=f(a+1)].$$

By replacing $f(a+1)$ in the first relation by the left-hand side of the second one, (9.11) follows after an elementary calculation that we leave to the reader.

For reasons of convenience, it is useful to agree that if the length of $[a, b]$ is 0 or 1, then all real-valued functions defined on $[a, b]$ are concave and linear. Moreover, we will say that a real-valued function f is concave or linear on a subinterval of its domain if the restriction of f to this subinterval is a concave or linear function.

If f is any function defined on $[a, b]$, then the set of concave functions g defined on $[a, b]$ and such that $g \geq f$ has a smallest element, termed the *concave envelope* of f.

Random Walks

Let $N \in \mathbb{N}$ be such that $N > 1$ and let E be the subinterval $[0, N]$ of $\mathbb{N}$. We will term *random walk* any Markov chain indexed by $\mathbb{N}$ with state space E and transition function P defined by

$$P(x, y) = \begin{cases} \frac{1}{2}, & \text{if } x \in]0, N[\text{ and } |x - y| = 1, \\ 0, & \text{if } x \in]0, N[\text{ and } |x - y| \neq 1, \\ \varepsilon(x, y), & \text{if } x \in \{0, N\}, \end{cases} \tag{9.12}$$

where $\varepsilon(x, y)$ is 0 if $x \neq y$ or 1 if $x = y$. It is clear from this definition that 0 and N are absorbing states of the chain. In the following we will assume that the underlying filtration of a random walk is its natural filtration.

Proposition (The superharmonic envelope in the case of a random walk). *Let f be a real-valued function defined on $\mathsf{E} = [0, N]$ and let P be the transition function defined in* (9.12). *In order that a real-valued function defined on E be superharmonic, it is necessary and sufficient that it be concave. The superharmonic envelope of f coincides therefore with the concave envelope of f. Moreover, if g denotes this envelope and if*

$$\mathsf{E}_0 = \{x \in \mathsf{E}: g(x) = f(x)\} = \{x_1, \ldots, x_n\},$$

where $x_1 < \cdots < x_n$, then $x_1 = 0$, $x_n = N$ and g is linear on $[x_k, x_{k+1}]$ for $k = 1, \ldots, n - 1$.

Proof. For all $x \in]0, N[$, the condition $Pg(x) \leq g(x)$ translates into

$$\tfrac{1}{2}g(x - 1) + \tfrac{1}{2}g(x + 1) \leq g(x).$$

In other words, a real-valued function g is superharmonic if and only if it is concave. It remains to prove that the concave envelope g of f is linear on $[x_k, x_{k+1}]$. By (9.5), $g(x) = \sup(f(x), \frac{1}{2}g(x - 1) + \frac{1}{2}g(x + 1))$ for all $x \in]0, N[$. Consequently, $g(x) = \frac{1}{2}g(x - 1) + \frac{1}{2}g(x + 1)$ for all $x \in \mathsf{E} - \mathsf{E}_0$, which means precisely that g is linear on each interval $[x_k, x_{k+1}]$.

Remark

If X denotes a random walk, then on the set $\{X_0 \in [x_k, x_{k+1}]\}$, the optimal choice in order to maximize the expected reward is to stop at the time of first visit to $\{x_k, x_{k+1}\}$. This choice does not depend on the law of X_0.

9.4 CONTROL AND STOPPING AT THE TIME OF FIRST VISIT TO A SET OF STATES

Consider a Markov chain X indexed by $\mathbb{I}$ with state space E and transition functions $P_1, \ldots, P_d$. Let D be a nonempty subset of E and f a nonnegative function defined on D.

Let $(\mathscr{G}_\gamma)$ be the filtration indexed by $\mathbb{G}_f$ defined by $\mathscr{G}_\gamma = \mathscr{F}_{\gamma_{l(\gamma)}}$ and let $\mathbf{G}_f$ be the set of finite controls relative to this filtration. Of course, these terms are to be interpreted according to the definition in Section 7.2. For all $\gamma \in \mathbb{G}_f$, let $\tau(\gamma)$ denote the time of first visit of $(X_{\gamma_n}, n \in \mathbb{N})$ to D and set

$$Y_\gamma = \begin{cases} f\left(X_{\gamma_{\tau(\gamma)}}\right) & \text{on } \{\tau(\gamma) < \infty\}, \\ 0 & \text{on } \{\tau(\gamma) = \infty\}. \end{cases} \tag{9.13}$$

Our objective is to establish that there exists a control Γ^o such that

$$\mathrm{E}(Y_{\Gamma^o}) = \sup_{\Gamma \in \mathbf{G}_f} \mathrm{E}(Y_\Gamma). \tag{9.14}$$

Such a control will be termed *optimal* for the process $Y = (Y_\gamma, \gamma \in \mathbb{G}_f)$.

In the remainder of this section, in order to avoid double subscripts, we will do as we did for increasing paths and write X_n^γ and $\mathscr{F}_n^\gamma$ instead of X_{γ_n} and $\mathscr{F}_{\gamma_n}$.

Reducing the Problem

In order to reduce this to a problem of optimal control of a Markov chain, we observe that if f is extended to all of E by setting $f(x) = 0$ for all $x \in \mathsf{E} - \mathsf{D}$, then for all $\gamma \in \mathbb{G}_f$,

$$Y_\gamma = f\left(\tilde{X}_\gamma\right), \tag{9.15}$$

where

$$\tilde{X}_\gamma = X^\gamma_{\tau(\gamma) \wedge l(\gamma)}. \tag{9.16}$$

This follows from the fact that $\tau(\gamma) \leq l(\gamma)$ on $\{\tau(\gamma) < \infty\}$ and $X^\gamma_{l(\gamma)} \in \mathsf{E} - \mathsf{D}$ on $\{\tau(\gamma) = \infty\}$.

The Markov Property of $\tilde{X}$

The process $\tilde{X} = (\tilde{X}_\gamma, \gamma \in \mathbb{G}_f)$ is a Markov chain relative to the filtration $(\mathscr{G}_\gamma)$ introduced above, with space state E and transition functions $\tilde{P}_1, \ldots, \tilde{P}_d$

defined by

$$\tilde{P}_i(x, y) = \begin{cases} P_i(x, y), & \text{if } x \in \mathsf{E} - \mathsf{D}, \\ \varepsilon(x, y), & \text{if } x \in \mathsf{D}, \end{cases} \tag{9.17}$$

where $\varepsilon(x, y)$ equals 0 if $x \neq y$ or 1 if $x = y$. Here is the proof of this assertion. Since $X^{\gamma}_{\tau(\gamma) \wedge l(\gamma)}$ is $\mathscr{F}^{\gamma}_{\tau(\gamma) \wedge l(\gamma)}$-measurable and $\mathscr{F}^{\gamma}_{\tau(\gamma) \wedge l(\gamma)} \subset \mathscr{F}^{\gamma}_{l(\gamma)} = \mathscr{G}_{\gamma}$, it follows that $\tilde{X}_{\gamma}$ is $\mathscr{G}_{\gamma}$-measurable. Fix $\gamma \in \mathbb{G}_{\mathrm{f}}$ and $y \in \mathsf{E}$, and show that

$$\mathrm{P}\left(\tilde{X}_{\delta_i(\gamma)} = y | \mathscr{G}_{\gamma}\right) = \tilde{P}_i\left(\tilde{X}_{\gamma}, y\right).$$

By the definition of $\tau(\gamma)$,

$$\tau(\delta_i(\gamma)) \wedge l(\delta_i(\gamma)) = \begin{cases} \tau(\gamma) & \text{on } \{\tau(\gamma) < \infty\}, \\ l(\gamma) + 1 & \text{on } \{\tau(\gamma) = \infty\}. \end{cases}$$

Taking $\delta_i(\gamma)$ instead of γ in (9.16) and observing that $\delta_i(\gamma)_{\tau(\delta_i(\gamma))} = \gamma_{\tau(\gamma)}$ on $\{\tau(\gamma) < \infty\}$ and $\delta_i(\gamma)_{l(\gamma)+1} = \gamma_{l(\gamma)} + e_i$, we deduce that

$$\tilde{X}_{\delta_i(\gamma)} = \begin{cases} X^{\gamma}_{\tau(\gamma)} = \tilde{X}_{\gamma} & \text{on } \{\tau(\gamma) < \infty\}, \\ X_{\gamma_{l(\gamma)} + e_i} & \text{on } \{\tau(\gamma) = \infty\}. \end{cases}$$

But X satisfies (9.1) and $\{\tau(\gamma) < \infty\} = \{\tau(\gamma) \leq l(\gamma)\} \in \mathscr{F}^{\gamma}_{l(\gamma)} = \mathscr{G}_{\gamma}$, which implies that

$$\begin{aligned} &\mathrm{P}\left(\tilde{X}_{\delta_i(\gamma)} = y | \mathscr{G}_{\gamma}\right) \\ &\quad = \mathrm{P}\left(\tilde{X}_{\gamma} = y, \tau(\gamma) < \infty | \mathscr{G}_{\gamma}\right) + \mathrm{P}\left(X_{\gamma_{l(\gamma)} + e_i} = y, \tau(\gamma) = \infty | \mathscr{F}^{\gamma}_{l(\gamma)}\right) \\ &\quad = \varepsilon\left(\tilde{X}_{\gamma}, y\right) 1_{\{\tau(\gamma) < \infty\}} + P_i\left(X^{\gamma}_{l(\gamma)}, y\right) 1_{\{\tau(\gamma) = \infty\}} \\ &\quad = \tilde{P}_i\left(\tilde{X}_{\gamma}, y\right) 1_{\{\tau(\gamma) < \infty\}} + \tilde{P}_i\left(\tilde{X}_{\gamma}, y\right) 1_{\{\tau(\gamma) = \infty\}} \\ &\quad = \tilde{P}_i\left(\tilde{X}_{\gamma}, y\right), \end{aligned}$$

where the next to the last equality is justified by (9.17) and the fact that $\tilde{X}_{\gamma} = X^{\gamma}_{\tau(\gamma)} \in \mathsf{D}$ on $\{\tau(\gamma) < \infty\}$ and $\tilde{X}_{\gamma} = X^{\gamma}_{l(\gamma)} \in \mathsf{E} - \mathsf{D}$ on $\{\tau(\gamma) = \infty\}$. It is thus established that $\tilde{X}$ is a Markov chain.

Existence of an Optimal Control for Y

The solution to our problem is now a direct consequence of the theorem of Section 9.2. Indeed, let us assume that $f(\tilde{X})$ is of class $\mathrm{D}_{\mathbf{G}_{\mathrm{f}}}$, or equivalently, that $f(X)$ is of class $\mathsf{D}_{\mathsf{A}_f}$, where f is the function given initially extended by the value 0 to $\mathsf{E} - \mathsf{D}$. Let g denote the superharmonic envelope of f relative to the transition functions $\tilde{P}_1, \dots, \tilde{P}_d$. Define $\mathsf{E}_0, \mathsf{E}_1, \dots, \mathsf{E}_d$ as in (9.8), but with $\tilde{P}_1, \dots, \tilde{P}_d$ instead of $P_1, \dots, P_d$. Moreover, define $(\Gamma(n),\ n \in \mathbb{N})$ by (9.9) and let τ denote the time of first visit of $(\tilde{X}_{\Gamma(n)},\ n \in \mathbb{N})$ to E_0. If τ is finite, then $\Gamma(\tau)$ is optimal for Y, in other words, the equality in (9.14) is satisfied by setting $\Gamma^o = \Gamma(\tau)$.

Optimization Using the Time of First Visit to D

It happens that $\Gamma(\tau)$ is also optimal in the case where τ is the time of first visit to D. The proof of this assertion is based on certain observations that will be useful further on. For any function h, it is obvious by (9.17) that $\tilde{P}_i h(x) = h(x)$ for all $x \in \mathsf{D}$. Define g' by

$$g'(x) = \begin{cases} f(x), & \text{if } x \in \mathsf{D}, \\ g(x), & \text{if } x \in \mathsf{E} - \mathsf{D}. \end{cases}$$

Clearly, $f \le g' \le g$. On the other hand, for $i = 1, \dots, d$, $\tilde{P}_i g'(x) = g'(x)$ if $x \in \mathsf{D}$ and $\tilde{P}_i g'(x) \le \tilde{P}_i g(x) \le g(x) = g'(x)$ if $x \in \mathsf{E} - \mathsf{D}$; therefore g' is a superharmonic majorant of f. Because g is the superharmonic envelope of f, we conclude that $g' = g$ and therefore that $g(x) = f(x)$ for all $x \in \mathsf{D}$. Denote now by τ the time of first visit of $(\tilde{X}_{\Gamma(n)},\ n \in \mathbb{N})$ to D and suppose that τ is finite. Then $g(\tilde{X}_{\Gamma(\tau)}) = f(\tilde{X}_{\Gamma(\tau)})$, which shows that condition a of Proposition 2 of Section 7.4 is satisfied. To conclude using this proposition, it remains to establish that condition b is also satisfied. This is equivalent to proving that $(g(\tilde{X}_{\Gamma(n \wedge \tau)}),\ n \in \mathbb{N})$ is a martingale, or equivalently, that

$$\mathrm{E}\Big(g\big(\tilde{X}_{\Gamma(n+1)}\big) | \mathscr{G}_{\Gamma(n)}\Big) = g\big(\tilde{X}_{\Gamma(n)}\big) \qquad \text{on } \{\tau \ge n\}. \tag{9.18}$$

By (9.9), (9.3), and (9.8),

$$\mathrm{E}\Big(g\big(\tilde{X}_{\Gamma(n+1)}\big) | \mathscr{G}_{\Gamma(n)}\Big) = \sup_i P_i g\big(\tilde{X}_{\Gamma(n)}\big).$$

On the other hand, since $f(x) = 0$ for all $x \in \mathsf{E} - \mathsf{D}$, (9.5) implies that $g(x) = \sup_i P_i g(x)$ for the same x. Because $\tilde{X}_{\Gamma(n)} \in \mathsf{E} - \mathsf{D}$ on $\{\tau > n\}$, equality (9.18) follows.

A Different Formulation

By a slight modification of the terminology in the definition of superharmonic envelope, the results of this section can be stated in an equivalent form without reference to the auxiliary chain $\tilde{X}$.

We say that a nonnegative function f defined on E is *superharmonic in* $\mathsf{E} - \mathsf{D}$ if $P_i f(x) \leq f(x)$ for all $x \in \mathsf{E} - \mathsf{D}$ and $i = 1, \ldots, d$. Given a nonnegative function f defined on E that vanishes on $\mathsf{E} - \mathsf{D}$, we will term *generalized envelope* of f the smallest element of the set of nonnegative functions defined on E that coincide with f on D and are superharmonic in $\mathsf{E} - \mathsf{D}$.

The considerations above establish the following theorem.

Theorem (Control and stopping at the time of first visit). *Let X be a Markov chain indexed by $\mathbb{I}$ with state space E and transition functions $P_1, \ldots, P_d$. Let D be a nonempty subset of E and f a nonnegative function defined on E that vanishes on $\mathsf{E} - \mathsf{D}$ and such that $f(X)$ is of class $\mathrm{D}_{\mathbf{A}_f}$. Let g denote the generalized envelope of f and set*

$$\mathsf{E}_0 = \mathsf{D} \text{ and } \mathsf{E}_i = \{x \in \mathsf{E} - \mathsf{D}: P_i g(x) = g(x)\}, \quad i = 1, \ldots, d. \tag{9.19}$$

Let Γ denote the predictable increasing path defined inductively by

$$\Gamma_0 = 0 \;\; and \;\; \Gamma_{n+1} = \begin{cases} \Gamma_n + e_1 & \text{on } \{X_n^\Gamma \in \mathsf{E}_0 \cup \mathsf{E}_1\}, \\ \Gamma_n + e_i & \text{on } \left\{X_n^\Gamma \in \mathsf{E}_i - \bigcup_{j:\, 1 \leq j < i} \mathsf{E}_j\right\}, \quad i = 2, \ldots, d. \end{cases} \tag{9.20}$$

Finally, for all predictable increasing paths Δ, let $\tau(\Delta)$ denote the time of first visit of X^Δ to D. If $\tau(\Gamma)$ is finite, then

$$\mathrm{E}\big(f\big(X^\Gamma_{\tau(\Gamma)}\big)\big) = \sup_{\Delta \in \mathbf{G}^\infty} \mathrm{E}\big(f\big(X^\Delta_{\tau(\Delta)}\big); \tau(\Delta) < \infty\big), \tag{9.21}$$

where $\mathbf{G}^\infty$ denotes the set of predictable increasing paths.

9.5 MARKOV STRUCTURES

In the theorem of Section 9.2, the construction of the optimizing sequence $(\Gamma(n),\ n \in \mathbb{N})$ and of the stopping time τ for which the control $\Gamma(\tau)$ is optimal (if τ is finite) uses at each stage $\Gamma(n)$ only a small portion of the information available in $\mathscr{G}_{\Gamma(n)}$, namely, the value of $X_{\Gamma(n)}$. It is natural to conclude from this that the search for an optimal control in the setting of a Markov chain can be limited to the elements of a comparatively small subset of $\mathbf{G}_\mathrm{f}$. Of course, such a possibility is all the more useful when the quantity of information represented by $\mathscr{G}_\gamma$ is large.

Generally, when trying to build an optimal control, it is interesting to know under which conditions it is sufficient to use only part of the available information. Such a simplification is attractive in applications, in particular when the storage and treatment of the information in its entirety are not feasible.

This section is the logical continuation of Sections 7.5 and 9.2. Our objective is to exhibit the conditions under which the type of simplification we have just mentioned is possible.

Reduced Information

In this section, $(\mathscr{H}_\gamma, \gamma \in \mathbb{G}_f)$ is a given family of sigma fields such that $\mathscr{H}_\gamma \subset \mathscr{G}_\gamma$ for all $\gamma \in \mathbb{G}_f$. We will denote this family more simply by $(\mathscr{H}_\gamma)$. We will not assume that $\mathscr{H}_\gamma \subset \mathscr{H}_\delta$ if $\gamma \prec \delta$, but we will suppose that each $\mathscr{H}_\gamma$ contains all null sets. Of course, $(\mathscr{G}_\gamma)$ is the basic filtration introduced in Section 7.2.

Intuitively, the sigma field $\mathscr{H}_\gamma$ describes the present, that is, the new information that becomes available just as we reach stage γ. Normally, $\mathscr{H}_\gamma$ contains only a small part of the information available at stage γ; in other words $\mathscr{H}_\gamma$ is small compared to $\mathscr{G}_\gamma$. For example, when $\mathscr{F}$ is finite, the cardinality of a generating partition of $\mathscr{G}_\gamma$ is typically an exponential function of $l(\gamma)$, whereas for $\mathscr{H}_\gamma$, this cardinality may only be a linear function of $l(\gamma)$.

We say that a process X indexed by $\mathbb{G}_f$ is *adapted* to $(\mathscr{H}_\gamma)$ if X_γ is $\mathscr{H}_\gamma$-measurable for all $\gamma \in \mathbb{G}_f$.

The Markov Property

We say that $(\mathscr{H}_\gamma)$ satisfies the *Markov property* if for all couples (γ, δ) of elements of $\mathbb{G}_f$ such that $\delta \in \mathbb{D}_\gamma$, $\mathscr{H}_\delta$ is conditionally independent of $\mathscr{G}_\gamma$ given $\mathscr{H}_\gamma$.

In the sequel, we will not use any particular consequence of this definition, but we nevertheless mention one here in order to provide a better understanding of the structure imposed by the Markov property: If $(\mathscr{H}_\gamma)$ satisfies this property, then for all couples (γ, δ) of elements of $\mathbb{G}_f$ such that $\gamma \prec \delta$, the sigma field $\bigvee_{n=l(\gamma)}^{l(\delta)} \mathscr{H}_{\delta[n]}$ is conditionally independent of $\mathscr{G}_\gamma$ given $\mathscr{H}_\gamma$ (cf. Exercise 6). On the other hand, it is easy to check that the Markov property does not in general imply conditional independence of the sigma fields $\bigvee_{\delta \in \mathbb{D}_\gamma} \mathscr{H}_\delta$ and $\mathscr{G}_\gamma$ given $\mathscr{H}_\gamma$ (cf. Exercise 7).

Having agreed that $\mathscr{H}_\gamma$ represents the information available in the present and $\mathscr{G}_\gamma$ the information contained in the (wide sense) past, the Markov property can be expressed by saying that the future along each control is conditionally independent of the past given the present.

Markov Controls

If Γ is a random variable with values in $\mathbb{G}_f$, we denote by $\mathscr{H}_\Gamma$ the sigma field generated by the sets of the form $\{\Gamma = \gamma\} \cap F_\gamma$, with $\gamma \in \mathbb{G}_f$ and $F_\gamma \in \mathscr{H}_\gamma$.

Clearly, H belongs to $\mathscr{H}_\Gamma$ if and only if for all $\gamma \in \mathbb{G}_f$, there exists $F_\gamma \in \mathscr{H}_\gamma$ such that $\{\Gamma = \gamma\} \cap H = \{\Gamma = \gamma\} \cap F_\gamma$.

Notice that if $\mathscr{H}_\gamma$ is generated by a random variable Y_γ for all $\gamma \in \mathbb{G}_f$, then $\mathscr{H}_\Gamma$ is generated by the couple (Y_Γ, Γ). Also notice that in the case where $(\mathscr{H}_\gamma)$ is a filtration and Γ is a finite control relative to this filtration, $\mathscr{H}_\Gamma$ is the usual sigma field associated with Γ according to the definition given in Section 7.3.

We say that an element $\Gamma \in \mathbf{G}_f$ is a *Markov control* if $\Gamma^{[n+1]}$ is $\mathscr{H}_{\Gamma^{[n]}}$-measurable for all $n \in \mathbb{N}$. By what was said above, $\Gamma^{[n+1]}$ is $\mathscr{H}_{\Gamma^{[n]}}$-measurable if for all couples (γ, δ) of elements of $\mathbb{G}_f$, there exists $F^n_{\gamma,\delta} \in \mathscr{H}_\gamma$ such that

$$\left\{\Gamma^{[n]} = \gamma\right\} \cap \left\{\Gamma^{[n+1]} = \delta\right\} = \left\{\Gamma^{[n]} = \gamma\right\} \cap F^n_{\gamma,\delta}. \tag{9.22}$$

Obviously, this condition only needs to be checked in the case where $\delta^{[n]} = \gamma$, because if $\delta^{[n]} \neq \gamma$, (9.22) is true with $F^n_{\gamma,\delta} = \varnothing$. Observe that in the case where $\delta^{[n]} = \gamma$, (9.22) reduces to

$$\left\{\Gamma^{[n+1]} = \delta\right\} = \left\{\Gamma^{[n]} = \gamma\right\} \cap F^n_{\gamma,\delta}. \tag{9.23}$$

The point to be noted is that (9.22) expresses the idea that once it is known that $\Gamma^{[n]} = \gamma$, the selection of $\Gamma^{[n+1]}$ is made using only information from the present.

Optimization Using Markov Controls

We are going to show that the existence of an optimal Markov control is closely related to the Markov property of the reduced information.

Theorem (Condition for existence of an optimal Markov control). *Let X be a process indexed by $\mathbb{G}_f$, adapted to $(\mathscr{H}_\gamma)$ and of class $\mathrm{D}_{\mathbf{G}_f}$. Let Z be Snell's envelope of X, $(\Gamma(n), n \in \mathbb{N})$ the optimizing sequence defined in* (7.16), *and τ the stopping time defined by*

$$\tau = \inf\left\{n \in \mathbb{N}: Z_{\Gamma(n)} = X_{\Gamma(n)}\right\}. \tag{9.24}$$

If $(\mathscr{H}_\gamma)$ satisfies the Markov property, then

$$\sup_{\Gamma \in \mathbf{G}_f} \mathrm{E}(X_\Gamma) = \sup_{\Gamma \in \mathbf{H}} \mathrm{E}(X_\Gamma), \tag{9.25}$$

where $\mathbf{H}$ denotes the set of Markov controls. Moreover, $\Gamma(n)$ is a Markov control, and if τ is finite, then $\Gamma(\tau)$ is an optimal Markov control. In other words, $\mathrm{E}(X_{\Gamma(\tau)})$ is equal to both sides of the equality in (9.25). *Conversely, if the equality in* (9.25) *is true for all processes X indexed by $\mathbb{G}_f$ that are bounded and adapted to $(\mathscr{H}_\gamma)$, then $(\mathscr{H}_\gamma)$ satisfies the Markov property.*

Proof. Assume that $(\mathscr{H}_\gamma)$ satisfies the Markov property. In order to establish (9.25), it suffices to prove that for all $\varepsilon > 0$, there exists an ε-optimal Markov control. We are going to prove that the ε-optimal control Γ^ε described at the end of Section 7.6 is a Markov control. To do this, we put ourselves in the setting of that section and assume that Z_γ, $D(\gamma)$, and $\Gamma(\gamma)$ are defined using (7.23), (7.24), and (7.25), which means that for the time being, we will not conform to the notation in the statement of the theorem. By (backwards) induction and using the Markov property, we see that $\mathscr{G}_\gamma$ can be replaced by $\mathscr{H}_\gamma$ in (7.23) and (7.24). Consequently, Z_γ and $D(\gamma)$ are $\mathscr{H}_\gamma$-measurable. Let us show by (backwards) induction that the control $\Gamma(\gamma)$ is Markov. Since this conclusion is clearly true if $l(\gamma) = n$, it suffices to prove that if $l(\gamma) < n$ and if $\Gamma(\delta')$ is Markov for all $\gamma' \in \mathbb{D}_\gamma$, then for all $m \in \mathbb{N}$ and all $\delta \in \mathbb{G}_n$, there exists $F \in \mathscr{H}_{\delta[m]}$ such that

$$\left\{\Gamma(\gamma)^{[m+1]} = \delta\right\} = \left\{\Gamma(\gamma)^{[m]} = \delta^{[m]}\right\} \cap F. \tag{9.26}$$

In the case where $m < l(\gamma)$, the relation $\gamma \prec \Gamma(\gamma)$ immediately implies that we can take $F = \Omega$ or $F = \varnothing$ according to whether $\delta = \gamma^{[m+1]}$ or $\delta \neq \gamma^{[m+1]}$. We now consider the case where $m \geq l(\gamma)$. Again thanks to the relation $\gamma \prec \Gamma(\gamma)$, we can suppose that $\gamma \prec \delta^{[m]}$, because otherwise $\{\Gamma(\gamma)^{[m+1]} = \delta\} \subset \{\Gamma(\gamma)^{[m]} = \delta^{[m]}\} = \varnothing$ and $F = \varnothing$ satisfies (9.26). By (7.25), $\{\Gamma(\gamma)^{[m+1]} = \gamma\} = \{\Gamma(\gamma)^{[m]} = \gamma\} \cap \{Z_\gamma = X_\gamma\}$, and taking into account the fact that $\gamma^{[m]} = \gamma$, we see that

$$\left\{\Gamma(\gamma)^{[m+1]} = \delta\right\} = \begin{cases} \left\{\Gamma(\gamma)^{[m]} = \delta^{[m]}\right\} \cap \{Z_\gamma = X_\gamma\}, & \text{if } \delta = \gamma, \\ \left\{\Gamma(D(\gamma))^{[m+1]} = \delta\right\} \cap \{Z_\gamma > X_\gamma\}, & \text{if } \delta \neq \gamma. \end{cases}$$

In the first case, it suffices to set $F = \{Z_\gamma = X_\gamma\}$. In the second case, by (7.24) and the hypothesis that stipulates that $\Gamma(\delta')$ is a Markov control for all $\delta' \in \mathbb{D}_\gamma$, we can write

$$\begin{aligned}
&\left\{\Gamma(D(\gamma))^{[m+1]} = \delta\right\} \\
&\quad = \bigcup_{\delta' \in \mathbb{D}_\gamma} \left(\left\{\Gamma(\delta')^{[m+1]} = \delta\right\} \cap \{D(\gamma) = \delta'\}\right) \\
&\quad = \bigcup_{\delta' \in \mathbb{D}_\gamma} \left(\left\{\Gamma(\delta')^{[m]} = \delta^{[m]}\right\} \cap F_{\delta'} \cap \{D(\gamma) = \delta'\}\right) \\
&\quad = \left\{\Gamma(D(\gamma))^{[m]} = \delta^{[m]}\right\} \cap \left(\bigcup_{\delta' \in \mathbb{D}_\gamma} (F_{\delta'} \cap \{D(\gamma) = \delta'\})\right),
\end{aligned}$$

where $F_{\delta'}$ denotes an element of $\mathscr{H}_{\delta[m]}$. Using (7.25), we see that (9.26) is satisfied by setting $F = (\bigcup_{\delta' \in \mathbb{D}_\gamma} (F_{\delta'} \cap \{D(\gamma) = \delta'\})) \cap \{Z_\gamma > X_\gamma\}$. Because $F \in \mathscr{H}_{\delta^{[m]}}$, the proof that $\Gamma(\gamma)$ is Markov is complete.

Let us return to the setting and notation of the theorem in order to prove the two assertions that follow (9.25). In the last paragraph of Section 7.6, we saw that $Z_\gamma = \lim_{n \to \infty} Z_\gamma^n$, where Z^n is Snell's envelope of the restriction of X to $\mathbb{G}_n$. From what has been established above, Z_γ^n is $\mathscr{H}_\gamma$-measurable. Consequently, Z is adapted to $(\mathscr{H}_\gamma)$. Notice incidentally that this conclusion can also be deduced from (7.22). If τ is finite, we know by the theorem of Section 7.5 that $\Gamma(\tau)$ is an optimal control. Let us show that this control is Markov. To this end, fix $m \in \mathbb{N}$ and consider a couple (γ, δ) of elements of $\mathbb{G}_f$ such that $\delta^{[m]} = \gamma$. We shall prove that

$$\left\{\Gamma(\tau)^{[m+1]} = \delta\right\} = \left\{\Gamma(\tau)^{[m]} = \gamma\right\} \cap F, \tag{9.27}$$

where

$$F = \begin{cases} \{Z_\gamma = X_\gamma\}, & \text{if } l(\delta) \le m, \\ \left\{\inf\left\{\delta' \in \mathbb{D}_\gamma : \mathrm{E}\left(Z_{\delta'} | \mathscr{H}_\gamma\right) = \sup\limits_{\eta \in \mathbb{D}_\gamma} \mathrm{E}\left(Z_\eta | \mathscr{H}_\gamma\right)\right\} = \delta\right\} \cap \{Z_\gamma > X_\gamma\}, & \\ & \text{if } l(\delta) > m. \end{cases}$$

This will establish our assertion because $F \in \mathscr{H}_\gamma$. By the way $(\Gamma(n), n \in \mathbb{N})$ is defined, it is clear that $l(\Gamma(n)) = n$ and that $\Gamma(n)^{[k]} = \Gamma(n \wedge k)$. Suppose that $l(\delta) \le m$. Then $\delta = \delta^{[m]} = \gamma$, and taking (9.24) into account, we can write

$$\begin{aligned} \left\{\Gamma(\tau)^{[m+1]} = \delta\right\} &= \{\Gamma(\tau) = \delta\} \\ &= \{\Gamma(\tau) = \gamma\} \cap \{Z_\gamma = X_\gamma\} \\ &= \{\Gamma(\tau \wedge m) = \gamma\} \cap \{Z_\gamma = X_\gamma\} \\ &= \left\{\Gamma(\tau)^{[m]} = \gamma\right\} \cap \{Z_\gamma = X_\gamma\}, \end{aligned}$$

which proves that (9.27) is true. On the other hand, consider the case where $l(\delta) > m$. If $l(\delta) > m + 1$, then (9.27) is trivially true, because $\{\Gamma(\tau)^{[m+1]} = \delta\} = \varnothing$ and $F = \varnothing$, since $\delta^{[m]} = \gamma$ and therefore $\delta \ne \mathbb{D}_\gamma$. If $l(\gamma) = m + 1$, then $\delta \in \mathbb{D}_\gamma$, again because $\delta^{[m]} = \gamma$. In this case, by (9.24) and (7.16), we see that

$$\begin{aligned} \left\{\Gamma(\tau)^{[m+1]} = \delta\right\} &= \{\tau > m\} \cap \{\Gamma(m) = \gamma\} \cap \{\Gamma(m+1) = \delta\} \\ &= \{\tau > m\} \cap \{\Gamma(m) = \gamma\} \cap \{Z_\gamma > X_\gamma\} \\ &\qquad \cap \left\{\inf\left\{\delta' \in \mathbb{D}_\gamma : \mathrm{E}\left(Z_{\delta'} | \mathscr{G}_\gamma\right) = \sup_{\eta \in \mathbb{D}_\gamma} \mathrm{E}\left(Z_\eta | \mathscr{G}_\gamma\right)\right\} = \delta\right\} \\ &= \left\{\Gamma(\tau)^{[m]} = \gamma\right\} \cap F, \end{aligned}$$

where the last equality is justified because it is possible to replace $\mathscr{G}_\gamma$ by $\mathscr{H}_\gamma$, thanks to the Markov property and the fact that Z is adapted to $(\mathscr{H}_\gamma)$. Thus (9.27) is established and $\Gamma(\tau)$ is Markov. In the same manner but more simply, it can be proved that $\Gamma(n)$ is Markov for all $n \in \mathbb{N}$.

We now prove the converse by assuming that $(\mathscr{H}_\gamma)$ does not satisfy the Markov property. In this case, there exists $\delta \in \mathbb{G}_{\mathrm{f}}$, $\eta \in \mathbb{D}_\delta$ and a bounded nonnegative and $\mathscr{H}_\eta$-measurable random variable Y such that

$$\mathrm{P}\{\mathrm{E}(Y|\mathscr{G}_\delta) \neq \mathrm{E}(Y|\mathscr{H}_\delta)\} > 0.$$

Set

$$F = \{\mathrm{E}(Y|\mathscr{G}_\delta) < \mathrm{E}(Y|\mathscr{H}_\delta)\}.$$

Since both conditional expectations have the same expectation, it is clear that $\mathrm{P}(F) > 0$. Moreover, F does not belong to $\mathscr{H}_\delta$, because if it did, we would conclude that

$$\mathrm{E}(Y; F) = \mathrm{E}(\mathrm{E}(Y|\mathscr{G}_\delta); F) < \mathrm{E}(\mathrm{E}(Y|\mathscr{H}_\delta); F) = \mathrm{E}(Y; F),$$

a contradiction. Set

$$X_\gamma = \begin{cases} \mathrm{E}(Y|\mathscr{H}_\delta), & \text{if } \gamma = \delta, \\ Y, & \text{if } \gamma = \eta, \\ 0, & \text{if } \gamma \in \mathbb{G}_{\mathrm{f}} - \{\delta, \eta\}. \end{cases}$$

This defines a process X indexed $\mathbb{G}_{\mathrm{f}}$ and adapted to $(\mathscr{H}_\gamma)$. Set

$$\Gamma = \begin{cases} \delta & \text{on } F, \\ \eta & \text{on } F^c. \end{cases}$$

Clearly, Γ is a control, and this control is not Markov because F does not belong to $\mathscr{H}_\delta$. On the other hand,

$$\mathrm{E}(X_\Gamma) = \mathrm{E}(\mathrm{E}(Y|\mathscr{H}_\delta)1_F + Y1_{F^c}) > \mathrm{E}(\mathrm{E}(Y|\mathscr{G}_\delta)1_F + Y1_{F^c}) = \mathrm{E}(Y).$$

Now let Δ be some arbitrary Markov control. Set $n = l(\delta)$, $F_1 = \{\Delta^{[n]} = \delta\}$ and let F_2 denote an element of $\mathscr{H}_\delta$ such that

$$\{\Delta^{[n+1]} = \eta\} = F_1 \cap F_2.$$

Then

$$\begin{aligned}
\mathrm{E}(X_\Delta) &= \mathrm{E}\big(\mathrm{E}(Y|\mathscr{H}_\delta)1_{\{\Delta=\delta\}} + Y1_{\{\Delta=\eta\}}\big)\\
&\le \mathrm{E}\big(\mathrm{E}(Y|\mathscr{H}_\delta)1_{\{\Delta^{[n]}=\delta\}\cap\{\Delta^{[n+1]}\neq\eta\}} + Y1_{\{\Delta^{[n+1]}=\eta\}}\big)\\
&= \mathrm{E}\big(\mathrm{E}(Y|\mathscr{H}_\delta)1_{F_1\cap F_2^c} + Y1_{F_1\cap F_2}\big)\\
&\le \mathrm{E}\big(\mathrm{E}(Y|\mathscr{H}_\delta)1_{F_2^c} + Y1_{F_2}\big)\\
&= \mathrm{E}(Y).
\end{aligned}$$

But $\mathrm{E}(Y) < \mathrm{E}(X_\Gamma)$, which implies that (9.25) is not true. The theorem is thus proved.

EXERCISES

1. Let X be a Markov chain indexed by $\mathbb{P}$ with state space E and transition functions $P_1, \ldots, P_d$. Show that for all $n > 0$, all nonnegative functions f defined on $\mathsf{E} \times \cdots \times \mathsf{E}$ (n times) and all sequences $(t_0, \ldots, t_n)$ of elements of $\mathbb{P}$ such that $t_0 = t$ and $t_j = \delta_{i_j}(t_{j-1})$, $j = 1, \ldots, n$,

$$\mathrm{E}\big(f(X_{t_1}, \ldots, X_{t_n})|\mathscr{F}_t\big) = (P_{i_1}P_{i_2} \cdots P_{i_n}f)(X_t),$$

where $P_{i_1}P_{i_2} \cdots P_{i_n}f$ denotes the function

$$x \mapsto \sum_{y_1\in\mathsf{E}} \cdots \sum_{y_n\in\mathsf{E}} P_{i_1}(x, y_1)P_{i_2}(y_1, y_2) \cdots P_{i_n}(y_{n-1}, y_n)f(y_1, \ldots, y_n).$$

2. Find a Markov chain X indexed by $\mathbb{N}$ and a function f defined on the state space E of this chain such that $(f(X_n), n \in \mathbb{N})$ is a supermartingale, but f is not superharmonic.

3. The data being that of the theorem of Section 9.2, deduce from Exercise 6 of Chapter 7 the following assertion: if $\mathrm{E}(\sup_{\gamma\in\mathbb{G}_f} f^+(X_\gamma)) < \infty$ and if $\lim_{l(\gamma)\to\infty} f(X_\gamma)$ exists (in $\overline{\mathbb{R}}$), then for all $\varepsilon > 0$, the time τ^ε of first visit of $(X_{\Gamma(n)}, n \in \mathbb{N})$ to $\{x \in \mathsf{E}: g(x) \le f(x) + \varepsilon\}$ is finite and $\Gamma(\tau^\varepsilon)$ is an ε-optimal control (that is, $\mathrm{E}(f(X_{\Gamma(\tau^\varepsilon)})) \ge \sup_{\Gamma\in\mathbf{G}_f} \mathrm{E}(f(X_\Gamma)) - \varepsilon$).

4. Suppose that $\mathsf{E} = \mathbb{N}_4$ and consider the transition function

$$P = (P(x,y), (x,y) \in \mathsf{E} \times \mathsf{E}) = \begin{pmatrix} 1 & 0 & 0 & 0 & 0 \\ \frac{1}{5} & 0 & \frac{4}{5} & 0 & 0 \\ \frac{3}{4} & 0 & 0 & 0 & \frac{1}{4} \\ 0 & \frac{2}{3} & \frac{1}{3} & 0 & 0 \\ 0 & 0 & 0 & 0 & 1 \end{pmatrix}.$$

Let f be the payoff function defined on E by $f(0) = 6$, $f(1) = 5$, $f(2) = 4$, $f(3) = 3$ and $f(4) = 0$. Suppose that X is a Markov chain index by $\mathbb{N}$ with state space E and transition function P.

a. Find the superharmonic envelope of f.

b. Find a finite stopping time τ^o such that $\mathrm{E}(f(X_{\tau^o})) = \sup_{\tau \in \mathbf{T}_f} \mathrm{E}(f(X_\tau))$, where $\mathbf{T}_f$ denotes the set of finite stopping times.

5. (An example in which there is no optimal finite stopping time). Fix $p \in]0, \frac{1}{2}[$ and let X be a Markov chain index by $\mathbb{N}$ with state space $\mathsf{E} = \mathbb{Z}$ and transition function P defined by

$$P(x,y) = \begin{cases} p, & \text{if } y = x+1, \\ 1-p, & \text{if } y = x-1, \\ 0, & \text{if } |x-y| \neq 1. \end{cases}$$

Assume further that $\mathrm{P}\{X_0 = x\} > 0$ for all $x \in \mathsf{E}$. Let f be the function defined on E by $f(x) = x^+$ and for all $n \in \mathbb{Z}$, set $\tau_n = \inf\{m \in \mathbb{Z}: X_m \geq n\}$.

a. Compute $\mathrm{E}(f(X_{\tau_n}) 1_{\{\tau_n < \infty\}} | X_0 = x)$. (*Hint.* Recalling b of Exercise 3 of Chapter 1, check that

$$P(\tau_n < \infty | X_0 = x) = \begin{cases} 1, & \text{if } x \geq n, \\ \left(\dfrac{p}{1-p}\right)^{n-x}, & \text{if } x < n.) \end{cases}$$

b. Show that there is $n_0 \in \mathbb{Z}$ such that

$$\mathrm{E}\big(f(X_{\tau_{n_0}}) 1_{\{\tau_{n_0} < \infty\}} | X_0 = x\big) = \sup_{n \in \mathbb{Z}} \mathrm{E}\big(f(X_{\tau_n}) 1_{\{\tau_n < \infty\}} | X_0 = x\big), \qquad \text{for all } x \in \mathsf{E}.$$

c. Define a function g on E by $g(x) = \mathrm{E}(f(X_{\tau_{n_0}}) 1_{\{\tau_{n_0} < \infty\}} | X_0 = x)$. Show that g is superharmonic, that $g \geq f$, and that g is the superharmonic envelope of f.

d. Conclude that $E(f(X_{\tau n0})\,1_{\{\tau n0<\infty\}}) = \sup_{\tau\in \mathbf{T}} E(f(X_\tau)\,1_{\{\tau n<\infty\}})$, where $\mathbf{T}$ denotes the set of stopping times relative to the natural filtration of X. Notice that $\lim_{n\to\infty} X_n = -\infty$, and therefore that $P\{\tau_{n_0} = \infty\} > 0$. (*Hint.* Define processes $Y = (Y_t,\ t \in \bar{\mathbb{I}})$ and $Z = (Z_t,\ t\in \bar{\mathbb{I}})$ by $Y_t = f(X_t)$ if $t \in \mathbb{I}$ and $Y_{\bowtie} = Z_{\bowtie} = 0$. Verify that Z is Snell's envelope of Y and conclude using the theorem of Section 3.3.)

e. Prove that if σ is a stopping time such that $P\{\sigma < \tau_{n0}\} > 0$ for some $x \in \mathsf{E}$, then $E(f(X_\sigma)\,1_{\{\tau<\infty\}})\ \sup_{\tau\in\mathbf{T}_f} E(f(X_\tau)\,1_{\{\tau<\infty\}})$, where $\mathbf{T}_f$ denotes the set of finite stopping times. (*Hint.* Show that the supremum over $\mathbf{T}_f$ equals $g(x)$, and that $P\{Y_\tau < Z_\tau\} > 0$. Conclude using Proposition 1 of Section 3.3.)

6. In the setting of Section 9.5, show that if $(\mathscr{H}_\gamma)$ satisfies the Markov property, then for all couples (γ, δ) of elements of $\mathbb{G}_f$ such that $\gamma \prec \delta$, the sigma field $\bigvee_{n=l(\gamma)}^{l(\delta)} \mathscr{H}_{\delta[n]}$ is conditionally independent of $\mathscr{G}_\gamma$ given $\mathscr{H}_\gamma$. More generally, show under the same condition that for all couples (Γ, Δ) of elements of $\mathbf{G}_f$ such that $\Gamma \prec \Delta$, the sigma field $\bigvee_{n\in\mathbb{N}} \mathscr{H}_{\Delta[l(\Gamma)+n]}$ is conditionally independent of $\mathscr{G}_\Gamma$ given $\mathscr{H}_\Gamma$.

7. In the setting of Section 9.5, give an example of a family of sigma fields $(\mathscr{H}_\gamma,\ \gamma \in \mathbb{G}_f)$ that satisfies the Markov property but for which there exists $\gamma \in \mathbb{G}_f$ such that $\bigvee_{\delta\in\mathbb{D}_\gamma} \mathscr{H}_\delta$ is not conditionally independent of $\mathscr{G}_\gamma$ given $\mathscr{H}_\gamma$.

8. Let $(\mathscr{F}^1_{t^1},\ t^1 \in \mathbb{N}),\ldots,(\mathscr{F}^d_{t^d},\ t^d\in\mathbb{N})$ be independent filtrations. For $t = (t^1,\ldots,t^d) \in \mathbb{I}\ (=\mathbb{N}^d)$, set $\mathscr{F}_t = \mathscr{F}^1_{t^1} \vee \cdots \vee \mathscr{F}^d_{t^d}$ and let $\mathbf{A}_f$ denote the set of finite accessible stopping points relative to the filtration $(\mathscr{F}_t,\ t\in\mathbb{I})$. For $i = 1,\ldots,d$, let $\mathbf{T}^i_f$ denote the set of finite stopping times relative to the filtration $(\mathscr{F}^i_{t^i})$ and let $(X^i_{t^i},\ t^i \in \mathbb{N})$ be a reward process (in the sense of Section 3.4) adapted to $(\mathscr{F}^i_{t^i})$. Define a process $(X_t,\ t\in\mathbb{I})$ by setting $X_t = \sum_{i=1}^d X^i_{t^i}$ for all $t = (t^1,\ldots,t^d)\in\mathbb{I}$. Show that

$$\sup_{T\in\mathbf{A}_f} E(X_T) = \sum_{i=1}^{d} \sup_{\tau_i\in\mathbf{T}^i_f} E\left(X^i_{\tau_i}\right).$$

(*Hint.* If $T = (T^1,\ldots,T^d)$ is a stopping point, notice that T^i is a stopping time for the filtration $(\tilde{\mathscr{F}}^i_{t^i})$, where for $i = 1,\ldots,d$ and all $t^i \in \mathbb{N}$,

$$\tilde{\mathscr{F}}^i_{t^i} = \mathscr{F}^i_{t^i} \vee \left(\bigvee_{j\neq i} \mathscr{F}^i_\infty \right).$$

Then use the theorem of Section 9.5)

9. (Application to the sampling problem of Chapter 6). In the context of Sections 6.1, 6.2, and 6.3, let $(\Omega, \mathscr{F}, \mathrm{P})$ and $(\mathscr{F}_t,\ t \in \mathbb{I})$ be the probability space and filtration defined in Section 6.2, let X be the process defined in (6.3), and for $t \in \mathbb{I}$, let $\pi^t = (\pi_1^t, \ldots, \pi_l^t)$, where for $j = 1, \ldots, l$, π_j^t is defined in (6.5). For all $\gamma \in \mathbb{G}_f$, set $\mathscr{G}_\gamma = \mathscr{F}_{\gamma_{l(\gamma)}}$ and $\tilde{X}_\gamma = X_{\gamma_{l(\gamma)}}$, and let $\mathscr{H}_\gamma$ be the sigma field generated by $\pi^{\gamma_{l(\gamma)}}$.

   a. Check that $\tilde{X}$ is adapted to $(\mathscr{H}_\gamma)$. (*Hint.* Use the proposition of Section 6.3.)

   b. Prove that $(\mathscr{H}_\gamma)$ satisfies the Markov property of Section 9.5 (*Hint.* Use part a along with Propositions 1 and 2 of Section 6.2.)

   c. Conclude that for the problem stated in Section 6.1, there exists an optimal strategy that is determined at each stage t by the value of π^t only. (Compare with the more precise statement of the theorem of Section 6.3.)

HISTORICAL NOTES

The concept of Markov chain introduced in this chapter is new, though families of several independent Markov chains, each indexed by $\mathbb{N}$, were studied for instance in Mandelbaum and Vanderbei (1981). In particular, the results of Section 9.2 extend results of these authors. On the other hand, the example and proposition of Section 9.3 are very classical, and were, for example, treated in Dynkin and Yushkevitch (1969). Section 9.4 is also new; the use of the index set $\mathbb{G}_f$ is crucial to the reduction to an optimal control problem carried out in this section. The problem considered in Section 9.5 is a formalization of the principle of sufficiency discussed by Bahadur (1954). The material presented there is taken from Lawler and Vanderbei (1983). These authors provided a partial extension to optimal control of a result of Irle (1981) for processes indexed by $\mathbb{N}$. The notion of Markov property used by these authors is sufficient for the existence of Markov controls but is not necessary, as becomes apparent when examining their definition along with the statement of Exercise 7.

Numerous examples and applications involving sequential stochastic optimization in a Markovian setup can be found in Puterman (1994).

CHAPTER 10

Optimal Switching Between Two Random Walks

In order to illustrate the theory presented in the previous chapter, we are going to study in detail an optimal control problem that involves switching back and forth between two random walks.

10.1 FORMULATING AND SOLVING THE PROBLEM

Let N^1 and N^2 be two integers such that $N^1 > 1$ and $N^2 > 1$. In the sequel, E will denote the *rectangle* $[0, N^1] \times [0, N^2]$, $\partial\mathsf{E}$ the *boundary* of E, that is, the union of the four *boundary segments* $[0, N^1] \times \{0\}$, $[0, N^1] \times \{N^2\}$, $\{0\} \times [0, N^2]$, and $\{N^1\} \times [0, N^2]$, and E° the *interior* of E, that is $\mathsf{E} - \partial\mathsf{E} =]0, N^1[\times]0, N^2[$.

Let $X^1 = (X^1_{t^1},\ t^1 \in \mathbb{N})$ and $X^2 = (X^2_{t^2},\ t^2 \in \mathbb{N})$ be two independent random walks as defined in Section 9.3, with respective state spaces $[0, N^1]$ and $[0, N^2]$ and respective transition functions P^1 and P^2. Imagine that an observer can control the evolution of X^1 and X^2 separately, that is, can leave t^1 fixed and let t^2 increase or leave t^2 fixed and let t^1 increase. This determines a process that evolves in E, termed a *switched process*. Assume that the observer can choose the switching strategy and the time at which the evolution ends, knowing that he will receive a reward that only depends on the state of the process at that time. This reward is represented by a nonnegative real-valued payoff function f defined on E that vanishes in E° (see Figure 2), and the objective is to maximize the expected reward. There is clearly no advantage to stopping the evolution in E°. On the other hand, the process cannot leave the boundary $\partial\mathsf{E}$ once this boundary has been reached, and so from that moment on, the problem reduces to the optimal stopping problem involving a random walk studied in Section 9.3. It is therefore convenient to eliminate this last phase of the evolution, which can be achieved simply by assuming that f is concave on each boundary segment of

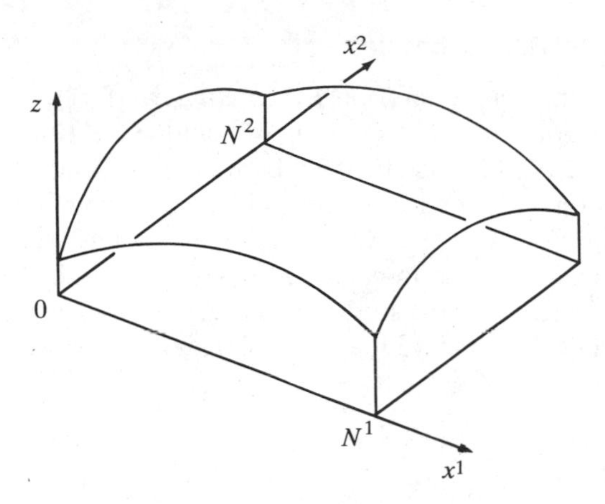

Figure 2. A possible graph of the payoff function f.

E, that is, $f(\cdot, 0)$, $f(\cdot, N^2)$, $f(0, \cdot)$, and $f(N^1, \cdot)$ are concave functions (cf. Exercise 1). Under this assumption, the evolution of the switched process terminates at the time of first visit to $\partial\mathsf{E}$. The problem we have just described is known as the *optimal switching problem*.

Solving the Problem

Let X denote the process $(X^1, X^2) = ((X^1_{t^1}, X^2_{t^2}), (t^1, t^2) \in \mathbb{I})$, where $\mathbb{I}$ denotes $\mathbb{N}^2$. It is clear that X is a Markov chain with state space E and transition functions P_1 and P_2 defined by

$$
\begin{aligned}
P_1(x, y) &= \begin{cases} P^1(x^1, y^1), & \text{if } x^2 = y^2, \\ 0, & \text{if } x^2 \neq y^2, \end{cases} \\
P_2(x, y) &= \begin{cases} P^2(x^2, y^2), & \text{if } x^1 = y^1, \\ 0, & \text{if } x^1 \neq y^1. \end{cases}
\end{aligned}
\tag{10.1}
$$

As already stated, let f be a real-valued nonnegative function defined on E that vanishes in E° and is concave on each boundary segment of E. Solving the optimal switching problem we have just described is equivalent to solving the problem of control and stopping at the time of first visit to $\mathsf{D} = \partial\mathsf{E}$ for the Markov chain $X = (X^1, X^2)$. For this, we only need to use the theorem of Section 9.4, which we can apply after observing that the stopping times $\tau(\Gamma)$ and $\tau(\Delta)$ in the statement of that theorem are finite, because the time τ^1 of first visit of X^1 to $\{0, N^1\}$ and τ^2 of X^2 to $\{0, N^2\}$ are finite and $\tau(\Delta) \leq \tau^1 + \tau^2$ for all $\Delta \in \mathbf{G}^\infty$.

Identifying the Generalized Envelope

A real-valued function $\tilde{f}$ defined on E is *biconcave* if $\tilde{f}(\cdot, x^2)$ and $\tilde{f}(x^1, \cdot)$ are concave functions for all $x^2 \in [0, N^2]$ and $x^1 \in [0, N^1]$. Since f is concave on each boundary segment of E, it is clear that if a nonnegative real-valued function $\tilde{f}$ defined on E coincides with f on $\partial\mathsf{E}$ and is superharmonic in E°, then $\tilde{f}$ is biconcave. On the other hand, any biconcave majorant of f remains biconcave if it is changed to coincide with f on $\partial\mathsf{E}$. In the context of the problem we are studying, the generalized envelope g of f is therefore the smallest biconcave majorant of f, in other words, the *biconcave envelope* of f.

Characterizing E_1 and E_2

Let g be the biconcave envelope of f. By (9.19) and (10.1), a point x in E° belongs to E_1 if and only if

$$\tfrac{1}{2}g\left(x^1 - 1, x^2\right) + \tfrac{1}{2}g\left(x^1 + 1, x^2\right) = g\left(x^1, x^2\right), \tag{10.2}$$

and it belongs to E_2 if and only if

$$\tfrac{1}{2}g\left(x^1, x^2 - 1\right) + \tfrac{1}{2}g\left(x^1, x^2 + 1\right) = g\left(x^1, x^2\right). \tag{10.3}$$

Therefore, in order that a horizontal segment $[x^1, y^1] \times \{x^2\}$ in E° be a subset of E_1, it is necessary and sufficient that the function $g(\cdot, x^2)$ be linear on $[x^1 - 1, y^1 + 1]$, and in order that a vertical segment $\{x^1\} \times [x^2, y^2]$ in E° be a subset of E_2, it is necessary and sufficient that the function $g(x^1, \cdot)$ be linear on $[x^2 - 1, y^2 + 1]$.

Describing the Solution

Once the sets E_1 and E_2 have been determined, the optimal control described in the theorem of Section 9.4 is achieved by applying the following rules:

- **a.** If $(X^1_{t^1}, X^2_{t^2}) \in \mathsf{E}_1$, leave t^2 fixed and let X^1 evolve.
- **b.** If $(X^1_{t^1}, X^2_{t^2}) \in \mathsf{E}_2$, leave t^1 fixed and let X^2 evolve.
- **c.** If $(X^1_{t^1}, X^2_{t^2}) \in \partial\mathsf{E}$, terminate the evolution of X.

In the sequel, we will express the rule formulated in a [b] above by saying that the horizontal [vertical] direction is optimal. The set $\mathsf{E}_1 \cap \mathsf{E}_2$ is an *indifference region* in which both directions are optimal.

10.2 SOME PROPERTIES OF THE SOLUTION

In order to reveal the structure of the solution of the optimal switching problem, it is necessary to prove several intermediate results. Recall that g is the biconcave envelope of f and that E_1 [E_2] is the subset of E° that consists of all points x that satisfy (10.2) [(10.3)]. Given two points $a = (a^1, a^2)$ and $b = (b^1, b^2)$ of E such that $a^1 \le b^1$ and $a^2 \le b^2$, the *rectangle* $[a^1, b^1] \times [a^2, b^2]$ will also be denoted $[a, b]$. If $a^2 = b^2$ [$a^1 = b^1$], this rectangle is a *horizontal* [*vertical*] *segment*. The *vertices* of $[a, b]$ are the points a, (b^1, a^2), b and (a^1, b^2). The *horizontal* [*vertical*] *boundary segments* of $[a, b]$ are the segments $[a^1, b^1] \times \{a^2\}$ and $[a^1, b^1] \times \{b^2\}$ [$\{a^1\} \times [a^2, b^2]$ and $\{b^1\} \times [a^2, b^2]$]. The union of the four boundary segments of $[a, b]$ constitutes the *boundary* $\partial[a, b]$ of $[a, b]$. The *interior* of $[a, b]$ is the set $[a, b]^{\circ} = [a, b] - \partial[a, b]$.

A subset F of E is *h-convex* [*v-convex*] if it contains all horizontal [vertical] segments $[x, y]$ whose extremities x and y belong to F; this subset is *biconvex* if it is both h-convex and v-convex. A real-valued function $\tilde{f}$ defined on E is *concave* or *linear* on a horizontal segment $[x^1, y^1] \times \{x^2\} \subset \mathsf{E}$ [on a vertical segment $\{x^1\} \times [x^2, y^2] \subset \mathsf{E}$] according as the function $\tilde{f}(\cdot, x^2)$ [$\tilde{f}(x^1, \cdot)$] is concave or linear on $[x^1, y^1]$ [on $[x^2, y^2]$]; this function is *h-concave* or *h-linear* [*v-concave* or *v-linear*] on an h-convex [v-convex] subset F of E if it is concave or linear on each horizontal [vertical] segment contained in F; it is *biconcave* [*bilinear*] on a biconvex subset F of E if it is both h-concave and v-concave [h-linear and v-linear] on F. When $\mathsf{F} = \mathsf{E}$, the indication "on F" will be omitted. We will say that $\tilde{f}$ is *strictly h-concave* at a point $(x^1, x^2) \in]0, N^1[\times [0, N^2]$ if

$$\tfrac{1}{2}\tilde{f}(x^1 - 1, x^2) + \tfrac{1}{2}\tilde{f}(x^1 + 1, x^2) < \tilde{f}(x^1, x^2)$$

and *strictly v-concave* at a point $(x^1, x^2) \in [0, N^1] \times]0, N^2[$ if

$$\tfrac{1}{2}\tilde{f}(x^1, x^2 - 1) + \tfrac{1}{2}\tilde{f}(x^1, x^2 + 1) < \tilde{f}(x^1, x^2).$$

Proposition 1 (Propagation of linearity and strict concavity). *Let $[a, b]$ be a subrectangle of E with a nonempty interior. If g is linear on the horizontal [vertical] boundary segments of $[a, b]$, then g is h-linear [v-linear] on $[a, b]$; in other words, the horizontal [vertical] direction is optimal in $[a, b]^{\circ}$. If in addition, g is strictly v-concave [strictly h-concave] at (a^1, x^2) or (b^1, x^2) for some $x^2 \in]a^2, b^2[$ [at (x^1, a^2) or (x^1, b^2) for some $x^1 \in]a^1, b^1[$], then g is strictly v-concave [strictly h-concave] at (x^1, x^2) for all $x^1 \in]a^1, b^1[$ [for all $x^2 \in]a^2, b^2[$].*

Proof. We will only consider the case where g is linear on the horizontal boundary segments of $[a, b]$, because the other case is similar. Let $\tilde{g}$ denote the function defined on E by

$$\tilde{g}(x^1, x^2) = \begin{cases} \dfrac{b^1 - x^1}{b^1 - a^1} g(a^1, x^2) + \dfrac{x^1 - a^1}{b^1 - a^1} g(b^1, x^2), & \text{if } (x^1, x^2) \in [a, b]^{\circ}, \\ g(x^1, x^2), & \text{if } (x^1, x^2) \in \mathsf{E} - [a, b]^{\circ}. \end{cases} \tag{10.4}$$

Since $g(\cdot, x^2)$ is concave, we see by (9.11) that $\tilde{g} \le g$. Let us prove that $\tilde{g}$ is h-concave. Since $\tilde{g}$ is h-linear on $[a, b]$ and coincides with g outside $[a, b]^{\circ}$, it suffices to check that if $x^2 \in]a^2, b^2[$, then

$$\tfrac{1}{2} g(a^1 - 1, x^2) + \tfrac{1}{2} \tilde{g}(a^1 + 1, x^2) \le g(a^1, x^2), \qquad \text{if } a^1 > 0,$$

and

$$\tfrac{1}{2} \tilde{g}(b^1 - 1, x^2) + \tfrac{1}{2} g(b^1 + 1, x^2) \le g(b^1, x^2), \qquad \text{if } b^1 < N^1.$$

But these two inequalities are obvious, because $\tilde{g} \le g$ and $g(\cdot, x^2)$ is concave. We now prove that $\tilde{g}$ is v-concave. Since $\tilde{g}$ coincides with g outside $[a, b]^{\circ}$, it suffices to establish that if $x^1 \in]a^1, b^1[$, then

$$\tfrac{1}{2} \tilde{g}(x^1, x^2 - 1) + \tfrac{1}{2} \tilde{g}(x^1, x^2 + 1) \le \tilde{g}(x^1, x^2) \tag{10.5}$$

for all $x^2 \in]a^2, b^2[$, and also for $x^2 = a^2$ if $a^2 > 0$ and for $x^2 = b^2$ if $b^2 < N^2$. By (10.4) and since $g(a^1, \cdot\,)$ and $g(b^1, \cdot\,)$ are concave, we see that for all $x^2 \in]a^2, b^2[$, the left-hand side of (10.5) is equal to

$$\begin{aligned} &\frac{b^1 - x^1}{b^1 - a^1} \left(\frac{1}{2} g(a^1, x^2 - 1) + \frac{1}{2} g(a^1, x^2 + 1) \right) \\ &\qquad + \frac{x^1 - a^1}{b^1 - a^1} \left(\frac{1}{2} g(b^1, x^2 - 1) + \frac{1}{2} g(b^1, x^2 + 1) \right) \\ &\le \frac{b^1 - x^1}{b^1 - a^1} g(a^1, x^2) + \frac{x^1 - a^1}{b^1 - a^1} g(b^1, x^2) \\ &= \tilde{g}(x^1, x^2). \end{aligned} \tag{10.6}$$

The verification for $x^2 = a^2$ and for $x^2 = b^2$ is done similarly using (9.11). We have therefore established that $\tilde{g}$ is biconcave. Because $g \ge \tilde{g} \ge f$ and g is the smallest biconcave majorant of f, it follows that $\tilde{g} = g$. Since $\tilde{g}$ is h-linear on $[a, b]$, the same is true of g. If g is strictly v-concave at (a^1, x^2) or (b^1, x^2) for some $x^2 \in]a^2, b^2[$, then the inequality in (10.6) and therefore in (10.5) is strict, which proves the second part of the proposition.

Remarks / Corollaries

Let $[a, b]$ be as in Proposition 1. By this proposition, and in particular by (10.4), if g is linear on the horizontal boundary segments of $[a, b]$, then for all $(x^1, x^2) \in [a, b]$,

$$g(x^1, x^2) = \frac{b^1 - x^1}{b^1 - a^1} g(a^1, x^2) + \frac{x^1 - a^1}{b^1 - a^1} g(b^1, x^2). \tag{10.7}$$

Similarly, if g is linear on the vertical boundary segments of $[a, b]$, then for all $(x^1, x^2) \in [a, b]$,

$$g(x^1, x^2) = \frac{b^2 - x^2}{b^2 - a^2} g(x^1, a^2) + \frac{x^2 - a^2}{b^2 - a^2} g(x^1, b^2). \tag{10.8}$$

If g is linear on all four boundary segments of $[a, b]$, then g is bilinear on $[a,b]$. In this case, by replacing $g(a^1, x^2)$ and $g(b^1, x^2)$ in (10.7) by the expression furnished by (10.8), we see after an elementary calculation that for all $(x^1, x^2) \in [a, b]$,

$$\begin{aligned} g(x^1, x^2) = g(a^1, a^2) &+ \frac{x^1 - a^1}{b^1 - a^1}(g(b^1, a^2) - g(a^1, a^2)) \\ &+ \frac{x^2 - a^2}{b^2 - a^2}(g(a^1, b^2) - g(a^1, a^2)) \\ &+ \frac{(x^1 - a^1)(x^2 - a^2)}{(b^1 - a^1)(b^2 - a^2)} \\ &\times (g(b^1, b^2) - g(b^1, a^2) - g(a^1, b^2) + g(a^1, a^2)). \end{aligned} \tag{10.9}$$

This result can be expressed by saying that the graph of g on $[a, b]$ is the restriction to $[a, b]$ of the hyperbolic paraboloid determined by the skew quadrilateral in $\mathbb{R}^3$ with vertices

$$\begin{aligned} A_1 &= (a^1, a^2, g(a^1, a^2)), \qquad & A_2 &= (b^1, a^2, g(b^1, a^2)), \\ A_3 &= (b^1, b^2, g(b^1, b^2)), \qquad & A_4 &= (a^1, b^2, g(a^1, b^2)), \end{aligned}$$

that is, the quadric surface generated by the straight lines parallel to the coordinate plane Ox^1x^3 of $\mathbb{R}^3$ that intersect the straight lines A_1A_4 and A_2A_3 (or by the straight lines parallel to Ox^2x^3 that intersect the straight lines A_1A_2 and A_3A_4 (cf. Figure 3)).

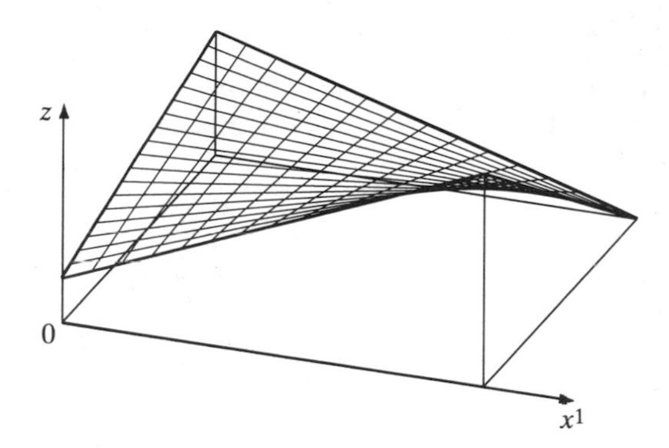

Figure 3. The graph of a bilinear function: a hyperbolic paraboloid.

Proposition 2 (Convexity property). E_1 *is v-convex and* E_2 *is h-convex.*

Proof. Let x and y be two points in E_1 such that $x^1 = y^1$ and $y^2 - x^2 > 1$. By the assertion that follows (10.3) (applied to the horizontal segments $[x, x]$ and $[y, y]$), g is h-linear on the horizontal boundary segments of the rectangle $[(x^1 - 1, x^2), (x^1 + 1, y^2)]$. By Proposition 1, we deduce that g is h-linear on this rectangle. Again by the assertion that follows (10.3), the vertical segment $[x, y]$ is therefore contained in E_1. It is thus established that E_1 is v-convex. The proof that E_2 is h-convex is similar.

Unimodal Functions

We will say that a function ξ defined on $[0, N^2]$ with values in $[0, N^1]$ is *unimodal of type* 1 [*unimodal of type* 2] if there exists $n \in [0, N^2]$ such that ξ is nondecreasing [nonincreasing] on $[0, n]$ and nonincreasing [nondecreasing] on $[n, N^2]$.

Proposition 3 (Existence of ξ_1 and ξ_2). *Assume that one of the following two hypotheses is satisfied*:

- **a.** *There exists* $x^1 \in]0, N^1[$ *such that* $(x^1, x^2) \in \mathsf{E}_2$ *for all* $x^2 \in]0, N^2[$.
- **b.** *For all* $x^2 \in]0, N^2[$, *there exists* $x^1 \in]0, N^1[$ *such that* $(x^1, x^2) \in \mathsf{E}_2 - \mathsf{E}_1$.

Then there exist two unimodal functions ξ_1 *and* ξ_2, *respectively of type* 1 *and type* 2, *such that* $\xi_1(0) = \xi_1(N^2) = 0$, $\xi_2(0) = \xi_2(N^2) = N^1$, *and* $0 < \xi_1(n) \leq \xi_2(n) < N^1$ *for all* $n \in]0, N^1[$, *which give rise to the inclusions*

$$\mathsf{F} \subset \mathsf{E}_1 \quad \text{and} \quad \mathsf{E}^{\circ} - \mathsf{F} \subset \mathsf{E}_2, \tag{10.10}$$

where F *denotes the set* $\{(x^1, x^2) \in \mathsf{E}^{\circ}\colon x^1 < \xi_1(x^2)$ or $x^1 > \xi_2(x^2)\}$.

Proof. For all $n \in [0, N^2]$, set

$$\mu_1(n) = \begin{cases} \inf\{m \in [0, N^1] : (m,n) \in \mathsf{E}_2\}, & \text{if } n \in]0, N^2[, \\ 0, & \text{if } n \in \{0, N^2\}, \end{cases}$$

and

$$\mu_2(n) = \begin{cases} \sup\{m \in [0, N^1] : (m,n) \in \mathsf{E}_2\}, & \text{if } n \in]0, N^2[, \\ N^1, & \text{if } n \in \{0, N^2\}. \end{cases}$$

Notice that

$$0 < \mu_1(n) \leq \mu_2(n) < N^1, \qquad \text{for all } n \in]0, N^2[.$$

Moreover, set

$$p^1 = \sup_{n \in [0, N^2]} \mu_1(n) \quad \text{and} \quad q^1 = \inf_{n \in [0, N^2]} \mu_2(n). \tag{10.11}$$

Then $p^1 < N^1$ and $q^1 > 0$. Choose p^2 and q^2 in $]0, N^2[$ such that

$$\mu_1(p^2) = p^1 \quad \text{and} \quad \mu_2(q^2) = q^1. \tag{10.12}$$

Let ξ_1 and ξ_2 denote the two functions defined on $[0, N^2]$, respectively, by

$$\xi_1(n) = \begin{cases} \sup_{k \in [0, n]} \mu_1(k), & \text{if } n \in [0, p^2], \\ \sup_{k \in [n, N^2]} \mu_1(k), & \text{if } n \in [p^2, N^2], \end{cases} \tag{10.13}$$

and

$$\xi_2(n) = \begin{cases} \inf_{k \in [0, n]} \mu_2(k), & \text{if } n \in [0, q^2], \\ \inf_{k \in [n, N^2]} \mu_2(k), & \text{if } n \in [q^2, N^2]. \end{cases} \tag{10.14}$$

Clearly, ξ_1 and ξ_2 are two unimodal functions, respectively of type 1 and type 2, independent of the choice of p^2 and q^2 and such that $\xi_1(0) = \xi_1(N^2) = 0$, $\xi_2(0) = \xi_2(N^2) = N^1$, $0 < \mu_1(n) \leq \xi_1(n)$, and $\xi_2(n) \leq \mu_2(n) < N^1$ for all $n \in]0, N^2[$. Moreover, these two functions satisfy

$$\begin{aligned} \xi_1(p^2) &= \sup_{n \in [0, N^2]} \xi_1(n) = \mu_1(p^2) = p^1, \\ \xi_2(q^2) &= \inf_{n \in [0, N^2]} \xi_2(n) = \mu_2(q^2) = q^1. \end{aligned} \tag{10.15}$$

Define F as in the statement of the proposition and let us prove that the inclusions in (10.10) hold. By Proposition 2, E_2 is h-convex; therefore

$$\mathsf{E}_2 = \{(x^1, x^2) \in \mathsf{E}^\circ : \mu_1(x^2) \le x^1 \le \mu_2(x^2)\}.$$

Consequently,

$$\mathsf{E}^\circ - \mathsf{F} = \{(x^1, x^2) \in \mathsf{E}^\circ : \xi_1(x^2) \le x^1 \le \xi_2(x^2)\} \subset \mathsf{E}_2,$$

which establishes the second inclusion in (10.10). In order to establish the first inclusion, it suffices to show that if (x^1, x^2) is a point of E° such that $\mu_1(x^2) \le x^1 < \xi_1(x^2)$ or $\xi_2(x^2) < x^1 \le \mu_2(x^2)$, then (x^1, x^2) belongs to E_1. Assume that the first alternative holds, as the proof in the other is analogous. Because $\mu_1(p^2) = \xi_1(p^2)$, we note that $x^2 \ne p^2$. Assume for example that $x^2 < p^2$. Then there exists $n \in]0, x^2]$ such that $\xi_1(x^2) = \xi_1(n) = \mu_1(n)$. But $x^1 < \xi_1(x^2) = \mu_1(n)$ and $x^1 < \xi_1(x^2) \le \xi_1(p^2) = \mu_1(p^2)$; therefore (x^1, n) and (x^1, p^2) belong to E_1. Now E_1 is v-convex by Proposition 2; therefore (x^1, x^2) also belongs to E_1.

It remains to prove that $\xi_1 \le \xi_2$. Assume that hypothesis a is satisfied. Then $\mu_1(x^2) \le x^1 \le \mu_2(x^2)$ for all $x^2 \in]0, N^2[$, and therefore $\xi_1 \le \xi_2$. Assume now that hypothesis b is satisfied. If there exists $x^2 \in]0, N^2[$ such that $\xi_1(x^1) > \xi_2(x^2)$, then (x^1, x^2) belongs to F, therefore to E_1, for all $x^1 \in]0, N^1[$, which contradicts the hypothesis; consequently, $\xi_1 \le \xi_2$.

10.3 THE STRUCTURE OF THE SOLUTION

Denote the optimal switching problem by (π). In order to explain the structure of the solution of (π), it is useful to be able to replace (π) by an equivalent problem $(\tilde{\pi})$ obtained by symmetry. Let us denote by Φ_1 and Φ_2, respectively, the reflection with respect to the vertical line whose equation is $x^1 = N^1/2$ and the reflection with respect to the line whose equation is $x^2 = x^1$. Let us also denote by $\tilde{\mathsf{E}}$, $\tilde{\mathsf{E}}_1$, and $\tilde{\mathsf{E}}_2$ the respective images of E, E_1, and E_2 [E, E_2, and E_1] by Φ_1 [Φ_2]. Obviously $\tilde{\mathsf{E}} = \mathsf{E}$ if $i = 1$ and $\tilde{\mathsf{E}} = [0, N^2] \times [0, N^1]$ if $i = 2$. Let $(\tilde{\pi})$ denote the optimal switching problem relative to the payoff function $\tilde{f} = f \circ \Phi_i$ on $\tilde{\mathsf{E}}$. It is easy to check that $\tilde{g} = g \circ \Phi_i$ is the biconcave envelope of $\tilde{f}$ and that $\tilde{\mathsf{E}}_1$ and $\tilde{\mathsf{E}}_2$ are the regions where the optimal directions for $(\tilde{\pi})$ are, respectively, horizontal and vertical. It follows that the structure of the solution of (π) can be deduced from the structure of the solution of $(\tilde{\pi})$, and conversely.

Two Types of Structures

The results that we are about to formulate rely on the conclusions furnished by Proposition 3 of Section 10.2. It is therefore necessary to make sure that

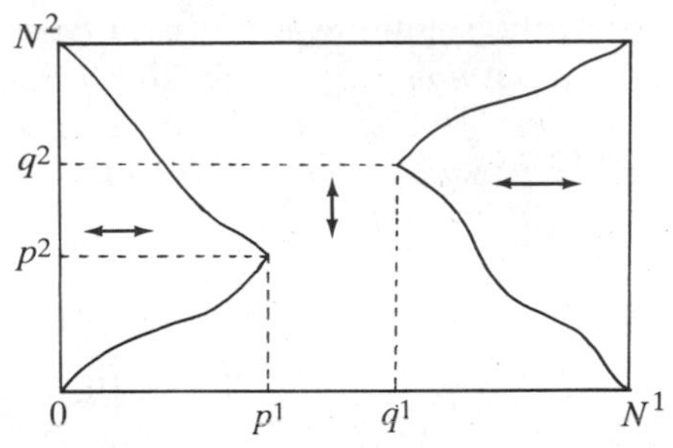

Figure 4. Solution of type 1.

one of the two hypotheses of this proposition is satisfied. If hypothesis b is not satisfied, that is, there exists $x^2 \in]0, N^2[$ such that $(x^1, x^2) \in \mathsf{E}_1$ for all $x^1 \in]0, N^1[$, then it suffices to transform (π) by applying Φ_2 to see that $(x^2, y^2) \in \tilde{\mathsf{E}}_2$ for all $y^2 \in]0, N^1[$, and therefore that hypothesis a is satisfied for the transformed problem. After possibly transforming (π) by applying Φ_2, we can therefore assume that one of the two hypotheses of Proposition 3 of Section 10.2 is satisfied and that the conclusions of this proposition apply. In particular, we will make use of the integers p^1, q^1, p^2, and q^2 defined in (10.11) and (10.12).

The inclusions in (10.10) can be expressed by saying that the horizontal direction is optimal in F and that the vertical direction is optimal in $\mathsf{E}^\circ - \mathsf{F}$. Of course, it is possible that in a portion of F or of $\mathsf{E}^\circ - \mathsf{F}$, both directions are optimal (cf. Exercise 2). In the case where $p^1 \leq q^1$, there is nothing to add to what has just been explained (cf. Figure 4). On the other hand, in the case where $p^1 > q^1$, we are going to show that there exists an indifference rectangle around which are four regions in which the horizontal and vertical directions are alternately optimal (cf. Figure 5).

In the sequel, we will say that the solution is *of type* 1 [*type* 2] if after possibly having transformed (π) by applying Φ_2, the inequality $p^1 \leq q^1$ [$p^1 > q^1$] holds.

The structure of the solution of type 2 will be described in the following theorem. Notice that this type of solution is only possible in the case where $N^1 > 2$ and $N^2 > 2$. Moreover, since $p^1 > q^1$, it is clear that $p^2 \neq q^2$. In order to avoid dealing with two cases, we will assume that $p^2 < q^2$. If this condition is not satisfied, then it will be after applying the transformation Φ_1.

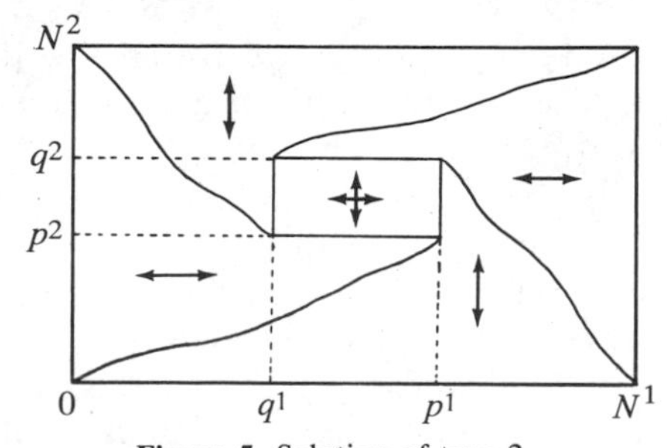

Figure 5. Solution of type 2.

Theorem (Structure of the solution when it is of type 2). *Assume that one of the two hypotheses of Proposition 3 of Section* 10.2 *is satisfied and define* p^1, q^1, p^2, *and* q^2 *by* (10.11) *and* (10.12), *with* p^2 *and* q^2 *chosen such that the value* $|p^2 - q^2|$ *is as small as possible. Moreover, define* ξ_1 *and* ξ_2 *by* (10.13) *and* (10.14). *If* $p^1 > q^1$ *and* $p^2 < q^2$, *then*

$$\begin{aligned} \xi_1(n) &\le q^1, \qquad \text{for all } n \in]p^2, N^2], \\ \xi_2(n) &\ge p^1, \qquad \text{for all } n \in [0, q^2[. \end{aligned} \tag{10.16}$$

Moreover, the structure of the solution to the optimal switching problem is described by the following conclusions: *The horizontal direction is optimal in*

$$\mathsf{F}_1 = \{(x^1, x^2) \in \mathsf{E}^{\mathrm{o}} : x^1 < \xi_1(x^2)\}$$

and

$$\mathsf{F}_3 = \{(x^1, x^2) \in \mathsf{E}^{\mathrm{o}} : x^1 > \xi_2(x^2)\},$$

the vertical direction is optimal in

$$\begin{aligned} \mathsf{F}_2 = \{(x^1, x^2) \in \mathsf{E}^{\mathrm{o}} &: x^2 < p^2 \text{ and } \xi_1(x^2) \le x^1 \le \xi_2(x^2) \\ &\text{or } p^2 \le x^2 < q^2 \text{ and } p^1 \le x^1 \le \xi_2(x^2)\} \end{aligned}$$

and

$$\begin{aligned} \mathsf{F}_4 = \{(x^1, x^2) \in \mathsf{E}^{\mathrm{o}} &: p^2 < x^2 \le q^2 \text{ and } \xi_1(x^2) \le x^1 \le q^1 \\ &\text{or } x^2 > q^2 \text{ and } \xi_1(x^2) \le x^1 \le \xi_2(x^2)\}, \end{aligned}$$

and both directions are optimal in the (*possibly empty*) *rectangle*

$$]q^1, p^1[\times]p^2, q^2[.$$

Proof. By Proposition 3 of Section 10.2, $[q^1, p^1[\times\{p^2\} \subset \mathsf{E}_1$ and $]q^1, p^1] \times \{q^2\} \subset \mathsf{E}_1$. But E_1 is v-convex by Proposition 2; therefore $]q^1, p^1[\times[p^2, q^2] \subset \mathsf{E}_1$. We now prove (10.16). Since $\xi_1(n) \le \xi_1(q^2) \le \xi_2(q^2) = q^1$ for all $n \in [q^2, N^2]$ and $\xi_2(n) \ge \xi_2(p^2) \ge \xi_1(p^2) = p^1$ for all $n \in [0, p^2]$, it is sufficient to prove that $\xi_1(n) \le q^1$ and $\xi_2(n) \ge p^1$ for all $n \in]p^2, q^2[$. If $q^2 - p^2 = 1$, there is nothing to prove. Assume that $q^2 - p^2 > 1$. If $p^1 - q^1 = 1$, then $\xi_1(n) \le q^1$ for all $n \in]p^2, q^2[$. Indeed, $\xi_1(n) < p^1$ for all $n \in]p^2, q^2[$ because $q^2 - p^2$ is minimal. Similarly, we see that $\xi_2(n) \ge p^1$ for all $n \in]p^2, q^2[$. If $p^1 - q^1 > 1$, assume that $\xi_1(n) > q^1$ for some $n \in]p^2, q^2[$. Since $\xi_1(q^2) \le \xi_2(q^2) = q^1$, there exists $k \in]p^2, q^2[$ such that $\xi_1(k) = \mu_1(k) > q^1$. Consequently $(q^1, k) \notin \mathsf{E}_2$; in other words, g is strictly v-concave at (q^1, k). Taking into account the fact that $\xi_1(k) < p^1$ due to the minimal character of

$q^2 - p^2$, by applying Proposition 1 of Section 10.2 to the rectangle $[q^1, p^1] \times [k-1, k+1]$, we deduce that $(\xi_1(k), k) \notin \mathsf{E}_2$. This contradicts the second inclusion in (10.10). We thus conclude that $\xi_1(n) \le q^1$ for all $n \in]p^2, q^2[$. Similarly, we see that $\xi_2(n) \ge p^1$ for all $n \in]p^2, q^2[$. Assertion (10.16) is therefore established. The final conclusions of the theorem follow directly from (10.10) and from the inclusion $]q^1, p^1[\times]p^2, q^2[\subset \mathsf{E}_1 \cap \mathsf{E}_2$.

Evolution of the Optimally Switched Process

Fix $x^2 \in]0, p^2[$, assume that $\xi_1(x^2) < p^1$, and let $y^2 = \inf\{n \in]x^2, p^2]: \xi_1(n) > \xi_1(x^2)\}$. If $z^2 \in]0, y^2[$, then on the set $\{X_0 = (\xi_1(x^2), z^2)\}$, the optimal strategy can be described as follows. Initially, leave t^1 fixed and equal to 0 and let t^2 evolve until either X visits $(\xi_1(x^2), 0)$, in which case the evolution stops and the payoff is $f(\xi_1(x^2), 0)$, or X visits $(\xi_1(x^2), y^2)$. In this case, leave t^2 fixed and let t^1 evolve until either X visits $(0, y^2)$, in which case the evolution stops and the payoff is $f(0, y^2)$, or X visits $(\xi_1(y^2), y^2)$. In the latter case, if $y^2 < p^2$, proceed as above as though the starting point were $(\xi_1(y^2), y^2)$. If X ever visits (p^1, p^2), then the optimal strategy depends on whether the solution is of type 1 or 2.

If the solution is of type 1, then from this time on, leave t^1 fixed and let t^2 evolve until X visits $(p^1, 0)$ or (p^1, N^2), at which time the evolution terminates and the payoff is, respectively, $f(p^1, 0)$ or $f(p^1, N^2)$.

If the solution is of type 2 and $p^2 < q^2$, then leave t^1 fixed and let t^2 evolve until either X visits $(p^1, 0)$, at which time the evolution terminates and the payoff is $f(p^1, 0)$, or X visits (p^1, q^2). In this case, leave t^2 fixed and let t^1 evolve until either X visits (N^1, q^2), at which time the evolution terminates and the payoff is $f(N^1, q^2)$, or X visits (q^1, q^2). The direction of evolution switches each time X visits one of the four states (q^1, q^2), (q^1, p^2), (p^1, p^2), or (p^1, q^2), until X finally visits one of the four states $(p^1, 0)$, (N^1, q^2), (q^1, N^2), or $(0, p^2)$, at which time the evolution terminates and the payoff is $f(p^1, 0)$, $f(N^1, q^2)$, $f(q^1, N^2)$, or $f(0, p^2)$, respectively.

10.4 CONSTRUCTING THE SWITCHING CURVES

The theorem in the previous section furnishes a complete description of the structure of the solution. Our objective is now to show how the solution can be determined from the payoff function f. This is a two-step process: The first step consists of determining whether the solution is of type 1 or 2 and integers p^1, q^1, p^2, and q^2 that satisfy (10.11) and (10.12), and the second aims at deriving a formula for evaluating the functions ξ_1 and ξ_2 (the graphs of which are the *switching curves* mentioned in this section's title). It turns out that it is convenient to proceed in the opposite order; that is, we shall assume in this section that p^1, q^1, p^2, and q^2 are given and will exhibit a formula for evaluating ξ_1 and ξ_2, whereas in the next two sections, we will determine the type of the solution and will compute p^1, q^1, p^2, and q^2.

The Functions φ^1 and φ^2

Denote by φ^1 and φ^2 the functions defined, respectively, on $[0, N^1[$ and $[0, N^2[$ by

$$\begin{aligned} \varphi^1(x^1) &= (x^1 + 1)f(x^1, 0) - x^1 f(x^1 + 1, 0), \\ \varphi^2(x^2) &= (x^2 + 1)f(0, x^2) - x^2 f(0, x^2 + 1). \end{aligned} \tag{10.17}$$

Clearly, $\varphi^1(0) = \varphi^2(0) = f(0,0)$. Moreover, for all $x^1 \in [0, N^1 - 1[$, $\varphi^1(x^1 + 1) - \varphi^1(x^1)$ is equal to

$$\begin{aligned} &(x^1 + 2)f(x^1 + 1, 0) - (x^1 + 1)f(x^1 + 2, 0) \\ &\qquad\qquad - (x^1 + 1)f(x^1, 0) + x^1 f(x^1 + 1, 0) \\ &= 2(x^1 + 1)f(x^1 + 1, 0) - (x^1 + 1)\big(f(x^1 + 2, 0) + f(x^1, 0)\big) \\ &\geq 2(x^1 + 1)f(x^1 + 1, 0) - 2(x^1 + 1)f(x^1 + 1, 0) \\ &= 0, \end{aligned}$$

where the inequality is justified by the fact that $f(\cdot, 0)$ is a concave function. Notice incidentally that this inequality is strict if $f(\cdot, 0)$ is strictly concave at $x^1 + 1$. Therefore φ^1 is nondecreasing, and one shows in the same way that φ^2 is also nondecreasing. It is also easy to check that for any fixed $x^1 \in [0, N^1[$, $\varphi^1(x^1)$ is the value at 0 of the unique linear function defined on $[0, N^1]$ whose value at x^1 is $f(x^1, 0)$ and whose value at $x^1 + 1$ is $f(x^1 + 1, 0)$. Similarly, for any fixed $x^2 \in [0, N^2[$, $\varphi^2(x^2)$ is the value at 0 of the unique linear function defined on $[0, N^2]$ whose value at x^2 is $f(0, x^2)$ and whose value at $x^2 + 1$ is $f(0, x^2 + 1)$. This property will in particular be useful in determining, under circumstances that will be explained further on, which of the values $\varphi^1(x^1)$ and $\varphi^2(x^2)$ is larger. Indeed, fix $x^1 \in [0, N^1[$ and $x^2 \in [0, N^2[$, and let h_1 and h_2 be two bilinear functions defined on E such that

$$\begin{aligned} h_1(x^1, 0) &= f(x^1, 0), & h_1(x^1 + 1, 0) &= f(x^1 + 1, 0), \\ h_2(0, x^2) &= f(0, x^2), & h_2(0, x^2 + 1) &= f(0, x^2 + 1). \end{aligned}$$

Then $h_1(0,0) = \varphi^1(x^1)$ and $h_2(0,0) = \varphi^2(x^2)$, which implies that $\varphi^1(x^1) < \varphi^2(x^2)$ if and only if $h_1(0,0) < h_2(0,0)$. This implication remains valid if the symbol $<$ is replaced by $\leq$, $>$, or $\geq$.

Comparing the Values of Two Bilinear Functions

If $a = (a^1, a^2)$ and $b = (b^1, b^2)$ are two points of E such that $a^1 < b^1$ and $a^2 < b^2$, and if c_1, c_2, c_3, and c_4 are real numbers, then there is a unique bilinear function h defined on E such that

$$\begin{aligned} h(a^1, a^2) &= c_1, & h(b^1, a^2) &= c_2, \\ h(b^1, b^2) &= c_3, & h(a^1, b^2) &= c_4. \end{aligned} \tag{10.18}$$

As has already been pointed out, the graph of h on E is the restriction to E of the hyperbolic paraboloid determined by the skew quadrilateral in $\mathbb{R}^3$ with vertices

$$\begin{aligned} A_1 &= (a^1, a^2, c_1), & A_2 &= (b^1, a^2, c_2), \\ A_3 &= (b^1, b^2, c_3), & A_4 &= (a^1, b^2, c_4), \end{aligned}$$

that is, the quadric surface generated by the straight lines parallel to Ox^1x^3 that intersect the straight lines A_1A_4 and A_2A_3 (or by the straight lines parallel to Ox^2x^3 that intersect the straight lines A_1A_2 and A_3A_4). It follows that if $\tilde{h}$ is the bilinear function defined on E by (10.18), but with $\tilde{c}_1$ instead of c_1, then $c_1 \le \tilde{c}_1$ $[c_1 < \tilde{c}_1]$ implies that $h(x^1, x^2) \le \tilde{h}(x^1, x^2)$ $[h(x^1, x^2) < \tilde{h}(x^1, x^2)]$ for all $(x^1, x^2) \in [0, b^1[\times[0, b^2[$.

Evaluating ξ_1 and ξ_2

As was already mentioned in Section 10.3, after possibly having transformed the problem by applying Φ_2, we may assume that the conclusions of Proposition 3 of Section 10.2 are satisfied. We also assume that p^1, q^1, p^2, and q^2 have been determined. Recall that $\xi_1(0) = \xi_1(N^2) = 0$ and $\xi_1(p^2) = p^1$. The theorem below shows how to compute the values of ξ_1 on the interval $]0, p^2[$. The values of ξ_1 on the interval $]p^2, N^2[$ as well as the values of ξ_2 can be computed analogously (cf. Exercise 4). Because we shall only consider the interval $]0, p^2[$, we can assume that $p^2 > 1$.

Theorem (Values of ξ_1 on $]0, p^2[$). *Let ξ_1 be the function defined in* (10.13) *and φ_1, φ_2 the functions defined in* (10.17). *Assume that $p^2 > 1$. Then for all $x^2 \in]0, p^2[$,*

$$\xi_1(x^2) = \begin{cases} \inf\{x^1 \in]0, p^1[: \varphi^1(x^1) \ge \varphi^2(x^2)\}, & \text{if } \{\ \} \neq \varnothing, \\ p^1, & \text{if } \{\ \} = \varnothing. \end{cases} \tag{10.19}$$

Proof. Throughout this proof, x^2 will denote a fixed but arbitrary element of $]0, p^2[$. If $p^1 = 1$, then $\xi_1(x^2) = 1$ and the conclusion (10.19) is obvious. We can therefore assume that $p^1 > 1$. It suffices to prove that

$$\varphi^1\big(\xi_1(x^2)\big) \geq \varphi^2(x^2), \qquad \text{if } \xi_1(x^2) < p^1, \tag{10.20}$$

and

$$\varphi^1\big(\xi_1(x^2) - 1\big) < \varphi^2(x^2), \qquad \text{if } \xi_1(x^2) > 1. \tag{10.21}$$

Assume that $\xi_1(x^2) < p^1$ and set

$$y^2 = \inf\{n \in]x^2, p^2]: \xi_1(n) > \xi_1(x^2)\}.$$

Let h_1 denote the bilinear function defined on E that takes the same values as g at each vertex of the rectangle $[(\xi_1(x^2), y^2 - 1), (\xi_1(x^2) + 1, y^2)]$, and h_2 the bilinear function defined on E that takes the value

$$\frac{1}{\xi_1(x^2) + 1} g\big(0, y^2 - 1\big) + \frac{\xi_1(x^2)}{\xi_1(x^2) + 1} g\big(\xi_1(x^2) + 1, y^2 - 1\big) \tag{10.22}$$

at $(\xi_1(x^2), y^2 - 1)$ and the same values as g at each of the three other vertices of the same rectangle. By Proposition 3 of Section 10.2, $[\xi_1(x^2), \xi_1(x^2) + 1] \times]0, y^2[$ is contained in E_2; in other words, $g(\xi_1(x^2), \cdot)$ and $g(\xi_1(x^2) + 1, \cdot)$ are linear on $[0, y^2]$. It follows that h_1, g, and f coincide at $(\xi_1(x^2), 0)$ and h_1, h_2, g, and f coincide at $(\xi_1(x^2) + 1, 0)$. Similarly, by the same proposition, $]0, \xi_1(y^2)[\times\{y^2\}$ is contained in E_1; therefore $g(\cdot, y^2)$ is linear on $[0, \xi_1(y^2)]$, which implies that h_1, h_2, g, and f coincide at $(0, y^2)$. Moreover, given the way in which h_2 is defined, it is clear that h_2, g, and f coincide at $(0, y^2 - 1)$. It follows that $h_1(0,0) = \varphi^1(\xi_1(x^2))$ and $h_2(0,0) = \varphi^2(y^2 - 1)$. Now $g(\cdot, y^2 - 1)$ is concave, so $h_1(\xi_1(x^2), y^2 - 1) \geq h_2(\xi_1(x^2), y^2 - 1)$ by (10.22) and (9.11). Consequently, by the comment that follows (10.18), $h_1(0,0) \geq h_2(0,0)$, and therefore $\varphi^1(\xi_1(x^2)) \geq \varphi^2(y^2 - 1)$, which establishes the assertion in (10.20), because $\varphi^2(y^2 - 1) \geq \varphi^2(x^2)$.

We now prove the assertion in (10.21). Consider first the case where $x^2 = 1$. In this case, the point $(\xi_1(1) - 1, 1)$ does not belong to E_2 by (10.13); therefore g is strictly v-concave at this point. Let h_1 denote the bilinear function defined on E that takes the value

$$\tfrac{1}{2} g\big(\xi_1(1) - 1, 0\big) + \tfrac{1}{2} g\big(\xi_1(1) - 1, 2\big)$$

at $(\xi_1(1) - 1, 1)$ and the same values as g at each of the three other vertices of the rectangle $[(\xi_1(1) - 1, 1), (\xi_1(1), 2)]$, and by h_2 the bilinear function that takes the same values as g at each vertex of the same rectangle. By Proposition 3 of Section 10.2, $g(\xi_1(1), \cdot)$ is linear on $[0, 2]$, whereas $g(\cdot, 1)$ and $g(\cdot, 2)$ are linear on $[0, \xi_1(1)]$. Taking into account the manner in which $h_1(\xi_1(1) - 1, \cdot)$ is defined, we conclude that $h_1(0,0) = \varphi^1(\xi_1(1) - 1)$ and $h_2(0,0) = \varphi^2(1)$. But because g is strictly v-concave at $(\xi_1(1) - 1, 1)$, we deduce that $h_1(\xi_1(1) - 1, 1) < h_2(\xi_1(1) - 1, 1)$ and therefore that $h_1(0,0) < h_2(0,0)$. It follows that $\varphi^1(\xi_1(1) - 1) < \varphi^2(1)$, which was to be proved.

Suppose now that the assertion in (10.21) is true for $x^2 = n - 1$, where $n \in]1, p^2[$, and show that it is then also true for $x^2 = n$. If $\xi_1(n-1) < \xi_1(n)$, then the point $(\xi_1(n) - 1, n)$ does not belong to E_2 by (10.13); therefore g is strictly v-concave at this point. By considering the vertices of the rectangle $[(\xi_1(n) - 1, n), (\xi_1(n), n+1)]$, we now conclude that (10.21) holds, exactly as in the case where $x^2 = 1$. If $\xi_1(n-1) = \xi_1(n)$, then the same conclusion follows directly from the hypothesis that postulates that $\varphi^1(\xi_1(n-1) - 1) < \varphi^2(n-1)$, given that $\varphi^2(n-1) \le \varphi^2(n)$.

Recognizing Elements of E_1 and E_2

An immediate consequence of the theorem is that for $(x^1, x^2) \in]0, p^1[\times]0, p^2[$,

$$\varphi^1(x^1) \ge \varphi^2(x^2) \text{ implies } (x^1, x^2) \in \mathsf{E}_2$$

and

$$\varphi^1(x^1) < \varphi^2(x^2) \text{ implies } (x^1, x^2) \in \mathsf{E}_1.$$

10.5 CHARACTERIZING THE TYPE OF THE SOLUTION

In this section, we will give a necessary and sufficient condition on the payoff function f for the solution of the optimal switching problem to be of type 1 or 2. This condition is geometric, and will be translated into an analytic condition in Section 10.6. Recall that f is a nonnegative real-valued function defined on E that vanishes in E°.

Bilinear Majorants

We will say that a real-valued function h defined on E is a *bilinear majorant* of f if h is bilinear and $h \ge f$. Notice that this inequality holds if it holds on $\partial \mathsf{E}$.

Contact Sets of a Bilinear Majorant

If h is a bilinear majorant of f, then the *contact sets* of h with f are the four sets

$$\begin{aligned}
\mathsf{C}_1 &= \{x^1 \in]0, N^1[: h(x^1, 0) = f(x^1, 0)\}, \\
\mathsf{C}_2 &= \{x^2 \in]0, N^2[: h(N^1, x^2) = f(N^1, x^2)\}, \\
\mathsf{C}_3 &= \{x^1 \in]0, N^1[: h(x^1, N^2) = f(x^1, N^2)\}, \\
\mathsf{C}_4 &= \{x^2 \in]0, N^2[: h(0, x^2) = f(0, x^2)\}.
\end{aligned}$$

Notice that these sets are intervals that can be singletons or empty.

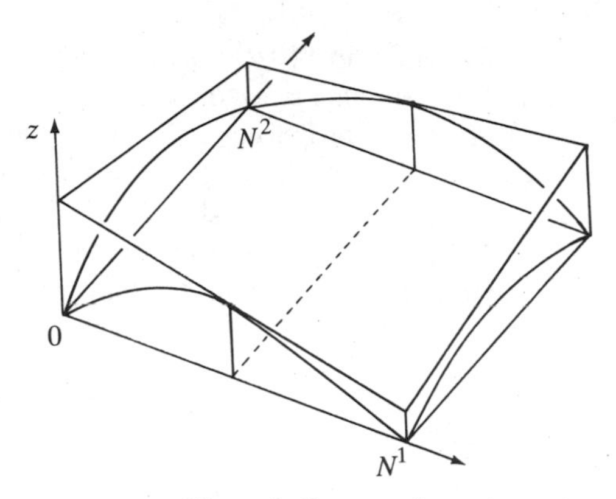

Figure 6. Contact of type 1.

Contact of Type 1 and of Type 2

Let h be a bilinear majorant of f and $\mathsf{C}_1, \ldots, \mathsf{C}_4$ the contact sets of h with f. We shall say that h has a *contact of type* 1 with f (see Figure 6) if

$$\text{either } \mathsf{C}_1 \cap \mathsf{C}_3 \neq \varnothing \quad \text{or} \quad \mathsf{C}_2 \cap \mathsf{C}_4 \neq \varnothing,$$

and that h has a *contact of type* 2 with f (see Figure 7) if $\mathsf{C}_1 \neq \varnothing, \ldots, \mathsf{C}_4 \neq \varnothing$ and either

$$\sup \mathsf{C}_3 < \inf \mathsf{C}_1 \quad \text{and} \quad \sup \mathsf{C}_4 < \inf \mathsf{C}_2,$$

or

$$\sup \mathsf{C}_1 < \inf \mathsf{C}_3 \quad \text{and} \quad \sup \mathsf{C}_2 < \inf \mathsf{C}_4.$$

Observe that the existence of a bilinear majorant with a contact of type 1 or a contact of type 2 with f is a property that is invariant under the

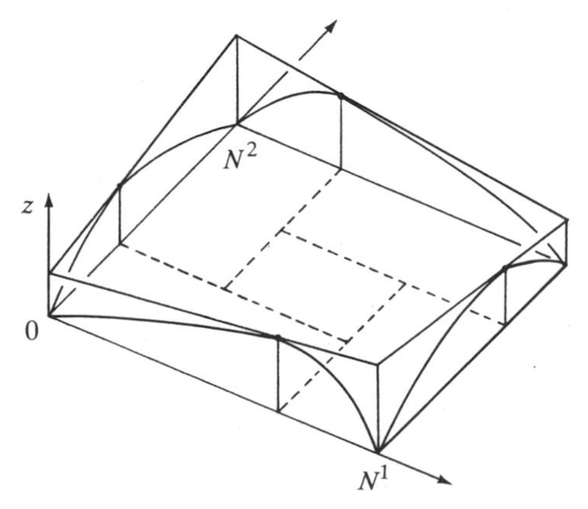

Figure 7. Contact of type 2.

reflections Φ_1 and Φ_2. Moreover, if h is a bilinear majorant of f, then

$$x^1 \in \mathsf{C}_1 \cap \mathsf{C}_3 \quad \text{implies} \quad \{x^1\} \times]0, N^2[\subset \mathsf{E}_2. \tag{10.23}$$

Indeed, if $x^1 \in \mathsf{C}_1 \cap \mathsf{C}_3$ and if g is the biconcave envelope of f, then $g \leq h$, because h is a biconcave majorant of f. Since $g(x^1, \cdot\,)$ is concave, $h(x^1, \cdot\,)$ is linear, and g and h agree at $(x^1, 0)$ and (x^1, N^2), $g(x^1, \cdot\,) \geq h(x^1, \cdot\,)$; therefore $g(x^1, \cdot\,) = h(x^1, \cdot\,)$, which proves (10.23).

Theorem (Criterion for the type of the solution). *The solution to the optimal switching problem is of type* 1 [*of type* 2] *if and only if there exists a bilinear majorant of f that has a contact of type* 1 [*of type* 2] *with f.*

Proof. We first establish that the condition concerning the type 1 solution is sufficient. Without loss of generality, we suppose that there exists a bilinear majorant h of f and $x^1 \in]0, N^1[$ such that $h(x^1, 0) = f(x^1, 0)$ and $h(x^1, N^2) = f(x^1, N^2)$. By (10.23), hypothesis a of Proposition 3 of Section 10.2 is verified, which establishes that the solution is of type 1.

We now establish that the condition is necessary. Assume that the solution is of type 1 and let g be the biconcave envelope of f. After possibly transforming the problem by applying Φ_2, we can assume that hypothesis a of Proposition 3 of Section 10.2 is satisfied.

The case where there exists $x^1 \in [0, N^1[$ such that both $g(x^1, \cdot)$ and $g(x^1 + 1, \cdot)$ are linear on $[0, N^2]$ is quite straightforward. In this case, let h be the bilinear function defined on E that agrees with g at each of the four vertices of the rectangle $[x^1, x^1 + 1] \times [0, N^2]$. Then h agrees with g on this rectangle, and since g is h-concave, we conclude that $h \geq g$ on E. Therefore h is a bilinear majorant of f with a type 1 contact with f, because either x^1 or $x^1 + 1$ belongs to $]0, N^1[$.

In the general case, there exists $x^1 \in]0, N^1[$ such that $g(x^1, \cdot)$ is linear on $[0, N^2]$. We shall construct a bilinear function h such that $h(x^1, \cdot) = g(x^1, \cdot)$ and $h(x^1 \pm 1, \cdot) \geq g(x^1 \pm 1, \cdot)$. Since g is h-concave, this will imply that $h \geq g$ on E, and therefore that h is a bilinear majorant of f with a type 1 contact with f.

Let k be the function defined on $[0, N^2]$ by

$$k(x^2) = \tfrac{1}{2}g(x^1 - 1, x^2) + \tfrac{1}{2}g(x^1 + 1, x^2), \qquad \text{for all } x^2 \in [0, N^2].$$

Then k is concave, and because g is h-concave, $g(x^1, x^2) \geq k(x^2)$ for all $x^2 \in [0, N^2]$. Let κ be the linear function defined on $[0, N^2]$ such that the following three properties are satisfied: $\kappa \geq k$, the function $g(x^1, \cdot) - \kappa(\cdot)$ is constant, and κ agrees with k at some element $x^2 \in [0, N^2]$.

In the case where $x^2 \in]0, N^2[$, define four numbers z'_-, z'_+, z''_-, and z''_+ by

$$z'_\pm = g(x^1 \pm 1, x^2 - 1), \qquad z''_\pm = 2g(x^1 \pm 1, x^2) - g(x^1 \pm 1, x^2 + 1).$$

For real numbers z_- and z_+ such that $z'_- \leq z_- \leq z''_-$ and $z'_+ \leq z_+ \leq z''_+$, let h_{z_-, z_+} be the bilinear function defined on E such that

$$h_{z_-, z_+}(x^1 \pm 1, x^2 - 1) = z_\pm \quad \text{and} \quad h_{z_-, z_+}(x^1 \pm 1, x^2) = g(x^1 \pm 1, x^2).$$

Because $z_- \leq z''_-$ and $z_+ \leq z''_+$,

$$\begin{aligned} h_{z_-, z_+}(x^1 \pm 1, x^2 + 1) &= 2g(x^1 \pm 1, x^2) - z_\pm \\ &\geq 2g(x^1 \pm 1, x^2) - z''_\pm = g(x^1 \pm 1, x^2 + 1). \end{aligned}$$

Observing that $g(x^1 \pm 1, \cdot\,)$ is concave, $h_{z_-, z_+}(x^1 \pm 1, x^2 - 1) = z_\pm \geq z'_\pm = g(x^1 \pm 1, x^2 - 1)$, and $h_{z_-, z_+}(x^1 \pm 1, x^2) = g(x^1 \pm 1, x^2)$, we conclude that

$$h_{z_-, z_+}(x^1 \pm 1, \cdot) \geq g(x^1 \pm 1, \cdot).$$

Moreover,

$$h_{z'_-, z'_+}(x^1, x^2 - 1) = k(x^2 - 1) \leq \kappa(x^2 - 1)$$

and

$$h_{z''_-, z''_+}(x^1, x^2 + 1) = k(x^2 + 1) \leq \kappa(x^2 + 1);$$

therefore

$$h_{z'_-, z'_+}(x^1, x^2 - 1) \leq \kappa(x^2 - 1) \quad \text{and} \quad h_{z''_-, z''_+}(x^1, x^2 - 1) \geq \kappa(x^2 - 1), \tag{10.24}$$

since $h_{z''_-, z''_+}(x^1, x^2) = \kappa(x^2)$. By the intermediate value theorem of calculus, there are real numbers z_- and z_+ such that $z'_- \leq z_- \leq z''_-$, $z'_+ \leq z_+ \leq z''_+$, and such that $h_{z_-, z_+}(x^1, x^2 - 1) = \kappa(x^2 - 1)$, or equivalently (since $h_{z_-, z_+}(x^1, x^2) = \kappa(x^2)$), such that $h_{z_-, z_+}(x^1, \cdot\,) = \kappa(\cdot)$. Now let h be the bilinear function equal to $h_{z_-, z_+} + c$ on E, where $c \geq 0$ is the constant value of $g(x^1, \cdot\,) - \kappa(\cdot)$. Then $h(x^1, \cdot\,) = g(x^1, \cdot\,)$, and $h(x^1 \pm 1, \cdot\,) \geq h_{z_-, z_+}(x^1 \pm 1, \cdot\,) \geq g(x^1 \pm 1, \cdot\,)$. Since g is h-concave, $h \geq g$ on E, and h is therefore a bilinear majorant of f with a type 1 contact with f.

In the case where $x^2 = N^2$, we let $z'_\pm$ be defined as above. For sufficiently large values of $z''_\pm$, (10.24) remains valid, as do the arguments that follow it. The case where $x^2 = 0$ can be handled in a similar way.

We now establish that the condition concerning the type 2 solution is necessary. Suppose that the solution is of type 2. Without loss of generality, we can assume that one of the two hypotheses of Proposition 3 of Section 10.2 is satisfied. Let p^1, q^1, p^2, and q^2 be as in the theorem of Section 10.3. By assumption, $p^1 > q^1$, and after possibly transforming the problem by applying Φ_1, we can assume that $p^2 < q^2$. Let g be the biconcave envelope of f and let h be the unique bilinear function that coincides with g at each of the four vertices of the rectangle $[q^1, p^1] \times [p^2, q^2]$. By the theorem of Section 10.3 and the observations in Section 10.4, h and g coincide on this rectangle. Since g is v-concave, $h \geq g$ on $[q^1, p^1] \times [0, p^2]$. But since $h = g$ on the segment $\{p^1\} \times [0, q^2]$ by the theorem of Section 10.3, the h-concavity of g implies $h \geq g$ on $[0, p^1] \times [0, p^2]$. Notice also that $h(p^1, 0) = g(p^1, 0) = f(p^1, 0)$. By a similar argument, one shows that $h \geq g$ on each of the rectangles $[p^1, N^1] \times [0, q^2]$, $[q^1, N^1] \times [q^2, N^2]$, and $[0, q^1] \times [p^2, N^2]$, which implies that $h \geq g$ and therefore that $h \geq f$. Since h and f agree at the four points $(p^1, 0)$, (N^1, q^2), (q^1, N^2), and $(0, p^2)$, h is a bilinear majorant of f and $\mathsf{C}_1 \neq \varnothing, \ldots, \mathsf{C}_4 \neq \varnothing$, where $\mathsf{C}_1, \ldots, \mathsf{C}_4$ are the contact sets of h with f. Because $p^1 > q^1$, hypothesis a of Proposition 3 of Section 10.2 is not satisfied. Together with (10.23), this implies that $\mathsf{C}_1 \cap \mathsf{C}_3 = \varnothing$. Since $p^1 \in \mathsf{C}_1$, $q^1 \in \mathsf{C}_3$ and both C_1 and C_3 are intervals, we conclude that $\sup \mathsf{C}_3 < \inf \mathsf{C}_1$. Since hypothesis b of the proposition just mentioned must be satisfied, one shows by a similar argument that $\mathsf{C}_2 \cap \mathsf{C}_4 = \varnothing$. Because $p^2 \in \mathsf{C}_4$ and $q^2 \in \mathsf{C}_2$, it follows that $\sup \mathsf{C}_4 < \inf \mathsf{C}_2$. Consequently, h has a contact of type 2 with f.

We now show that the condition concerning the type 2 solution is sufficient. Assume that there exists a bilinear majorant h of f that has a contact of type 2 with f and let $\mathsf{C}_1, \ldots, \mathsf{C}_4$ be the contact sets of h with f. Set $x^1 = \inf \mathsf{C}_1$, $y^1 = \sup \mathsf{C}_3$, $x^2 = \sup \mathsf{C}_4$, and $y^2 = \inf \mathsf{C}_2$. Notice that $x^1, y^1 \in]0, N^1[$ and $x^2, y^2 \in]0, N^2[$. Without loss of generality, we can assume that

$$y^1 < x^1 \quad \text{and} \quad x^2 < y^2. \tag{10.25}$$

Notice that $h > f$ at the two points $(x^1 - 1, 0)$ and $(y^1 + 1, N^2)$. In order to establish that the solution is of type 2, we first show that

$$(\{x^1\} \times]0, y^2[) \cup (\{y^1\} \times]x^2, N^2[) \subset \mathsf{E}_2. \tag{10.26}$$

Let F denote the set on the left-hand side of the inclusion and set $a_1 = (x^1, 0)$, $a_2 = (N^1, y^2)$, $a_3 = (y^1, N^2)$, $a_4 = (0, x^2)$, $b_1 = (x^1, x^2)$, $b_2 = (x^1, y^2)$, $b_3 = (y^1, y^2)$, $b_4 = (y^1, x^2)$, and $b_5 = b_1$. Because g is a biconcave majorant of f, the numbers $\lambda_j = g(b_j)$, $j = 1, \ldots, 5$ satisfy the system of inequalities

$$\lambda_j \geq r_j f(a_j) + \rho_j \lambda_{j+1}, \qquad j = 1, \ldots, 4, \tag{10.27}$$

where $r_j = d(b_j, b_{j+1})/d(a_j, b_{j+1})$, $\rho_j = 1 - r_j$, and $d(\cdot, \cdot)$ denotes Euclidean distance. On the other hand, it is easy to check that the system of equations

$$\alpha_j = r_j f(a_j) + \rho_j \alpha_{j+1}, \qquad j = 1, \ldots, 4, \tag{10.28}$$

where $\alpha_1, \ldots, \alpha_4$ are the unknowns and $\alpha_5 = \alpha_1$, by convention, has a unique solution, namely, $\alpha_j = h(b_j)$ (cf. Exercise 3). Subtracting (10.28) from (10.27), we see that

$$\lambda_j - \alpha_j \geq \rho_j(\lambda_{j+1} - \alpha_{j+1}), \qquad j = 1, \ldots, 4,$$

from which it follows that $\lambda_j \geq \alpha_j$, $j = 1, \ldots, 4$, since $\rho_1 \rho_2 \rho_3 \rho_4 < 1$ by (10.25). Therefore, $g(b_j) \geq h(b_j)$, $j = 1, \ldots, 4$, which implies, because $g \leq h$, that $g = h$ on F and thus that $\mathsf{F} \subset \mathsf{E}_2$.

For future reference, we point out that we have also shown that

$$(]0, x^1[\times \{x^2\}) \cup (]y^1, N^1[\times \{y^2\}) \subset \mathsf{E}_1, \tag{10.29}$$

and by Proposition 2 of Section 10.2, that

$$h = g \qquad \text{on } [y^1, x^1] \times [x^2, y^2]. \tag{10.30}$$

We now show that hypothesis b of Proposition 3 of Section 10.2 is verified. If it were not, then there would exist $n \in]0, N^2[$ such that $]0, N^1[\times \{n\} \subset \mathsf{E}_1$, that is,

$$g(m, n) = \frac{N^1 - m}{N^1} f(0, n) + \frac{m}{N^1} f(N^1, n), \qquad \text{for all } m \in [0, N^1].$$

Suppose that $n \in]0, x^2]$. Then $f(0, n) \leq h(0, n)$ because h is a majorant of f, and $f(N^1, n) < h(N^1, n)$ because $n \leq x^2 < y^2 = \inf \mathsf{C}_2$. Therefore, since $(x^1, n) \in \mathsf{F}$,

$$g(x^1, n) < \frac{N^1 - x^1}{N^1} h(0, n) + \frac{x^1}{N^1} h(N^1, n) = h(x^1, n) = g(x^1, n),$$

which is a contradiction. The cases where $n \in]x^2, y^2[$ and $n \in [y^2, N^2[$ lead in a similar way to a contradiction.

Consider now the functions μ_1 and μ_2 defined in the proof of Proposition 3 of Section 10.2 and the integers p^1, q^1, p^2, and q^2 defined in (10.11) and (10.12). Because $\mathsf{F} \subset \mathsf{E}_2$, $\mu_1(n) \leq x^1$ and $\mu_2(n) \geq y^1$ for all $n \in [0, N^2]$. We are going to show that

$$\sup \mu_1 = x^1 \quad \text{and} \quad \inf \mu_2 = y^1, \tag{10.31}$$

which will complete the proof of the assertion concerning the type 2 solution, since $x^1 > y^1$. By hypothesis, $h(x^1 - 1, 0) > f(x^1 - 1, 0) = g(x^1 - 1, 0)$, whereas $h(x^1 - 1, x^2) = g(x^1 - 1, x^2)$ by (10.30). Because $h \geq g$, there exists therefore $n \in]0, x^2]$ such that g is strictly v-concave at $(x^1 - 1, n)$; in other words $(x^1 - 1, n) \notin \mathsf{E}_2$. Since E_2 is h-convex by Proposition 2 of Section 10.2, and since $(x^1, n) \in \mathsf{E}_2$, it follows that $(m, n) \notin \mathsf{E}_2$ for all $m \in]0, x^1[$. Consequently $\mu_1(n) = x_1$ and therefore the supremum in (10.11) is x^1. Similarly, one shows that the infimum in (10.11) is y^1; therefore (10.31) is proved.

10.6 DETERMINING THE TYPE OF THE SOLUTION

In this section, we shall give analytic conditions on the payoff function f that ensure the existence of a bilinear majorant of f having a contact of type 1 or a contact of type 2 with f. By the theorem of Section 10.5, these conditions will ensure that the solution is of type 1 or 2. We shall also show how to construct this bilinear majorant along with integers p^1, q^1, p^2, and q^2 that satisfy (10.11) and (10.12).

It will be convenient to have at hand symbols for the vertices and the boundary segments of E: We set $V_1 = (0, 0)$, $V_2 = (N^1, 0)$, $V_3 = (N^1, N^2)$, $V_4 = (0, N^2)$, and by convention, $V_5 = V_1$ and $V_0 = V_4$; for $j = 0, \ldots, 4$, we shall also denote by S_j the boundary segment of E with extremities V_j and V_{j+1}.

The Functions φ_j^i

We are going to define eight functions φ_j^i, $i = 1, 2$, $j = 1, \ldots, 4$, as follows. First of all, φ_1^1 and φ_1^2 are, respectively, the functions φ^1 and φ^2 defined in (10.17). Next, we define φ_j^i, $i = 1, 2$, $j = 2, 3, 4$, by transforming E by reflections in order to bring V_j to V_1 and then by applying (10.17) to the image of E. For example, φ_2^1 and φ_3^2 are the functions defined, respectively, on $]0, N^1]$ and $]0, N^2]$ by

$$\varphi_2^1(x^1) = (N^1 - x^1 + 1) f(x^1, 0) - (N^1 - x^1) f(x^1 - 1, 0),$$

$$\varphi_3^2(x^2) = (N^2 - x^2 + 1) f(N^1, x^2) - (N^2 - x^2) f(N^1, x^2 - 1).$$

By convention, φ_0^i will be φ_4^i. Notice that the lower index j of φ_j^i refers to the vertex V_j, whereas the upper index i indicates the direction (horizontal if $i = 1$ or vertical if $i = 2$). Notice also the interpretation that has already been explained for φ_1^1 and φ_1^2: For instance, for $x^2 \in]0, N^2]$ fixed, $\varphi_3^2(x^2)$ is the value at N^2 of the unique linear function defined on $[0, N^2]$ whose value at x^2 is $f(N^1, x^2)$ and at $x^2 - 1$ is $f(N^1, x^2 - 1)$.

The Sets D_j

We now define four sets $\mathsf{D}_1, \dots, \mathsf{D}_4$ as follows. The set D_1 [D_3] is the set of all $x^1 \in [0, N^1[$ [$x^1 \in]0, N^1]$] such that the linear function defined on $[0, N^2]$ with value $\varphi_1^1(x^1)$ at 0 and value $\varphi_4^1(x^1)$ at N^2 [value $\varphi_2^1(x^1)$ at 0 and value $\varphi_3^1(x^1)$ at N^2] is larger than $f(0, \cdot)$ [$f(N^1, \cdot)$] (see Figure 8). The set D_2 [D_4] is the set of all $x^2 \in [0, N^2[$ [$x^2 \in]0, N^2]$] such that the linear function defined on $[0, N^1]$ with value $\varphi_1^2(x^2)$ at 0 and value $\varphi_2^2(x^2)$ at N^1 [value $\varphi_4^2(x^2)$ at 0 and value $\varphi_3^2(x^2)$ at N^1] is larger than $f(\cdot, 0)$ [$f(\cdot, N^2)$].

Notice that if $x^1 \in \mathsf{D}_1$, then $y^1 \in \mathsf{D}_1$ whenever $y^1 > x^1$. An analogous property holds for D_2, D_3, and D_4. We use the convention $\inf \mathsf{D}_1 = N^1$ if $\mathsf{D}_1 = \varnothing$, $\inf \mathsf{D}_2 = N^2$ if $\mathsf{D}_2 = \varnothing$, $\sup \mathsf{D}_3 = 0$ if $\mathsf{D}_3 = \varnothing$, and $\sup \mathsf{D}_4 = 0$ if $\mathsf{D}_4 = \varnothing$. Given the payoff function f, the infima and suprema of $\mathsf{D}_1, \dots, \mathsf{D}_4$ are not difficult to determine.

Let $x^1 = \inf \mathsf{D}_1$ and suppose that $x^1 > 1$. Then the set $\{x^2 \in]0, N^2[: \varphi_1^2(x^2) > \varphi_1^1(x^1 - 1)\}$ is not empty and the infimum y^2 of this set satisfies

$$\varphi_4^2(y^2) > \varphi_4^1(x^1 - 1). \tag{10.32}$$

Indeed, the set in question contains $N^2 - 1$, because otherwise $\varphi_1^2(N^2 - 1) \leq \varphi_1^1(x^1 - 1)$ and therefore the linear function h defined on $[0, N^2]$ with value $\varphi_1^1(x^1 - 1)$ at 0 and value $\varphi_4^1(x^1 - 1)$ at N^2 would dominate $f(\cdot, 0)$, contradicting the fact that $x^1 - 1 \notin \mathsf{D}_1$. Now if $\varphi_4^2(y^2) \leq \varphi_4^1(x^1 - 1)$, then since $\varphi_1^2(y^2 - 1) \leq \varphi_1^1(x^1 - 1)$, the function h would dominate $f(\cdot, 0)$, a contradiction.

Given the four sets $\mathsf{D}_1, \dots, \mathsf{D}_4$, we consider two cases.

Case A: $\inf \mathsf{D}_1 < \sup \mathsf{D}_3$ or $\inf \mathsf{D}_2 < \sup \mathsf{D}_4$.

Case B: $\inf \mathsf{D}_1 > \sup \mathsf{D}_3$ and $\inf \mathsf{D}_2 > \sup \mathsf{D}_4$.

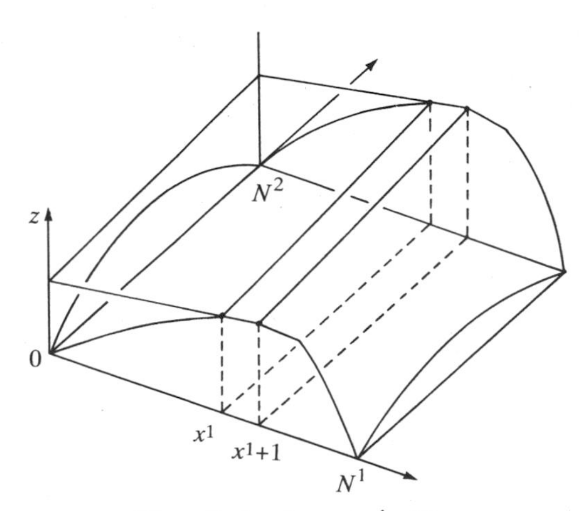

Figure 8. An element x^1 of D_1.

In either Case A or B, we shall show how to determine the type of the solution. If neither Case A nor Case B holds, then $\inf \mathsf{D}_1 \geq \sup \mathsf{D}_3$, $\inf \mathsf{D}_2 \geq \sup \mathsf{D}_4$, and one or both of these two inequalities is an equality. This is a critical case in which determining the type of the solution requires additional work.

Theorem 1 (Recognizing and computing the solution of type 1). *If Case* A *holds, then the solution of the optimal switching problem is of type* 1. *In this case, after possibly having transformed the problem by applying* Φ_2, *the inequality* $\inf \mathsf{D}_1 < \sup \mathsf{D}_3$ *holds and hypothesis* a *of Proposition* 3 *of Section* 10.2 *is satisfied. In addition, the integers* p^1, q^1, p^2, *and* q^2 *defined by*

$$\begin{aligned} &p^1 = \inf \mathsf{D}_1, \quad p^2 = \inf\{x^2 \in]0, N^2[: \varphi_1^2(x^2) > \varphi_1^1(p^1 - 1)\}, && \text{if } \inf \mathsf{D}_1 > 0, \\ &p^1 = 1, \qquad\quad p^2 = 1, && \text{if } \inf \mathsf{D}_1 = 0, \end{aligned} \tag{10.33}$$

$$\begin{aligned} &q^1 = \sup \mathsf{D}_3, \quad q^2 = \sup\{x^2 \in]0, N^2[: \varphi_3^2(x^2) > \varphi_3^1(q^1 + 1)\}, && \text{if } \sup \mathsf{D}_3 < N^1, \\ &q^1 = N^1 - 1, \quad q^2 = 1, && \text{if } \sup \mathsf{D}_3 = N^1, \end{aligned} \tag{10.34}$$

satisfy (10.11) *and* (10.12).

Proof. If $\inf \mathsf{D}_2 < \sup \mathsf{D}_4$, then $\inf \mathsf{D}_1 < \sup \mathsf{D}_3$ for the transformation of the problem obtained by applying Φ_2. So let $\tilde{p}^1 = \inf \mathsf{D}_1$, $\tilde{q}^1 = \sup \mathsf{D}_3$ and assume that $\tilde{p}^1 < \tilde{q}^1$. For $x^1 \in [\tilde{p}^1, \tilde{q}^1[$, let h be the unique bilinear function that agrees with f at the four vertices of the rectangle $[x^1, x^1 + 1] \times [0, N^2]$. Since $x^1 \in \mathsf{D}_1$, h dominates f on the rectangle $[0, x^1] \times [0, N^2]$. Because $x^1 + 1 \leq \tilde{q}^1$ and $\tilde{q}^1 \in \mathsf{D}_3$, we conclude that $x^1 + 1 \in \mathsf{D}_3$, which means that h also dominates f on the rectangle $[x^1 + 1, N^1] \times [0, N^2]$. Since either x^1 or $x^1 + 1$ belongs to $]0, N^1[$, h is a bilinear majorant of f that has a contact of type 1 with f, so the solution is of type 1 by the theorem of Section 10.5.

By (10.23), these considerations also show that the biconcave envelope g of f is v-linear on $[\tilde{p}^1, \tilde{q}^1] \times [0, N^2]$. In particular, hypothesis a of Proposition 3 of Section 10.2 is satisfied, and the integer p^1 defined in (10.11) satisfies $1 \leq p_1 \leq \tilde{p}^1$ if $\tilde{p}^1 \geq 1$ and $p_1 = 1$ if $\tilde{p}^1 = 0$. When $p^1 = 1$, $p^2 = 1$ satisfies (10.12). In order to prove that $\tilde{p}^1 = p^1$ when $\tilde{p}^1 \geq 1$, we only need to show that if $\tilde{p}^1 > 1$, then there is $x^2 \in]0, N^2[$ such that $(\tilde{p}^1 - 1, x^2) \in \mathsf{E}_1 - \mathsf{E}_2$. Let h be the bilinear function that agrees with f at the four vertices of the rectangle $[\tilde{p}^1 - 1, \tilde{p}^1] \times [0, N^2]$. Notice that h does not dominate g on this rectangle, for otherwise, h would agree with g there, and since g is h-concave, h would dominate g and therefore f on S_4, which would contradict the fact that $\tilde{p}^1 - 1 \notin \mathsf{D}_1$. It follows that $h < g$ at some point of the segment $\{\tilde{p}^1 - 1\} \times]0, N^2[$, and so $g(\tilde{p}^1 - 1, \cdot\,)$ is not linear. This means that there is $x^2 \in]0, N^2[$ such that $(\tilde{p}^1 - 1, x^2) \notin \mathsf{E}_2$, that is, such that $(\tilde{p}^1 - 1, x^2) \in \mathsf{E}_1 - \mathsf{E}_2$.

To complete the proof of (10.33), assume $\inf \mathsf{D}_1 > 0$ and let $\tilde{p}^2$ denote the right-hand side of the second equality in (10.33). Notice that the set over which the infimum is taken is not empty by the comment that precedes (10.32). All we now need to show is that $\mu_1(\tilde{p}^2) = p^1$, where μ_1 is the function defined in the proof of Proposition 3 of Section 10.2. Let p^2 be chosen as in this proof. If $\tilde{p}^2 = p^2$, what must be shown is true by the choice of p^2. If $\tilde{p}^2 < p^2$, then since $\varphi_1^2(\tilde{p}^2) > \varphi_1^1(p^1 - 1)$ and $\varphi_1^2(\tilde{p}^2 - 1) \le \varphi_1^1(p^1 - 1)$, we see by the theorem of Section 10.4 that $\xi_1(\tilde{p}^2) = p^1$ and $\xi_1(\tilde{p}^2 - 1) < p^1$. By (10.13), it follows that $\mu_1(\tilde{p}^2) = p^1$. In order to complete the proof, it therefore suffices to establish that $\tilde{p}^2 \le p^2$. Since $p^1 - 1 \notin \mathsf{D}_1$, we conclude from (10.32) that

$$\varphi_4^2(\tilde{p}^2) > \varphi_4^1(p^1 - 1). \tag{10.35}$$

Assume now that $p^2 < \tilde{p}^2$. Applying the statement analogous to (10.19) for the rectangle $[0, p^1] \times [p^2, N^2]$, we see from (10.35) that $\xi_1(\tilde{p}^2) = p^1$. Therefore g would be h-linear on $[0, p^1] \times [p^2, \tilde{p}^2]$. But then the bilinear function h that agrees with g on $[0, p^1] \times [\tilde{p}^2 - 1, \tilde{p}^2]$ would dominate g on $[0, p^1] \times [0, \tilde{p}^2]$, because g is υ-concave. As g is strictly υ-concave at $(p^1 - 1, p^2)$ by (10.12), the inequality $f(p^1 - 1, 0) < h(p^1 - 1, 0)$, or, equivalently, $h(0,0) > \varphi_1^1(p^1 - 1)$ (since $f(p^1, 0) = h(p^1, 0)$) would hold. Because $h(0,0)$ is also equal to $\varphi_1^2(\tilde{p}^2 - 1) \le \varphi_1^1(p^1 - 1)$, we have derived a contradiction. The proof of (10.34) is similar and is left to the reader.

The Functions F_j

For $j = 1, \ldots, 4$, we define a real-valued function F_j whose domain is the interval $[f(V_j), \infty[$ of $\mathbb{R}$. In fact, we only consider the case where $j = 1$, since F_2, F_3, and F_4 are defined analogously. For $z \in [f(V_1), \infty[$, let h denote the linear function defined on $[0, N^1]$ such that $h \ge f(\cdot, 0)$, $h(0) = z$, and $h(x^1) = f(x^1, 0)$ for at least one $x^1 \in]0, N^1]$ (if $z = f(V_1)$, then x^1 can be taken equal to 1). Then $F_1(z)$ is by definition equal to $h(N^1)$ (see Figure 9).

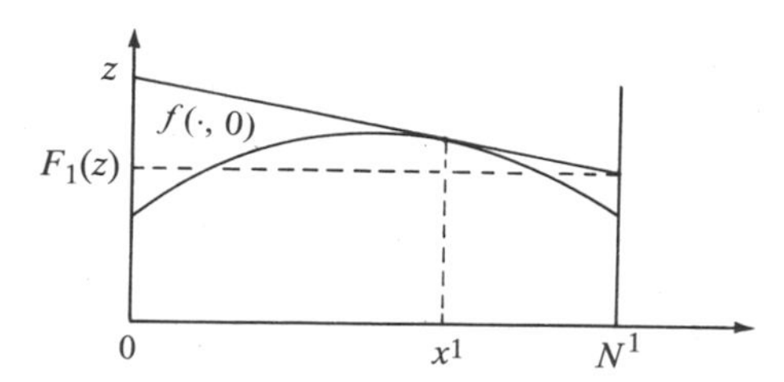

Figure 9. Illustration of the function F_1.

For $j = 1, \dots, 4$, it is clear that F_j is continuous, nonincreasing, and even strictly decreasing on $[f(V_j), z_j]$, where $z_1 = \varphi_1^1(N^1 - 1)$, $z_2 = \varphi_2^2(N^2 - 1)$, $z_3 = \varphi_3^1(1)$, and $z_4 = \varphi_4^2(1)$. Moreover,

$$\begin{aligned} F_1\big(\varphi_1^1(x^1)\big) &= \varphi_2^1\big(x^1 + 1\big), && \text{for all } x^1 \in \big[0, N^1\big[, \\ F_2\big(\varphi_2^2(x^2)\big) &= \varphi_3^2\big(x^2 + 1\big), && \text{for all } x^2 \in \big[0, N^2\big[, \\ F_3\big(\varphi_3^1(x^1)\big) &= \varphi_4^1\big(x^1 - 1\big), && \text{for all } x^1 \in \big]0, N^1\big], \\ F_4\big(\varphi_4^2(x^2)\big) &= \varphi_1^2\big(x^2 - 1\big), && \text{for all } x^2 \in \big]0, N^2\big]. \end{aligned} \tag{10.36}$$

Another property of the function F_1 is that

$$0 \le x^2 < \inf \mathsf{D}_2 \quad \text{implies} \quad \varphi_1^2(x^2) < F_1^{-1}\big(\varphi_2^2(x^2)\big). \tag{10.37}$$

Indeed, if $\varphi_1^2(x^2)$ were not less than $F_1^{-1}(\varphi_2^2(x^2))$, then $F_1(\varphi_1^2(x^2)) \le \varphi_2^2(x^2)$, which would imply that $x^2 \in \mathsf{D}_2$, a contradiction.

The Function F

The function $F = F_4 \circ F_3 \circ F_2 \circ F_1$ will play an important role below. If z_1 is a fixed point of F and if $z_{j+1} = F_j(z_j)$, $j = 1, 2, 3$, then the unique bilinear function h whose value at V_j is z_j, $j = 1, \dots, 4$, is called the *bilinear function associated with* this fixed point. It is a bilinear majorant of f, and if $z_j > f(V_j)$, $j = 1, \dots, 4$, then all four contact sets of h with f are nonempty.

Theorem 2 (Recognizing and computing the solution of type 2). *In Case* B, *the solution of the optimal switching problem is of type* 2. *Let* $d_1^1 = \inf \mathsf{D}_1$ *and* $d_2^2 = \inf \mathsf{D}_2$. *The function* F *has a unique fixed point* z *in the interval* $[\varphi_1^1(d_1^1 - 1), F_1^{-1}(\varphi_2^2(d_2^2 - 1))]$ *of* $\mathbb{R}$, *and the bilinear function* h *associated with* z *is a bilinear majorant of* f *that has a contact of type* 2 *with* f. *The fixed point* z *is characterized by the following property: for* z' *in this interval,* $z' < z$ *implies* $z' < F(z') < z$, *and* $z < z'$ *implies* $z < F(z') < z'$. *If* $\mathsf{C}_1, \dots, \mathsf{C}_4$ *are the contact sets of* h *with* f, *then after possibly transforming the problem by applying* Φ_1, *the inequality* $\inf \mathsf{C}_1 > \sup \mathsf{C}_3$ *holds and hypothesis* b *of Proposition* 3 *of Section* 10.2 *is verified. In addition, the integers* p^1, q^1, p^2, *and* q^2 *defined by*

$$p^1 = \inf \mathsf{C}_1, \quad q^1 = \sup \mathsf{C}_3, \quad p^2 = \inf \mathsf{C}_4, \quad q^2 = \sup \mathsf{C}_2$$

satisfy (10.11) *and* (10.12).

Proof. Assume Case B holds and define four subsets $\mathsf{F}_1, \ldots, \mathsf{F}_4$ of E by setting

$$\mathsf{F}_1 = \{(x^1, x^2): \varphi_4^2(x^2) > \varphi_4^1(x^1) \text{ and } \varphi_1^2(x^2) > \varphi_1^1(x^2)\},$$
$$\mathsf{F}_2 = \{(x^1, x^2): \varphi_1^1(x^1) > \varphi_1^2(x^2) \text{ and } \varphi_2^1(x^1) > \varphi_2^2(x^2)\},$$
$$\mathsf{F}_3 = \{(x^1, x^2): \varphi_2^2(x^2) > \varphi_2^1(x^1) \text{ and } \varphi_3^2(x^2) > \varphi_3^1(x^1)\},$$
$$\mathsf{F}_4 = \{(x^1, x^2): \varphi_3^1(x^1) > \varphi_3^2(x^2) \text{ and } \varphi_4^1(x^1) > \varphi_4^2(x^2)\}.$$

Notice that

$$\mathsf{F}_1 \cap \mathsf{F}_2 = \varnothing, \quad \mathsf{F}_2 \cap \mathsf{F}_3 = \varnothing, \quad \mathsf{F}_3 \cap \mathsf{F}_4 = \varnothing, \quad \text{and} \quad \mathsf{F}_4 \cap \mathsf{F}_1 = \varnothing. \quad (10.38)$$

Since each function φ_j^i is monotone, it easy to check that

$$\text{each set } \mathsf{F}_j \text{ is enclosed by the graph of a unimodal function} \quad (10.39)$$

(and by a segment contained in S_{j-1}). Let $d_3^1 = \sup \mathsf{D}_3$ and $d_4^2 = \sup \mathsf{D}_4$. Notice that because $\inf \mathsf{D}_1 > \sup \mathsf{D}_3$, $N^1 \geq d_1^1 > d_3^1 \geq 0$, so in particular, $d_1^1 > 0$ and $d_3^1 < N^1$. Similarly, $d_2^2 > 0$ and $d_4^2 < N^2$. Set

$$d_1^2 = \inf\{x^2 \in]0, N^2[: \varphi_1^2(x^2) > \varphi_1^1(d_1^1 - 1)\},$$
$$d_2^1 = \sup\{x^1 \in]0, N^1[: \varphi_2^1(x^1) > \varphi_2^2(d_2^2 - 1)\},$$
$$d_3^2 = \sup\{x^2 \in]0, N^2[: \varphi_3^2(x^2) > \varphi_3^1(d_3^1 + 1)\},$$
$$d_4^1 = \inf\{x^1 \in]0, N^1[: \varphi_4^1(x^1) > \varphi_4^2(d_4^2 + 1)\}.$$

For reasons similar to the one that showed that the set defined just before (10.32) is nonempty, all four sets on the right-hand side above are nonempty. Notice that

$$(d_1^1 - 1, d_1^2) \in \mathsf{F}_1, \quad (d_2^1, d_2^2 - 1) \in \mathsf{F}_2,$$
$$(d_3^1 + 1, d_3^2) \in \mathsf{F}_3, \quad (d_4^1, d_4^2 + 1) \in \mathsf{F}_4.$$

Indeed, the first relation is true because $\varphi_1^2(d_1^2) > \varphi_1^1(d_1^1 - 1)$, by definition of d_1^2, and $\varphi_4^2(d_1^2) > \varphi_4^1(d_1^1 - 1)$, by (10.32). The other three relations can be checked in a similar way.

Figure 10 shows a possible position of the sets $\mathsf{F}_1, \ldots, \mathsf{F}_4$. Since $d_2^2 > d_4^2$, F_1 cannot intersect F_3 by (10.38) and (10.39). Similarly, F_2 cannot intersect F_4, and so $\mathsf{F}_1, \ldots, \mathsf{F}_4$ are disjoint. From the figure, notice that we can assume that $d_1^2 \leq d_3^2$, because otherwise this inequality would hold after applying Φ_1.

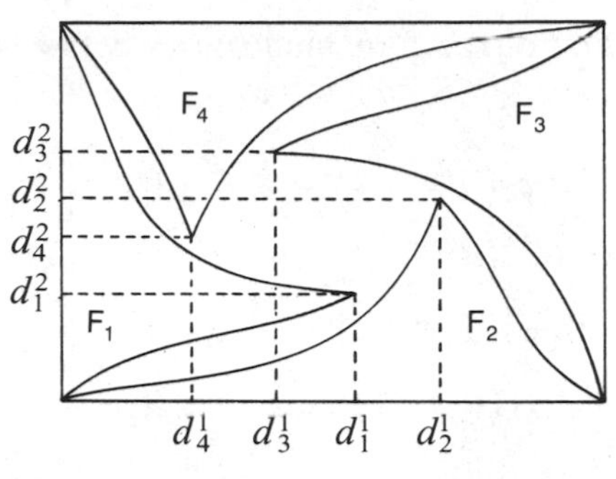

Figure 10.

Also from the figure, notice that $d_3^1 < d_1^1$ and $d_4^2 < d_2^2$ imply that

$$d_4^1 \le d_3^1, \quad d_1^1 \le d_2^1, \quad d_1^2 \le d_4^2 \quad \text{and} \quad d_2^2 \le d_3^2,$$

and therefore that

$$d_4^1 \le d_3^1 < d_1^1 \le d_2^1 \quad \text{and} \quad d_1^2 \le d_4^2 < d_2^2 \le d_3^2. \tag{10.40}$$

Since $d_1^1 \le d_2^1$ and $\varphi_2^1(d_2^1) > \varphi_2^2(d_2^2 - 1)$, it follows from (10.36) that

$$F_1\big(\varphi_1^1(d_1^1 - 1)\big) = \varphi_2^1\big(d_1^1\big) \ge \varphi_2^1\big(d_2^1\big) > \varphi_2^2\big(d_2^2 - 1\big),$$

and therefore that

$$\varphi_1^1\big(d_1^1 - 1\big) < F_1^{-1}\big(\varphi_2^2\big(d_2^2 - 1\big)\big),$$

and one checks in a similar way that

$$\varphi_2^2\big(d_2^2 - 1\big) < F_2^{-1}\big(\varphi_3^1\big(d_3^1 + 1\big)\big),$$
$$\varphi_3^1\big(d_3^1 + 1\big) < F_3^{-1}\big(\varphi_4^2\big(d_4^2 + 1\big)\big),$$
$$\varphi_4^2\big(d_4^2 + 1\big) < F_4^{-1}\big(\varphi_1^1\big(d_1^1 - 1\big)\big).$$

Define four intervals $I_1, \ldots, I_4$ of $\mathbb{R}$ by setting

$$I_1 = \big[\varphi_1^1\big(d_1^1 - 1\big), F_1^{-1}\big(\varphi_2^2\big(d_2^2 - 1\big)\big)\big], \quad I_2 = \big[\varphi_2^2\big(d_2^2 - 1\big), F_2^{-1}\big(\varphi_3^1\big(d_3^1 + 1\big)\big)\big],$$
$$I_3 = \big[\varphi_3^1\big(d_3^1 + 1\big), F_3^{-1}\big(\varphi_4^2\big(d_4^2 + 1\big)\big)\big], \quad I_4 = \big[\varphi_4^2\big(d_4^2 + 1\big), F_4^{-1}\big(\varphi_1^1\big(d_1^1 - 1\big)\big)\big].$$

Moreover, set $I_5 = I_1$. By (10.40), $d_3^1 < d_1^1$, and therefore, using (10.36) and the property of F_2 analogous to (10.37), we see that

$$F_1\big(\varphi_1^1\big(d_1^1 - 1\big)\big) \le F_1\big(\varphi_1^1\big(d_3^1\big)\big) = \varphi_2^1\big(d_3^1 + 1\big) < F_2^{-1}\big(\varphi_3^1\big(d_3^1 + 1\big)\big).$$

It follows that the first of the four inequalities below is established and the others can be established in a similar way:

$$\begin{aligned} F_1\big(\varphi_1^1(d_1^1 - 1)\big) &< F_2^{-1}\big(\varphi_3^1(d_3^1 + 1)\big), \\ F_2\big(\varphi_2^2(d_2^2 - 1)\big) &< F_3^{-1}\big(\varphi_4^2(d_4^2 + 1)\big), \\ F_3\big(\varphi_3^1(d_3^1 + 1)\big) &< F_4^{-1}\big(\varphi_1^1(d_1^1 - 1)\big), \\ F_4\big(\varphi_4^2(d_4^2 + 1)\big) &< F_1^{-1}\big(\varphi_2^2(d_2^2 - 1)\big). \end{aligned} \tag{10.41}$$

Since F_j is nonincreasing, we conclude that $F_j(I_j) \subset I_{j+1}$ for $j = 1, \ldots, 4$. Consequently, $F(I_1) \subset I_1$, and so F admits a fixed point z in I_1. Let h be the bilinear function associated with z, and set $z_1 = z$ and $z_{j+1} = F_j(z_j)$ for $j = 1, 2, 3$. Notice that $z_1 \in F_4(I_4)$, so by (10.41),

$$z_1 \leq F_4\big(\varphi_4^2(d_4^2 + 1)\big) < F_1^{-1}\big(\varphi_2^2(d_2^2 - 1)\big),$$

and therefore

$$z_2 = F_1(z_1) > \varphi_2^2(d_2^2 - 1).$$

Similarly, one shows that

$$z_2 < F_2^{-1}\big(\varphi_3^1(d_3^1 + 1)\big), \quad z_3 < F_3^{-1}\big(\varphi_4^2(d_4^2 + 1)\big), \quad z_4 < F_4^{-1}\big(\varphi_1^1(d_1^1 - 1)\big)$$

and

$$z_3 > \varphi_3^1(d_3^1 + 1), \quad z_4 > \varphi_4^2(d_4^2 + 1), \quad z_1 > \varphi_1^1(d_1^1 - 1).$$

It follows that h and f do not agree at V_j, $j = 1, \ldots, 4$, and because h is the bilinear function associated with z, the contact sets $\mathsf{C}_1, \ldots, \mathsf{C}_4$ of h with f must be nonempty. The inequalities just established show that $\mathsf{C}_1, \ldots, \mathsf{C}_4$ satisfy

$$\inf \mathsf{C}_1 \geq d_1^1, \quad \inf \mathsf{C}_2 \geq d_2^2, \quad \sup \mathsf{C}_3 \leq d_3^1, \quad \text{and} \quad \sup \mathsf{C}_4 \leq d_4^2. \tag{10.42}$$

By (10.40), this implies that h has a contact of type 2 with f, therefore, by the theorem of Section 10.5, that the solution is of type 2. Using the argument that follows (10.30), we see that hypothesis b of Proposition 3 of Section 10.2 is verified.

We shall now prove that F has at most one fixed point in I_1 (in Exercise 5, it is shown that any fixed point of F outside of I_1 does not correspond to a bilinear function that has a contact of type 2 with f). Fix x_1 in the interior of I_1, and for $j = 1, 2, 3$, let $x_{j+1} = F_j(x_j)$. Let $[\alpha_1(x_1), \beta_1(x_1)] \times \{0\}$ be the segment of S_1 on which f coincides with the linear function defined on S_1

that takes the value x_1 at V_1 and the value x_2 at V_2 (this segment may be reduced to a single point, in which case $\alpha_1(x_1) = \beta_1(x_1)$). For $j = 2, 3, 4$, define the interval $[\alpha_j(x_j), \beta_j(x_j)]$ analogously. It is not difficult to deduce from the definition of F_1 that

$$F_1(x_1) = x_1 + N^1 \frac{f(\beta_1(x_1), 0) - x_1}{\beta_1(x_1)}.$$

Denote by D_+ and D_- the operations that consist of taking the derivative on the right and left, respectively. Since for some $\varepsilon > 0$, $\beta_1(\cdot)$ is constant on the interval $[x_1, x_1 + \varepsilon]$ of $\mathbb{R}$,

$$D_+F_1(x_1) = 1 - \frac{N^1}{\beta_1(x_1)}.$$

Similarly,

$$D_-F_1(x_1) = 1 - \frac{N^1}{\alpha_1(x_1)},$$

and analogous formulas hold for $D_\pm F_j(x_j)$, $j = 2, 3, 4$ (cf. Exercise 6). Because the F_j are decreasing, the chain rule for derivatives applied to F yields

$$\begin{aligned} D_+F(x_1) &= D_-F_4(x_4)D_+F_3(x_3)D_-F_2(x_2)D_+F_1(x_1) \\ &= \left(1 - \frac{N^2}{N^2 - \beta_4(x_4)}\right)\left(1 - \frac{N^1}{N^1 - \alpha_3(x_3)}\right) \\ &\quad \times \left(1 - \frac{N^2}{\alpha_2(x_2)}\right)\left(1 - \frac{N^1}{\beta_1(x_1)}\right), \end{aligned}$$

which, after reducing each factor to a common denominator, is equal to

$$\frac{\alpha_3(x_3)}{\beta_1(x_1)} \frac{N^1 - \beta_1(x_1)}{N^1 - \alpha_3(x_3)} \frac{\beta_4(x_4)}{\alpha_2(x_2)} \frac{N^2 - \alpha_2(x_2)}{N^2 - \beta_4(x_4)}.$$

Since x_1 is in the interior of I_1, $\beta_1(x_1) \geq d_1^1$, $\alpha_2(x_2) \geq d_2^2$, $\alpha_3(x_3) \leq d_3^1$, and $\beta_4(x_4) \leq d_4^2$. It follows that the product above is not greater than

$$\frac{d_3^1}{d_1^1} \frac{(N^1 - d_1^1)}{(N^1 - d_3^1)} \frac{d_4^2}{d_2^2} \frac{(N^2 - d_2^2)}{(N^2 - d_4^2)} < 1$$

by (10.40). Consequently, $D_+F < 1$ in the interior of I_1, and so F has at most one fixed point in I_1. Because F is non-decreasing, this also shows that for $z' \in I_1$, $z' < F(z') < z$ if $z' < z$, and $z < F(z') < z'$ if $z < z'$.

Set $p^1 = \inf \mathsf{C}_1$, $q^1 = \sup \mathsf{C}_3$, $p^2 = \inf \mathsf{C}_4$, and $q^2 = \sup \mathsf{C}_2$. If $p^2 > 1$ and $q^2 < N^2 - 1$, then

$$h(0, p^2 - 1) > f(0, p^2 - 1) \quad \text{and} \quad h(N^1, q^2 + 1) > f(N^1, q^2 + 1). \tag{10.43}$$

By (10.26) and (10.29),

$$(\{p^1\} \times]0, q^2[) \cup (\{q^1\} \times]p^2, N^2[) \subset \mathsf{E}_2, \tag{10.44}$$

and

$$(]0, p^1[\times \{p^2\}) \cup (]q^1, N^1[\times \{q^2\}) \subset \mathsf{E}_1. \tag{10.45}$$

Let μ_1 and μ_2 be the functions defined in the proof of Proposition 3 of Section 10.2. By (10.31), $p^1 = \sup \mu_1$ and $q^1 = \inf \mu_2$, so (10.11) is satisfied. We now show that p^2 satisfies (10.12). By (10.44), $\mu_1(n) \le q^1$ for $n > p^2$, and so there is $n \in]0, p^2]$ such that $\mu_1(n) = p^1$. If $n = p^2$, then $\mu_1(p^2) = p^1$; in other words, p^2 satisfies (10.12). Suppose that $p^2 > 1$ and $n \in]0, p^2[$. To conclude that $\mu_1(p^2) = p^1$, we must show that $(p^1 - 1, p^2) \notin \mathsf{E}_2$. Indeed, because $(p^1, p^2) \in \mathsf{E}_2$ and E_2 is h-convex by Proposition 2 of Section 10.2, we will be able to conclude that $(m, p^2) \notin \mathsf{E}_2$ for $m \in]0, p^1[$, which implies that $\mu_1(p^2) = p^1$. Suppose that $(p^1 - 1, p^2) \in \mathsf{E}_2$. Then $g(p^1 - 1, p^2 - 1) = h(p^1 - 1, p^2 - 1)$. Since $(m, n) \in \mathsf{E}_1$ for $m \in]0, p^1[$ and E_1 is v-convex by Proposition 2 of Section 10.2, we conclude from (10.45) that $(m, p^2 - 1) \in \mathsf{E}_1$ for $m \in]0, p^1[$. But then g is h-linear on $[0, p^1] \times \{p^2\}$ and agrees with h at $(p^1 - 1, p^2 - 1)$ and at $(p^1, p^2 - 1)$, so g agrees with h at $(0, p^2 - 1)$. Therefore, h and f agree at this point, contradicting (10.43).

The proof that q^2 satisfies (10.12) is similar and is left to the reader.

EXERCISES

1. Let $\tilde{f}$ be a nonnegative real-valued function defined on E that vanishes in $\mathsf{E} - \partial\mathsf{E}$, and let f be the function defined on E as follows: $f(\cdot, 0)$ and $f(\cdot, N^2)$ [$f(0, \cdot)$ and $f(N^1, \cdot)$] are the concave envelopes of $\tilde{f}(\cdot, 0)$ and $\tilde{f}(\cdot, N^2)$ [$\tilde{f}(0, \cdot)$ and $\tilde{f}(N^1, \cdot)$], respectively, and f vanishes in $\mathsf{E} - \partial\mathsf{E}$. Prove that the biconcave envelopes of f and $\tilde{f}$ coincide.

2. (Indifference regions) In this exercise, the presence of indifference regions other than the central rectangle that appears in the solution of type 2 is examined. Let ξ_1 be the function defined in (10.13) and let p^1 and p^2 be the integers defined in (10.11) and (10.12).

a. Assume that $p^1 > 1$, fix $x^1 \in]0, p^1[$, and let $x^2 = \inf\{n \in]0, p^2]: \xi_1(n) \geq x^1\}$. Show that if $x^2 > 1$ and $f(\cdot, 0)$ is strictly concave at x^1, then $\{x^1\} \times]0, x^2[\subset \mathsf{E}_2 - \mathsf{E}_1$. (*Hint*. Check that by construction, $[x^1 - 1, x^1 + 1] \times]0, x^2[\subset \mathsf{E}_2$, then use Proposition 1 of Section 10.2.)

b. Assume $p^1 > 2$ and fix $x^1, y^1 \in]0, p^1[$ such that $y^1 < x^1$. Assume that $f(\cdot, 0)$ is linear on $[y^1, x^1]$ and define x^2 as above. Show that if $x^2 > 1$, then $\xi_1(x^2 - 1) \leq y^1$ and $]y^1, x^1[\times]0, x^2[\subset \mathsf{E}_1 \cap \mathsf{E}_2$. (*Hint*. Use the theorem of Section 10.4 to prove the inequality, then the theorem of Section 10.3 and Proposition 2 of Section 10.2 to prove the inclusion.)

3. (Computing the biconcave envelope of the payoff function) Assume that one of the two hypotheses of Proposition 3 of Section 10.2 is satisfied and let p^1, q^1, p^2, and q^2 be the integers defined in (10.11) and (10.12).

a. If the solution is of type 1, determine the values of $g(p^1, p^2)$ and $g(q^1, q^2)$. (*Hint*. Show that they are interpolations of $f(p^1, 0)$ and $f(p^1, N^2)$, and $f(q^1, 0)$ and $f(q^1, N^2)$, respectively.)

b. If the solution is of type 2 and $p^2 < q^2$, use the notation introduced after (10.26) and for $j = 5, 6, 7$, set $a_j = a_{j-4}$, $\rho_j = \rho_{j-4}$, $r_j = r_{j-4}$. For $j = 1, \ldots, 4$, let

$$\alpha_j = \frac{r_j f(a_j) + \rho_j r_{j+1} f(a_{j+1}) + \rho_j \rho_{j+1} r_{j+2} f(a_{j+2}) + \rho_j \rho_{j+1} \rho_{j+2} r_{j+3} f(a_{j+3})}{1 - \rho_1 \rho_2 \rho_3 \rho_4}.$$

Show that $g(p^1, p^2) = \alpha_1$, $g(p^1, q^2) = \alpha_2$, $g(q^1, q^2) = \alpha_3$, and $g(q^1, p^2) = \alpha_4$. (*Hint*. Check that the formula above yields the solution of the system of equations defined in (10.28); the determinant of this system is $1 - \rho_1 \rho_2 \rho_3 \rho_4$.)

c. Let $N = p^1 + p^2$ and let ξ_1 be the function defined in (10.13). Consider the functions η_1 and η_2 defined on $[0, N]$, with values in $[0, p^1]$ and $[0, p^2]$, respectively, by the following properties: η_1 and η_2 are nondecreasing, $\eta_1(n) + \eta_2(n) = n$ and $\eta_1(n) \leq \xi_1(\eta_2(n))$ for $n \in [0, N]$, and the graph of η contains the graph of ξ_1 restricted to $[0, p^2]$. Let φ be the function defined on $[0, N]$ by $\varphi(n) = f(\eta_1(n), \eta_2(n))$ if $\eta_1(n)\eta_2(n) = 0$, $\varphi(N) = g(p^1, p^2)$, and for all $n \in [0, N]$ such that $\eta_1(n) > 0$ and $\eta_2(n) > 0$, by setting $\varphi(n)$ equal to

$$\eta_1(n)\eta_2(n)\left\{\frac{g(p^1, p^2)}{p^1 p^2} + \sum_{k=n}^{N-1}\left(\frac{f(0, \eta_2(k))}{\eta_2(k)}\left(\frac{1}{\eta_1(k)} - \frac{1}{\eta_1(k+1)}\right) + \frac{f(\eta_1(k), 0)}{\eta_1(k)}\left(\frac{1}{\eta_2(k)} - \frac{1}{\eta_2(k+1)}\right)\right)\right\}.$$

Prove that $g(\eta_1(n), \eta_2(n)) = \varphi(n)$ for all $n \in [0, N]$. (*Hint*. Proceed by backwards induction on n. Assume that $n < N$ and for instance, that

$\eta_1(n-1) = \eta_1(n) - 1$ and $\eta_2(n-1) = \eta_2(n)$. The properties of η_1 and η_2 and the theorem of Section 10.3 imply that $g(\cdot, \eta_2(n))$ is linear on $[0, \eta_1(n)]$, which shows that

$$g(\eta_1(n-1), \eta_2(n-1)) = \frac{1}{\eta_1(n)} f(0, \eta_2(n)) + \frac{\eta_1(n) - 1}{\eta_1(n)} g(\eta_1(n), \eta_2(n)).$$

Now use the induction hypothesis and the observation that

$$\frac{1}{\eta_1(n)} f(0, \eta_2(n)) = \eta_1(n-1)\eta_2(n-1)\left(\frac{f(0, \eta_2(n-1))}{\eta_2(n-1)} \frac{1}{\eta_1(n)\eta_1(n-1)}\right)$$

and

$$\frac{1}{\eta_1(n)\eta_1(n-1)} = \frac{1}{\eta_1(n-1)} - \frac{1}{\eta_1(n)},$$

because $\eta_1(n-1) = \eta_1(n) - 1$.)

d. Use the result of part c to compute $g(x^1, x^2)$ for any $(x^1, x^2) \in [0, p^1] \times [0, p^2]$. (*Hint*. Distinguish the two cases $\eta_1^{-1}(x^1) \geq \eta_2^{-1}(x^2)$ and $\eta_1^{-1}(x^1) < \eta_2^{-1}(x^2)$, and use the theorem of Section 10.3.)

4. (Describing the switching curves) Consider the functions φ_j^i defined in Section 10.6 and assume that one of the hypotheses of Proposition 3 of Section 10.2 is satisfied. Let p^1, q^1, p^2, and q^2 be the integers defined in (10.11) and (10.12).

a. Show that if the solution is of type 1, then for all $x^2 \in]p^2, N^2[$,

$$\xi_1(x^2) = \begin{cases} \inf\{x^1 \in]0, p^1[: \varphi_4^1(x^1) \geq \varphi_4^2(x^2)\}, & \text{if } \{\ \} \neq \varnothing, \\ p^1, & \text{if } \{\ \} = \varnothing. \end{cases}$$

b. Show that if the solution is of type 2 and $p^2 < q^2$, then $\xi_1(x^2)$ satisfies the same equality as above but with p^1 replaced by q^1.

5. Consider the function F defined in Section 10.6 and the interval I_1 defined in the proof of Theorem 2 of that section. Let z be a fixed point of F that does not belong to I_1 and let h be the bilinear function associated with z. Show that h does not have a contact of type 2 with f.

(*Hint*. Using the notation defined in the theorem and its proof, assume for instance that $z < \varphi_1^1(d_1^1 - 1)$ and set $z_1 = z$ and $z_{j+1} = F_j(z_j)$ for $j = 1, 2, 3$. Show that $\beta_1(z) < d_1^1$ and that $\beta_4(z_4) < d_1^2$. Use (10.32) to conclude that $z_4 > \varphi_4^1(d_1^1 - 1)$, and therefore that $\alpha_3(z_3) \geq d_1^1$. Since $\beta_1(z) < d_1^1$,

$$z_2 = F_1(z_1) \geq \varphi_2^1(d_1^1) \geq \varphi_2^1(d_2^1) > \varphi_2^2(d_2^2 - 1);$$

therefore $\alpha_2(z_2) \geq d_2^2 > d_1^2 > \beta_4(z_4)$. Conclude that h does not have a contact of type 2 with f.)

6. Let $F_1, \ldots, F_4$ be the functions defined in Section 10.6 and I_1 and $\alpha_j(\cdot), \beta_j(\cdot)$, $j = 1, \ldots, 4$, be the interval and functions defined in the proof of Theorem 2 of that section. Let $z_1 \in I_1$, and for $j = 1, 2, 3$, set $z_{j+1} = F_j(z_j)$. Show that

$$D_+F_2(z_2) = 1 - \frac{N^2}{\beta_2(z_2)}, \qquad D_-F_2(z_2) = 1 - \frac{N^2}{\alpha_2(z_2)},$$

$$D_+F_3(z_3) = 1 - \frac{N^1}{N^1 - \alpha_3(z_3)}, \qquad D_-F_3(z_3) = 1 - \frac{N^1}{N^1 - \beta_3(z_3)},$$

$$D_+F_4(z_4) = 1 - \frac{N^2}{N^2 - \alpha_4(z_4)}, \qquad D_-F_4(z_4) = 1 - \frac{N^2}{N^2 - \beta_4(z_4)}.$$

7. Solve the optimal switching problem on $\mathsf{E} = [0, 20] \times [0, 20]$ with the payoff function f defined by

$$f(m, 0) = f(m, 20) = 40m - 2m^2, \qquad \text{for } m \in [0, 20],$$

$$f(0, n) = f(20, n) = 20n - n^2, \qquad \text{for } n \in [0, 20],$$

and $f(m, n) = 0$ if $(m, n) \in \mathsf{E}^\circ$. (*Hint*. In order to conclude that the solution is of type 1, find a constant function h that is a bilinear majorant of f with a type 1 contact with f. Then apply the formulas in (10.17) to see that $\varphi_1^1(m) = 2m(m + 1)$ and $\varphi_1^2(n) = n(n + 1)$. Tabulate these two functions, then check that $\inf \mathsf{D}_1 = 7$, and due to the symmetry properties of f, that $\sup \mathsf{D}_3 = 13$, where the sets D_1 and D_3 are defined in Section 10.6. Apply Theorem 1 of Section 10.6 to determine the values of p^1, q^1, p^2 and q^2, and the theorem of Section 10.4 (along with Exercise 4) to construct the switching curves.)

8. Solve the optimal switching problem on $\mathsf{E} = [0, 20] \times [0, 20]$ with the payoff function f defined by

$$f(m,0) = 26m - m^2, \qquad f(m,20) = f(20-m,0), \qquad \text{for } m \in]0,20[,$$
$$f(20,n) = 26n - n^2, \qquad f(0,n) = f(20,20-n), \qquad \text{for } n \in]0,20[,$$

$f(0,0) = f(20,0) = f(20,20) = f(0,20) = 0$, and $f(m,n) = 0$ if $(m,n) \in \mathsf{E}^{\circ}$. (*Hint*. Find a constant function h that is a bilinear majorant of f with a type 2 contact with f, in order to conclude that the solution is of type 2. Then apply Theorem 2 of Section 10.6 to determine p^1, q^1, p^2 and q^2. Use the formulas in (10.17) to check that $\varphi_1^1(m) = m(m+1)$ and $\varphi_1^2(n) = n^2 + n + 120$. Tabulate these functions and use the symmetry properties of f to determine the switching curves.)

9. Solve the optimal switching problem on $\mathsf{E} = [0, 20] \times [0, 20]$ with the payoff function f defined by

$$f(m,0) = 26m - m^2, \qquad f(m,20) = f(20-m,0), \qquad \text{for } m \in]0,20[,$$
$$f(20,n) = 0.9(26n - n^2), \qquad f(0,n) = f(20,20-n), \qquad \text{for } n \in]0,20[,$$

$f(0,0) = f(20,0) = f(20,20) = f(0,20) = 0$, and $f(m,n) = 0$ if $(m,n) \in \mathsf{E}^{\circ}$. (*Hint*. The first step is to determine whether the solution is of type 1 or 2. Using (10.17) and analogous formulas for φ_j^i, show that

$$\varphi_1^1(m) = m(m+1), \qquad \varphi_2^1(m) = m^2 - 41m + 540,$$
$$\varphi_2^2(n) = 0.9n(n+1), \qquad \varphi_3^2(n) = 0.9\left(n^2 - 41n + 540\right).$$

Using the symmetry properties of f, tabulate the eight functions φ_j^i, $i = 1, 2$, $j = 1, 2, 3, 4$. Then check that $\inf \mathsf{D}_1 = 11$, $\sup \mathsf{D}_3 = 9$, $\inf \mathsf{D}_2 = 12$, and $\sup \mathsf{D}_4 = 8$, and therefore, by Theorem 2 of Section 10.6, the solution is of type 2. Let F be the function defined in Section 10.6. Apply the theorem just mentioned to see that F has a unique fixed point z in the interval $I = [110, 402.8]$. Let $\alpha_j(x_j)$ and $\beta_j(x_j)$ be the functions defined in the proof of Theorem 2 of Section 10.6. Notice that $\varphi_1^1(11) = 132$ and show that $F_1(132) = 192$, $F_2(192) = 134$, $F_3(134) = 190.67$, $F_4(190.67) = 134.44$, and therefore $F(132) = 134.44$. Similarly, show that $\varphi_1^1(12) = 156$ and $F(156) = 136.17$. Conclude from the theorem just mentioned that $134.44 < z < 136.17$. Observe that $\alpha_1(134.44) = \beta_1(136.17) = 12$, and that $\beta_2(F_1(134.44)) = \alpha_2(F_1(136.17)) = 15$. Deduce that $p^1 = 12$ and $q^2 = 15$. Complete the solution of the problem using the symmetry properties of f and the theorem of Section 10.4.)

HISTORICAL NOTES

The optimal switching problem described in this chapter is a special case of an optimization problem for several Markov chains first mentioned by Mandelbaum and Vanderbei (1981). A related class of continuous-time problems involving Brownian motions had been studied earlier by Walsh (1968), though the problem exactly analogous to the one presented here was first considered by Mandelbaum (1988). This article contained an explicit solution to the problem for a very restricted class of payoff functions. Mandelbaum, Shepp, and Vanderbei (1990) then solved the continuous-time problem under the assumption that the boundary data are concave and smooth, uncovering the structure of the solution using the so-called principle of smooth fit. They also provided explicit formulas for the value function and the switching curves. The solution to the discrete problem discussed in this chapter is taken from Cairoli and Dalang (1995a and 1995b), who showed that the discrete problem can in fact be solved explicitly and that the continuous-time formulas, properly interpreted, remain valid in the discrete case, yielding the theorems of Section 10.3 and 10.4 These articles also provided necessary and sufficient conditions on the payoff function for the solution to be of type 1 or of type 2 (Section 10.5) and the explicit formulas given in Section 10.6.

Some extensions to more than two diffusions of the results of Mandelbaum, Shepp, and Vanderbei (1990) are studied in Vanderbei (1992). A general class of related control problems has been considered, for instance, by Krylov (1972) and Evans and Friedman (1979).

The formulas for the biconcave envelope given in Exercise 3 are the discrete analogs of formulas given in Mandelbaum, Shepp, and Vanderbei (1990). Exercises 7, 8 and 9 are taken from Cairoli and Dalang (1995b).

Bibliography

Armitage, P. (1985), "The Search for Optimality in Clinical Trials," *Int. Statist. Rev.*, **53-1**, 15–24.

Arrow, K. J., Blackwell, D., and Girshick, M.A. (1949), "Bayes and Minimax Solutions of Sequential Decision Problems," *Econometrica*, **17**, 213–244.

Bahadur, R. R. (1954), "Sufficiency and Statistical Decision Functions," *Ann. Math. Statist.*, **25**, 423–462.

Berge, C. (1973), *Graphes et Hypergraphes*, North-Holland, Amsterdam.

Berge, C. and Duchet, P. (1984), "Strongly Perfect Graphs," in Berge, C. and Chvátal, V. (eds.), *Topics on Perfect Graphs*, North-Holland Mathematics Studies Vol 88, Annals of Discrete Math. No. 21, North-Holland, Amsterdam. pp. 57–62.

Berry, D. A., and Fristedt, B. (1985), *Bandit Problems*, Chapman and Hall, New York.

Blackwell, D., and Girshick, M. A. (1954), *Theory of Games and Statistical Decisions*, Wiley, New York.

Bochner, S. (1955), "Partial Ordering in the Theory of Martingales," *Ann. Math.*, **62**(1), 162–169.

Bolle, Y. (1993), "Sommes de variables aléatoires réelles indépendantes multi-indéxées et généralisations du lemme de Kronecker." Travail de diplôme, Institut de Mathematiques, Université de Lausanne, Switzerland.

Bradt, R. N., and Karlin, S. (1956), "On the Design and Comparison of Certain Dichotomous Experiments," *Ann. Math. Statist.*, **27**, 390–409.

Breiman, L. (1968), *Probability*, Addison-Wesley, Reading, Massachusetts.

Brown, L. D. (1977), "Closure Theorems for Sequential-Design Processes," in Gupta, S. S., and Moore, D. S. (eds.), *Statistical Decision Theory and Related Topics II*, Academic Press, New York, pp. 57–91.

Cairoli, R. (1970), "Une Inégalité pour Martingales à Indices Multiples et ses Applications," in *Sém. de Probab. IV*, Lecture Notes in Mathematics Vol. 124, Springer Verlag, New York, pp. 1–27.

Cairoli, R., and Dalang, R. C. (1995a), "Optimal Switching between two Brownian Motions," in Cranston, M. C. and Pinsky, M. A. (eds.), *Stochastic Analysis*, Proc. of Symposia in Pure Math. Vol. 57, American Mathematical Society, Providence, Rhode Island.

Cairoli, R., and Dalang, R. C. (1995b), "Optimal Switching between Two Random Walks," *Ann. Probab.*, to appear.

Cairoli, R., and Gabriel, J.-P. (1978), "Arrêt Optimal de Certaines Suites de Variables Aléatoires Indépendantes," in *Sém. de Probab. XIII*, Lecture Notes in Mathematics Vol. 721, Springer Verlag, New York, pp. 174–198.

Chernoff, H. (1959), "Sequential Design of Experiments," *Ann. Math. Statist.*, **30**(3), 755–770.

Chow, Y. S. (1960), "Martingales in a σ-Finite Measure Space Indexed by Directed Sets," Trans. Am. Math. Soc., **97**, 254–285.

Chow, Y. S., and Robbins, H. (1965), "On Stopping Rules for S_n/n," *Illinois J. Math.*, **9**, 444–454.

Chow, Y. S., Robbins, H., and Siegmund, D. (1971), *Great Expectations: The Theory of Optimal Stopping*, Houghton-Mifflin, Boston.

Chow, Y. S., and Teicher, H. (1988), *Probability Theory: Independence, Interchangeability, Martingales* (2nd ed.), Springer Verlag, New York.

Chung, K. L. (1974), *A Course in Probability Theory* (2nd ed.), Academic Press, New York.

Chvátal, V. (1975), "On certain polytopes associated with graphs," *J. Combinatorial Th. (B)*, **18**, 138–154.

Chvátal, V. (1984), "Perfectly ordered graphs," in Berge, C. and Chvátal, V. (eds.), *Topics on Perfect Graphs*, North-Holland Mathematics Studies Vol. 88, Annals of Discrete Math No. 21, North-Holland, Amsterdam, pp. 63–66.

Dalang, R. C. (1987), "Randomization in the Two-Parameter Optimal Stopping Problem: Combinatorial, Probabilistic and Infinitesimal Methods," Thèse de doctorat No. 688, Ecole Polytechnique Fédérale de Lausanne, Switzerland.

Dalang, R. C. (1988a), "On Infinite Perfect Graphs and Randomized Stopping Points on the Plane," *Probab. Theory Related Fields*, **78**(3), 357–378.

Dalang, R. C. (1988b), "On Stopping Ponts in the Plane that Lie on a Unique Optional Increasing path," *Stochastics*, **24**, 245–268.

Dalang, R. C. (1989), "Optimal Stopping of Two-Parameter Processes on Nonstandard Probability Spaces," *Trans. Am. Math. Soc.*, **313**(2) 697–719.

Dalang, R. C. (1990), "Randomization in the Two-Armed Bandit Problem," *Ann. Probab.*, **18**(1), 218–225.

Dalang, R. C., Trotter, L. E., and de Werra, D. (1988), "On Randomized Stopping Points and Perfect Graphs", *J. Combinatorial Theory* (*B*), **45**(3), 320–344.

Davis, B. (1971), "Stopping Rules for S_n/n and the Class L log L," *Z. Wahrscheinlichkeitstheorie v. Gebiete*, **17**, 147–150

Davis, B. (1973), "Moments of Random Walks Having Infinite Variance and the Existence of Certain Optimal Stopping Rules for S_n/n," *Ill. J. Math.*, **17**, 75–81.

Dellacherie, C., and Meyer, P. A. (1978), *Probabilities and Potential*, North-Holland, New York, Chaps. I–IV.

Dellacherie, C., and Meyer, P. A., (1982), *Probabilities and Potential B Theory of Martingales*, North-Holland, New York, Chaps. V–VIII.

Doob, J. L. (1953). *Stochastic Processes*, Wiley, New York.

Dubins, L., and Savage, J. (1965), *How to Gamble if You Must*: *Inequalities for Stochastic Processes*, McGraw Hill, New York.

Dvoretzky, A. (1967), "Existence and Properties of Certain Optimal Stopping Rules," in *Fifth Berkeley Symposium Math. Statist. Probab.*, University of California Press, Berkeley, pp. 441–452.

Dynkin, E. B., and Yushkevitch, A. A. (1969), *Markov Processes Theorems and Problems*, Plenum Press, New York.

El Karoui, N., and Karatzas, I. (1993), "General Gittins Index Processes in Discrete Time," *Proc. Nat. Acad. Sci.*, **90**, 1232–1236.

El Karoui, N., and Karatzas, I. (1994), "Dynamic Allocation Problems in Continuous time," *Ann. Appl. Probab.*, **4**(2), 255–286.

Etemadi, N. (1991), "Almost Sure Convergence for Partial sums of Independent Random Vectors with Multi-Dimensional Time Parameter," *Commun. Statist.*, **20**, 3909–3923.

Evans, L. C., and Friedman, A. (1979), "Optimal Stochastic Switching and the Dirichlet Problem for the Bellman Equation," *Trans. Am. Math. Soc.*, **253**, 365–389.

Fleming, W. H., and Soner, H. M. (1993), *Controlled Markov Processes and Viscosity Solutions*, Springer Verlag, New York.

Fouque, J. P. (1983), "The Past of a Stopping Point and Stopping for Two-Parameter Processes," *J. Multivariate Anal.*, **11**, 561–577.

Fulkerson, D. R. (1973), "On the Perfect Graph Theorem," in Hu, T. C. and Robertson, S. M. (eds.), *Mathematical Programming*, Academic Press, New York, pp. 69–77.

Gabriel, J.-P. (1975), "Loi des Grands Nombres, Séries et Martingales Indexées par un Ensemble Filtrant," Thèse de doctorat No. 225, Ecole Polytechnique Fédérale de Lausanne, Switzerland.

Gabriel, J.-P. (1977), "Martingales with a Countable Filtering Index Set," *Ann. Probab.*, **5**, 888–898.

Gittins, J. C. (1979), "Bandit Processes and Dynamic Allocation Indices (with Discussion)," *J. R. Statist. Soc. Ser. B*, **41**, 148–177.

Gittins, J. C. (1989), *Multi-Armed Bandit Allocation Indices*, Wiley, New York.

Gittins, J. C., and Jones, D. M. (1974), "A Dynamic Allocation Index for the Sequential Design of Experiments," in Gani, J. (ed.), *Progress in Statistics*, North Holland, Amsterdam, pp. 241–266.

Golumbic, M. C. (1980), *Algorithmic Graph Theory and Perfect Graphs*, Academic Press, New York.

Grillenberger, C., and Krengel, U. (1981), "Remark on Strong Martingales and the Linear Embedding of Tactics," *J. Multi. Anal.*, **11**, 568–571.

Gut, A. (1976), "Convergence of Reversed Martingales with Multidimensional Indices," *Duke Math. J.*, **43**, 269–275.

Gut, A. (1978), "Marcinkiewicz Laws and Convergence Rates in the Laws of Large Numbers for Random Variables with Multidimensional Indices," *Ann. Probab.*, **6**, 469–482.

Gut, A. (1979), "Moments of the Maximum of Normed Partial Sums of Random Variables with Multidimensional Indices," *Z. Wahrscheinlichskeitstheorie v. Gebiete*, **46**, 205–220.

Haggstrom, G. W. (1966), "Optimal Stopping and Experimental Design," *Ann. Math. Statist.*, **37**, 7–20.

Hardy, G. H. (1974), On the Convergence of Certain Multiple Series, *Collected Papers of G. H. Hardy*, Clarendon Press, Oxford. Vol. VI.

Hayre, L. S. (1982), "A Note on the Asymptotic Optimality of a Two Population Sequential Probability Ratio Test," *J. Statist. Planning Inf.*, **6**, 127–130.

Helms, L. L. (1958), "Mean Convergence of Martingales," *Trans. Am. Math. Soc.*, **87**, 439–446.

Hewitt, E., and Stromberg, K. (1965), *Real and Abstract Analysis*, Springer Verlag, New York.

Hoffmann-Jorgensen, J. (1974), "Sums of Independent Banach Space Valued Random Variables," *Studia Math. T. L.*, **II**, 159–186.

Irle, A. (1981), "Transitivity in Problems of Optimal Stopping," *Ann. Probab.*, **9**(4), 642–647.

Ishikida, T., and Varaiya, P. (1993), "Multi-armed Bandit Problem Revisited" (preprint).

Karatzas, I. (1984), "Gittins Indices in the Dynamic Allocation Problem for Diffusion Processes," *Ann. Probab.*, **12**, 173–192.

Keener, R. W. (1980), "Renewal Theory and the Sequential Design of Experiments with Two States of Nature," *Commun. Statist.-Theor. Meth.*, **A9** (16), 1699–1726.

Keener, R. (1986), "Multi-Armed Bandits with Simple Arms," *Adv. Appl. Math.*, **7**, 199–204.

Krengel, U., and Sucheston, L. (1981), "Stopping Rules and Tactics for Processes Indexed by a Directed Set," *J. Mult. Anal.*, **11**, 199–229.

Krickeberg, K. (1956), "Convergence of Martingales with a Directed Index Set," *Trans. Am. Math. Soc.*, **83**, 313–337.

Krylov, V. (1972), "Control of the Solution of a Stochastic Differential Equation," *Th. Probab. Appl.*, **14**, 114–130.

Krylov, V. (1980), *Controlled Diffusion Processes*, Springer Verlag, New York.

Kullback, S. (1968), *Information Theory and Statistics*, Dover Publications, Inc., New York.

Kurtz, T. G. (1980), "The Optional Sampling Theorem for Martingales Indexed by a Directed Set," *Ann. Probab.*, **8**, 675–681.

Kushner, H. J., and Dupuis, P. G. (1992), *Numerical Methods for Stochastic Control Problems in Continuous Time*, Springer Verlag, New York.

Lai, T. L. (1987), "Adaptive Treatment Allocation and the Multi-Armed Bandit Problem," *Ann. Statist.*, **15**(3), 1091–1114.

Lalley, S. P., and Lorden, G. (1986), "A Control Problem Arising in the Sequential Design of Experiments," *Ann. Probab.*, **14**(1), 136–172.

Lawler, G. F., and Vanderbei, R. J. (1983), "Markov Strategies for Optimal Control Problems Indexed by a Partially Ordered Set," *Ann. Probab.*, **11**(3), 642–647.

Lehman, E. (1959), *Testing Statistical Hypotheses*, Wiley, New York.

Louis, T. A. (1975), "Optimal Allocation in Sequential Tests Comparing the Means of Two Gaussian Populations," *Biometrika*, **62**(2), 359–369.

Louis, T. A. (1977), "Sequential Allocation in Clinical Trials Comparing Two Exponential Survival Curves," *Biometrics*, **33**, 627–634.

Mandelbaum, A. (1986), "Discrete Multi-Armed Bandits and Multi-Parameter Processes," *Probab. Theory and Related Fields*, **71**, 129–147.

Mandelbaum, A. (1987), "Continuous Multi-Armed Bandits and Multi-Parameter Processes," *Ann. Probab.*, **15**, 1527–1556.

Mandelbaum, A. (1988), "Navigating and Stopping Multi-Parameter Processes," in Fleming, W., and Lions, P. L. (eds.), *Stochastic Differential Systems, Stochastic Control Theory and Applications*, Springer Verlag, New York, pp. 339–372.

Mandelbaum, A., Shepp, L., and Vanderbei, R. J. (1990), "Optimal Switching between a Pair of Brownian Motions," *Ann. Probab.*, **18**(3), 1010–1033.

Mandelbaum, A., and Vanderbei, R. J. (1981), "Optimal Stopping and Supermartingales over Partially Ordered Sets," *Z. Wahrscheinlichkeitstheorie v. Gebiete*, **57**, 253–264.

Mazziotto, G., and Millet, A. (1986), "Points, Lignes et Systèmes d'Arrêt Flous et Problème d'Arrêt Optimal," Lecture Notes in Mathematics, vol. 1204, Springer Verlag, New York, pp. 81–94.

Mazziotto, G., and Millet, A. (1987), "Stochastic Control of Two-Parameter Processes Application: The Two-Armed Bandit Problem," *Stochastics*, **22**, 251–288.

Mazziotto, G., and Szpirglas, J. (1983), "Arrét Optimal Sur le Plan," *Z. Wahrscheinlichskeitstheorie v. Gebiete*, **62**, 215–233.

McCabe, B. J., and Shepp, L. A. (1970), "On the Supremum of S_n/n," *Ann. Math. Statist.*, **41**, 2166–2168.

Moore, Ch.N. (1973), "*Summable Series and Convergence Factors*," Dover Publications, Inc., New York.

Neveu, J. (1975), *Discrete-Parameter Martingales*, North Holland, New York.

Nualart, D. (1992), "Randomized Stopping points and Optimal Stopping on the Plane," *Ann. Probab.*, **20**(2), 883–900.

Presman, E. L., and Sonin, I.N. (1990), *Sequential Control with Incomplete Information: The Bayesian Approach to Multi-Armed Bandit Problems*, Academic Press, New York.

Puterman, M., (1994), *Markov Decision Processes: Discrete Stochastic Dynamic Programming*, Wiley, New York.

Robbins, H. (1952), "Some Aspects of the Sequential Design of Experiments," *Bull. Am. Math. Soc.*, **58**(5), 527–535.

Robbins, H., and Siegmund, D. O. (1974), "Sequential Tests Involving Two Populations," *J. Am. Statist. Assoc.*, **69**, 132–139.

Shiryayev, A. N. (1978), *Optimal Stopping Rules*, Springer Verlag, New York.

Siegmund, D. (1985), *Sequential Analysis: Tests and Confidence Intervals*, Springer Verlag, New York.

Smythe, R. T. (1973), "Strong Law of Large Numbers for r-Dimensional Arrays of Random Variables," *Ann. Probab.*, **1**(1), 164–170.

Snell, J. L. (1952), "Application of Martingale System Theorems," *Trans. Am. Math. Soc.*, **73**, 293–312.

Vanderbei, R. J. (1992), "Optimal Switching among Several Brownian Motions," *SIAM J. Control Optim.*, **30**, 1150–1162.

Varaiya, P. P., Walrand, J. C., and Buyukkoc, C. (1985), "Extensions of the Multi-armed Bandit Problem: The Discounted Case," *IEEE Trans. Autom. Control.* **AC-30**(5), 426–439.

Wald, A. (1947), *Sequential Analysis*, Wiley, New York.

Wald, A. (1950), *Statistical Decision Functions*, Wiley, New York.

Wald, A., and Wolfowitz, J. (1948), "Optimum Character of the Sequential Probability Ratio Test," *Ann. Math. Statist.*, **19**, 326–339.

Wald, A., and Wolfowitz, J. (1950), "Bayes Solutions of Sequential Decision Problems," *Ann. Math. Statist.*, **21**, 82–99.

Walsh, J. B. (1968), "Probability and a Dirichlet problem for Multiply Superharmonic Functions," *Ann. Inst. Fourier* (*Grenoble*), **18**(2), 221–279.

Walsh, J. B. (1981), "Optional Increasing Paths," in *Proc. Aléatoires à Deux Indices*, Lecture Notes in Mathematics Vol. 863, Springer Verlag, New York, pp. 265–439.

Washburn, Jr., R. B., and Willsky, A. S. (1981), "Optional Sampling of Submartingales Indexed by Partially Ordered Sets," *Ann. Probab.*, **9**, 957–970.

Whittle, P. (1965), "Some General Results in Sequential Design (with Discussion)," *J. R. Statist. Soc. Ser. B.*, **27**, 371–388.

Whittle, P. (1982). *Optimization over Time*: *Dynamic Programming and Stochastic Control* Wiley, New York, Vol. 1.

Index of Notation

In this printing, the reader may notice 2 slight variations in the following characters: ∂, ε, g, p, q, x, and y.

Index of Terms

WILEY SERIES IN PROBABILITY AND STATISTICS

ESTABLISHED BY WALTER A. SHEWHART AND SAMUEL S. WILKS

Editors

Vic Barnett, Ralph A. Bradley, Nicholas I. Fisher, J. Stuart Hunter, J. B. Kadane, David G. Kendall, David W. Scott, Adrian F. M. Smith, Jozef L. Teugels, Geoffrey S. Watson

Probability and Statistics

ANDERSON · An Introduction to Multivariate Statistical Analysis, *Second Edition*
*ANDERSON · The Statistical Analysis of Time Series
ARNOLD, BALAKRISHNAN, and NAGARAJA · A First Course in Order Statistics
BACCELLI, COHEN, OLSDER, and QUADRAT · Synchronization and Linearity: An Algebra for Discrete Event Systems
BARNETT · Comparative Statistical Inference, *Second Edition*
BERNARDO and SMITH · Bayesian Statistical Concepts and Theory
BHATTACHARYYA and JOHNSON · Statistical Concepts and Methods
BILLINGSLEY · Convergence of Probability Measures
BILLINGSLEY · Probability and Measure, *Second Edition*
BOROVKOV · Asymptotic Methods in Queuing Theory
BRANDT, FRANKEN, and LISEK · Stationary Stochastic Models
CAINES · Linear Stochastic Systems
CAIROLI and DALANG · Sequential Stochastic Optimization
CHEN · Recursive Estimation and Control for Stochastic Systems
CONSTANTINE · Combinatorial Theory and Statistical Design
COOK and WEISBERG · An Introduction to Regression Graphics
COVER and THOMAS · Elements of Information Theory
*DOOB · Stochastic Processes
DUDEWICZ and MISHRA · Modern Mathematical Statistics
ETHIER and KURTZ · Markov Processes: Characterization and Convergence
FELLER · An Introduction to Probability Theory and Its Applications, Volume I, *Third Edition*, Revised; Volume II, *Second Edition*
FULLER · Introduction to Statistical Time Series
FULLER · Measurement Error Models
GIFI · Nonlinear Multivariate Analysis
GUTTORP · Statistical Inference for Branching Processes
HALD · A History of Probability and Statistics and Their Applications before 1750
HALL · Introduction to the Theory of Coverage Processes
HANNAN and DEISTLER · The Statistical Theory of Linear Systems
HEDAYAT and SINHA · Design and Inference in Finite Population Sampling
HOEL · Introduction to Mathematical Statistics, *Fifth Edition*
HUBER · Robust Statistics
IMAN and CONOVER · A Modern Approach to Statistics
JUREK and MASON · Operator-Limit Distributions in Probability Theory
KAUFMAN and ROUSSEEUW · Finding Groups in Data: An Introduction to Cluster Analysis
LARSON · Introduction to Probability Theory and Statistical Inference, *Third Edition*
LESSLER and KALSBEEK · Nonsampling Error in Surveys
LINDVALL · Lectures on the Coupling Method
MANTON, WOODBURY, and TOLLEY · Statistical Applications Using Fuzzy Sets
MORGENTHALER and TUKEY · Configural Polysampling: A Route to Practical Robustness
MUIRHEAD · Aspects of Multivariate Statistical Theory
OLIVER and SMITH · Influence Diagrams, Belief Nets and Decision Analysis

*Now available in a lower priced paperback edition in the Wiley Classics Library.

Probability and Statistics (Continued)

*PARZEN • Modern Probability Theory and Its Applications
PILZ • Bayesian Estimation and Experimental Design in Linear Regression Models
PRESS • Bayesian Statistics: Principles, Models, and Applications
PUKELSHEIM • Optimal Experimental Design
PURI and SEN • Nonparametric Methods in General Linear Models
PURI, VILAPLANA, and WERTZ • New Perspectives in Theoretical and Applied Statistics
RENCHER • Methods of Multivariate Analysis
RAO • Asymptotic Theory of Statistical Inference
RAO • Linear Statistical Inference and Its Applications, *Second Edition*
ROBERTSON, WRIGHT, and DYKSTRA • Order Restricted Statistical Inference
ROGERS and WILLIAMS • Diffusions, Markov Processes, and Martingales, Volume II: Îto Calculus
ROHATGI • A Introduction to Probability Theory and Mathematical Statistics
ROSS • Stochastic Processes
RUBINSTEIN • Simulation and the Monte Carlo Method
RUZSA and SZEKELY • Algebraic Probability Theory
SCHEFFE • The Analysis of Variance
SEBER • Linear Regression Analysis
SEBER • Multivariate Observations
SEBER and WILD • Nonlinear Regression
SERFLING • Approximation Theorems of Mathematical Statistics
SHORACK and WELLNER • Empirical Processes with Applications to Statistics
SMALL and McLEISH • Hilbert Space Methods in Probability and Statistical Inference
STAPLETON • Linear Statistical Models
STAUDTE and SHEATHER • Robust Estimation and Testing
STOYANOV • Counterexamples in Probability
STYAN • The Collected Papers of T. W. Anderson: 1943–1985
WHITTAKER • Graphical Models in Applied Multivariate Statistics
YANG • The Construction Theory of Denumerable Markov Processes

Applied Probability and Statistics

ABRAHAM and LEDOLTER • Statistical Methods for Forecasting
AGRESTI • Analysis of Ordinal Categorical Data
AGRESTI • Categorical Data Analysis
ANDERSON and LOYNES • The Teaching of Practical Statistics
ANDERSON, AUQUIER, HAUCK, OAKES, VANDAELE, and WEISBERG • Statistical Methods for Comparative Studies
*ARTHANARI and DODGE • Mathematical Programming in Statistics
ASMUSSEN • Applied Probability and Queues
*BAILEY • The Elements of Stochastic Processes with Applications to the Natural Sciences
BARNETT • Interpreting Multivariate Data
BARNETT and LEWIS • Outliers in Statistical Data, *Second Edition*
BARTHOLOMEW, FORBES, and McLEAN • Statistical Techniques for Manpower Planning, *Second Edition*
BATES and WATTS • Nonlinear Regression Analysis and Its Applications
BELSLEY • Conditioning Diagnostics: Collinearity and Weak Data in Regression
BELSLEY, KUH, and WELSCH • Regression Diagnostics: Identifying Influential Data and Sources of Collinearity
BHAT • Elements of Applied Stochastic Processes, *Second Edition*
BHATTACHARYA and WAYMIRE • Stochastic Processes with Applications
BIEMER, GROVES, LYBERG, MATHIOWETZ, and SUDMAN • Measurement Errors in Surveys
BIRKES and DODGE • Alternative Methods of Regression
BLOOMFIELD • Fourier Analysis of Time Series: An Introduction

*Now available in a lower priced paperback edition in the Wiley Classics Library.

Applied Probability and Statistics (Continued)

BOLLEN · Structural Equations with Latent Variables
BOULEAU · Numerical Methods for Stochastic Processes
BOX · R. A. Fisher, the Life of a Scientist
BOX and DRAPER · Empirical Model-Building and Response Surfaces
BOX and DRAPER · Evolutionary Operation: A Statistical Method for Process Improvement
BOX, HUNTER, and HUNTER · Statistics for Experimenters: An Introduction to Design, Data Analysis, and Model Building
BROWN and HOLLANDER · Statistics: A Biomedical Introduction
BUCKLEW · Large Deviation Techniques in Decision, Simulation, and Estimation
BUNKE and BUNKE · Nonlinear Regression, Functional Relations and Robust Methods: Statistical Methods of Model Building
CHATTERJEE and HADI · Sensitivity Analysis in Linear Regression
CHATTERJEE and PRICE · Regression Analysis by Example, *Second Edition*
CLARKE and DISNEY · Probability and Random Processes: A First Course with Applications, *Second Edition*
COCHRAN · Sampling Techniques, *Third Edition*
*COCHRAN and COX · Experimental Designs, *Second Edition*
CONOVER · Practical Nonparametric Statistics, *Second Edition*
CONOVER and IMAN · Introduction to Modern Business Statistics
CORNELL · Experiments with Mixtures, Designs, Models, and the Analysis of Mixture Data, *Second Edition*
COX, BINDER, CHINNAPPA, CHRISTIANSON, COLLEDGE, and KOTT · Business Survey Methods
COX · A Handbook of Introductory Statistical Methods
*COX · Planning of Experiments
CRESSIE · Statistics for Spatial Data, *Revised Edition*
DANIEL · Applications of Statistics to Industrial Experimentation
DANIEL · Biostatistics: A Foundation for Analysis in the Health Sciences, *Sixth Edition*
DAVID · Order Statistics, *Second Edition*
*DEGROOT, FIENBERG, and KADANE · Statistics and the Law
*DEMING · Sample Design in Business Research
DILLON and GOLDSTEIN · Multivariate Analysis: Methods and Applications
DODGE and ROMIG · Sampling Inspection Tables, *Second Edition*
DOWDY and WEARDEN · Statistics for Research, *Second Edition*
DRAPER and SMITH · Applied Regression Analysis, *Second Edition*
DUNN · Basic Statistics: A Primer for the Biomedical Sciences, *Second Edition*
DUNN and CLARK · Applied Statistics: Analysis of Variance and Regression, *Second Edition*
ELANDT-JOHNSON and JOHNSON · Survival Models and Data Analysis
EVANS, PEACOCK, and HASTINGS · Statistical Distributions, *Second Edition*
FISHER and VAN BELLE · Biostatistics: A Methodology for the Health Sciences
FLEISS · The Design and Analysis of Clinical Experiments
FLEISS · Statistical Methods for Rates and Proportions, *Second Edition*
FLEMING and HARRINGTON · Counting Processes and Survival Analysis
FLURY · Common Principal Components and Related Multivariate Models
GALLANT · Nonlinear Statistical Models
GLASSERMAN and YAO · Monotone Structure in Discrete-Event Systems
GROSS and HARRIS · Fundamentals of Queueing Theory, *Second Edition*
GROVES · Survey Errors and Survey Costs
GROVES, BIEMER, LYBERG, MASSEY, NICHOLLS, and WAKSBERG · Telephone Survey Methodology
HAHN and MEEKER · Statistical Intervals: A Guide for Practitioners
HAND · Discrimination and Classification
*HANSEN, HURWITZ, and MADOW · Sample Survey Methods and Theory, Volume I: Methods and Applications

*Now available in a lower priced paperback edition in the Wiley Classics Library.

*HANSEN, HURWITZ, and MADOW · Sample Survey Methods and Theory, Volume II: Theory
HEIBERGER · Computation for the Analysis of Designed Experiments
HELLER · MACSYMA for Statisticians
HINKELMAN and KEMPTHORNE: Design and Analysis of Experiments, Volume I: Introduction to Experimental Design
HOAGLIN, MOSTELLER, and TUKEY · Exploratory Approach to Analysis of Variance
HOAGLIN, MOSTELLER, and TUKEY · Exploring Data Tables, Trends and Shapes
HOAGLIN, MOSTELLER, and TUKEY · Understanding Robust and Exploratory Data Analysis
HOCHBERG and TAMHANE · Multiple Comparison Procedures
HOEL · Elementary Statistics, *Fifth Edition*
HOGG and KLUGMAN · Loss Distributions
HOLLANDER and WOLFE · Nonparametric Statistical Methods
HOSMER and LEMESHOW · Applied Logistic Regression
HØYLAND and RAUSAND · System Reliability Theory: Models and Statistical MethodsL
HUBERTY · Applied Discriminant Analysis
IMAN and CONOVER · Modern Business Statistics
JACKSON · A User's Guide to Principle Components
JOHN · Statistical Methods in Engineering and Quality Assurance
JOHNSON · Multivariate Statistical Simulation
JOHNSON and KOTZ · Distributions in Statistics
Continuous Univariate Distributions—2
Continuous Multivariate Distributions
JOHNSON, KOTZ, and BALAKRISHNAN · Continuous Univariate Distributions, Volume 1, *Second Edition*
JOHNSON, KOTZ, and BALAKRISHNAN · Continuous Univariate Distributions, Volume 2, *Second Edition*
JOHNSON, KOTZ, and KEMP · Univariate Discrete Distributions, *Second Edition*
JUDGE, GRIFFITHS, HILL, LÜTKEPOHL, and LEE · The Theory and Practice of Econometrics, *Second Edition*
JUDGE, HILL, GRIFFITHS, LÜTKEPOHL, and LEE · Introduction to the Theory and Practice of Econometrics, *Second Edition*
KALBFLEISCH and PRENTICE · The Statistical Analysis of Failure Time Data
KASPRZYK, DUNCAN, KALTON, and SINGH · Panel Surveys
KISH · Statistical Design for Research
*KISH · Survey Sampling
LANGE, RYAN, BILLARD, BRILLINGER, CONQUEST, and GREENHOUSE · Case Studies in Biometry
LAWLESS · Statistical Models and Methods for Lifetime Data
LEBART, MORINEAU, and WARWICK · Multivariate Descriptive Statistical Analysis: Correspondence Analysis and Related Techniques for Large Matrices
LEE · Statistical Methods for Survival Data Analysis, *Second Edition*
LePAGE and BILLARD · Exploring the Limits of Bootstrap
LEVY and LEMESHOW · Sampling of Populations: Methods and Applications
LINHART and ZUCCHINI · Model Selection
LITTLE and RUBIN · Statistical Analysis with Missing Data
MAGNUS and NEUDECKER · Matrix Differential Calculus with Applications in Statistics and Econometrics
MAINDONALD · Statistical Computation
MALLOWS · Design, Data, and Analysis by Some Friends of Cuthbert Daniel
MANN, SCHAFER, and SINGPURWALLA · Methods for Statistical Analysis of Reliability and Life Data
MASON, GUNST, and HESS · Statistical Design and Analysis of Experiments with Applications to Engineering and Science
McLACHLAN · Discriminant Analysis and Statistical Pattern Recognition
MILLER · Survival Analysis

Applied Probability and Statistics (Continued)

MONTGOMERY and PECK · Introduction to Linear Regression Analysis, *Second Edition*
NELSON · Accelerated Testing, Statistical Models, Test Plans, and Data Analyses
NELSON · Applied Life Data Analysis
OCHI · Applied Probability and Stochastic Processes in Engineering and Physical Sciences
OKABE, BOOTS, and SUGIHARA · Spatial Tesselations: Concepts and Applications of Voronoi Diagrams
OSBORNE · Finite Algorithms in Optimization and Data Analysis
PANKRATZ · Forecasting with Dynamic Regression Models
PANKRATZ · Forecasting with Univariate Box-Jenkins Models: Concepts and Cases
PORT · Theoretical Probability for Applications
PUTERMAN · Markov Decision Processes: Discrete Stochastic Dynamic Programming
RACHEV · Probability Metrics and the Stability of Stochastic Models
RÉNYI · A Diary on Information Theory
RIPLEY · Spatial Statistics
RIPLEY · Stochastic Simulation
ROSS · Introduction to Probability and Statistics for Engineers and Scientists
ROUSSEEUW and LEROY · Robust Regression and Outlier Detection
RUBIN · Multiple Imputation for Nonresponse in Surveys
RYAN · Statistical Methods for Quality Improvement
SCHUSS · Theory and Applications of Stochastic Differential Equations
SCOTT · Multivariate Density Estimation: Theory, Practice, and Visualization
SEARLE · Linear Models
SEARLE · Linear Models for Unbalanced Data
SEARLE · Matrix Algebra Useful for Statistics
SEARLE, CASELLA, and McCULLOCH · Variance Components
SKINNER, HOLT, and SMITH · Analysis of Complex Surveys
STOYAN · Comparison Methods for Queues and Other Stochastic Models
STOYAN, KENDALL, and MECKE · Stochastic Geometry and Its Applications
STOYAN and STOYAN · Fractals, Random Shapes and Point Fields
THOMPSON · Empirical Model Building
THOMPSON · Sampling
TIERNEY · LISP-STAT: An Object-Oriented Environment for Statistical Computing and Dynamic Graphics
TIJMS · Stochastic Modeling and Analysis: A Computational Approach
TITTERINGTON, SMITH, and MAKOV · Statistical Analysis of Finite Mixture Distributions
UPTON and FINGLETON · Spatial Data Analysis by Example, Volume I: Point Pattern and Quantitative Data
UPTON and FINGLETON · Spatial Data Analysis by Example, Volume II: Categorical and Directional Data
VAN RIJCKEVORSEL and DE LEEUW · Component and Correspondence Analysis
WEISBERG · Applied Linear Regression, *Second Edition*
WESTFALL and YOUNG · Resampling-Based Multiple Testing: Examples and Methods for *p*-Value Adjustment
WHITTLE · Optimization Over Time: Dynamic Programming and Stochastic Control, Volume I and Volume II
WHITTLE · Systems in Stochastic Equilibrium
WONNACOTT and WONNACOTT · Econometrics, *Second Edition*
WONNACOTT and WONNACOTT · Introductory Statistics, *Fifth Edition*
WONNACOTT and WONNACOTT · Introductory Statistics for Business and Economics, *Fourth Edition*
WOODING · Planning Pharmaceutical Clinical Trials: Basic Statistical Principles
WOOLSON · Statistical Methods for the Analysis of Biomedical Data

Tracts on Probability and Statistics

BILLINGSLEY · Convergence of Probability Measures
TOUTENBURG · Prior Information in Linear Models